Fatigue of Materials and Structures

Fatigue failure of engineering materials and structures has long been a great challenge for structural integrity, reliability, and safety in mechanical, civil, and aerospace engineering. These failure mechanisms and their modeling are critical concerns for managing aging structures, and they directly affect sustainability across society.

In this context, the fundamental theories and methods of fatigue failure of engineering materials and structures are explored in detail. Fatigue damage accumulation, crack initiation, and crack growth analysis are presented from materials to structures, deterministic to probabilistic fatigue, physics to data science, uniaxial to multiaxial fatigue, and ultra low-cycle fatigue to very high cycle fatigue. The focus is on mechanical understanding and risk management for design, maintenance, and operation. Some recent advancements include fatigue of additive manufactured (AM) metals and advanced materials, which could potentially transform fatigue analysis and offer new perspectives on fatigue failure mechanisms and reliability design. Both experimental supporting evidence and simulation benefits are demonstrated. It integrates recent developments in artificial intelligence with fatigue in AM metals and advanced materials.

It provides case studies, and future research challenges for the fusion of fatigue physics modeling with data analytics, for graduate students and advanced practitioners.

Structural Damage, Fatigue and Fracture

This series promotes scientific issues associated with damage, fatigue, and fracture phenomena applied to all kinds of materials, structural components and structures, as well as their engineering practical applications.

Series Editors:
José A.F.O. Correia, *Department of Civil Engineering, University of Porto*
Shun-Peng Zhu, *School of Mechanical and Electrical Engineering University of Electronic Science and Technology of China*
Matthew Hebdon, *Department of Civil, Architectural and Environmental Engineering, Cockrell School of Engineering, The University of Texas at Austin*

Honorary Editor:
Filippo Berto, *Department of Chemical Engineering, Materials and Environment, La Sapienza University*

Fatigue of Materials and Structures: Physics and Data Science
Qingyuan Wang, Shun-Peng Zhu, José A.F.O. Correia, Abilio M.P. De Jesus, and Grzegorz Lesiuk

Fatigue of Materials and Structures

Physics and Data Science

Edited by
Qingyuan Wang, Shun-Peng Zhu,
José A.F.O. Correia, Abílio M.P. De Jesus,
and Grzegorz Lesiuk

CRC Press
Taylor & Francis Group
Boca Raton London New York

CRC Press is an imprint of the
Taylor & Francis Group, an **informa** business

Designed cover image: Cover artwork and figures by Qingyuan Wang and Shun-Peng Zhu, used with permission.

First edition published 2025
by CRC Press
2385 NW Executive Center Drive, Suite 320, Boca Raton FL 33431

and by CRC Press
4 Park Square, Milton Park, Abingdon, Oxon, OX14 4RN

CRC Press is an imprint of Taylor & Francis Group, LLC

ISBN: 9781032497556 (hbk)
ISBN: 9781032520834 (pbk)
ISBN: 9781003405122 (ebk)

DOI: 10.1201/9781003405122

Typeset in Times
by codeMantra

Contents

Foreword

Fatigue is a discipline with a long history, whose origin can be traced back to the 1830s, when German mining engineer Albert conducted the first fatigue test on metal conveyor chains and subsequently published the first article on fatigue. Initially, "Tired" was used to describe the phenomenon, until Braithwaite from Britain coined the term "fatigue" in 1854. According to various statistics, fatigue-induced failures account for up to 80% of all structural failures. Particularly, with the advancement of modern industrial design and analysis software, failures due to insufficient static strength can be avoided. Instead, the fatigue strength (or fatigue resistance) under cyclic loadings generally directly determines the fatigue life of the structure. Although research on this issue has been ongoing for nearly 200 years, it is hard to describe it as "well-studied."

Fatigue fracture is a type of low-stress brittle failure, occurring under loads far below the static material's strength limit, or even its yield limit. As a result, there is often no noticeable plastic deformation in most cases. Stress concentration near structural discontinuities or internal defects are the primary causes of fatigue failure, which induces initial microcracks. Under alternating stress, microcracks continuously propagate and coalesce, eventually forming a macroscopic crack that leads to sudden fracture without any warning. Due to the severe impact of fatigue on the service life of engineering structures or components, it is often referred to as the "cancer of industry."

Improper fatigue assessment techniques can result either in overly conservative safety margins or in insufficient safety in fatigue design. In the past, through the continuous efforts by generations of scholars and engineers, significant achievements have been made in the study of fatigue across experimental research, theoretical development, and engineering applications. With the increasing focus on structural integrity and service reliability, fatigue strength design theory has evolved, from infinite life design, safe life design, and damage tolerance design to defect tolerance design. Presently, as structural design in modern engineering trends toward extreme dimensions, loads, and parameters, researchers dedicated to fatigue study will face increasingly greater difficulties.

The development of fatigue theory and engineering applications requires not only researchers' efforts but also the advancement of related disciplines, technologies, and industries. Since this century, the rapid progress of advanced testing, monitoring and analysis instruments, as well as computer technology has enabled insights into failure mechanisms of materials and structures and comprehensive integration with data science. It led to new findings and subfields, bringing considerable opportunities and challenges.

To systematically report on the latest research advancements in fatigue design for materials and structures, the authors from three countries have collaborated to write this book, combining their expertise and research accumulation in respective subfields. In addition to traditional fatigue research topics and a series of engineering case studies, this book further supplements chapters on very high cycle fatigue, fatigue reliability, artificial intelligence in fatigue, as well as fatigue in additive manufacturing and advanced materials.

This book is of significant reference for engineering and technical personnel engaged in fatigue design and assessment of materials and structures, as well as faculty and students in related disciplines at higher education institutions, to gain a comprehensive understanding of the overall methodologies in materials and structural fatigue. Its publication strongly promotes the progress and high-level development of the global manufacturing industry in the field of fatigue and reliability engineering.

Tatsuo Sakai
Research Organization of Science and Technology
Ritsumeikan University, Kusatsu, Shiga, Japan
September 24, 2024

Preface

In engineering practice, fatigue failure of materials and structures has always been a key challenge threatening the safety of main structures and equipment. Fatigue failure is often caused by accumulated damage under repeated loads, which eventually leads to failure. With the complexity of engineering structures and the diversification of application environments, it is increasingly important to understand and predict their fatigue behavior. However, uncertainty in material properties, changes in loading conditions, deviations in the manufacturing process, and the influence of environmental factors bring challenges to fatigue design requiring models that are more reliable to keep accuracy in fatigue life predictions.

In this context, the combination of physical understanding of fatigue damage and data science provides a new perspective for solving material and structural fatigue problems. In-depth research on physical mechanisms can reveal the microscopic behavior of materials during fatigue, while data science helps us more accurately and cost-effectively predict fatigue life and optimize structural design through the analysis of a large amount of experimental data and simulation results. By comprehensively considering multiple influencing factors and combining physical models with data-driven methods, we can more comprehensively evaluate the fatigue performance of structures, thereby improving the reliability and safety of designs. This fusion can not only extend the service life of structures but also significantly reduce maintenance costs and potential risks caused by fatigue failures.

This book, "Fatigue of Materials and Structures: Physics and Data Science," is a comprehensive overview and in-depth discussion of the latest research advances in the field of fatigue from both physics and data science aspects. It not only provides experts with cutting-edge theories and methods but also provides beginners with a solid foundation for entering the field. Through this book, readers can gain a deep understanding of the basic principles, latest technologies, and application examples of fatigue in different engineering fields.

This book consists of 11 chapters, constructing a comprehensive framework from theory to practice and exploring the fatigue of materials and structures from physics and data science aspects. Chapter 1 is an introduction that clarifies the background and progress of fatigue, laying a foundation for the subsequent chapters. Chapter 2 elaborates on variable amplitude fatigue as well as several case studies of engineering components. Chapter 3 introduces multiaxial fatigue as well as case studies of aeroengine components. Section 4 comprehensively introduces the behavior of fatigue crack growth and relevant fatigue life prediction methods. Chapter 5 provides a systematical introduction on ultra low-cycle fatigue and case studies of a full-scale pipe. Chapter 6 turns into very high cycle fatigue as well as its recent advances and future challenges. Chapter 7 introduces the fatigue reliability issue and its application to fatigue design. Chapter 8 presents the fatigue issues of additively manufactured materials. Chapter 9 introduces the fatigue of joints and bridges as well as several case studies of bridges and methodologies for multi-scale structural analysis.

Chapter 10 elaborates fatigue modeling methods based on artificial intelligence and machine learning. Chapter 11 introduces fatigue issues of advanced materials, especially in ceramics and hydrogels.

In conclusion, this book serves as a vital resource for those seeking to deepen their knowledge of fatigue in engineering contexts. By merging the foundational principles of physical mechanisms with the innovative approaches of data science, this book not only bridges the gap between theory and application but also empowers readers with the tools necessary to tackle the complex challenges of fatigue in modern engineering. Whether you are an expert looking to stay informed of the latest advancements or a beginner eager to build a strong foundation, this book offers valuable insights that will enhance your understanding and contribute to the ongoing evolution of fatigue analysis and design. Through its comprehensive coverage and interdisciplinary approach, this book promises to inspire new strategies for improving the reliability, safety, and longevity of materials and structures across various engineering domains.

Qingyuan Wang
Shun-Peng Zhu
José A.F.O. Correia
Abílio M.P. De Jesus
Grzegorz Lesiuk
September 2024

About the Authors

Qingyuan Wang is a Full Professor of Sichuan University since 2001. He was the President of Chengdu University between 2014 and 2024. He received his Ph.D. in 1998 from Ecole Centrale Paris (University Paris-Saclay), France. He was a visiting Professor at the University of Paris East (France), Technical University of Claustal (Germany), University of New Hampshire (USA), University of Liverpool (UK), and Kyushu University (Japan).

His research activity is focused on materials, mechanics and structures, super long life durability, 3D printing, green buildings, and low carbon materials. He has published over 300 papers in SCI index journals, including *Nature, Nature Commun, Prog Mater Sci, Adv Mater, Adv Func Mater, Adv Science, Acta Mater, IJ Fatigue, etc.* He was named on the list of Chinese Most Cited Researchers from 2014 to 2023.

He served as an Editor-in-Chief for *Buildings and Materials*, a co-chief editor for *Chinese Journal of Experimental Mechanics*, an associate editor for *J Micrometer & Molecular Physics*, an editorial member for *FFEMS, IJ Fatigue*, etc. He has chaired or co-chaired a number of international or national conferences, including the 6th International Conference on Very High Cycle Fatigue (VHCF-6, 2014), 2021+1 Chinese Congress on Mechanics, and The 55th of Advanced Materials Congress (2023, Sweden), 22nd National Conference of Fatigue & Fracture (2024, Chengdu), etc. He has received a number of awards, including the Distinguished Chinese Scholar Award from the NSF of China (2009), The Second Prize of National Natural Science Award (PI, China, 2018), The First Prize of Sci & Tech Progress (PI, The Chinese Society of Mechanics, 2021), European Advanced Materials Award of the Year 2022 (International Association of Advanced Materials), Outstanding Paper Award of 2013–2023 (*Materials Research Letters*), Senior Career Award of 2024 (*Inter J Structural Integrity*), ICCES Distinguished Fellow Award of 2024 (the International Conference on Computational & Experimental Engineering and Sciences and the Tech Science Press) and JSPS Fellowship Award of 2024 (the Japan Society for the Promotion of Science). He is an Effective Fellow candidate of Chinese Academy of Engineering (2019), Fellow of European Academy of Sciences and Arts (2023), and Fellow of European Academy of Sciences (2024).

Shun-Peng Zhu is a Full Professor in the School of Mechanical and Electrical Engineering at the University of Electronic Science and Technology of China, China. He received his Ph.D. in Mechanical Engineering from UESTC in 2011, worked as a PIF Fellow of Politecnico di Milano, Italy from 2016 to 2018 and was a research associate at the University of Maryland, United States from 2010 to 2011. His research, published in scholarly journals and edited volumes, over 150 peer-reviewed book chapters, journals, and proceedings papers explores the aspects of fatigue design, structural reliability, probabilistic failure modeling, multi-physics damage modeling, and life prediction for engineering materials/components.

He received the Second Prize of Natural Science Awards of the Sichuan Province and Chinese Society of Aeronautics and Astronautics in 2022, four Most Cited Paper Awards of *FFEMS Journal* since 2018, Award of Merit (2019) and Robert Moskovic Award (2024) of ESIS TC12, FEUP Medal of University of Porto in 2024, Top 100 Most Influential International Academic Paper Award of China in 2018, Polimi International Fellowship in 2016, Hiwin Doctoral Dissertation Award in 2012, Best Paper Awards of five international conferences. He has been featured as the Most Cited Chinese Researcher in the field of Safety, Risk, Reliability, and Quality since 2018, World's Top 2% Scientists since 2020, and serves as a Co-Editor-in-Chief of the *International Journal of Structural Integrity and ICE-Transport*, an editorial board member of seven international journals, like *IJ Fatigue*, *FFEMS*, *EFA*, etc. and Taylor & Francis book series.

José A.F.O. Correia is an Assistant Professor at the Department of Mechanical Engineering and a Researcher at the CONSTRUCT Institute, Faculty of Engineering, University of Porto, Portugal. Since March 2018, he has been a Guest Professor at the Delft University of Technology (Netherlands). Also, he was an Invited Professor at the University of Porto (2018–2019) and at the University of Coimbra (2016/09–2021/08), and he was an Associate Professor by invitation at UTAD (2022/03–2023/02).

In 2015, He received a Ph.D. in Civil Engineering from the University of Porto. Recently, in July 2024, he obtained the Habilitation in Civil Engineering – Fatigue and Structural Integrity from the University of Porto (Portugal). Since 2019, he has been included in the world's top 2% scientists list of the Stanford University ranking – Engineering, Materials area. He is the co-author of journal papers in the most relevant scientific journals, proceedings in conferences, book chapters, and also the co-author of books. He also edited special issues of scientific journals as a Guest Editor. Since 2015, he has been an editorial board member, editor-in-chief, and section editor of various scientific journals and book series.

He is a supervisor of doctoral and master's students in the areas of Civil and Mechanical Engineering. He is the (co)coordinator at FEUP of eight R&D projects (e.g., FADEST, FiberBridge, SOS-WindEnergy, Hyperloop-Verne, PT-Chinese Project, NEXUS, and SuRe3-OW-GH2) and consulting services (university extension) budgeted at more than 1000k€ and 190k€, respectively.

Abílio M.P. De Jesus is a Full Professor at the Department of Mechanical Engineering, Faculty of Engineering, University of Porto (FEUP), Portugal. With an academic career spanning approximately 28 years, he is also a research member of the Institute of Science and Innovation in Mechanical and Industrial Engineering (INEGI) and an integrated member of the Associated Laboratory for Energy, Transports, and Aeronautics (LAETA). He earned his Ph.D. in Mechanical Engineering from the University of Trás-os-Montes and Alto Douro (UTAD) in 2004 while serving as a Professor there.

In the past 10 years, his research has increasingly focused on manufacturing processes, particularly subtractive and additive methods. He is working to consolidate a research group in this field, with a focus on process simulation, material characterization, hybrid manufacturing (additive and subtractive), and their impact on fatigue and fracture behaviors. He has (co)authored over 250 papers in national and international scientific journals and presented more than 370 scientific works at conferences. His contributions to the fields of fatigue and fracture of materials and structures, as well as manufacturing, are well-recognized, with 355 documents indexed in SCOPUS and an H-index of 46, reflecting over 7000 citations. He is the editor of the Structural Integrity Book series, serves as editor-in-chief of two journals (JMBM, DeGruyter, ICE-Transport), and has coordinated or participated in numerous European projects such as ULCF, FASTCOLD, IN2TRACK2, IN2TRACK3.

Currently, he serves as the Director of the Bachelor's program in Mechanical Engineering at FEUP, Deputy Director of the Master's program in Mechanical Engineering, co-chair of the Technical Line on Advanced Production at LAETA, and co-chair of the Technical Division on Technologies for Materials Processing of the Portuguese Materials Society and Group Leader of LAETA on Materials and Manufacturing. He is also the FEUP coordinator for a Recovery and Resilience Project on railways at FEUP, with a budget of €4.0 million.

Grzegorz Lesiuk is a Professor at the Wroclaw University of Science and Technology (WUST), Poland. He is the Member of the Committee on Mechanics and Civil Engineering of the Polish Academy of Sciences branch Wroclaw, President of the Polish Group of Crack Mechanics (PGMP), and President of the Wroclaw Branch of PTMTS (Polish Society for Theoretical and Applied Mechanics). He received his doctoral degree in technical sciences in 2013 and received his habilitation in 2020 at the Faculty of Mechanical Engineering of Wroclaw University of Technology. His scientific activities are in the field of mechanics, fracture mechanics, and fatigue of materials and structures, as well as new materials and modeling of their durability under cyclic and static loads. In this field, he has more than 120 scientific papers published in renowned national and international scientific and technical journals. He is the co-author of an academic mechanics coursebook and the co-author of a monograph on the problems of degradation of materials and structures. He is a co-organizer of many scientific events of international as well as national scope, i.e., XXICMFM (International Colloquium Mechanical Fatigue of Metals), XVIII KKMP (National Conference on Fracture Mechanics), and the creator of a new initiative under ERASMUS BIP courses for summer and winter schools of fracture mechanics and fatigue of materials. He has been repeatedly awarded for his scientific activities: Admission as a member of the Academy of Young Scientists and Artists operating at the Mayor of the City of Wroclaw – nomination received from the Mayor of the City of Wroclaw (2018), Award of the Rector of Wroclaw University of Technology (2018, 2020), International ESIS Award – TC12 (Merit Award) 2019 and two international TOP 1% Reviewers (Publons&Web of Science) awards – 2019, Lierati Award (2019) – Emerald Journals Group, 2023 – ESIS Award of Merit – Robert Moskovic Award (Serbia, IV, 2023), – FEUP medal University of Porto – 2023. In addition, in his career, he has managed numerous projects of national as well as international scope funded by NCBiR, NCN, and NAWA.

Acknowledgments

For this book, we acknowledge the support and funding provided by Key Program of the National Natural Science Foundation of China (Grant No. 12232004 and 12332012). The authors are grateful to several people who have contributed to this book in various ways: **Changqi Luo, Hao Ma, Zijian Xu, Pengyu Chen, Borui Wu, Mengli Yan,** and **Xinya You** who are membered in the *Structural Integrity & Intelligent Design Group* at School of Mechanical and Electrical Engineering, University of Electronic Science and Technology of China, China.

Additionally, this book was supported by the following grants: base funding – UIDB/04708/2020 and programmatic funding – UIDP/04708/2020 of the CONSTRUCT – Instituto de I&D em Estruturas e Construções – funded by national funds through the FCT/MCTES (PIDDAC); and, individual project grant (2020.03856.CEECIND), awarded to José A.F.O. Correia, by national funds (PIDDAC) through the Fundação para a Ciência e Tecnologia (FCT/MCTES). This work is also a result of the Agenda "NEXUS: Innovation Pact Digital and Green Transition – Transports, Logistics and Mobility", nr. C645112083-00000059, investment project nr. 53, financed by the Recovery and Resilience Plan (PRR) and European Union – NextGeneration EU. The publication has been prepared as a part of the Support Program of the Partnership between Higher Education and Science and Business Activity Sector financed by City of Wroclaw (Poland) and supporting of Wroclaw University of Science and Technology through the program "Academia Professorum Iuniorum." Professor Grzegorz Lesiuk would like to acknowledge the International Visegrad Fund Fellowship project no #62310197.

Also, this book has been partially supported by FCT – Foundation for Science and Technology, through the project PTDC/EME-EME/7678/2020, Giga-Cycle Fatigue Behaviour of Engineering Metallic Alloys. The Associated Laboratory for Energy, Transports and Aeronautics, through the grant UIDB/50022/2020, is fully acknowledged.

Contributors

Rui Calçada
CONSTRUCT-LESE, Faculty of Engineering
University of Porto
Porto, Portugal

Chao He
Failure Mechanics and Engineering Disaster Prevention and Mitigation Key Laboratory of Sichuan Province
College of Architecture and Environment, Sichuan University
Chengdu, China

Jin-Chao He
School of Mechanical and Electrical Engineering
University of Electronic Science and Technology of China
Chengdu, China

Cláudio S. Horas
CONSTRUCT-LESE, Faculty of Engineering
University of Porto
Porto, Portugal

Wentao Huang
Failure Mechanics and Engineering Disaster Prevention and Mitigation Key Laboratory of Sichuan Province
College of Architecture and Environment, Sichuan University
Chengdu, China

Ding Liao
School of Mechanical and Electrical Engineering
University of Electronic Science and Technology of China
Chengdu, China

Yongjie Liu
Failure Mechanics and Engineering Disaster Prevention and Mitigation Key Laboratory of Sichuan Province
College of Architecture and Environment, Sichuan University
Chengdu, China

Changqi Luo
School of Mechanical and Electrical Engineering
University of Electronic Science and Technology of China
Chengdu, China

Xiaopeng Niu
School of Mechanical and Electrical Engineering
University of Electronic Science and Technology of China
Chengdu, China

Kayode Olaleye
Faculty of Mechanical Engineering
Wroclaw University of Science and Technology
Wroclaw, Poland

João Carlos Pereira
Faculty of Engineering
INEGI, University of Porto
Porto, Portugal

Dariusz Rozumek
Faculty of Mechanical Engineering
Opole University of Technology
Opole, Poland

Mieczysław Szata
Faculty of Mechanical Engineering
Wroclaw University of Science and Technology
Wroclaw, Poland

Shaoxiong Xie
Failure Mechanics and Engineering Disaster Prevention and Mitigation Key Laboratory of Sichuan Province
College of Architecture and Environment, Sichuan University
Chengdu, China

Kun Yang
Institute for Advanced Study
Chengdu University
Chengdu, China

1 Introduction to Fatigue

José A.F.O. Correia and Abílio M.P. De Jesus

1.1 BRIEF INTRODUCTION

Today, scientists and designers/engineers have a comprehensive understanding of the fatigue damage phenomenon and its failure mechanisms. This collective awareness is reflected in the definition of fatigue provided by ASTM E1823 standard. This standard describes fatigue as the process of permanent and localized structural change that occurs at the material level when this structure, and naturally the critical locations, is subjected to constant amplitude or variable amplitude loading conditions, which may produce cracks or final fracture after a sufficient number of loading cycles. The incorporation of approaches, methods, and methodologies for fatigue design into contemporary design codes further underscores this awareness, highlighting the importance of addressing fatigue to ensure the safety of structural systems, including mechanical ones.

Although noteworthy advancements were made in fatigue understanding, as evidenced by numerous specialized scientific journals dedicated to the topic, the phenomenon remains not fully comprehended, indicating that further research and progress are yet required.

This chapter provides a review about fatigue methodologies currently accepted. It begins with a short history of fatigue main contributions, followed by an overview of fatigue damage mechanism and the main fatigue methodologies available for life prediction. This chapter then explores probabilistic fatigue approaches, offering insights into current research and developments in this area. Afterward, the main design criteria adopted in the industry are disclosed. This chapter ends with a brief reference to the usage of data science in fatigue assessment and perspectives on the challenges in this promising field.

1.2 HISTORICAL OVERVIEW OF FATIGUE

The modern understanding of fatigue has been shaped by numerous contributions over the last 185 years. While it is not feasible to acknowledge all these contributions here, certain milestones are noteworthy in the history of fatigue [1].

The first studies of fatigue are often credited to the German mining engineer Wilhelm Albert, who in 1837 carried out pioneering tests on conveyor chains, publishing the first documented fatigue test data. Rankine (1842) pointed out and discussed the fatigue resistance of stock axles. Furthermore, life-safe design was explored for the axles of horse-drawn coaches in Morin's (1853) book on the strength of materials. This included maintenance strategies such as crack repairs and identifying critical areas prone to cracks. The word "fatigue" was reportedly coined

DOI: 10.1201/9781003405122-1

by Braithwaite, in 1854, when he documented failures in various machinery due to fatigue, also discussing permissible stresses for components under fatigue load.

From 1837 to 1958, numerous incidents involving fatigue failure in railway car axles were recorded, leading to many casualties. From 1858 to 1870, Wöhler's work became a cornerstone in fatigue research. He experimented on railway axles and was among the first to be aware of the importance of amplitude and mean stress in material fatigue degradation. Wöhler yet designed early testing machines and contributed to service load measurements and finite life design. His data were initially presented in tables, but Spangenberg later introduced graphical S-N curves to represent this data, which since 1936 have been known as Wöhler curves, honouring his contributions. Basquin further refined this by using bi-logarithmic axes to depict these curves, introducing a power relationship to describe fatigue experimental data.

Bauschinger's research, between 1870 and 1905, was crucial for understanding metal cyclic plasticity, foundational for the low-cycle fatigue (LCF) theory, which was later developed independently by Coffin and Manson. Another relevant aspect was the introduction of the stress concentration factor associated with fatigue, by Kirsch (1898), due to its impact on fatigue damage at critical locations. In 1903, the first descriptions of the metallurgical aspects of fatigue, observing slip bands on the sample surfaces, were introduced by Ewing and Humfrey.

The period from 1905 to 1920 saw the pioneering full-scale fatigue tests on full-scale aircraft components, focusing on improving fatigue strength. The concept of "notch effect" emerged in literature, and authors like Smith, Haigh, Gough, Griffith, and others became prominent. Between 1920 and 1945, significant fatigue research areas were developed, and many are still under investigation today. These include studies on variable amplitude loading, with key contributions by Gassner, and the development of fatigue damage accumulation theories called linear damage rules by Palmgren and Miner, who introduced the widely known Palmgren-Miner rule. Griffith's work laid the foundations for fracture mechanics, and Forest conducted the initial tests for the crack propagation characterization. In 1951, statistical methods applied to the materials strength were pioneered by Waloddi Weibull, and Müller-Stock evaluated large data sets of fatigue experiments.

Investigations by Coffin and Manson between 1945 and 1956 led to the development of strain-life models, specifically for the LCF regime, stimulated by the needs of the aircraft industry, which also contributed to the discussion of safe-life and safe-failure design philosophies that were later incorporated into design codes and standards. The validity of the Palmgren-Miner rule was extensively studied, and Irwin's 1958 work on the stress intensity factor, SIF or K, further contributed to the topic of linear elastic fracture mechanics. Post-1960, fracture mechanics and its application to fatigue advanced significantly, particularly with Paris's 1962 work connecting fatigue and fracture mechanics. Paris's relation has since been a fundamental model, though subsequent research has sought to refine it. Elber's concept of "crack closure" under variable amplitude loading was another key development. The evolution of testing machines, including servo-hydraulic systems, has enabled extensive research into variable amplitude loading and multiaxial fatigue loading. Resonance testing technology has appeared more recently facilitating the research into the very high cycle fatigue regimes.

Despite these developments, several areas require further research, such as predicting fatigue lifetime under variable loading, transferring specimen-based fatigue data to real components, corrosion fatigue, and the combined effects of high temperatures and fatigue stresses [2–5]. Multiaxial stress histories, especially those arising from non-proportional variable amplitude loading, also remain a challenging field. Large structures (e.g., bridges, offshore structures), which often face multiaxial, variable amplitude loads, and potential corrosion, pose complex problems that necessitate probabilistic approaches due to the high variability in the fatigue properties of old materials (fatigue resistance) or due to uncertainty in the fatigue actions.

1.3 GENERAL CONCEPTS CONCERNING FATIGUE

1.3.1 Fatigue Damage Evolution

Early in the 20th century, microscopic studies revealed that fatigue cracks initially form as microscopic cracks in consistent slip bands, which are not observable to the naked eye. As research progressed, more detailed observations about the progression of these small cracks emerged. Investigations have suggested that crack growth can begin almost immediately if cyclic stresses above the yield strength (or endurance limit) are imposed. The fatigue strength limit (or fatigue endurance limit) is defined as the cyclic stress threshold above which fatigue failure will occur in the material or mechanical/structural component. Even though crack nucleation starts early, these micro-cracks often continue undetectable for a significant portion of the fatigue lifetime. Once cracks become detectable, the residual lifetime of a material sample or structural component is typically a small fraction of its entire life [6]. However, this fraction may represent a larger quantity when dealing with real-world structural systems or engineering structures (e.g., vessels, airframe structures, and bridges) due to the size of components and redundancy in real structures.

After a micro-crack has formed, the growth process can remain slow and irregular, conditioned by the material's microstructure, including grains interfaces. However, once the crack has propagated beyond the initial nucleation site, its growth typically becomes more consistent. This marks the onset of the primary crack propagation phase. As illustrated in Figure 1.1, the fatigue life leading up to failure embraces two distinct stages: the fatigue crack incubation (micro-cracks) and the macroscopic fatigue crack growth (macro-cracks). It is crucial to distinguish between these two stages because various surface conditions can significantly impact the initiation phase but have little effect on the propagation one. Notably, different methods are

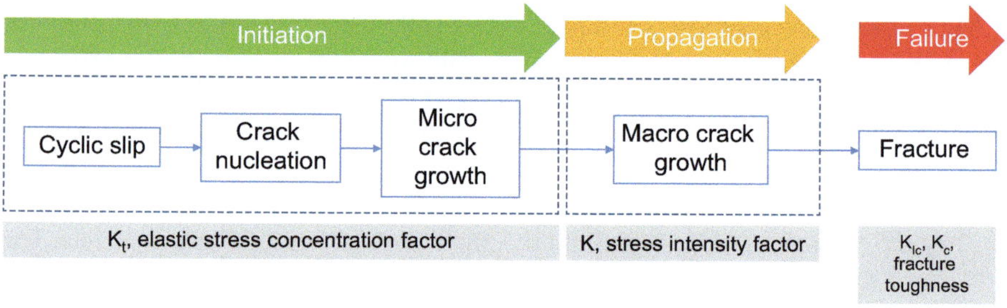

FIGURE 1.1 Stages of fatigue damage evolution with influencing parameters. (Adapted from Ref. [17].)

used for the prediction of each phase. For fatigue crack initiation period, the elastic stress concentration factor, K_t, is the governing mechanical factor, while the stress intensity factor, K or SIF, is the most important and most used leading factor for the fatigue crack propagation period.

Fatigue crack nucleations and subsequent crack macroscopic fatigue crack growth result from cyclic slip bands, which are generated by cyclic plasticity due to dislocations movement. Fatigue damage manifests even at stress under the yield strength, where plastic deformation affects only a few material grains, in the form of micro-plasticity. This micro phenomenon is more clearly observed at the surface, as grains are only constrained by material on one side, making them more susceptible to deformation.

Cyclic slip is dependent on the application of cyclic shear stresses. At a microscopic level, this shear stress is unevenly applied throughout the gains of the material. Variations in shear stresses occur across different crystallographic slip planes due to differences in grain size, shape, orientation, and the material's elastic anisotropy. Some surface grains experience more favourable conditions for cyclic slip compared to others. Furthermore, as the load increases, the slip induces strain hardening within the slip band. Consequently, larger shear stress will be applied in the reverse direction within the same slip band after unloading. This reversed slip, despite having occurred in the same slip band, mobilizes the neighbour parallel slip planes.

Reduced constraint at the material-free surface is often claimed as a factor favouring crack nucleation at the surface. Nevertheless, additional reasons support the existence of crack nucleation at the free surface. Practically, inhomogeneous stress distribution caused by notches, holes, or other geometric discontinuities leads to peak stresses at the surface, creating stress concentrations. Additionally, surface roughness/finishing further facilitates crack nucleation. These combined factors suggest that the fatigue crack nucleation phase is predominantly a surface-related problem.

When a micro-crack remains on the scale of a single grain of a crystalline structure with multiple slip systems, it causes a non-uniform stress field at the micro scale, with a crack tip stress concentration. This localized stress can activate multiple slip systems. As the crack extends into neighbour grains, the constraint on slip increases. Consequently, it becomes more challenging for slip displacements to be accommodated by a single slip system alone, necessitating the activation of slip planes in various directions. As a result, the direction of the micro-crack path often deviates from the original slip band orientation and tends to align closely perpendicular to the loading, as illustrated in Figure 1.2 [6].

Micro-crack propagation is influenced by cyclic plasticity, and obstacles to slip can create a barrier for crack advancement. Bibliography reveals that micro-crack growth often starts with higher rates, which afterward decreases or even arrests due to material microstructural obstacles. At this stage, surface-specific factors become less relevant. For example, the reduced constraint on cyclic sliding at the surface of the free material does not apply internally, where inclusions or other irregularities can trigger crack initiation. Additionally, surface finishing and other surface features have minimal impact on crack growth once it has penetrated the material. Thus, the resistance to crack growth within most materials depends on their inherent properties.

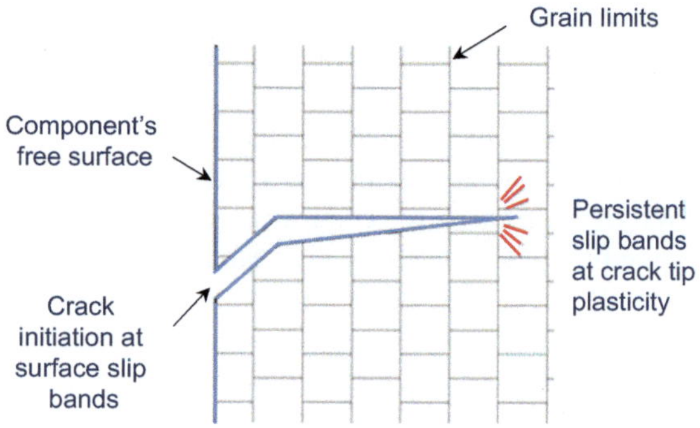

FIGURE 1.2 Cross section of micro-crack growth. (Adapted from Ref. [17].)

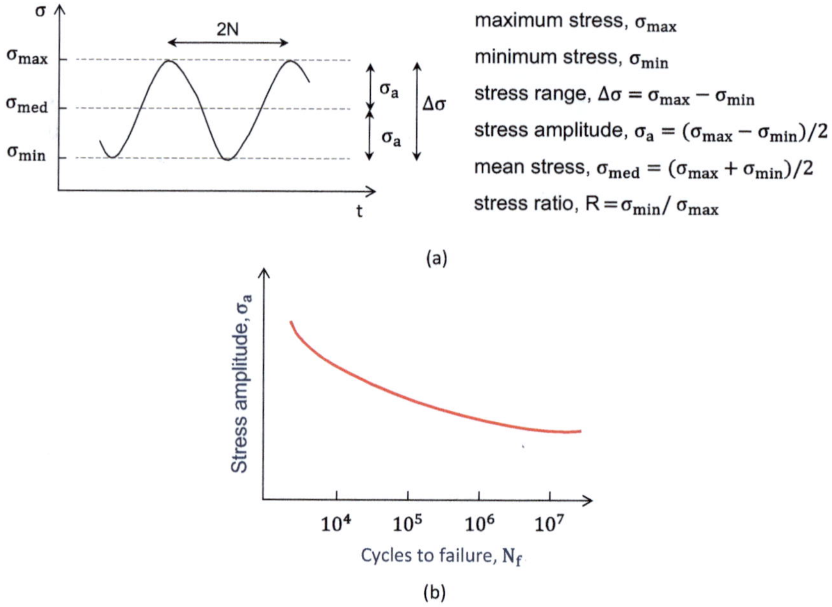

FIGURE 1.3 Fatigue loading and fatigue resistance representations. (a) Constant stress amplitude; (b) Stress-life S-N diagram. (Adapted from Ref. [17].)

1.3.2 Main Factors Influencing Fatigue Damage

Since Wöhler's groundbreaking work on metals, it is well established that fatigue damage is linked to the time-fluctuation of stresses or strains. Increased variation in these stresses or strains accelerates the rate of fatigue damage. Fatigue damage assessment is commonly described through fatigue curves, which are generally derived for cyclic experiments of constant stress or strain amplitude, performed on materials and mechanical/structural details. Figure 1.3 illustrates common S-N curves for constant stress amplitude, showing the relationship between a stress variation parameter (e.g., stress amplitude), known as fatigue loading, and the fatigue lifetime (cycles to rupture).

A distinct horizontal threshold at long fatigue lives (high-cycle fatigue) may be exhibited for some materials, normally, representing the conventional fatigue endurance limit. However, other materials exhibit a continuous progressive reduction in fatigue resistance at high fatigue lives, demonstrating that some materials do not exhibit a well-defined endurance limit. This is often observed in certain high-strength steels and aluminium alloys. In these cases, the endurance limit is agreed as the stress range or amplitude corresponding to a sufficiently high fatigue lifetime.

S-N curves are usually developed using the stress parameter (e.g., stress amplitude, stress range, and maximum stress) as a function of the number of cycles, as the primary fatigue parameter. Then, each S-N curve is typically constructed for a specific secondary stress parameter, such as the mean stress, the maximum stress, the minimum stress, or the stress ratio. Generally, the fatigue resistance diminishes as the mean stress increases. Figure 1.4a illustrates a set of S-N curves (amplitude of stress *vs.* fatigue lifetime) highlighting how different mean stresses affect the curves. This figure shows how the fatigue strength limit is influenced by the mean stress. The influence of mean stress on fatigue resistance can be illustrated using constant life diagrams, such as those suggested by Goodman, Gerber, and Soderberg, which are established in the literature and design criteria (refer to Figure 1.4b). These diagrams help to model the mean stress effects at the conventional fatigue endurance limits or the fatigue resistance assumed for the specific fatigue life to be defined by the designer or engineer. In addition to these relationships, several other methods can be found in the literature and used to explain the average stress effects [7].

Furthermore to stress variation (e.g., stress range, stress amplitude), which are primary fatigue parameters, and mean stress cycle, which are secondary parameters, several other factors influence fatigue behaviour. For steels (e.g., carbon, low-alloy), the following influencing factors may be referred [8]:

i. Materials monotonic strength;
ii. Materials ductility;
iii. Fabrication defects within the steel;
iv. Surface finishing;

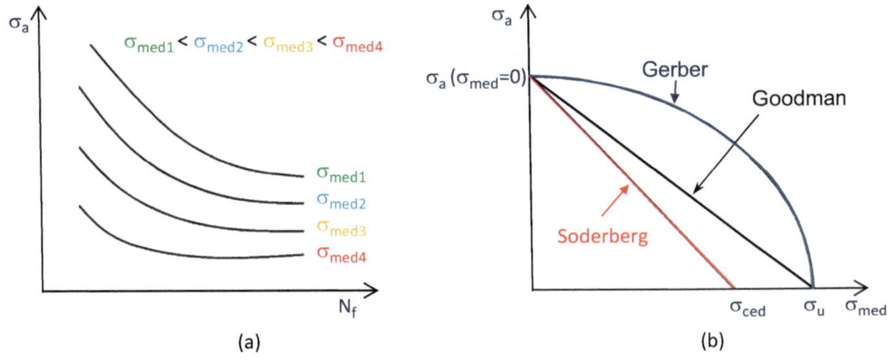

FIGURE 1.4 Mean stress effects on fatigue strength. (a) S-N diagrams; (b) constant life diagrams. (Adapted from Ref. [17].)

v. Size effects;
vi. Notch effects;
vii. Corrosive environments (e.g., H_2);
viii. Self-equilibrated internal residual stresses.

Mechanical and structural components and details often exhibit stress concentrations, K_t, due to surface geometric discontinuities, which leads to a reduction in fatigue resistance. Nevertheless, the reduction in fatigue resistance is not typically proportional to the elastic stress concentration. Instead, a lower decrease in fatigue resistance is observed, characterized by the fatigue notch reduction factor (K_f). This K_f factor is therefore defined as the ratio between the fatigue strengths of smooth and notch specimens, at a given fatigue endurance; K_f is generally influenced by both the fatigue endurance and stress levels, being reduced as stress levels increase and fatigue lives shorten, primarily as a result of notch root plasticity. When the material's sensitivity to notch effects increases, the fatigue notch reduction factor (K_f) approaches the elastic stress concentration factor (K_t).

The fatigue design approaches for components with notches, submitted to dominant elastic fatigue, were initially developed by Neuber [9,10] and Peterson [11]. More recently, the theory of critical distances (TCD) has provided additional insights into this area [12,13].

1.4 FATIGUE LIFE PREDICTION

1.4.1 Global Fatigue Approaches

A wealth of research has been carried out to understand the fatigue behaviour of components or structural details and to contribute to the development of fatigue design procedures. Fatigue damage is inherently complex and challenging to model accurately due to the many interacting factors that influence how structures respond to repeated loading. As a result, design and assessment methods often rely on empirical approaches. Since a universal theory capable of consistently predicting fatigue responses does not yet exist, experimental testing remains essential for evaluating fatigue resistance. Both small-scale (e.g., material smooth specimens) and full-scale (e.g., components) tests have been implemented to study the fatigue characteristics of these components.

Depending on the specifics of the research and the examined region of the S-N diagram, various empirical models have been proposed to describe the experimental data (e.g., linear, piecewise linear, parabolic, and hyperbolic) [14]. Some common relations are listed in Table 1.1. Nevertheless, not all of these relations find a physical meaning.

S-N curves, typically represented as straight lines on log-log scales when plotting stress range ($\Delta\sigma$) against fatigue life (N_f), are commonly used to describe the cyclic behaviour of structural details, as shown in Figure 1.5. The cyclic lifetime (N_f) is determined based on a constant amplitude (σ_a) or stress range ($\Delta\sigma$). To determine the representative strength of a specific structural detail, multiple specimens must be

TABLE 1.1
Literature-Based Relations Proposed to Describe the S-N Mean Lines [14]

Model	Function Form
Wöhler (1870)	$\log N = A - B \cdot \Delta\sigma;\ \Delta\sigma \geq \Delta\sigma_0$
Basquin (1910)	$\log N = A - B \cdot \log \Delta\sigma;\ \Delta\sigma \geq \Delta\sigma_0$
Strohmeyer (1914)	$\log N = A - B \cdot \log(\Delta\sigma - \Delta\sigma_0)$
Palmgren (1924)	$\log(N + D) = A - B \cdot \log(\Delta\sigma - \Delta\sigma_0)$
Palmgren (1924)	$\log N = A - B \cdot \log(\Delta\sigma - \Delta\sigma_0)$
Weibull (1949)	$\log(N + D) = A - B \cdot \log\left(\frac{(\Delta\sigma - \Delta\sigma_0)}{(\Delta\sigma_{st} - \Delta\sigma_0)}\right)$
Stüssi (1955)	$\log N = A - B \cdot \log\left(\frac{(\Delta\sigma - \Delta\sigma_0)}{(\Delta\sigma_{st} - \Delta\sigma)}\right)$
Bastenaire (1972)	$(\log N - B)\cdot(\Delta\sigma - \Delta\sigma_0) = A \cdot \exp\left[-C \cdot (\Delta\sigma - \Delta\sigma_0)\right]$
Spindel-Haibach (1981)	$\log\left(\frac{N}{N_0}\right) = \frac{\lambda + \delta \cdot \left(-\log(1-p)\right)^{1/\beta}}{\log(\Delta\sigma / \Delta\sigma_0)}$
Castillo et al. (1985)	$\log(N/N_0) = A \cdot \log(\Delta\sigma/\Delta\sigma_0) - B \cdot \log(\Delta\sigma/\Delta\sigma_0)$ $+ B \cdot \left\{(1/\alpha) \cdot \log\left[1 + (\Delta\sigma/\Delta\sigma_0)^{-2\alpha}\right]\right\}$
Pascual and Meeker (1999)	$\log N = A - B \cdot \log(\Delta\sigma - \Delta\sigma_0)$
Kohout and Vechet (2001)	$\log\left(\frac{\Delta\sigma}{\Delta\sigma_\infty}\right) = \log\left(\frac{(N + N_1)}{(N + N_2)}\right)$
Castillo and Canteli (2001)	$N_i = \exp\left[\log N_0 + \frac{\lambda + \delta\left(-\log(1-p)\right)^{1/\beta}}{\log \Delta\sigma - \log \Delta\sigma_0}\right]$
Correia et al. [15]	$\psi = q(N_f)$ $p = 1 - \exp\left\{-\left[\frac{\log(N_f/N_0)\cdot\log(\psi/\psi_0) - \lambda}{\delta}\right]^{\beta}\right\}$ $\log(N_f/N_0)\cdot\log(\psi/\psi_0) \geq \lambda$
Correia et al. [16]	$\psi(N) = \psi_e\left(\frac{N + N_u}{N + N_e}\right)^{b'}$

fatigue tested. Deviations between predicted mean strengths and actual test results frequently arise due to many influencing factors [17].

It is common in fatigue recommendations adopted in design practice [18] to follow the Basquin proposal for representing the S-N relation:

$$\Delta\sigma^m \cdot N_f = C \tag{1.1}$$

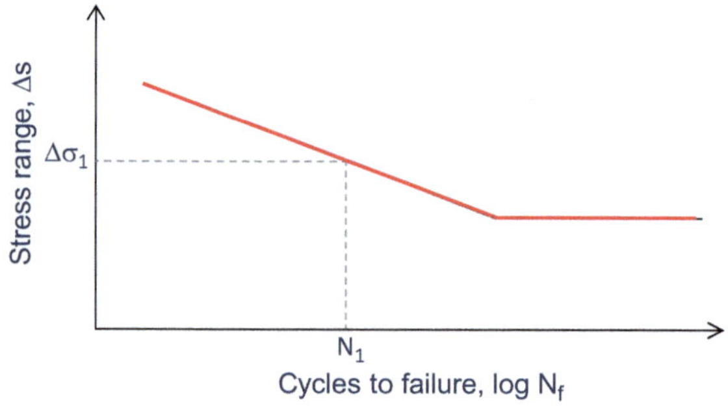

FIGURE 1.5 Linear representation of the S-N curve on bi-logarithm representation. (Adapted from Ref. [17].)

where exponent m and coefficient C are material or detail constants. This S-N relation may alternatively be described by the equivalent linearized model, which is more appropriate for the constants identification through a linear regression analysis [19]:

$$Y = A + B \cdot X \tag{1.2}$$

X (= $\log(\Delta\sigma)$) is the independent variable and Y (= $\log(N_f)$) is the dependent variable; A and B are constants to be evaluated from linear regression analysis. Therefore, it is viable to translate the S-N curves through the relations [19].

$$\begin{cases} \log(N_f) = A + B \cdot \log(\Delta\sigma) \\ \log(\Delta\sigma) = -\dfrac{A}{B} + \dfrac{1}{B} \cdot \log(N_f) \end{cases} \tag{1.3}$$

where parameters A and B are related to C and m, as follows:

$$\begin{cases} C = 10^A \\ m = -B \end{cases} \tag{1.4}$$

For example, to evaluate the fatigue residual life of old metallic bridges, offshore structures, energy transmission towers, and others, it is essential to identify and categorize critical structural details. The management of fatigue lifetime in structural elements is outlined by design regulations from various countries or geographic regions (e.g., Europe, North America). Eurocode 3 [20], BS 5400 [21], and AASHTO [22] are some references for structures, including bridges. Eurocode 3 does not address the fatigue resistance of riveted shear connections which are frequently found in old steel bridges and towers [20]. A practical approach is to analyse riveted joints as equivalent to those using non-preloaded high-strength bolts, where Eurocode 3 specifies the use of Detail Category 112 (refer to Figure 1.6). Both Åkesson and Edlund [23] and Helmerich et al. [24] recommended Eurocode 3, Detail Category 71, for riveted joints, considering it a more suitable option than Detail Category 112, despite its conservative nature for stress ranges below approximately 60 MPa.

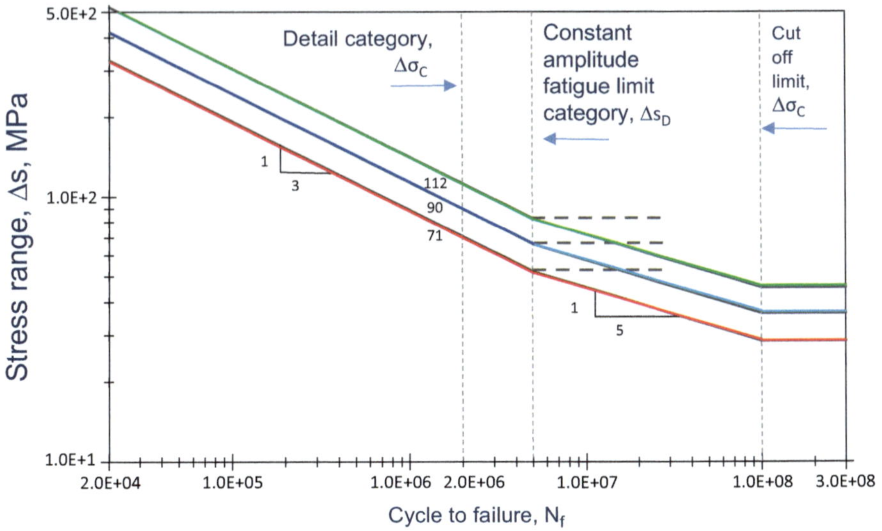

FIGURE 1.6 Fatigue design curves for nominal stress range, according to EC3, part 1–9 [20]. (Adapted from Ref. [17].)

Taras and Greiner [25] conducted a statistical analysis of extensive fatigue data from the literature on riveted connections in ancient bridges. They recommended categorizing the fatigue resistance of riveted connections into five classes, based on a design curve with a slope $m=5$, instead of $m=3$, as suggested in design codes like EC3-1-9 [20]. Among these classes are single and double shear riveted joints under tension. For double shear riveted splices, they recommend a design S-N curve with a slope of $m=5$ and a fatigue strength of 90 MPa at 2 million cycles, and for single shear riveted splices, a curve with a slope of $m=5$ and a fatigue strength of 71 MPa at 2 million cycles. Additionally, they highlighted the importance of accounting for mean stress effects in riveted joints and suggested the use of normalized stress.

The implementation of the Eurocodes has resulted in a halt in the development of national codes in Europe. Nevertheless, riveted joints are addressed in BS 5400, allowing for further treatment within this framework. BS 5400 classifies the riveted shear splice detail as Class D and distinguishes between riveted and bolted shear connections, unlike EC3. Nevertheless, fatigue prediction provided for riveted joints in this standard is considered unsatisfactory [17].

AASHTO code also includes a specific fatigue prediction method for riveted shear joints, categorized under Detail Category D, which is extensively used by railroad authorities in North America and is the most reflective of recent test data among the reviewed standards [17].

1.4.2 Local Fatigue Approaches

The philosophy behind design against fatigue has transitioned from criteria based on fatigue limit and infinite life to methodologies that account for finite life behaviour [26]. Currently, local approaches to fatigue, which apply local fatigue parameters to link fatigue test results, particularly those related to fatigue crack initiation, are prevalent.

To predict fatigue life for specific conditions, various local fatigue parameters have been proposed. These local approaches are typically classified into three main types: stress-, strain-, and energy-based methods. Each type relies on stress, strain, or energy parameters, respectively. Models have been proposed to address the effects of mean stress levels in these approaches, which are discussed in the subsequent paragraphs. Furthermore, fatigue approaches that allow various local fatigue criteria have also been proposed [15,16].

1.4.2.1 Stress-Life Methods

The stress-life approach predicts the fatigue life based on a fatigue stress parameter (e.g., amplitude, range). This approach compares the fatigue stress parameter with a known S-N curve. The S-N diagrams are constructed based on test data, by a fitting procedure of an existing relation.

The stress-life approach is primarily applied to high-cycle fatigue scenarios, as it assumes a linear stress-strain relationship. In low-cycle fatigue conditions, where the material undergoes a nonlinear stress-strain relation, this method is less applicable [26]. In the context of S-N diagrams, the relationship between the fatigue stress parameter (e.g., stress amplitude, $\Delta\sigma/2$) and the number of cycles to failure (N_f) is often approximated by a linear representation if both the stress parameter and life (cycles) are plotted on a log-log scales. This linear approximation facilitates the prediction of fatigue life and can be defined for stress amplitude as the fatigue stress parameter, as:

$$\frac{\Delta\sigma}{2} = \sigma'_f \left(2N_f\right)^b \quad (1.5)$$

where σ'_f is the fatigue strength coefficient, and b is the fatigue strength exponent.

In numerous realistic loading scenarios, non-null mean stresses are common. The impact of mean stress is frequently disregarded in fatigue life calculations, as for example for welded components the substantial residual stresses present in welds generally overshadow any effects of the applied stress level. Conversely, for non-welded components, it is essential to consider the mean stress effects in fatigue assessments to ensure more realistic predictions. Morrow [27,28] introduced a correction method to model the impact of mean stress on fatigue. This correction is formulated as follows:

$$\frac{\Delta\sigma}{2} = \left(\sigma'_f - \sigma_m\right)\left(2N_f\right)^b \quad (1.6)$$

where σ_m represents the mean stress, while the other variables are consistent with those in Eq. (1.5). The influence of a tensile (positive) mean stress is effectively modelled as a decrease in fatigue strength. This approach assumes that there is a combination of stress amplitude, $\Delta\sigma/2$, and mean stress, σ_m, which will result in the same cycles to failure as a fully reversed (null mean stress) stress amplitude, $(\Delta\sigma/2)_{-1}$:

$$\left(\frac{\Delta\sigma}{2}\right)_{-1} = \frac{\Delta\sigma/2}{1-\sigma_m/\sigma'_f} \quad (1.7)$$

The correction by Morrow has been shown to be quite effective for structural steel grades [29].

1.4.2.2 Strain-Life Methods

The strain-life approach is extensively used in fatigue, particularly for predicting fatigue crack initiation from notches. Unlike the stress-based approaches, which primarily consider stress as a fatigue parameter, the strain-based approach takes into account the plastic deformation that can occur in localized regions (e.g., notches) where cracks start. This method depends on the assumption that the behaviour of the material at the root of a notch (stress concentrator) is similar to that given by a smooth specimen subjected to strain-controlled cyclic loading of equivalent strain [28].

As a substitute for using the S-N curve as in the stress-based approach, the strain-based approach relies on a strain *versus* life relation and the cyclic curve of the material, the stress amplitude *versus* stress amplitude curve. These curves are characterized by a set of constants (coefficients and exponents) that define the fatigue properties of the material. Initially, there was a lack of sufficient fatigue data to determine these properties accurately for many engineering metals, leading to various empirical equations that correlated fatigue properties with tensile properties [28]. However, the advancement of servo-controlled mechanical testing machines and the progress of strain-based approaches have reduced the dependence on these empirical equations, providing a wealth of data on fatigue properties for numerous materials [17].

For materials at room temperature, cyclic hardening or softening typically occurs at first cycles stabilizing afterwards. The stabilized cyclic stress *versus* strain curve is commonly represented through the Ramberg-Osgood mathematical relation [29,30]. This curve is often derived from multiple test specimens, each one tested at different strain-controlled amplitudes, or from a unique specimen, tested under an incremental step test scheme [28]. Equation (1.8) relates the elastic strain amplitude, $\Delta\varepsilon_E/2$, with the number of cycles to fatigue crack initiation:

$$\frac{\Delta\varepsilon_E}{2} = \frac{\sigma'_f}{E}\left(2N_f\right)^b \tag{1.8}$$

where E is Young's modulus.

The relationship between the plastic strain amplitude and fatigue crack initiation life is given by the following relation, which can be linearized using logarithmic transformations:

$$\frac{\Delta\varepsilon_P}{2} = \varepsilon'_f\left(2N_f\right)^c \tag{1.9}$$

where ε'_f and c stand respectively for the fatigue ductility coefficient and exponent. Adding both components of the elastic and plastic strain amplitude, as shown by Eqs. (1.8) and (1.9), results in one of the most important relations for the strain-based approaches (refer to Figure 1.7):

$$\varepsilon_a = \frac{\Delta\varepsilon_E}{2} + \frac{\Delta\varepsilon_P}{2} = \frac{\sigma'_f}{E}\left(2N_f\right)^b + \varepsilon'_f\left(2N_f\right)^c \tag{1.10}$$

To support the application of the strain-based approaches, the Ramberg-Osgood equation, has been playing an essential role, which is defined as follows:

$$\frac{\Delta\varepsilon}{2} = \frac{\Delta\sigma/2}{E} + \left(\frac{\Delta\sigma/2}{K'}\right)^{1/n'} \tag{1.11}$$

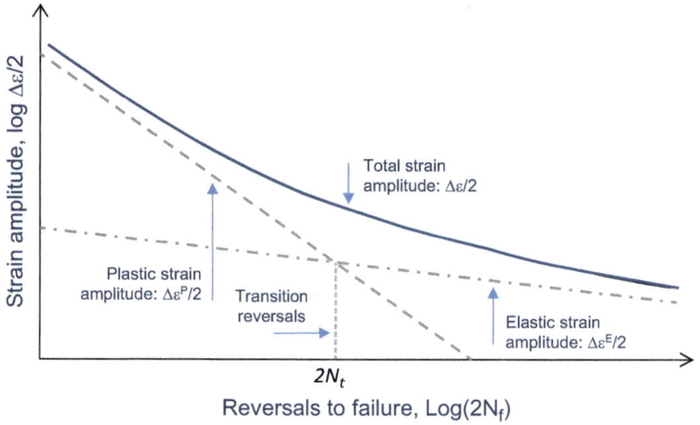

FIGURE 1.7 Schematic illustration of the strain-life relations. (Adapted from Ref. [17].)

where K' and n' are Ramberg-Osgood constants, i.e., the cyclic strain-hardening strength coefficient and exponent, respectively.

The effect of mean stress can be included using the Smith, Watson, and Topper (SWT) model, for the strain-based method [31]. This model is effective for a vast number of materials [29]. The SWT model postulates that the fatigue lifetime, considering mean stress, is dependent on the product of the maximum stress with the strain amplitude, $\Delta\varepsilon/2 \cdot \sigma_{\max}$. The model is formulated as:

$$\mathrm{SWT} = \varepsilon_a \cdot \sigma_{\max} = \frac{\left(\sigma'_f\right)^2}{E}\left(2N_f\right)^{2b} + \sigma'_f \cdot \varepsilon'_f \left(2N_f\right)^{b+c} \tag{1.12}$$

1.4.2.3 Energy-Life Methods

Experimental evidence demonstrated the crucial role that plastic deformation plays in the fatigue phenomenon. Cyclic plasticity, which involves slip along specific crystallographic planes and dislocations movement, is closely related to cyclic stress at the microscopic level. During these irreversible deformations, strain energy is dissipated [26,32–34].

The energy-based approach, which uses energy as a fatigue parameter, highlights the interrelation between stress, strain, and fatigue damaging. This method unifies high and LCF and is promising for correlating fatigue data from specimens of varying geometries and sizes tested under different conditions [17]. Although it has been primarily developed over the last decades and is not yet widely used, the method shows significant potential.

The total dissipated energy up to fatigue failure is generally considered to be dependent on the cumulative number of cycles endured. The area under stress-strain hysteresis loops corresponding to a cycle is directly linked to the fatigue damage imparted by that cycle [26]. Various energy-related parameters have been proposed, including variations to account for the mean stress of the cycle. For example, the plastic strain energy density per cycle (ΔW^P) is more suitable for large plastic strains (low-cycle fatigue) but is less effective for small strain magnitudes because, in high-cycle fatigue, the plastic strain energy is minimal and challenging to measure accurately. To overcome the previous limitations, the combined plastic and elastic tensile strain energy density per cycle (ΔW^t) was suggested to model the mean stress effects [35] and is advisable for deformation-controlled conditions.

Ellyin and Kujawski [36] proposed that the total strain energy per reversal, ΔW, can be defined as:

$$\Delta W = \frac{1}{2}\Delta W^P + \frac{1}{2}\Delta\sigma\Delta\varepsilon \tag{1.13}$$

A more appropriate criterion to account for the mean stress effect is established by setting the parameter, ΔW^t as:

$$\Delta W^t = \Delta W^P + \frac{(\sigma_{\max})^2}{2E} \tag{1.14}$$

In the previous two equations, $\Delta\sigma$ and $\Delta\varepsilon$ are the stress and strain ranges, and $\sigma_{\max}$ is the maximum stress of the cycle. The above parameters of energy fatigue can be correlated with the number of cycles to failure, through a power relation. For example, using ΔW^t as a fatigue parameter, it results as:

$$\Delta W^t = \kappa_t \left(2N_f\right)^{\alpha_t} + \Delta W_0^t \tag{1.15}$$

where $\alpha_t < 0$ and $\kappa_t > 0$ are material parameters and ΔW_0^t is the total strain energy range threshold so-called the energy-based conventional fatigue limit. Ellyin [26] proposed a more elaborated energy fatigue relation, accounting explicitly for the mean stress:

$$D = \Delta W^P + f\left(\frac{\sigma_m}{\Delta\sigma/2}\right)\frac{(\Delta\sigma/2)^2}{2E} = \kappa_t\left(2N_f\right)^{\alpha_t} + \Delta W_0^t \tag{1.16}$$

$$f\left(\frac{\sigma_m}{\Delta\sigma/2}\right) = 1 + \eta\left(\frac{\sigma_m}{\Delta\sigma/2}\right) \tag{1.17}$$

where η is a coefficient, considered a material property, varying between 0 and 1. This coefficient portrays the sensitivity of the material to the cycle mean stress, to be calibrated from high-cycle experimental data representing different mean stresses [15,17].

1.4.2.4 Fatigue-Fracture Approaches

Methods based on fracture mechanics have brought new possibilities for fatigue-fracture modelling. A significant milestone in this area is credited to Paris [37], who first identified the relationship between the stress intensity factor range of an existing crack and its propagation rate:

$$\frac{da}{dN} = f(\Delta K) \tag{1.18}$$

where da/dN is the fatigue crack propagation rate, and ΔK is the stress intensity factor range. The function f relates to the ΔK and needs to be defined. Paris proposed a power-law relationship between the fatigue crack propagation rate and the variation (range) of the stress intensity factor [37]:

$$\frac{da}{dN} = C \cdot (\Delta K)^m$$

$$\Delta K = K_{\text{max}} - K_{\text{min}} \tag{1.19}$$

$$\Delta K_{\text{th}} \leq \Delta K \leq K_c$$

where C and m are the material parameters, K_{max} is the maximum stress intensity factor, K_{min} is the minimum stress intensity factor, ΔK_{th} is the threshold value of the stress intensity factor range, and, K_c is the fracture toughness. This relationship for fatigue crack propagation was validated for a specific regime of fatigue crack growth, particularly the so-called regime II. This regime is between near-threshold fatigue crack propagation (regime I) and near-unstable crack propagation (regime III), as depicted in Figure 1.8.

Although the Paris relation has made a significant impact, it has notable limitations that have been addressed by more complex fatigue crack propagation models. Beden et al. [38] provided a comprehensive literature survey of various fatigue crack growth models extensible to both constant and variable amplitude loading. One key parameter affecting the fatigue crack growth, which the basic Paris model does not consider, is the stress ratio (mean stress) effect. To address this shortcoming, Walker [17,38] introduced an alternative relation:

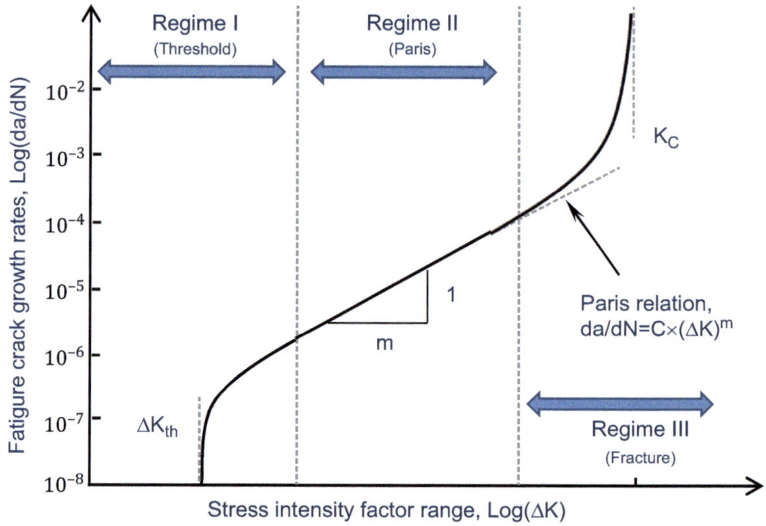

FIGURE 1.8 Different fatigue crack growth regimes. (Adapted from Ref. [17].)

$$\frac{da}{dN} = C \cdot \left(\frac{\Delta K}{(1-R)^{1-\gamma}} \right)^m = C \cdot \left(\overline{\Delta K} \right)^m \tag{1.20}$$

This model adds an additional parameter, γ, in addition to Paris's constants, C and m, $\overline{\Delta K}$ is the Walker stress intensity factor range. When $R = 0$, this relation delivers the Paris relation. The γ parameter is determined as the best-fitting value to experimental data – a unique straight line on the log-log map of *da/dN versus* ΔK, where ΔK is defined as:

$$\overline{\Delta K} = \frac{\Delta K}{(1-R)^{1-\gamma}} \tag{1.21}$$

A two-parameter approach was introduced by Walker for modelling fatigue crack growth. The Walker fatigue crack growth model is then reformulated as follows:

$$\frac{da}{dN} = C \cdot \left[(\Delta K)^{\gamma} \cdot (K_{max})^{1-\gamma} \right]^m = C \cdot \left[(\Delta K)^{1-p} \cdot (K_{max})^p \right]^m \tag{1.22}$$

Dinda and Kujawski [17,38] suggested a modification of the previous model (Eq. 1.22) by filtering the negative (compressive) fraction of the range of the stress intensity factor:

$$\frac{da}{dN} = C \cdot \left[\left(\Delta K^{+} \right)^{1-p} \cdot (K_{max})^p \right]^m \tag{1.23}$$

This approach to fatigue crack propagation, which utilizes a two-parameter fatigue crack driving force, was also adopted by other researchers. Notable examples include the work of Donald and Paris [39] and Vasudevan et al. [40]. The latter group proposed that crack growth only occurs when both ΔK and K_{max} thresholds are surpassed.

While both Paris and Walker models are applicable for crack growth in regime II, Forman proposed an extension to address regime III crack propagation [17,38]:

$$\frac{da}{dN} = \frac{C(\Delta K)^m}{(1-R)(K_c - K_{max})} \tag{1.24}$$

Hartman and Schijve [17,38] expanded the Forman model to encompass all fatigue crack growth regimes:

$$\frac{da}{dN} = \frac{C(\Delta K - \Delta K_{th})^m}{(1-R)K_c - \Delta K} \tag{1.25}$$

This latter model resembles a sigmoidal function with two asymptotes at $K_{max} = K_c$ and $\Delta K = \Delta K_{th}$. A notable limitation of this relation is its lack of direct consideration for the stress ratio dependency of ΔK_{th}. Numerous other fatigue crack growth models exist in the bibliography, but only a selection of the more historically significant ones are mentioned here. The aforementioned relations utilize the ΔK parameter, which is supported by linear elastic fracture mechanics (LEFM), making them applicable primarily to cracks with minimal plasticity at the tip. For more complex elastoplastic conditions, some researchers have suggested using elastoplastic fracture mechanics (EPFM) parameters to better describe fatigue crack propagation. An example by Dowling and Begley [41] considers the application of the J-integral:

$$\frac{da}{dN} = C \cdot \Delta J^m \quad (1.26)$$

This power relation is akin to the Paris model (effective in region II) but is particularly useful in scenarios involving large-scale yielding. The previously mentioned relations have been proposed for constant amplitude fatigue loading conditions and may not be suitable for variable amplitude loading due to load interactions that can considerably affect crack growth rates. For example, it is known that after an overload, crack propagation can suffer retardation. To address this, Wheeler [42] introduced a propagation law with a retardation parameter that simulates the retardation effects after single overloads. This parameter considers the interaction between the plasticity zones at the crack tip caused by the overload and the steady-state propagation conditions. However, Wheeler's model shows some limitations (e.g., not considering underloads; not account for interactions between multiple underloads and overloads). Willenborg [43] evolved the Wheeler work, including considerations for underloads. The NASGRO model, which is globally recognized, builds upon the Willenborg model and extends it to account for the reduction in retardation effects due to underloads. Correia et al. [44] presented a research work on the fatigue crack growth behaviour of a puddle iron from the Fão bridge, using the CT specimens under constant and variable amplitude loading conditions, where the crack closure effects were also measured.

Elber [45,46] made a significant contribution to the understanding of fatigue crack propagation, presenting the crack opening or closure assumption which was used to discuss crack propagation behaviour under variable amplitude loading. A load cycle only contributes to crack growth only when the near-tip crack faces are open, as stated by Elber. To quantify this, the effective value of the stress intensity factor range, U, was introduced as:

$$U = \frac{\Delta K_{\text{eff}}}{\Delta K} = \frac{K_{\max} - K_{\text{op}}}{K_{\max} - K_{\min}} \quad (1.27)$$

where $K_{\max}$, $K_{\min}$, and K_{op} respectively represent the maximum stress intensity factor, minimum stress intensity factor, and opening stress intensity factor. The previously mentioned fatigue crack growth relations can be reformulated by adopting the effective stress intensity factor range, ΔK_{eff}. For instance, the Paris relation can be presented as:

$$\frac{da}{dN} = C \cdot (\Delta K_{\text{eff}})^m = C \cdot (U \Delta K)^m \quad (1.28)$$

In this approach, crack closure is considered responsible for the effects of load interactions. Accurate definition of the parameters U and ΔK_{eff} is essential for applying the crack closure model, as these parameters must be determined for materials specificity, fatigue loading conditions, and regimes on the fatigue propagation process. The crack opening/closure model provides another understanding to the residual stresses perspective, induced by crack tip plasticity, with the former becoming widespread.

Under variable amplitude loading, defined by approximately random stress spectra, loading sequential effects are less significant. In such cases, simpler models neglecting load interaction effects can be adequate for damage assessment. These models can be applied on a cycle-by-cycle basis. Also, some researchers [47] have proposed using the Miner model to define a constant amplitude damage equivalent (effective) stress intensity factor, defined as:

$$\Delta K_{\text{eff}} = \left(\sum \frac{n_i \Delta K_i^m}{N_T} \right)^{1/n} \tag{1.29}$$

where ΔK_{eff} represents the equivalent stress intensity factor for the constant amplitude damage, N_T is the total number of cycles, n_i the number of cycles with a stress intensity factor range of ΔK_i, and m is the exponent from the Paris relation.

The root mean square (RMS) has been suggested by Hudson [48] to compute the mean of minimum and maximum stresses to calculate ΔK_{eff} and ΔK_i, which are required for fatigue modelling. This approach provides an average estimate of the crack growth for the stress spectrum that was used.

Castillo, Fernández-Canteli, and Siegele (CCS) [49] suggested a novel fatigue crack growth correlation model using the idea that the crack propagation rates can be modelled as a cumulative distribution function. This new approach addresses the limitations of dimensional parameters in existing previous models by employing dimensional analysis to derive dimensionless normalized parameters. Additionally, the Gumbel cumulative distribution function was explored as an alternative. The proposed model is represented by the following distribution (details available in reference [49]):

$$\frac{\log \Delta K^* - \log \Delta K_{\text{th}}^*}{\log \Delta K_{\text{up}}^* - \log \Delta K_{\text{th}}^*} = \exp\left[-\exp\left(\frac{\alpha - \log \frac{da^*}{dN^*}}{\gamma} \right) \right] \tag{1.30}$$

This relation depends on the following four constants, α, γ, ΔK_{th}^* and ΔK_{up}^* which may be evaluated by the least-square method [49]. The superscript '*' in the variables of the model aims to represent normalized/ non-dimensional parameters, which are defined as follows:

$$a^* = \frac{a}{W}; \quad N^* = \frac{N}{N_0}; \quad \Delta K^* = \frac{K_{\max} - K_{\min}}{K_c}; \quad \Delta K_{\text{th}}^* = \frac{\Delta K_{\text{th}}}{K_c}; \quad \Delta K_{\text{up}}^* = \frac{\Delta K_{\text{up}}}{K_c} \tag{1.31}$$

where the following nomenclature should apply:

a = crack size;
W = characteristic size (e.g., specimen dimension);
N = fatigue cycles;
N_0 = reference fatigue cycles;

K_{min}, K_{max} = minimum and maximum stress intensity factors;
ΔK_{th} = threshold for the fatigue crack propagation;
ΔK_{up} = limit for the stress intensity factor range;
K_c = fracture toughness.

A proposal for modification of this CCS crack growth model considering the values of the cyclic J-integral was suggested by Correia et al. [50] and given by the following equation:

$$\frac{da^*(N^*)}{dN^*} = \exp\left[F^{-1}\left(\frac{\log \Delta J^* - \log \Delta J^*_{th}}{\log \Delta J^*_{up} - \log \Delta J^*_{th}}\right)\right]$$

$$a^* = \frac{a}{W};\quad N^* = \frac{N}{N_0};\quad \Delta J^* = \frac{J_{max} - J_{min}}{J_c};\quad \Delta J^*_{th} = \frac{\Delta J_{th}}{J_c};\quad \Delta J^*_{up} = \frac{\Delta J_{up}}{J_c} \tag{1.32}$$

where J_{max} is the maximum value of the cyclic J-integral, J_{min} is the minimum value of the cyclic J-integral, J_c is the fracture toughness in values of J-integral, ΔJ_{th} is the threshold value of the cyclic J-integral, ΔJ_{up} is the limit value of the cyclic J-integral, and, ΔJ^*, ΔJ^*_{th}, ΔJ^*_{up} are the normalized/non-dimensional parameters for the applied range, threshold range, and limit values of the cyclic J-integral, respectively. Furthermore, Correia et al. [51] incorporated the crack closure effects into the CCS fatigue crack growth model.

1.4.2.5 Fatigue Assessment Supported by Fracture Mechanics

A brief overview of fatigue crack propagation models was introduced in the previous section. The large number of models available in the literature is motivated by the search for precise fatigue crack propagation models which is crucial for reliable fatigue predictions of components. Typically, fracture mechanics supports these propagation laws, which through proper integration are used to assess the residual fatigue lifetime of components or structures with known initial defects acting as cracks. This procedure is expressed as follows:

$$\frac{da}{dN} = f(\Delta K, R, \ldots) \Rightarrow N_f = \int_{a_i}^{a_f} \frac{1}{f(\Delta K, R, \ldots)} da \tag{1.33}$$

where a_i and a_f represent the initial and critical crack sizes, the later corresponding to unstable crack propagation, determined by material limited toughness, or the plastic collapse.

In addition to residual lifetime assessments, which form the basis of damage tolerance design, integrating fatigue crack propagation models can also be applied to evaluate the complete fatigue life. In these situations, crack nucleation is ignored, and fatigue life is viewed solely as a fatigue crack propagation process, starting in existing defects. The primary challenge in this approach consists of determining the starting defect dimension for the propagation assessment. A common practice for metals involves the postulation of an initial crack length within the range 0.25–1 mm [52–54]. However, this hypothesis might underestimate the component's fatigue life.

To address this, some researchers combine this method with a local strain-life approach to evaluate the number of cycles required to develop such an engineering crack [55]. This method, however, has the drawback of requiring a precise definition of the initial crack size within the assumed range, often necessitating a tuning process.

Another approach is to support the initial defect size distribution definition through non-destructive inspection techniques (NDI) [56]. Nevertheless, this method is limited if the defects distribution falls under the NDI technique detection threshold. In such cases, a safe approach involves assuming that the initial defect size is equivalent to the NDI detection limit [57].

An additional practice consists of estimating an equivalent initial flaw size (EIFS) compatible with a fracture mechanics-based fatigue modelling. The EIFS considers both the quality of the material, including manufacturing defects and inherent material properties. Calculating EIFS typically involves a trial-and-error inverse method. This method, based on an initial guess of the defect geometry and location, runs a fatigue crack growth analysis, looking to match the material failure data (stress-life) in an iterative way [58]. For instance, Yang and Manning [59] applied this inverse back-extrapolation technique to determine the EIFS distribution for 2024-T351 aluminium alloy. White et al. [60] utilized a probabilistic fracture mechanics-based approach based on the same method to evaluate the equivalent pre-crack size. Molent et al. [61] employed a back-extrapolation of experimental crack growth curves to time zero to estimate the EIFS for the 7050 aluminium alloy.

A significant issue with the inverse fitting method is the stress level dependency of the EIFS [62]. Ideally, the EIFS should reflect a material-process property indicating the initial quality independent of the applied stress levels, making it applicable across various stress conditions. However, it is important to note that the EIFS is not a physical property but rather a model parameter derived from test data aiming to facilitate lifetime predictions using long cracks growth analysis, overcoming the complexities of short crack propagation modelling. The generalization of the EIFS concept has been hampered by the extensive test data required to develop a reliable EIFS statistical distribution [63].

Incorporating uncertainties associated with EIFS adds complexity to the problem. Given the intrinsic variability in fatigue crack propagation data, as well as the modelling approximations, it is crucial to account for these sources of uncertainties through probabilistic lifetime predictions. The back-extrapolation inverse method, which involves Monte Carlo simulation combined with iterative fatigue crack propagation analysis, can make computing a probabilistic EIFS distribution very resource-intensive [64,65].

To address these challenges, reference [58] proposes a new methodology for calculating probabilistic EIFS supported by the Kitagawa–Takahashi relation. This approach determines the EIFS by comparing the fatigue limit of a component with and without an assumed defect, using only the fatigue limit data and the fatigue crack propagation threshold. This method allows for direct calculation of EIFS statistics without the need for inverse back-extrapolation fatigue crack propagation modelling.

Additionally, the fatigue approach based on fracture mechanics methods has also been used for modelling the engineering cracks initiation, which according to

Savaidis et al. [66] consists of a 1 mm. Resorting to J-integral Elastoplastic Fracture Mechanics, postulated micro-cracks where integrated. Using a back-extrapolation procedure, the model was applied to simulate the formation of technical cracks at notches. Both the elastoplastic and crack closure conditions were accounted for in this methodology.

1.5 RANDOM NATURE OF FATIGUE

Ignoring the stochastic nature of fatigue damaging processes by relying solely on deterministic approaches for lifetime predictions refutes the basic principles of structural integrity analysis and could have severe implications for industries such as metallic bridges, aerospace, and machinery. Such deterministic methods are only justifiable when current knowledge and technology do not yet support the development and application of more effective probabilistic models.

Fatigue results can be influenced by numerous sources of variability, including material differences and production processes. To establish appropriate safety margins for mechanical components and structural details. It is essential to explain these different sources of scatter [67–69]. Probabilistic approaches to fatigue offer a solution to this challenge, providing a more comprehensive understanding and management of the inherent uncertainties.

1.5.1 Probabilistic Fatigue Models

Probabilistic methods in fatigue analysis are the need to consider the variability in S-N curves, enabling the establishment of suitable reliability levels for fatigue assessments. According to reference [70], the common variability in the S-N diagram was discussed. The variability increases with longer fatigue lives and lower stress amplitudes (refer to Figure 1.9).

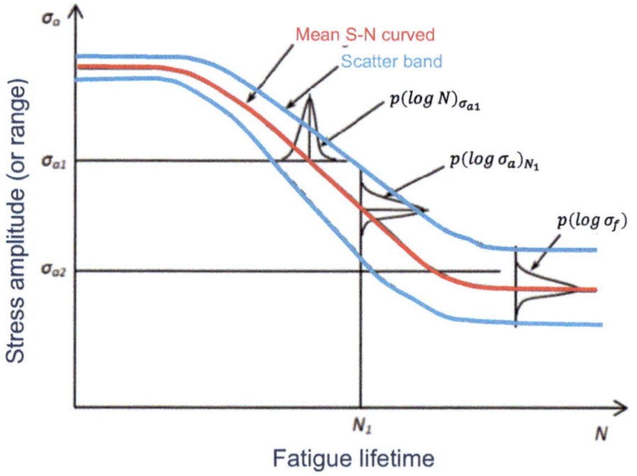

FIGURE 1.9 Common dispersion in an S-N field [70]. (Adapted from Ref. [17].)

In engineering, the variability of S-N curves is often addressed using a simple method: a linear regression analysis on $\log(N_f)$ *versus* $\log(\sigma_a)$ data is implemented to derive a mean S-N curve; then its standard deviation, under the assumption of a lognormal distribution of the fatigue life at a specified stress level, is settled. Safety is assured by applying the lower two-sigma or three-sigma design curve approach. This approach shifts the median strain-life curve leftward by two or three standard deviations in logarithmic coordinates. This technique is employed by main design codes, such as EC3, BS5400, AASHTO, and DNV. However, it does not consider the statistical distribution of the results based on sample size or the necessary reliability/confidence levels [71–73]. Additionally, it fails to address the variable dispersion throughout the different fatigue regimes, which may not be suitable for very-high cycle fatigue lives. This procedure is currently standardized in materials testing standards like ASTM E739 [74] and ISO 12107 [75].

Zhao et al. [76] introduced a probabilistic S-N band that extends it up to the giga-life fatigue life of the railway axle. Using test data from mid-life to long-life regimes, fatigue curves, and conventional fatigue limits were estimated, which are linked to competing probability levels. The aim of the proposed field is to improve the current S-N curves proposed in the design codes, which are considered by the authors as excessively conservative for very high cycle fatigue regimes. In their study, the fatigue lives related to the stress amplitude were assumed to follow a lognormal distribution.

Schijve [70] discussed popular statistical distribution functions used in the fatigue assessment of structures. He noted that, despite the considerable qualitative understanding of material fatigue damage, it is impossible to derive statistical distributions for fatigue from only the physical understanding. Instead, a statistical distribution function must be postulated. In his perspective, Schijve examined the three-parameters Weibull, the lognormal, and the three-parameter $\log(N - N_0)$ normal statistical distributions using available experimental fatigue data, using a stress parameter and measuring the cycles to failure. The $\log(N - N_0)$ normal and three-parameter Weibull distributions showed a good fit for the test results within the data range, whereas the $\log(N)$ normal distribution did not fit well if the data indicated a skew distribution.

Furthermore, extrapolation of the statistical distribution function to very low failure probabilities (extremes) remains unreliable, implying that suitable statistical distributions are unknown. This type of research assesses the goodness of fit of known distributions for parts of the S-N field. Still, it only provides a partial picture of part of the probabilistic P-S-N field.

Significant research has been focused on modelling the complete probabilistic S-N field. Notable contributions include the ones by Basternaire, Spindel, and Haibach, and Pascual and Meeker [17]. Castillo and Fernández-Canteli revitalized a new class of probabilistic models to describe the complete S-N field, incorporating both physical and statistical conditions. This latter approach will be discussed in more detail in the next subsection due to its most recent proposition.

One widely adopted probabilistic approach for integrating variability into existing deterministic models for fatigue analysis involves using appropriate probabilistic distributions to account for the variability of individual parameters or properties [77–79]. These parameters are treated as stochastic variables obeying to specific distributions. The model's output (e.g., p-S-N field of a mechanical component or a structural detail), can be calculated using sampling techniques (e.g., Monte Carlo simulations). This method allows for utilizing complex fatigue models previously validated through deterministic approaches, deriving probabilistic outputs that would otherwise be analytically unattainable. This probabilistic approach has been applied to model fatigue crack initiation and propagation [80–82]. Correia et al. [13] similarly to Sanches et al. [83] employed this method for probabilistically modelling the fatigue strength of riveted connections in old bridges. They used strain-life approaches (i.e. local approaches) for fatigue crack initiation simulation and fracture mechanics-based approaches for macroscopic fatigue crack growth. In their models, uncertainties in parameters were handled by assuming a stochastic nature and described by selected distributions. Monte Carlo simulations were applied to compute the p-S-N field for riveted shear splices. This approach also included uncertainties in mechanical parameters that affect joint behaviour, such as sliding coefficient and preload of rivets (clamping stresses).

Kandarpa et al. [84] utilized Morrow's strain-life local approach to predict the fatigue performance of notched samples. They treated the material parameters in the strain-life relation as stochastic ones. The lognormal statistical distribution was applied to describe the fatigue ductility and fatigue strength and the normal distribution for the corresponding strength and ductility exponents; the stress concentration factor of the notched details was also assumed a random variable following a lognormal distribution. They performed sensitivity analysis to identify the more influencing factors and identify which uncertainties had the most impact. The probability of fatigue failure was then calculated using the following procedures: (1) Monte Carlo simulation (MCS), (2) first-order reliability method (FORM), and (3) second-order reliability method (SORM). Their analysis of the predictions revealed that SORM and MCS produced identical outcomes, whereas FORM tended to overestimate the failure probabilities.

Meggiolaro and Castro [85] conducted a statistical evaluation of constants from the Coffin-Manson (ε-N) and Ramberg-Osgood (σ-ε) relations, utilizing extensive data from steels and aluminium alloys. They tested various statistical distributions for each parameter involved, looking for exploring local strain-based probabilistic approaches.

Ni and Mahadevan [86] suggested a probabilistic method for the fatigue life prediction of spot-welded joints used in the automotive industry. Using the local approach, based on Morrow's stress-life relationship, it was possible to convert it into a probabilistic model, treating the ductility strength and fatigue parameters as stochastic variables after statistical distributions with best fit to the experimental data. As a result, they derived a probabilistic fatigue p-ε-N field. They also suggested a probabilistic cumulative damage accumulation approach, based on Miner's rule, to handle variable amplitude loading as:

$$\sum_i \frac{n_{\mathrm{pi}}}{N_{\mathrm{fpi}}} = 1 \tag{1.34}$$

where N_{fpi} is derived from the fatigue p-ε-N field.

Zhao et al. [87] developed a fatigue reliability analysis framework based on the strain-life method. They began by analysing experimental methods and test data. They then proposed random models where material constants were treated as dependent stochastic variables using the Morrow relation and the modified Ramberg-Osgood material description for random cyclic stress and stress-strain relationships. Their approach considered survival probability and sample size confidence. They addressed the limitations of existing methods, including the assumption of incomplete independent random variables and the lack of consideration for random cyclic stress-strain relationships.

Williams et al. [88] introduced a methodology for developing probabilistic strain-life curves from test data under strain-controlled loading. Their methodology includes (1) criteria specification to ensure data significance and representativeness, (2) a statistical analysis methodology aiming at accurately representing each domain of material performance, and (3) an approximate Owen's tolerance limit method to take into account the scatter in fatigue data, allowing the establishment of reliability/confidence levels based on sample size.

Probabilistic fatigue simulations supported by fracture mechanics need yet important developments, mainly due to the lack of probabilistic analytical solutions for fatigue crack growth. Recognizing that fatigue crack growth is a random process, one alternative solution involves the selection of a deterministic crack growth model and considering that its parameters are stochastic variables following a given statistical distribution, based on test data and empirical approximations. Liu and Mahadevan [89] applied this approach to derive fatigue p-S-N fields. They used the equivalent initial flaw size (EIFS) concept together with a deterministic fatigue crack growth model and assumed a lognormal distribution for the EIFS. Treating the coefficient and fatigue crack propagation threshold as stochastic variables, the fatigue p-S-N field was simulated through Monte Carlo sampling.

Several studies have analysed variability in the parameters of existing deterministic crack propagation laws. Virkler et al. [90] investigated the statistical distribution of fatigue crack growth in centre-cracked plates made of 2024-T3 aluminium alloy under constant amplitude loading. Their data demonstrated considerable scatter in crack propagation rates, with the 3-parameter lognormal distribution describing fatigue life best at a given crack length. Annis [91] suggested a Monte Carlo simulation that is compared with the results of Virkler et al.'s experimental data. He found that selecting random variables for both C and m parameters in the Paris equation led to more significant variation in Monte Carlo simulation results than experimental data. Annis proposed that C and m should not be chosen independently, advocating for a bivariate normal distribution to reduce error to approximately 1% [17]. Consequently, some researchers now consider deterministic just one of the parameters of the Paris law and treat the remaining as random.

Ellyin and Fakinlede [92] developed a probabilistic model to describe fatigue crack propagation assuming a nonlinear damage accumulation that accounts for loading sequential effects. Based on the contributions of a deterministic model by Ellyin and Kujawshi's [26], their model considers simultaneous damage to multiple representative material elements within the plastic zone of the inverted crack tip, rather than just one element. The loading history of each material element is evaluated using a nonlinear damage rule supported by an energy-based model. Failure of elements within the cyclic plastic zone is determined using a probabilistic form of Miner's linear damage accumulation rule, where the threshold value for failure is assumed to follow a random variable.

Recently, some authors such as Correia et al. [93] and Zhang et al. [94] explored the determination of the Wöhler S-N equation coefficients using the least-squares and orthogonal regression methods applied for the fatigue S-N data of old riveted steel connections and concrete, respectively. Liu et al. [95] evaluated the probabilistic S-N fields stochastically based on conventional fatigue models and their coefficients calculated by the Bayesian method.

1.5.2 Recently Developed Probabilistic Fatigue Models

Castillo and Fernández-Canteli and their collaborators have formulated probabilistic fatigue models based on extreme value distributions such as Weibull and Gumbel distributions. The proposed contributions include de following possibilities [14,15,96]:

- Probabilistic S-N bands for a given mean stress;
- Probabilistic S-N bands for varying mean stresses;
- Probabilistic ε-N bands;
- Correspondence between p-S-N bands and p-ε-N bands;
- Generalization of the probabilistic S-N bands to the conventional fatigue regimes;
- Probabilistic understanding of the Palmgren-Miner's number;
- Inclusion of statistical size effects in fatigue assessment;
- Generalization of the probabilistic S-N bands to other more complete fatigue parameters.

Those developments were performed supported by a base Weibull/Gumbel probabilistic model derived under physical and statistical assumptions, formulating a new description of probabilistic S-N bands. Such base probabilistic field may be explored to correlate the fatigue data from smooth or notched specimens, representing mechanical details or structural components, and including run-outs. This model, which was originally stated in the stress domain, is particularly suited for medium to very high cycle fatigue conditions. The model's formulation adheres to physical constraints (such as the identification of relevant variables and dimensionless analysis) and statistical criteria (including the weakest link principle, stability, range limitations, and boundary behaviour). Additionally, by ensuring compatibility between the lifetime distribution for a given stress range and the stress range distribution for a specific lifetime, the model derivation culminates with a functional equation, which has the following distribution solution that defines the probabilistic S-N field [14]:

$$F\left(\log(\Delta\sigma);\log(N)\right) = p = 1-\exp\left\{-\left[\frac{\left(\log(N)-B\right)\left(\log(\Delta\sigma)-C\right)-\lambda}{\delta}\right]^{\beta}\right\} \tag{1.35}$$

$$\left(\log(N)-B\right)\left(\log(\Delta\sigma)-C\right)\geq\lambda$$

In this model, N represents the lifetime; $\Delta\sigma$ denotes the stress range; $F()$ is the cumulative distribution function (CDF) of N for a given $\Delta\sigma$; $B=\log(N_0)$, where

N_0 is the threshold lifetime; $C = \log(\Delta\sigma_0)$, with $\Delta\sigma_0$ being the threshold fatigue strength so-called conventional fatigue limit; and λ, β, and δ are the Weibull statistical model parameters – β is the shape parameter, δ is the scale parameter, and λ is the location parameter – that defines the position of the 0th percentile curve. Equation (1.36) has been analysed as well as effectively utilized in various lifetime evaluations [14]. Equation (1.34) can also be expressed in terms of normalized cycle numbers (N^*) and normalized stress ranges ($\Delta\sigma^*$):

$$F\left(\log(\Delta\sigma);\log(N)\right) = p = 1 - \exp\left\{-\left[\frac{\log\left(N^*\right)\log\left(\Delta\sigma^*\right) - \lambda}{\delta}\right]^{\beta}\right\} \tag{1.36}$$

$$\log\left(N^*\right)\log\left(\Delta\sigma^*\right) \geq \lambda$$

where $\Delta\sigma^* = \sigma_M^* - \sigma_m^*$ denotes the normalized stress range; $\sigma_M^* = \sigma_{max}/\sigma_0$ and $\sigma_m^* = \sigma_{min}/\sigma_0$ are the normalized maximum and minimum stresses, respectively, and $N^* = N/N_0$ represents the normalized number of cycles; σ_{max} and σ_{min} are the maximum and minimum stresses of the stress cycle. $\sigma_0 = 10^B$ and $N_0 = 10^C$ are model parameters. Figure 1.10 illustrates the Weibull probabilistic S-N field, highlighting several constant probability S-N lines. These lines are hyperbolas limited by the two asymptotes $\log(N_0) = B$, representing the conventional fatigue limit which mathematically is the horizontal asymptote, and $\log(\Delta\sigma_0) = C$, representing the lifetime threshold which mathematically is the vertical asymptote. The null percentile S-N line represents the minimum number of cycles required to achieve failure for various values of $\Delta\sigma$. The asymptotes of the S-N bands limit its application to specific subdomains. Already the vertical asymptote is contentious due to plastic effects in the LCF region, which cause actual percentile curves to bend downward in this area. Conversely, the horizontal asymptote is less disputed, even for materials that

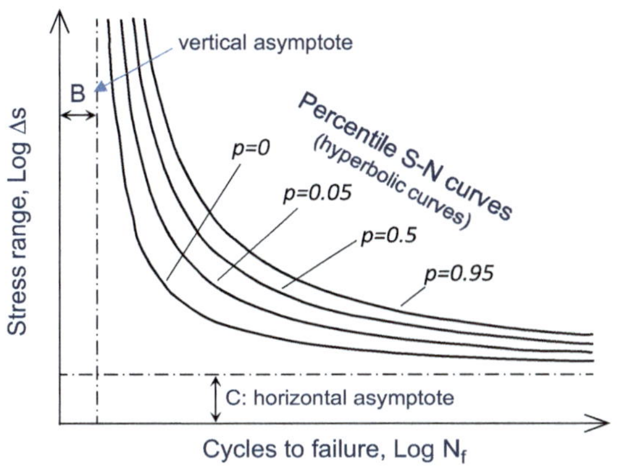

FIGURE 1.10 Base probabilistic Weibull p-S-N model representation by Castillo and Fernández-Canteli. (Adapted from Ref. [17].)

do not have a distinct fatigue limit, such as some aluminium alloys, since the model accommodates the case where $C = 0$. Additionally, this probabilistic fatigue model proposed by Castillo and Fernández-Canteli [14] allows the use of several uniaxial and multiaxial fatigue criteria as suggested by Correia et al. [15].

Besides the Weibull distribution, Gumbel distribution was also demonstrated to be a solution of the probabilistic S-N bands. In fact, the Gumbel distribution is considered a limiting case of the Weibull statistical distribution when β approaches infinity. In practice for β values greater than 8, Gumbel distribution resembles the Weibul. In this case, the probabilistic S-N bands can be described by:

$$F\left(\log(\Delta\sigma);\log(N)\right) = p = 1 - \exp\left\{-\exp\left[\frac{\left(\log(N)-B\right)\left(\log(\Delta\sigma)-C\right)-\lambda}{\delta}\right]\right\} \tag{1.37}$$

or using non-dimensional variables:

$$F\left(\log(\Delta\sigma);\log(N)\right) = p = 1 - \exp\left\{-\exp\left[\frac{\log\left(N^*\right)\log\left(\Delta\sigma^*\right)-\lambda}{\delta}\right]\right\} \tag{1.38}$$

The parameters can be identified using a two-stage procedure: (1) estimating the threshold B and C parameters using the constrained least-squares method; then (2) estimation of λ, β, and δ Weibull parameters using the maximum likelihood method [14].

One major finding by Castillo and Fernández-Canteli was the new variable, combining a fatigue stress parameter and fatigue lifetime

$$V = \left[\log(N) - B\right]\left[\log(\Delta\sigma) - C\right] \tag{1.39}$$

which must follow the Weibull or Gumbel extreme value distributions to fulfil the statistical and physical essential assumptions, including the compatibility requirement between the life and strength distributions. The formal structure of the probabilistic S-N bands has been exploited by using different fatigue parameters, allowing for example the use of complex parameters accounting for mean stress effects, multiaxial fatigue, etc.; also, the constants of the model can be adjusted to cover different p-S-N fields for different mean stress levels (stress ratios). The scale Weibull parameter can be applied to cover the statistical size effects. Mohabeddine et al. [97] developed a comparative study to obtain the coefficients of the conventional S-N curve using the ISO12107 standard approaches and linear regression together with the Weibull and Gumbel distributions applied to steel details. Also in this study, the author recommends the use of the probabilistic fatigue model proposed by Castillo and Fernández-Canteli incorporating in this model the stress reduction factor (SRF).

1.6 FATIGUE DESIGN CRITERIA

The criteria for a proper fatigue design have progressed over the years from the earlier concept of infinite-life design to a focus on damage tolerance design. Despite this evolution, each criterion retains relevance depending on the specific application.

Each design criteria is based on specific fatigue approaches previously discussed. The fatigue design criteria can be categorized as follows [98]:

- Design for infinite life
- Design for safe life
- Design aiming fail-safe
- Design for damage tolerance

The design for the infinite life corresponds to the oldest criterion, which grants unlimited safety, necessitating local fatigue parameters (e.g., stresses or strains) to remain predominantly elastic and well below the relevant fatigue endurance limit. This approach is yet practical for components experiencing millions of cycles (e.g., engine components). However, it is often challenging to define or obtain the relevant fatigue limits for most parts due to significant variable amplitude loading. Moreover, this design criterion may be impractical or uneconomical in many design scenarios (e.g., aircraft overweight). Therefore, in a globally competitive market, there is a need for alternative cost-effectiveness fatigue design criteria [98].

Infinite-life design was initially suitable for applications like rolling stock axles as studied by Wöhler. However, car designers found that designing parts to withstand hundreds of thousands of cycles, rather than millions, was sufficient for components like suspension springs or reverse gears, where maximum loads occur infrequently. This approach, called safe-life design, is also adopted in other industries such as pressure vessels and aircraft engines. Safe-life design includes a safety margin to account for variability in fatigue results and other uncertainties. Design calculations may be supported by stress-life, strain-life, or fracture mechanics-based crack growth relationships and rely on field and simulated testing. For instance, the USA Air Force requires fatigue test life to be four times the expected service life for aircraft parts. Bearings are rated based on a reference load that 90% are expected to endure for a given time. Safety margins in safe-life criterion can be specified in terms of life, loading, or both, as outlined in standards like Boiler and Pressure Vessel Code by ASME [98].

A structure, component, or transportation vehicle must be retired from service when it reaches its designated safe life. However, this approach can be inefficient and costly, necessitating retiring the entire fleet before the average calculated or tested life is reached. Additionally, testing and analysis cannot always foresee all potential collapses. To address these issues, aircraft engineers developed fail-safe fatigue design criteria. This method avoids the excessive weight and cost associated with significant safety factors and the risks of smaller safety factors. Fail-safe design ensures that the overall system remains operational if one part fails. It acknowledges that fatigue cracks can develop but ensures that these cracks do not cause catastrophic failure before they are detected and repaired. Strategies for achieving fail-safe design include:

- Alternative load paths;
- Load redistribution between components;
- Incorporating crack arresters at intervals;
- Regular inspections.

Initially focused on airframes such as fuselages, wings, and control surfaces, this philosophy is currently applied in various fields. For instance, engines are designed with fail-safe principles in multi-engine aircraft, while landing gears are designed with a safe-life approach rather than fail-safe features [98].

The damage-tolerant design philosophy refines the fail-safe approach by assuming that cracks will form due to manufacturing or fatigue. Fracture mechanics is applied to predict whether assumed cracks will propagate to a size that causes failure before they are detected through inspections. Key aspects of this approach include:

- **Residual strength:** the remaining resistance of a cracked component decreases as the crack grows;
- **Fatigue crack growth behaviour:** the rate at which cracks propagate under cyclic loading is influenced by material properties and other factors.
- **Crack detection:** non-destructive testing methods are able to identify cracks before they lead to failure.

This design methodology also incorporates environmental conditions, load history, and safety margins. It is mandatory by the USA Air Force and applied in pressure vessel design, exemplified by the "leak before break" principle. For jet engine turbine blades, which are often retired based on their safe-life predictions, damage-tolerant methods allow for extended service life by combining analytical assessments with additional testing.

1.7 DATA SCIENCE IN FATIGUE ASSESSMENT

Fatigue assessment involves evaluating the durability of materials and components under cyclic loading to predict their lifespan and ensure safety. Traditionally, this process relied heavily on empirical testing and analytical models. However, the integration of data science has the potential to enhance these methods, in particular in the following ways [99–104]:

i. **Data collection and integration:** data science techniques facilitate the collection and integration of large datasets from various sources, such as sensor data from testing equipment, historical performance data, and operational conditions. This comprehensive data helps in understanding fatigue behaviour more accurately.

ii. **Predictive modelling:** machine learning algorithms are being employed to develop predictive models that estimate fatigue life based on historical data and real-time inputs. Techniques such as regression analysis, neural networks, and ensemble methods improve the accuracy of life predictions by identifying patterns and relationships in complex datasets.

iii. **Pattern recognition:** data science tools can identify patterns and anomalies in fatigue data that might not be apparent through traditional analysis. For example, clustering algorithms can group similar fatigue failure patterns, aiding in identifying common failure modes.

iv. **Real-time monitoring and analysis:** advanced data analytics enable real-time monitoring of components using sensors and IoT devices. This real-time data can be analysed to detect early signs of fatigue and predict potential failures before they occur, allowing for timely maintenance or replacements.

v. **Optimization and simulation:** data-driven simulations use historical data to optimize material selection and component design. By integrating data science with simulation tools, engineers can model different loading scenarios and assess how changes in design or materials impact fatigue performance.

vi. **Big Data and statistical analysis:** big data techniques are employed to analyse vast amounts of fatigue test data. Statistical methods are used to quantify uncertainties and improve the reliability of fatigue life predictions, handling complex datasets involving multiple variables and interactions.

vii. **Failure analysis and root cause identification:** data science aids in understanding the root causes of fatigue failures by analysing failure data and identifying contributing factors. Techniques such as causal analysis and data mining help in pinpointing underlying issues and preventing future occurrences.

The following sectors and applications are currently benefitting from data science techniques:

i. **Aerospace and automotive industries:** data science enhances fatigue assessment by improving predictive maintenance schedules and optimizing design processes, leading to increased safety and reduced costs.

ii. **Structural health monitoring:** data science techniques are used in structural health monitoring systems to continuously assess the condition of infrastructure like bridges and buildings, identifying fatigue-related issues before they become critical.

iii. **Materials science:** advanced analytics are used to evaluate the fatigue properties of new materials and composites, facilitating the development of more durable and reliable materials.

Nevertheless, the development and exploitation of data science-based techniques assisting the solution of structural integrity problems, including fatigue and fracture, is still at infancy. There are still significant challenges to overcome:

i. **Data quality and management:** ensuring the accuracy and completeness of data is crucial. Effective data management practices are necessary to handle the vast amounts of data being generated.

ii. **Integration with traditional methods:** combining data science with traditional fatigue assessment methods requires careful integration to leverage the strengths of both approaches.

iii. **Scalability and adaptability:** developing scalable solutions that can adapt to different types of components and loading conditions remains a challenge.

Data science has been transforming fatigue assessment by enhancing predictive accuracy, improving real-time monitoring, and optimizing design and maintenance strategies. As technology advances, the integration of data science in fatigue assessment is expected to further improve safety, efficiency, and cost-effectiveness in various industries.

ABBREVIATIONS

AASHTO American Association of State Highway and Transportation Officials
ASME American Society of Mechanical Engineers
ASTM American Society for Testing and Materials
BS British Standards
CCS Castillo, Fernández-Canteli and Siegele
CDF Cumulative Distribution Function
CEN Comité Européen de Normalisation
CT Compact Tension Specimen
DNV Det Norske Veritas
EC3 Eurocode 3: Design of steel structures
EIFS Equivalent Initial Flaw Size
EPFM Elastoplastic Fracture Mechanics
EUROCODE European Standards managed by CEN
FORM First-Order Reliability Method
IoT Internet of Things
ISO International Standardization Organization
LCF Low-Cycle Fatigue
LEFM Linear Elastic Fracture Mechanics
MCS Monte Carlo Simulation
NASGRO Fracture Mechanics and Fatigue Crack Growth Analysis Software
NDI Non-destructive Inspection Techniques
RMS Root Mean Square
SIF Stress Intensity Factor
SORM Second-Order Reliability Method
SRF Stress Reduction Factor
SWT Smith-Watson-Topper
TCD Theory of Critical Distances
USA United States of America

REFERENCES

1. Schütz W. A history of fatigue. *Eng. Fract. Mech.*, 1996; 54: 263–300.
2. Liao D, Zhu SP, Correia JA, De Jesus AM, Berto F. Recent advances on notch effects in metal fatigue: A review. *Fatigue Fract. Eng. Mater. Struct.*, 2020; 43(4): 637–659.
3. Zhu SP, Ai Y, Liao D, Correia JA, De Jesus AM, Wang Q. Recent advances on size effect in metal fatigue under defects: A review. *Int. J. Fract.*, 2022; 234(1–2): 21–43.

4. Zhu SP, Ai Y, Liao D, Correia JA, De Jesus AM, Wang Q. Stress gradient effect in metal fatigue: Review and solutions. *Theor. Appl. Fract. Mech.*, 2022; 121: 103513.
5. Niu X, He C, Zhu SP, Foti P, Berto F, Wang L, Liao D, Wang Q. Defect sensitivity and fatigue design: Deterministic and probabilistic aspects in additively manufactured metallic materials. *Prog. Mater. Sci.*, 2024; 144: 101290.
6. Schijve J. *Fatigue of Structures and Materials.* Kluwer Academic Publishers, New York, NY, 2004.
7. Dowling NE. *Mean Stress Effects in Stress-Life and Strain-Life Fatigue.* Second SAE Brasil International Conference on Fatigue, Sao Paulo, Brasil, June 2004.
8. Boardman B. Fatigue resistance of steels. In: *ASM Handbook, Volume 1: Properties and Selection: Irons, Steels, and High-Performance Alloys*, R. Steiner, ed., ASM Handbook Committee, OH, USA, 1990; 673–688.
9. Neuber H. *Theory of Notch Stresses: Principles for Exact Calculation of Strength with Reference to Structural Form and Material* (2nd edition). Springer-Verlag, Berlin, 1958.
10. Neuber H. Theory of stress concentration for shear-strained prismatic bodies with arbitrary nonlinear stress–strain law. *Trans. ASME J. Appl. Mech.,* 1961; 28: 544–551.
11. Peterson RE. Notch sensitivity. In: *Metal Fatigue*, G. Sines and J. L. Waisman, eds., McGraw Hill, New York, NY, 1959; 293–306.
12. Taylor D. *The Theory of Critical Distances: A New Perspective in Fracture Mechanics.* Elsevier, Oxford, 2007.
13. Correia, JAFO, De Jesus, AMP, Da Silva, ALL, Da Silva, JFN. A procedure to derive probabilistic fatigue strength data for riveted joints. *Bridge Maintenance, Safety, Management and Life-Cycle Optimization - Proceedings of the 5th International Conference on Bridge Maintenance, Safety and Management*, 2010; 3086–3093.
14. Castillo E, Fernández-Canteli A. *A Unified Statistical Methodology for Modeling Fatigue Damage.* Springer, Dordrecht, 2009.
15. Correia, JAFO, Apetre N, Arcari A, De Jesus AMP, Muñiz-Calvente M, Calçada R, Berto F, Fernández-Canteli A. Generalized probabilistic model allowing for various fatigue damage variables. *Int. J. Fatigue*, 2017; 100(Part 1): 187–194.
16. Correia, JAFO, Raposo P, Calvente M, Blasón S, Lesiuk G, De Jesus AMP, Moreira PMGP, Calçada RAB, Fernández-Canteli A. A generalization of the fatigue Kohout-Věchet model for several fatigue damage parameters. *Eng. Fract. Mech.*, 2017; 185: 284–300.
17. Correia JAFO. An integral probabilistic approach for fatigue lifetime prediction of mechanical and structural components. PhD Thesis, FEUP, 2014.
18. APK. *Association pour la Promotion de l'Enseignement de la Construction Acier. Construction métallique et mixte acier-béton: Calcul et dimensionnement selon les Eurocodes 3 et 4.* Éditions Eyrolles, Paris, 1996.
19. American Society for Testing and Materials. ASTM E739-91: standard practice for statistical analysis of linear or linearized stress-life and strain-life fatigue data. In: *Annual Book of ASTM Standards*, J. Brozzetti and P. Bourrier, eds., ASTM International, West Conshohocken, PA, 2004; 03.01: 1–7.
20. CEN-TC 250. *EN 1993-1-9: Eurocode 3, Design of Steel Structures-Part 1–9: Fatigue.* European Committee for Standardization, Brussels, 2003.
21. British Standards Institution. *BS 5400 Steel, Concrete and Composite Bridges: Part 10: Code of Practice for Fatigue.* BSI, London, 1980.
22. AASHTO. *AASHTO LRFD: Bridge Design Specification.* AASHTO, Washington, D.C., 1995.
23. Åkesson B, Edlund B. Remaining fatigue life of riveted railway bridges. *Stahlbau*, 1996; 65(11): 429–436.

24. Helmerich R, Brandes K, Herter J. Full scale laboratory fatigue tests on riveted railway bridges. *Proceedings IABSE Workshop*, Lausanne. IABSE, Zurich, 1997; 191–200.
25. Taras A, Greiner R. Statistical background to the proposed fatigue class catalogue for riveted components. Report: Contribution to WG6.1 – Assessment of Existing Steel Structures, ECCS TC6–2010, Spring Meeting – Lausanne – March 22–23, 2010; 20(1): 91–103.
26. Ellyin F. *Fatigue Damage, Crack Growth and Life Prediction.* Chapman & Hall, London, 1997.
27. Morrow JD. Cyclic plastic strain energy and fatigue of metals. *Int. Frict. Damp. Cyclic Plast. ASTM STP,* 1965; 378: 45–87.
28. Morrow JD, Socie DF. Evolution of fatigue crack initiation life prediction methods. In: *Materials, Experimentation and Design in Fatigue: Proceedings of Fatigue,* Warwick University, England, E Sherratt and JB Sturgeon, eds., Westbury House, Guildford, 1981; 3–21.
29. Dowling NE. *Mechanical Behaviour of Materials: Engineering Methods for Deformation, Fracture and Fatigue* (2nd edition). Prentice Hall, Inc., Upper Saddle River, NJ, 1999.
30. Ramberg W, Osgood WR. Description of stress–strain curves by three parameters. Technical Note No. 902, NACA tech, 1943.
31. Smith KN, Watson P, Topper TH. A stress-strain function for the fatigue of metals. *J. Mater.*, 1970; 5(4): 767–778.
32. Zhu SP, Liu Y, Liu Q, Yu ZY. Strain energy gradient-based LCF life prediction of turbine discs using critical distance concept. *Int. J. Fatigue*, 2018; 113: 33–42.
33. Zhu SP, Yu ZY, Liu Q, Ince A. Strain energy-based multiaxial fatigue life prediction under normal/shear stress interaction. *Int. J. Damage Mech.*, 2019; 28(5): 708–739.
34. Zhu SP, Yang YJ, Huang HZ, Lv Z, Wang HK. A unified criterion for fatigue-creep life prediction of high temperature components. *Proc. Inst. Mech. Eng. Pt. G J. Aerosp. Eng.*, 2017; 231(4): 677–688.
35. Kujawski D, Ellyin F. A unified approach to mean stress effect on fatigue threshold conditions. *Int. J. Fatigue*, 1995; 17(2): 101–106.
36. Ellyin F, Kujawski D. An energy-based fatigue failure criterion. *Microstruct. Mech. Behav. Mater.*, 1985; 2: 591–600
37. Paris PC, Gomez M, Anderson WE. A rational analytic theory of fatigue. *Trend Eng.*, 1961; 13: 9–14.
38. Beden SM, Abdullah S, Ariffin AK. Review of fatigue crack propagation models for metallic components. *Eur. J. Sci. Res.*, 2009; 28: 364–397.
39. Donald K, Paris PC. An evaluation of ΔK_{eff} estimation procedure on 6061-T6 and 2024-T3 aluminium alloys. *Int. J. Fatigue*, 1999; 21: S47–S57.
40. Vasudevan AK, Sadananda K, Louat N. A review of crack closure, fatigue crack threshold and related phenomena. *Mater. Sci. Eng.*, 1994; A188(1–2): 1–22.
41. Dowling NE, Begley JA. Fatigue crack growth during gross plasticity and the J-integral. In: Rice J.R. and Paris P.C., *Mechanics of Crack Growth.* ASTM STP 590, American Society for Testing and Materials, Philadelphia, PA, 1976; 82–105.
42. Wheeler OE. Spectrum loading and crack growth. *J. Basic Eng.*, 1972; 94: 181–186.
43. Borg J, Engle RM, Wood HA. A crack growth retardation model using an effective stress concept. Report AFFDL-TR71-1, Air Force Flight Dynamic Laboratory, Dayton, 1971.
44. Correia JAFO, de Jesus AMP, Carvalho H, Lesiuk G, Mourão A, Grilo LF, Calçada RAB. Fatigue crack growth modelling of Fão Bridge puddle iron under variable amplitude loading. *Int. J. Fatigue*, 2020; 136: 105588.
45. Elber W. Fatigue crack closure under cyclic tension. *Eng. Fract. Mech.*, 1970, 2: 37–45.

46. Elber W. The significant of fatigue crack closure. In: *Damage Tolerance in Aircraft Structures*, ASTM International, eds. ASTM STP 486. American Society for Testing and Materials, Philadelphia, PA, 1971; 230–242.
47. Correia JAFO, Blasón S, Arcari A, Muñiz-Calvente M, Apetre N, Moreira PMGP, De Jesus AMP, Fernández-Canteli A. Modified CCS fatigue crack growth model for the AA2019-T851 based on plasticity-induced crack closure. *Theor. Appl. Fract. Mech.*, 2016; 85: 26–36.
48. Varma D. *Fatigue Crack Growth.* New Age International (P) Ltd., New Delhi, 1990.
49. Hudson CM. A root-mean-square approach for predicting fatigue crack growth under random loading. In: *Methods and Models for Predicting Fatigue Crack Growth under Random Loading.* ASTM STP 748. JB Chang and CM Hudson, eds., American Society for Testing and Materials, Philadelphia, PA, 1981; 41–52.
50. Castillo E, Fernández-Canteli A, Siegele D. Obtaining S-N curves from crack growth curves: An alternative to self-similarity. *Int. J. Fracture*, 2014; 187: 159–172.
51. Correia JAFO, Blasón S, De Jesus AMP, Fernández-Canteli A, Moreira PMGP, Tavares PJS. Fatigue life prediction based on an equivalent initial flaw size approach and a new normalized fatigue crack growth model. *Eng. Fail. Anal.*, 2016; 69: 15–28.
52. Department of Defense. Joint service specification guide aircraft structures, JSSG-2006. Department of Defense, United States of America, 1998.
53. Gallagher JP, Berens AP, Engle Jr RM. *USAF Damage Tolerant Design Handbook: Guidelines for the Analysis and Design of Damage Tolerant Aircraft Structures.* Final Report, Flight Dynamics Laboratory, Air Force Wright Aeronautical Laboratories, Air Force Systems Command, Wright Patterson Air Force Base, Ohio, 1984.
54. Merati A, Eastaugh G. Determination of fatigue related discontinuity state of 7000 series of aerospace aluminum alloys. *Eng. Fail. Anal.*, 2007; 14(4): 673–685.
55. De Jesus AMP, Da Silva ALL, Correia JAFO. Fatigue of riveted and bolted joints made of puddle iron—A numerical approach. *J. Construct. Steel Res.*, 2014; 102: 164–177.
56. Krasnowski BR, Rotenberger KM, Spence WW. A damage-tolerance method for helicopter dynamic components. *J. Am. Helicopter Soc.,* 1991; 36(2): 52–60.
57. Forth SC, Everett Jr RA, Newman JA. A novel approach to rotorcraft damage tolerance. *6th Joint FAA/DoD/NASA Aging Aircraft Conference*, San Francisco, CA. 2002.
58. Liu Y, Sankaran Mahadevan S. Probabilistic fatigue life prediction using an equivalent initial flaw size distribution. *Int. J. Fatigue*, 2009; 31: 476–487.
59. Yang JN, M.S. Distribution of equivalent initial flaw size. *Proceedings of the Annual Reliability and Maintainability Symposium*, San Francisco, CA. 1980.
60. White P, Molent L, Barter S. Interpreting fatigue test results using a probabilistic fracture approach. *Int. J. Fatigue*, 2005; 27(7): 752–767.
61. Molent L, Sun Q, Green A. Charcterisation of equivalent initial flaw sizes in 7050 aluminium alloy. *J. Fatigue Fract. Eng. Mater. Struct.*, 2006; 29: 916–937.
62. Moreira PMGP, de Matos PFP, de Castro PMST. Fatigue striation spacing and equivalent initial flaw size in Al 2024-T3 riveted specimens. *Theor. Appl. Fract. Mech.*, 2005; 43(1): 89–99.
63. Fawaz SA. *Equivalent Initial Flaw Size Testing and Analysis.* Air Force Research Laboratory, Ohio, United States, 2000.
64. Cross R, Makeev A, Armanios E. Simultaneous uncertainty quantification of fracture mechanics based life prediction model parameters. *Int. J. Fatigue,* 2007; 29(8): 1510–1515.
65. Makeev A, Nikishkov Y, Armanios E. A concept for quantifying equivalent initial flaw size distribution in fracture mechanics based life prediction models. *Int. J. Fatigue*, 2007; 29(1): 141–145.
66. Savaidis G, Savaidis A, Zerres P, Vormwald M. Mode I fatigue crack growth at notches considering crack closure. *Int. J. Fatigue*, 2010; 35: 1543–1558.

67. Zhu SP, Niu X, Keshtegar B, Luo C, Bagheri M. Machine learning-based probabilistic fatigue assessment of turbine bladed disks under multisource uncertainties. *Int. J. Struct. Integr.*, 2023; 14(6): 1000–1024.
68. Zhu SP, Liu Q, Lei Q, Wang Q. Probabilistic fatigue life prediction and reliability assessment of a high pressure turbine disc considering load variations. *Int. J. Damage Mech.*, 2018; 27(10): 1569–1588.
69. Zhu SP, Huang HZ, Peng W, Wang HK, Mahadevan S. Probabilistic Physics of Failure-based framework for fatigue life prediction of aircraft gas turbine discs under uncertainty. *Reliab. Eng. Syst. Saf.*, 2016; 146: 1–12.
70. Schijve J. Statistical distribution functions and fatigue of structures. *Int. J. Fatigue*, 2005; 27: 1031–1039.
71. He JC, Zhu SP, Liao D, Niu XP. Probabilistic fatigue assessment of notched components under size effect using critical distance theory. *Eng. Fract. Mech.*, 2020; 235: 107150.
72. Niu XP, Wang RZ, Liao D, Zhu SP, Zhang XC, Keshtegar B. Probabilistic modeling of uncertainties in fatigue reliability analysis of turbine bladed disks. *Int. J. Fatigue*, 2021; 142: 105912.
73. Niu XP, Zhu SP, He JC, Ai Y, Shi K, Zhang L. Fatigue reliability design and assessment of reactor pressure vessel structures: Concepts and validation. *Int. J. Fatigue*, 2021; 153: 106524.
74. American Society for Testing and Materials. *ASTM. E739-91: Standard Practice for Statistical Analysis of Linear or Linearized Stress-Life (S-N) and Strain Life (ε-N) Fatigue Data.* Annual Book of ASTM Standards. American Society for Testing and Materials, Philadelphia, PA, 1991; 597–603.
75. International Organization for Standardization. *ISO 12107: Metallic Materials-Fatigue Testing-Statistical Planning and Analysis of Data, Standard.* International Institute of Standardization, Geneva, Switzerland, 2012.
76. Zhao YX, Yang B, Feng MF, Wang H. Probabilistic fatigue S–N curves including the super-long life regime of a railway axle steel. *Int. J. Fatigue*, 2009; 31: 1550–1558.
77. Zhu SP, Foletti S, Beretta S. Probabilistic framework for multiaxial LCF assessment under material variability. *Int. J. Fatigue*, 2017; 103: 371–385.
78. Zhu SP, Liu Q, Peng W, Zhang XC. Computational-experimental approaches for fatigue reliability assessment of turbine bladed disks. *Int. J. Mech. Sci.*, 2018; 142–143: 502–517.
79. Zhu SP, Liu Q, Zhou J, Yu ZY. Fatigue reliability assessment of turbine discs under multi-source uncertainties. *Fatigue Fract. Eng. Mater. Struct.*, 2018; 41(6): 1291–1305.
80. Ai Y, Zhu SP, Liao D, Correia JA, Souto C, De Jesus AM, Keshtegar B. Probabilistic modeling of fatigue life distribution and size effect of components with random defects. *Int. J. Fatigue*, 2019; 126: 165–173.
81. Ai Y, Zhu SP, Liao D, Correia JA, De Jesus AM, Keshtegar B. Probabilistic modelling of notch fatigue and size effect of components using highly stressed volume approach. *Int. J. Fatigue*, 2019; 127: 110–119.
82. Liao D, Zhu SP, Keshtegar B, Qian G, Wang Q. Probabilistic framework for fatigue life assessment of notched components under size effects. *Int. J. Mech. Sci.*, 2020; 181: 105685.
83. Sanches RF, De Jesus AMP, Correia JAFO, Da Silva ALL. A probabilistic fatigue approach for riveted joints using Monte Carlo simulation. *J. Construct. Steel Res.*, 2015; 110: 149–162.
84. Kandarpa S, Spencer Jr BF, Kirkner DJ. Reliability analysis of structural components utilizing the strain-life method. *Eng. Fract. Mech.*, 1996; 53(5): 761–774.
85. Meggiolaro MA, Castro JT. Statistical evaluation of strain-life fatigue crack initiation predictions. *Int. J. Fatigue*, 2004; 26: 463–476.

86. Ni K, Mahadevan S. Strain-based probabilistic fatigue life prediction of spot-welded joints. *Int. J. Fatigue*, 2004; 26(7): 763–772.
87. Zhao YX, Yang B, Zhai ZY. The framework for a strain-based fatigue reliability analysis. *Int. J. Fatigue*, 2008; 30(3): 493–501.
88. Williams CR, Lee YL, Rilly JT. A practical method for statistical analysis of strain–life fatigue data. *Int. J. Fatigue*, 2003; 25(5): 427–436.
89. Liu Y, Mahadevan S. Probabilistic fatigue life prediction using an equivalent initial flaw size distribution. *Int. J. Fatigue*, 2009; 31(3): 476–487.
90. Virkler DA, Hillberry B, Goel PK, The statistical nature of fatigue crack propagation. *Eng. Mater. Technol.*, 1979; 101(2): 148–153.
91. Annis C. Probabilistic life prediction isn't as easy as it looks. *Probabilistic Asp. Life Predict.*, 2004; 1(2): 11557.
92. Ellyin F, Fakinlede COA. Probabilistic crack growth by nonlinear damage accumulation. *Int. J. Fract.*, 1988; 36: 137–149.
93. Correia J, Mourão A, Xin H, De Jesus A, Bittencourt T, Calçada R, Berto F. Fatigue strength assessment of riveted railway bridge details based on regression analyses combined with probabilistic models. *J. Mater. Res. Tech.*, 2023; 23: 3257–3271.
94. Zhang Y, Sun Q, Xin H, Liu Y, Correia JAFO, Berto F. Probabilistic flexural fatigue strength of ultra-lightweight cement concrete and high strength lightweight aggregate concrete. *Int. J. Fatigue*, 2022; 158: 106743.
95. Liu J, Correia JAF, Xin H, Gao Q, Sun Q, Zhao B, Berto F. Stochastic analysis on fatigue life of butt-welded joints with uneven thickness considering mean stress effects. *Mech. Adv. Mater. Struct.*, 2023; 30(11): 2173–2184.
96. Fernández Canteli A, Castillo E, Blasón S, Correia JAFO, De Jesus AMP. Generalization of the Weibull probabilistic compatible model to assess fatigue data into three domains: LCF, HCF and VHCF. *Int. J. Fatigue*, 2022; 159: 106771.
97. Mohabeddine A., Correia JAFO, Montenegro PA, De Jesus A, Castro JM, Berto F. Probabilistic S-N curves for CFRP retrofitted steel details. *Int. J. Fatigue*, 2021; 148: 106205.
98. Stephens RI, Fatemi A., Stephens RR, Fuchs HO. *Metal Fatigue in Engineering* (2nd edition). John Wiley & Sons, Inc., New York, 2001.
99. Zhang J, Zhu J, Guo W, Guo W. A machine learning-based approach to predict the fatigue life of three-dimensional cracked specimens. *Int. J. Fatigue*, 2022; 159: 106808.
100. Nasiri S., Khosravani MR. Applications of data-driven approaches in prediction of fatigue and fracture. *Mater. Today Commun.*, 2022; 33: 104437.
101. Meng D, Yang S, Zhang Y, Zhu SP. Structural reliability analysis and uncertainties-based collaborative design and optimization of turbine blades using surrogate model. *Fatigue Fract. Eng. Mater. Struct.*, 2019; 42(6): 1219–1227.
102. Zhu SP, Hao YZ, Liao D. Probabilistic modeling and simulation of multiple surface crack propagation and coalescence. *Appl. Math. Model.*, 2020; 78: 383–398.
103. Li XK, Zhu SP, Liao D, Correia JA, Berto F, Wang Q. Probabilistic fatigue modelling of metallic materials under notch and size effect using the weakest link theory. *Int. J. Fatigue*, 2022; 159: 106788.
104. Wang L, Zhu SP, Luo C, Niu X, He JC. Defect driven physics-informed neural network framework for fatigue life prediction of additively manufactured materials. *Philos. Trans. R. Soc. A: Math. Phys. Eng. Sci.*, 2023; 381: 20220386.

2 Variable Amplitude Fatigue

Jin-Chao He, Shun-Peng Zhu, and Qingyuan Wang

2.1 INTRODUCTION

Fatigue is the main failure mode of engineering components, and most reported fatigue research studies in the literature are conducted under constant amplitude fatigue (CAF) loadings. The CAF loading is explained as the loading with constant stress amplitude and mean value, such as the classical example of sinusoidal loading. However, in engineering practice, few structures are totally serviced under CAF loadings during the whole service period. Most of these engineering structures are operated under variable amplitude loadings where the load amplitude and mean value change with time, which brings the issue of life prediction under variable amplitude fatigue (VAF) loadings. Specifically, VAF issues could be found in many engineering applications, e.g. aerospace, offshore, bridges, wind turbines, railway, automotive, etc. Therefore, compared with fatigue design under CAF loadings, fatigue modeling under VAF loadings is much more complicated.

In view of the significance of the VAF issue, various approaches have been developed for life predictions of engineering components/structures under VAF loadings, which can be mainly sorted into the time-domain approaches and the frequency-domain approaches. For the frequency-domain approaches, the stress power spectrum density function is obtained by using Fourier transform. As for the time-domain approaches, three main aspects need to be considered for life predictions. The first is discretizing the continuous loading history into loading cycles based on cycle counting methods, such as the rain-flow cycle counting method. The second is evaluating the fatigue damage by using a proper fatigue damage parameter. The third is calculating the total fatigue damage with damage accumulation rules, such as the most commonly used linear fatigue damage rule, i.e., the Miner's rule.

To provide an in-depth understanding of the life prediction issue under VAF loadings, this chapter provides an introduction to the commonly focused aspects of the VAF issue. Specifically, to deal with the complex VAF issues, the commonly applied methods to describe and determine the loading spectrum are presented in Section 2.2. After that, fatigue test criteria under VAF loadings and the differences between CAF tests are shown in Section 2.3. Then, commonly adopted methods for modeling the fatigue damage accumulation are presented in Section 2.4. Then, commonly applied approaches for life predictions under VAF loadings are presented and compared in Section 2.5. Finally, several cases for life predictions of structures under VAF loadings are conducted in Section 2.6.

DOI: 10.1201/9781003405122-2

2.2 FATIGUE LOADING SPECTRUM

2.2.1 Different Types of Loading Spectra

The fatigue loading spectrum is defined as the cyclic loads on a structure in service. Under CAF loadings, the knee point stress range can be found where fatigue lives are infinite. For fatigue under VAF loadings, situations get more complicated, where fatigue damage of each loading cycle through loading history should be clearly defined. Hence, fatigue life prediction under VAF loadings is much more complicated than that under CAF loadings, and characterization of the loading spectrum is significant for accurate fatigue life prediction. In addition, due to the complexity of the VAF issue, the fatigue life is highly related to various factors. First of all, the residual stress (if exists) which is a significant influence factor of fatigue behavior is highly affected by VAF loadings with beneficial/detrimental consequences. Secondly, the interaction between tensile overloads with other mechanisms e.g. creep, yielding, or fracture as well as the interaction between compressive overloads with local buckling are highly related to the fatigue performance. Therefore, in view of the significant differences between CAF and VAF loadings, it is of vital importance to clearly define and describe the real loading spectrum to reach an accurate fatigue life prediction.

Examples of typical CAF and VAF loadings are presented in Figure 2.1 [1]. Actually, the VAF loading indicates single overload or underload cycles on the basis of the CAF loading history. As shown in Figure 2.1a, a typical example of CAF loading is presented where the mean value and the amplitude are constants. Specifically, CAF loadings as presented in Figure 2.1a can be found in service period of rotating, reciprocating, or process machinery. Sometimes, variation of mean stress or periods of overload might occur during stable CAF loadings, as shown in Figure 2.1b. Actually, periodic variation of the mean stress is not induced by the operation of equipment, but induced by some external factors. For instance, the mean stress of stable CAF loadings might change due to the thermal stress related to the operation environment. Figure 2.1c presents the variation of the loads at intervals between the maximum and the minimum stresses, which can be found in bridge structures, lifting devices, pressure equipment, etc. Variation of the large block-wise mean stress is another type of VAF loading, as shown in Figure 2.1d, which can be normally found in aircraft components with an air-to-ground shift or vehicles with full-to-empty random loads. Figure 2.1e presents a load case whose mean stress and loading amplitude vary in each cycle, which is normally found in components affected by random events, e.g. wind, wave, etc.

Typical cases shown in Figure 2.1 illustrate a variety of VAF loadings. Specifically, the VAF loadings can be recognized as two categories, i.e. deterministic and stochastic VAF loadings. The VAF loading can be treated as deterministic if its amplitude in a time can be accurately estimated. For example, the loading history of the pressure vessel can be treated as deterministic. Instead, the stochastic VAF loading presents unavoidable statistical characterization. Specifically, the amplitude of the stochastic VAF loading at a given time can not be represented by a certain value. For example, the wind loads faced by the wind turbine is a stochastic VAF loading. To describe the stochastic VAF loading, mathematical statistical theory is needed.

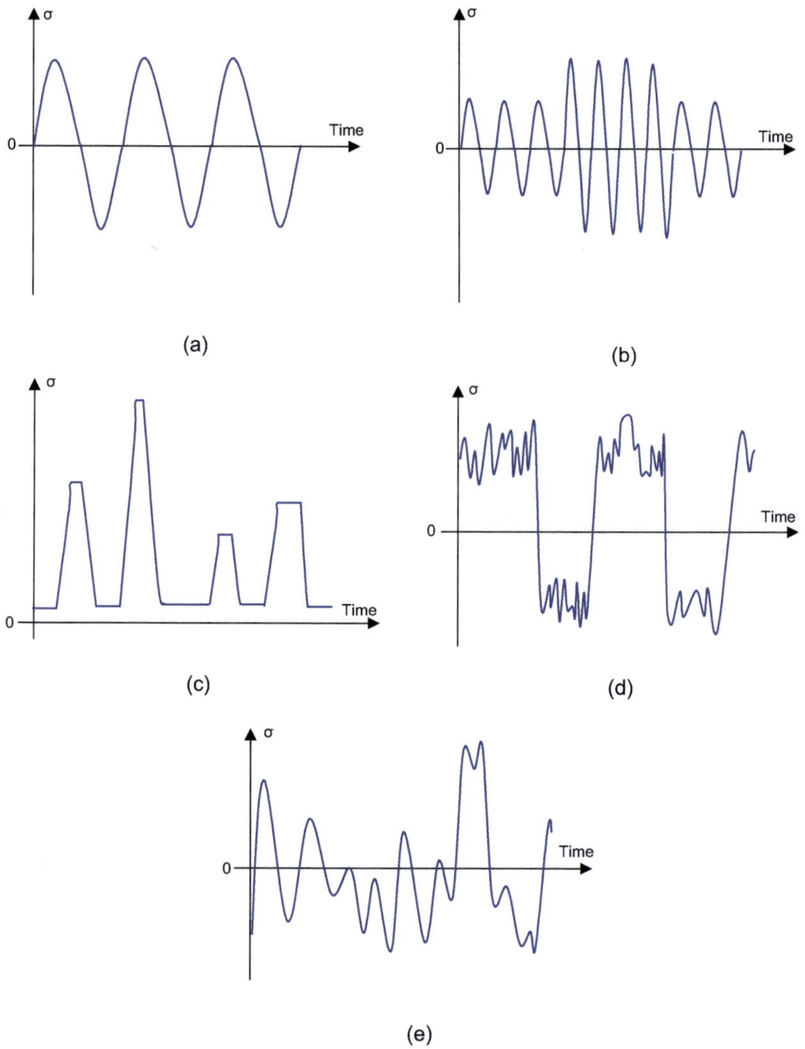

FIGURE 2.1 Typical cases of CAF and VAF loadings. (a) Constant amplitude loading; (b) overload phenomenon; (c) variation of peak stress; (d) variation of mean stress; (e) random loading. (The figure is created based on the work of Ref. [1].)

2.2.2 Description of the Loading Histories

Note from Figure 2.1a–c, the individual loading cycles can be easily identified. However, for the VAF loadings as shown in Figure 2.1d and e, the loading cycles cannot be easily obtained. A lot of load variation points exist among the whole loading history, and during fatigue life prediction or fatigue tests under VAF loadings, it is not possible to apply the whole load history as an input. Therefore, reasonable quantification and simplification of the load history are of vital significance. The process of simplifying the loading history to several cycles is called "cycle counting". Until now, more than ten cycle counting methods have been developed abroad. The accuracy of fatigue life estimation and fatigue test results is largely affected by loading spectrum, and the compilation of loading spectrum is related to the used cycle counting method. The loading spectrum compiled using different cycle counting methods for the same loading history can sometimes differ greatly. Of course, life estimation or tests based on the loading spectrum will also provide different results. From a statistical perspective, the description of the loading histories is broadly divided into three categories: one-parameter cycle counting, two-parameters cycle counting, and energy spectrum approach. The so-called one-parameter cycle counting method refers to considering only one variable per loading cycle, such as range (the difference between adjacent peaks and valleys). The two-parameter cycle counting method considers both variables simultaneously. Considering only one certain parameter is generally not sufficient to describe the load cycle characteristics, indicating that the one-parameter cycle counting method presents significant drawbacks. To provide an understanding of the commonly applied cycle counting methods, both one-parameter and two-parameters cycle counting methods are systematically presented and compared in this section.

2.2.2.1 One-Parameter Cycle Counting

Recently, one-parameter cycle counting has been widely applied for cycle counting, which can be mainly classified into range counting, peak valley and level crossing, etc. However, the one-parameter cycle counting method cannot accurately describe the loading spectrum which links to the final fatigue failure. In view of this, the one-parameter cycle counting method is introduced in this section as a comparison to the one-parameter cycle counting method.

2.2.2.1.1 Level-Crossing Counting

Example of level-crossing counting is shown in Figure 2.2a [2]. Specifically, when the positive sloped position of the load exceeds the reference load, or when the negative sloped position of the load is less than the reference load, one count is recorded. The reference loading crossing is counted on positive sloped part. It's worth mentioning that it does not matter whether the counted counts are negative or positive, and the distinction of the negative and positive is conducted to reduce the number of events. In addition, to eliminate possible amplitude variations which will bring the increase of number of events, the traditional level-crossing counts should be restricted. Specifically, the restriction can be realized by filtering the small load excursion before conducting cycle counting. Another method to realize this restriction is described as follows: Firstly, no counts should be conducted at the reference loading; Secondly, for successive crossings of the secondary lower level, only one count is needed, see Figure 2.2b. Specifically, the differences between these two methods are that the second method applies the same secondary levels for all counted levels above the reference load, and the first method applies the same secondary levels for all counting levels below the reference load. In addition, for fatigue life prediction under VAF loadings, the most crucial cycle count is obtained by constructing largest, second largest until all level crossings based on the level-crossing counting method, as shown in Figure 2.2c.

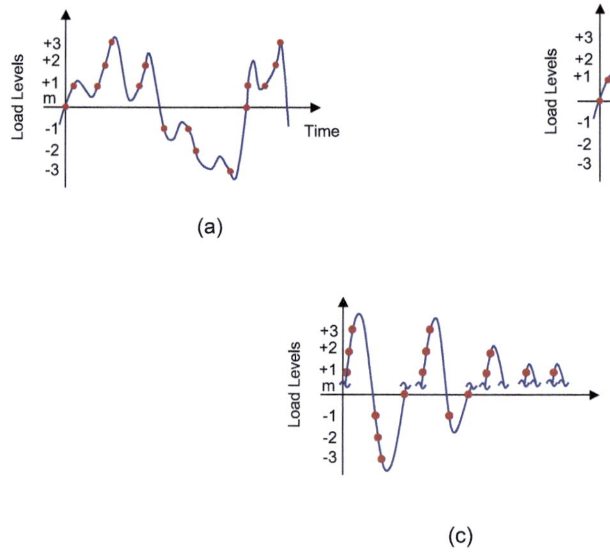

FIGURE 2.2 Example of level-crossing cycle counting. (a) Level-crossing counting; (b) restricted level-crossing counting; (c) cycle derived from level-crossing counting [2].

2.2.2.1.2 Peak Counting

The peak counting method is performed by detecting the occurrence of the maximum load or the minimum load at location position. Specifically, the relative maximum load is counted when it exceeds the reference load level, or when it is lower than the reference loading, as presented in Figure 2.3a [2]. Normally, results obtained from peaks and valleys should be recorded separately. In addition, all the peaks and valleys can be counted without considering the reference load. Specifically, the small amplitude loadings issues can be avoided when mean crossing peak counting is applied, which is performed by counting the largest peak/valley, as presented in Figure 2.3b. In addition, for life prediction under VAF loadings, the most crucial cycle count is obtained by constructing largest, second largest until all level crossings based on the peak counting method, as shown in Figure 2.3c.

2.2.2.2 Two-Parameter Cycle Counting

As the most commonly used two-parameters cycle counting method, the rain-flow counting method was developed by assuming that the damage is related to plastic deformation. In general, although the nominal stress is within the elastic range, plastic deformation still exists from a local and microscopic perspective. The strain-time history shown in Figure 2.4a [3] corresponds to the stress-strain curves, see Figure 2.4b. It can be seen that two small cycles 2-3-2′, 5-6-5′ and one systemic circulation 1-4-7 form two small hysteresis loops and one large hysteresis loop respectively. If fatigue damage is marked by this and supposed that fatigue damage induced by a large range is not related to truncation to complete a small hysteresis loop, the smaller cycles that make up the smaller hysteresis loop can be extracted from the entire strain-time history one by one and recombined. In this way, the strain-time history in Figure 2.4a is

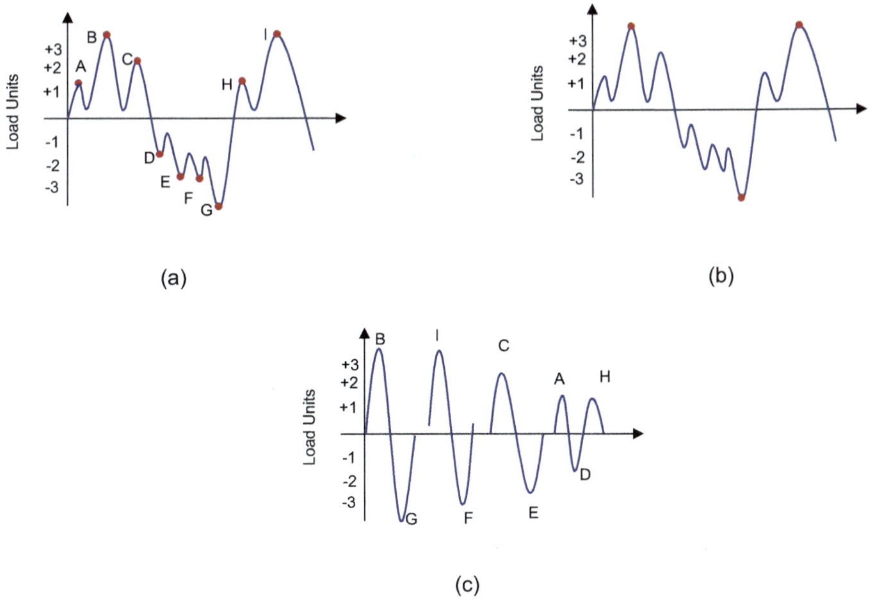

FIGURE 2.3 Example of peak counting. (a) Peak counting; (b) mean crossing peak counting; (c) cycles derived from peak counting [2].

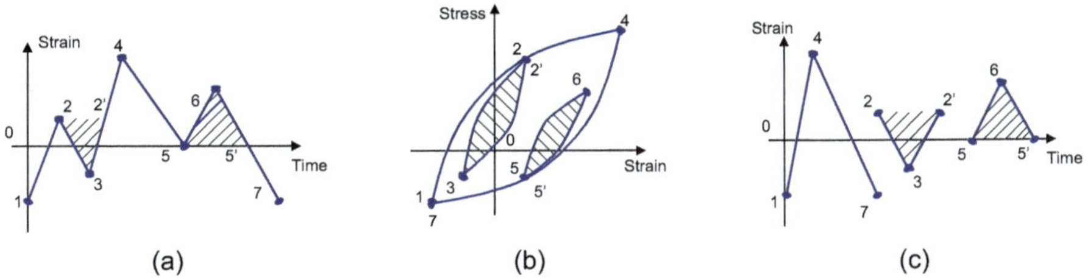

FIGURE 2.4 Strain-time history and stress-strain curve. (a) Strain-time history; (b) stress-strain curve; (c) simplified strain-time history. (The figure is created based on the work of Ref. [3].)

simplified to Figure 2.4c, and it is believed that both of them cause equivalent fatigue damage to the material.

Based on the above principles, the strain-time data is recorded by turning it over 90° see Figure 2.4. The time axis is just vertically downwards, and recorded data are in the form of several roofs. Specifically, rainwater just flows down the roof and hence this method is named as "rain-flow counting method", see Figure 2.5 [3], which is explained as:

1. Rain-flow begins from starting point of test record and sequentially at inner edge of each peak, i.e. starting from 1, 2, …;
2. Rain drops vertically at the peak point (i.e. the eaves) until there is a corrected maximum on the opposite side compared to the initial maximum;

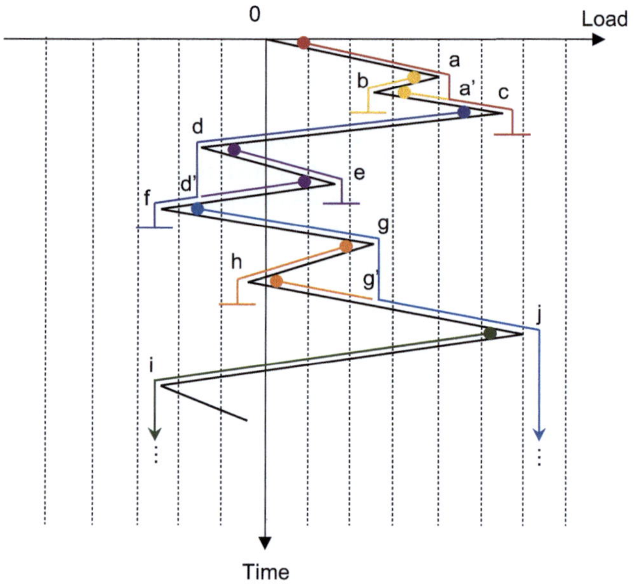

FIGURE 2.5 Illustration of the rain-flow counting method. (The figure is created based on the work of Ref. [3].)

3. As the rain stream encounters rain, the flowing stops and one loading cycle forms;
4. Draw each cycle based on the starting and ending points of raindrop flow, take out all cycles gradually, and the peak valley value is recorded;
5. Amplitude of the loading cycle is obtained by using horizontal length of each flow.

Although there are more than ten reported counting approaches, the rain-flow counting is recognized as an advanced counting method, because its counting process is in line with the stress-strain characteristics of materials and has a more reliable mechanical basis. There are three calculation solutions for the rain-flow counting method: four peak-valley counting method, three peak-valley counting method, and real-time counting method.

The four peak-valley counting method has been widely applied in practices, whose procedure is described as:

1. **Data compression:** Data compression is the process of processing raw experimental data into an array that facilitates the calculation of cycle numbers, including equivalent number compression and peak valley value detection. The purpose of performing equivalent number compression is to avoid confusion during peak valley value detection and to retain only one identical value of the phase. Peak valley detection is one of the main contents of data compression, which extracts the peak valley values from the data as an array for the next step of data processing.

2. **Once rain-flow counting:** Once rain-flow counting is an iterative process of extracting cycles from compressed data and recording their characteristic values. According to the specifications of the four peak-valley counting model.
3. **Closed processing of compressed data:** The waveform formed by the closed processing of compressed data and the remaining points in the rain-flow counting is a standard divergence convergence type. At this time, based on the rain-flow counting rule, the entire cycle cannot be formed anymore. It can only be cut off at the point of peak (or valley) value and then connected end-to-end to form a converged divergence type loading spectra that starts with the peak (or valley) value and ends with the peak (or valley) value.
4. **Secondary rain-flow counting:** Secondary rain-flow counting means that continuing the extraction cycle of the completed waveform until the last three points form the maximum integer cycle.

2.2.2.3 Energy Spectrum Approach

Energy spectrum approach is a more efficient method than the cycle counting method for describing the load history. Specifically, the calculated loading cycles are not able to relate with the hysteresis loops. In addition, the energy spectrum approach can only be used under the situation of stationary loading condition. The energy spectrum approach is developed by splitting the stationary time series into an infinite number of harmonic components, as:

$$\sigma(t) = \int_0^{\infty} \sigma_\omega \sin(\omega t + \varphi_\omega) d\omega \tag{2.1}$$

where σ_ω denotes the harmonic stress component amplitude with the frequency ω; φ_ω denotes the random phase distortion. Equation (2.1) needs to locate the amplitude of σ_ω as a function of ω which can be realized with the help of Fourier transformation technique. Specifically, the narrow frequency band energy which is proportional to $0.5\sigma_\omega^2$ can be expressed as:

$$S(\omega)d\omega = 0.5\sigma_\omega^2 d\omega \tag{2.2}$$

Based on the obtained loading spectrum, the general moment function can be defined as:

$$m_n = \int_0^{\infty} \omega^n S(\omega) d\omega \tag{2.3}$$

where m_n denotes the moment function of the loading spectrum. Specifically, when n equals to zero, m_0 denotes the area under the spectrum curve. Then, the averaging up-crossing time period is expressed as:

$$T_Z \approx 2\pi\sqrt{\frac{m_0}{m_2}} \tag{2.4}$$

where m_2 denotes second-order moment of energy spectrum. To describe stress level deviation, normal distribution can be applied, as:

$$f(\sigma)=\frac{1}{2\pi s_\sigma}\exp\left(-\frac{1}{2}\left(\frac{\sigma}{s_\sigma}\right)^2\right) \tag{2.5}$$

The stress history can be expressed as a Gaussian process with standard deviation of s_σ. Specifically, s_σ can be related to the energy spectrum, as:

$$s_\sigma^2 = m_0 \tag{2.6}$$

To describe the loading history, only the amplitude should be focused rather than the arbitrary stress level given in Eq. (2.5). It can be seen that the stress range obeys the Rayleigh distribution for narrow energy spectrum, as:

$$f(\Delta\sigma)=\frac{\Delta\sigma}{4m_0}\exp\left(-\left(\frac{\Delta\sigma}{2\sqrt{2m_0}}\right)^2\right) \tag{2.7}$$

Specifically, the mean value of 1/3 largest loading cycle can be expressed as the significant loading cycle, as:

$$\Delta\sigma_{\text{sign}} = 4\sqrt{m_0} \tag{2.8}$$

2.3 FATIGUE TESTS UNDER VAF LOADINGS

Fatigue testing under VAF loadings is a method to simulate the actual working conditions of the component of interest and provide data for characterizing and assessing its fatigue behavior. VAF test can provide references for structural and material fatigue design. Actually, fatigue performances of main structures are normally assessed by testing S-N curves. Early in 1939, Gassner [4] indicated that the applied loading amplitude in fatigue tests should be the same as the observed loading spectrum during the service process. In addition, the so-called block-program VAF test is designed and the first VAF loading spectrum was established. Later, the electro-hydraulic servo fatigue test machine was developed, which enables the implementation of block-program VAF loading in fatigue test. The emergence of the electro-hydraulic servo fatigue test machine brings a great breakthrough in VAF aspect. However, since the electro-hydraulic servo fatigue test machine is mainly applied for strength detections and life prediction, it is difficult to perform VAF simulation test for the actual VAF phenomenon of main components. In addition, due to the fact that the block-program VAF test ignores the impact of the loading sequence, the block-program VAF test still needs to be further improved and refined.

Until now, although research studies have been conducted on a large number of fatigue tests, the vast majority are limited to CAF tests. And relevant VAF test criterion is not very mature. In the relevant research fields abroad, although there is much literature on the VAF test, most of them on VAF test does not have a clear description of the loading history, test requirements, and test information. The lack of this information makes it difficult to evaluate the VAF test well. The key factor in implementing the VAF test is the acquisition of test loading control signals and the adherence to specific test requirements. In view of the existing problems in fatigue tests under VAF loadings, the loading history, test principles, and the important influencing factors on the fatigue test result are summarized in this section.

2.3.1 Loading History

The basic principle of the VAF test is to apply the load by using control signals obtained from accumulation frequency plots (program sequence blocks) or random sampling. Specifically, the accumulation frequency map of the loading amplitude or range (spectrum) is constructed from the data obtained by processing the original loading history using relevant cycle counting methods. Therefore, the most important prerequisite is the loading history (i.e. a sequence of loading cycles with amplitude changes from one cycle to the next). The original loading history can generally be obtained in two ways. For one thing, the loading history applied in fatigue test is normally obtained from the measurement of the in-service components, but this real-time measurement is only possible if the structure is easy to be measured. Once the entire real-time loading history is recorded, it can be applied to a fatigue test machine. However, when the original loading history is obtained by directly measuring the load of in-service components, filtering may be necessary to eliminate electronic mechanical noise or other interference vibrations not caused by fatigue processes. In addition, during the filtering, to avoid ignoring the real damage cycle that plays a major role in the fatigue damage process of structural components, it is necessary to set a critical value or threshold for the loading amplitude. For another, typical standard loading history in the industrial field can also be cited. Although these loading histories are feasible for general research on specified structures, the verification of fatigue characteristics should not only use one of the standard loading histories; on the contrary, the loading spectrum and load sequence must be related to the actual load conditions that the structure is subjected to in service.

In addition, due to the fact that obtaining the loading cycles to failure from the original loading history is difficult, the original loading history normally needs to be simplified for further VAF tests. The simplification of the original loading history is usually achieved by loading signal analysis into two analog loading control signals. These two types of loading analog control signals are obtained by reconstructing signals randomly extracted from block-programs or conversion matrices, as shown in Figure 2.6 [5].

Actually, only one portion of the "random" extraction appears to be truly random. The random sampling program is explained as follows (taking $i=3$ as an example):

1. Obtain a_{39} using pseudo-random extraction with $j>9$;
2. Due to the alternating upward and downward transitions in random sampling, and to avoid discontinuous loading signals, the next transformation should be conducted from level 9 to another a_{9j}. Therefore, here j will be extracted as a pseudo-random extraction with $j<9$, such as obtaining a_{96} when $j=6$;

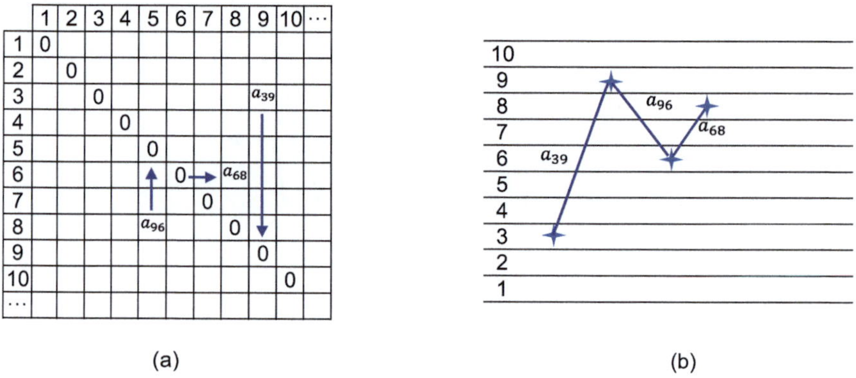

FIGURE 2.6 Loading control signals analysis. (a) Example of transformation matrix; (b) random draw procedure. (The figure is created based on the work of Ref. [5].)

3. Similar to step (2), i should be pseudo-random extracted in the range $i=6$ and $j>6$, such as obtaining a_{68} when $j=8$;
4. If the transformation from i to j has been calculated for m times, now $a_{ij} = m$. When reconstructing the control signal, once the first transformation is added to the signal, $a_{ij} = m-1$. Specifically, the reconstruction of the signal is finished once the value of all a_{ij} equal to zero.

2.3.2 Test Principles and Schemes

The VAF test can be approximated as a CAF test conducted at different loading levels. The biggest difference between CAF and VAF tests is that the VAF test is conducted at different loading levels, repeating the loading sequence until failure (limited crack depth or length, deformation, or complete fracture), while the loading amplitude remains unchanged under the CAF test. It should be noted that the loading sequence in the VAF test must be repeated at least 5–10 times. Due to the interaction between sequence effects and load cycles, different load cycle sequences in the VAF loading history have a significant impact on the $\sum n/N$ value at fracture, especially for the block-program fatigue load sequence. Therefore, during the VAF tests, application of the block-program VAF test should be avoided.

For a given amplitude signal with its own frequency feature, it should be ensured that the test configuration is suitable for this signal. If not applicable, the components, fixtures, loading units, or any components used for test should be adjusted to adapt to the frequency, amplitude, and all other characteristics of the signal. In addition, when the loading frequency effect is ignored, the frequency is increased to shorten the test time. Nevertheless, due to the interaction between sample stiffness and test machine, the test frequency is restrained. In this case, acceleration is achieved through amplitude and frequency adaptive control, as shown in Figure 2.7 [5]. However, when components with frequencies greater than the upper limit f of the loading bandwidth are filtered in different ways, the applied load will lose its authenticity. In view of this, it is deeded to reduce the test speed until the frequency of all important harmonic components input to the fatigue test machine is less than f.

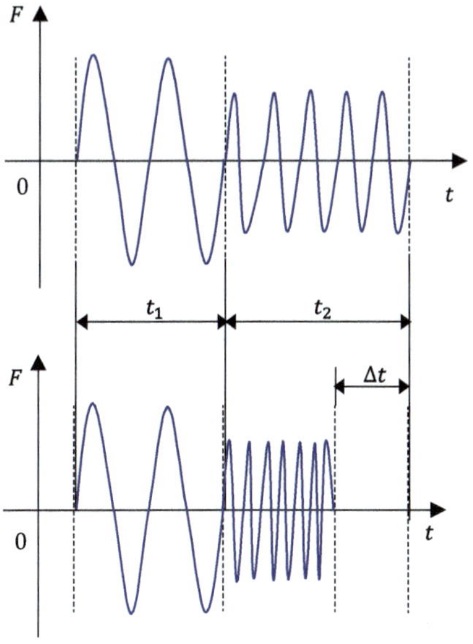

FIGURE 2.7 Acceleration of tests by frequency adaptation. (The figure is created based on the work of Ref. [5].)

The accumulation frequency plots (program modules) or randomly selected control signals are used for the application of VAF loading during tests. Specifically, the response of the tested specimen is monitored through the measurement values given by the load sensor, force sensor, or strain gauge, and these output data are used for closed-loop control. The specific experimental test principle is shown in Figure 2.8.

The main parameter records by the test spectrum include: the spectrum of the loading history, maximum and minimum amplitudes in the spectrum, the maximum stress ratio R, the length and shape of the loading sequence. In addition, during the test, the turning point, amplitude distribution, conversion matrix, and power spectrum density need to be recorded and compared as a reference for the changes made to the test setup to adapt to the signal characteristics (including frequency). Normally, the frequency spectrum can not include the applied frequency, and the test frequency is related to the interaction between the stiffness of the tested component and test machine. However, for VAF tests of dynamic and nonlinear systems, the recorded loading history must contain information about the frequency spectrum.

2.3.3 Important Influencing Factors

Considering that the fatigue tests under VAF loadings is very complex, tested lives may be affected by many factors, e.g. loading level, loading frequency, etc. Therefore, before conducting fatigue test under VAF loadings, all these factors affecting factors to the tested results should be clarified and checked, including the following:

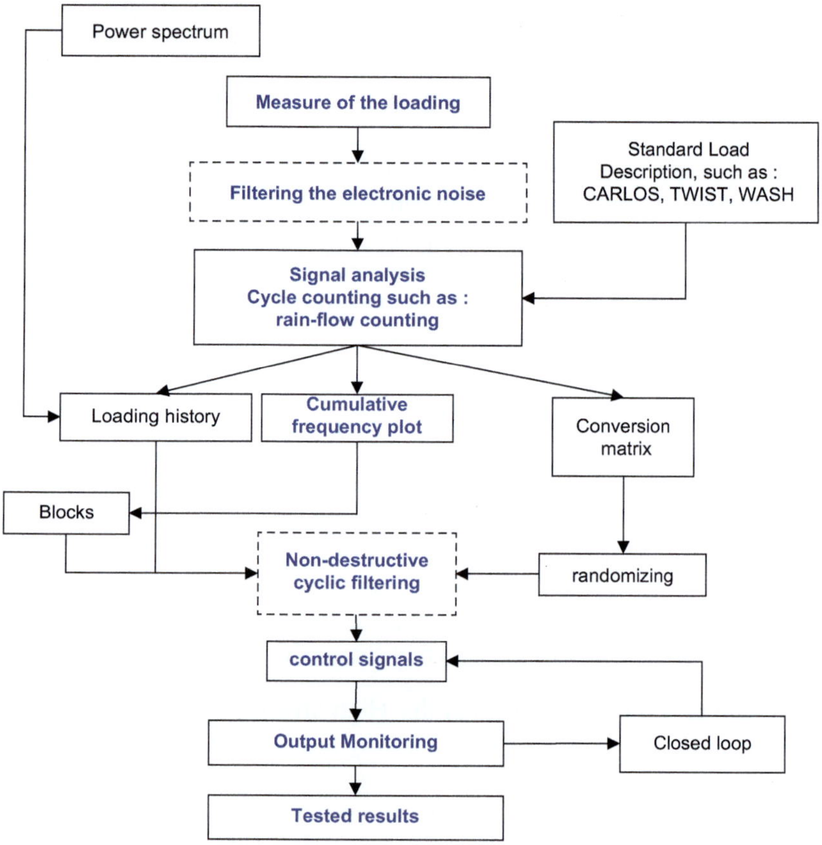

FIGURE 2.8 Procedure for the VAF test.

2.3.3.1 Cycle Counting Method

Various cycle counting methods have been systematically introduced in Section 2.2.2, which are able to be applied for dealing with the loading spectrum. However, which method provides the best estimations to the real service loading needs to be further investigated and verified. Some researchers have studied the behavior of fatigue crack propagation by using five different types of cycle counting method. Specifically, a series of VAF tests under Gaussian loading sequence with different irregularity factors were conducted. Finally, results indicated that the predicted fatigue lives show strong dependence on the applied cycle counting method.

2.3.3.2 Number of Loading Levels

Taking block loading as an example, this method is normally applied with eight different loading levels, while in the engineering practice different number of loading levels can be applicable. Things are the same for the number of levels applied for the estimation of the accumulation frequency function. As loading levels decrease, the impact of higher loading amplitude blocks on the lower one need to be considered.

In addition, the risk of overloading/underloading effect is more significant when the duration of fatigue crack propagation is non-negligible.

2.3.3.3 Loading Levels Application Order

The influence of loading levels application order normally happens along with the block loading, especially when applying in less than eight loading levels. To eliminate the influence of loading levels application order, it is suggested to increase the loading levels and repeat the loading sequence before crack initiation. However, for several block-program applications, some researchers revealed a strong relationship between fatigue with order of loading level. In addition, the influence of loading levels application order is important during crack initiation tests as well.

2.3.3.4 Loading Frequency

As reported by some researchers, the loading frequency effect on fatigue lives of metallic materials can be ignored under the frequency remains lower than 250 Hz. Later, it was found that this conclusion is only suitable for several types of metallic materials. For example, with the increase of the loading frequency, the temperature of the specimens made from ductile materials increase, which will bring unavoidable influence to the fatigue life. Specifically, during fatigue tests on some materials (such as thermo-couples, infrared camera, etc.), the temperature has to be measured since it can not be ignored, which will bring great changes to the fatigue performance.

2.3.3.5 Limitation of the Signals under High Stresses

The maximum loading level under VAF testing is a very significant index. Indeed, due to the fact that the residual stress is related to local plastic deformation near the notch root, the highest loading cycles on the notched specimens pose a great influence on the fatigue lives. In addition, the highest cycles also pose a great influence on the crack propagation curve. The physical phenomena in the simulation test should not exceed the stress-strain range of material under actual usage conditions.

2.3.3.6 Irregularity Factor

For the VAF tests under narrow band loading, the irregularity effect on fatigue lives could be ignored. However, for VAF tests with narrow band loading, the tested fatigue life is highly dependent on the irregularity factor. Mainly two reasons lead to this: (1) the influence of the mean value of cycles; (2) the less damaging effect of low amplitude cycles. Therefore, it is suggested by some researchers that the narrow band signals should be chosen during VAF tests.

2.4 FATIGUE DAMAGE ACCUMULATION: FUNDAMENTALS

2.4.1 Evolution of Fatigue Damage

As aforementioned, fatigue failure is the key failure mode threatening the safety of engineering components/structures. Hence, effective strategies for fatigue damage quantification have always been the research hotspot of fatigue studies. During fatigue analysis of these components, the concept “damage” was developed a long time ago, whose physical explanation is based on structural integrity, such as initiation of micro-crack, degradation of the structural strength, etc. Recently, a series of methods have been developed for defining fatigue damage, which brings difficulties for fatigue strength assessment and fatigue life prediction. Mainly two ways can be found for the definition of fatigue damage, one from the micro or physical perspective (such as the crack density in fatigue damage zone, acoustic emission, etc.) and another from the macro perspective (such as the Miner's rule, residual strength rule, etc.). Specifically, no matter how the fatigue damage is defined, it shows the following two features, namely irreversibility and randomness, which are explained as follows:

1. **Irreversibility:** Fatigue damage is a continuous variable in space and time, which is a general expression of the effect of micro defects on mechanical performance. Therefore, there is a certain functional relationship between real damage and the defined damage state variables. According to continuum damage mechanics, damage accumulation is a macro continuous variable in the irreversible change process of the internal structure of materials under alternating loads. From the energy perspective, fatigue failure is an irreversible dissipation process of the energy. Damage mechanics characterizes the internal damage variable of materials from the macroscopic perspective by the number of micro cracks and micropores, or the change of elastic modulus, yield limit, mass, density, etc. These variables are monotonous with the time process; therefore, fatigue damage is an irreversible process from both macroscopically and microscopically.
2. **Randomness:** The randomness of fatigue damage mainly originates from two aspects. First, uncertainties of the microstructure. Microscopic observations indicate that the fatigue damage is anisotropic, and the damage evolution process is also random and discontinuous. Macroscopic fatigue damage is the collective contribution of a large amount of micro damage caused by the uncertainty of microstructure. In addition, the uncertainty of microstructure also leads to the stochastic distribution of the macroscopic properties, e.g. fracture toughness, yield limit, tensile strength, elastic modulus, etc. Secondly, uncertainties from external factors. External conditions such as external loads, machining processes, and service environments are accompanied with randomness in the process of fatigue damage, which leads to significant dispersions of the fatigue damage evolution. In the laboratory, life dispersion can be minimized by controlling experimental conditions. However, even if the test loading spectrum, test conditions, and test pieces are strictly controlled, the fatigue damage still exhibits considerable dispersions.

When the material bears a load above the fatigue limit, fatigue damage will accumulate in each loading cycle, which is averaged as $1/N$. In addition, the fatigue damage will accumulate with the increase of the loading cycles. Specifically, the damage induced by n CAF cycles is n/N. For VAF loadings, damage under n loading cycles is $\sum_{i=1}^{l} n_i/N_i$, where l is the total loading levels, n_i denotes loading cycles of ith loading level, N_i

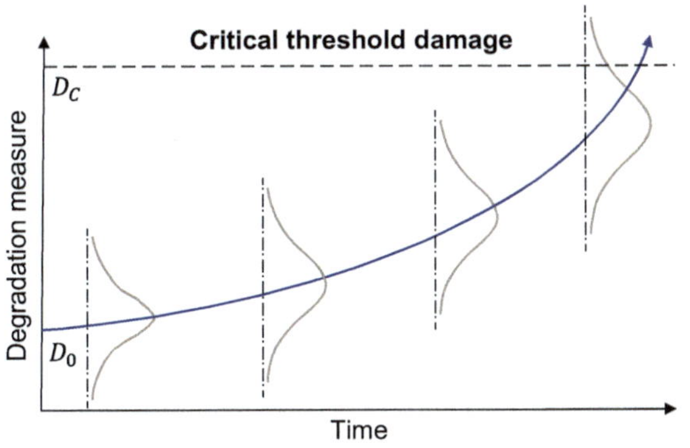

FIGURE 2.9 Degradation process of fatigue damage. (The figure is created based on the work of Ref. [6].)

denotes the fatigue life of *i*th loading level. Fatigue fracture of the material happens as the fatigue damage obtains critical threshold damage D_C. As an irreversible damage accumulation process, fatigue damage leads to fatigue fracture eventually. Moreover, due to the uncertainties, probabilistic models are commonly utilized for fatigue damage quantification. Figure 2.9 [6] depicts the degradation process graphically where D_0 and D_C denote initial damage and critical threshold damage, respectively.

In practical engineering, main components are normally subjected to VAF loadings. However, existing life prediction theories and methods are normally developed based on the assumption of CAF loadings. Considering that fatigue life prediction theory for CAF loadings has been developed into a complete analysis system after years of development, transferring the fatigue life analysis issue under VAF loadings into a combination of multiple fatigue life analysis issues under CAF loadings can greatly save research time and costs. In view of this, the theory of fatigue accumulation damage has gradually emerged.

To establish a reasonable damage accumulation rule, definition of the fatigue damage D is critical. In other words, the fatigue damage D needs to be defined with a clear physical explanation. In addition, the defined fatigue damage should present the same damage evolution rule with the experimental data. Specifically, the defined fatigue damage needs to reply to the following three questions: (1) What is the damage caused to materials or structures by each cycle? (2) How does the fatigue damage accumulate during multiple loading cycles? (3) How to express the critical threshold damage at failure time?

After extensive engineering experiments and theoretical research, a large amount of fatigue accumulation damage rules have been developed, as shown in Table 2.1 [7]. Among them, Miner's rule has been widely applied for damage modeling due to its high calculation efficiency. However, due to the fact that the Miner's rule ignores the interaction between loadings as well as the load sequence effect [8], its prediction accuracy is normally unacceptable. In view of the unavoidable shortcomings of the Miner's rule, a series of damage accumulation rules were gradually developed. Until now, the defined fatigue damage accumulation theories are sorted into three types,

TABLE 2.1
Commonly Used Fatigue Damage Accumulation Rules

Type	Authors	Fatigue Damage Model		Material Parameters
Linear damage accumulation rules	Palmgren Miner	Fatigue damage	$D_i = 1/N_i$	N_i
		Failure criterion	$D_c = 1$	
	Manson	Fatigue damage	$D_1 = 1/N_1 \quad D_2 = 1/N_2$	$N_1 \quad N_2$
		Failure criterion	$D_c = 1$	
	Shanley	Fatigue damage	$D_i = CS^{kb}n$	$C \quad k$
		Failure criterion	$D_c = 1$	
	Buch	Fatigue damage	$D_i = 1/N_i$	N_i
		Failure criterion	$D_c = D_f$	
	Heller	Fatigue damage	$D_i = W_i/N_i \quad W_i = (\bar{\sigma}/\sigma_i)^d$	d
		Failure criterion	$D_c = 1$	
	Holycross	Fatigue damage	$D_i = \left(A\Delta W_i^d\right)^{-1}$	$A \quad d$
		Failure criterion	$D_c = 1$	
	Grover	Fatigue damage	$D_i = 1/\alpha_i N_i$	α_i
		Failure criterion	$D_c = 1$	
	Kramar	Fatigue damage	$D_i = \sigma_{si}/\sigma_s^*$	σ_s^*
		Failure criterion	$D_c = 1$	
Nonlinear damage accumulation rules	Macro Starkey	Fatigue damage	$D \propto (n_i/N_i)^C$	C
		Failure criterion	$D_c = 1$	
	Corten Dolan	Fatigue damage	$D = mrn^a$	a
		Failure criterion	$D_c = 1$	
	Henry	Fatigue damage	$D = (\sigma_{e0} - \sigma_e)/\sigma_{e0}$	σ_e
		Failure criterion	$D_c = 1$	
	Bui-Quoc	Fatigue damage	$D = (\sigma_{e0} - \sigma_e)/(\sigma_{e0} - \sigma_{ec})$	$\sigma_e \quad \sigma_{ec}$
		Failure criterion	$D_c = 1$	
	Morrow	Fatigue damage	$D = (n_i/N_i)^C$	C
		Failure criterion	$D_c = 1$	

including the linear-based, nonlinear-based, and the probabilistic-based damage accumulation rules, which will be systematically introduced in the next subsections.

2.4.2 Linear Fatigue Damage Accumulation Rule

The linear fatigue damage accumulation rule means that under cyclic loadings, fatigue damage accumulates linearly, and each loading cycle is independent and uncorrelated with others. Fatigue failure occurs in the specimen or component as the accumulated fatigue damage obtains the critical value. Recently, several linear damage accumulation rules were developed, as shown in Table 2.1. The most widely applied Miner's rule and the double linear fatigue damage accumulation rule are introduced herein.

2.4.2.1 Miner's Rule

The typical linear accumulation damage theory is Miner's rule, abbreviated as Miner's rule. For a loading sequence containing m loading blocks, if n_i is applied for expression of the loading cycles of ith loading block under the CAF loading Δ_{σ_i}, and N_i is applied for expression of fatigue life under the CAF loading Δ_{σ_i}, then Miner's rule can be expressed as:

$$\sum_{i=1}^{m} \frac{n_i}{N_i} = 1 \tag{2.9}$$

It's worth mentioning that the application of Eq. (2.9) for fatigue damage calculation is based on three assumptions: (1) applied loads in each loading block are symmetric cyclic loading, namely, no mean stress can be observed; (2) At arbitrary given loading level, rate of accumulation damage is no related to previous loading histories; (3) Order of loading block does not affect the fatigue lives.

Due to its simplicity in form and high efficiency in calculation, the Miner's rule is widely utilized for damage calculation and life predictions. However, strong dependence of Miner's rule on the aforementioned three assumptions affects its prediction accuracy. A large number of experiments have demonstrated that for a complex loading sequence, $\sum_{i=1}^{m} n_i/N_i$ is not equivalent to 1, which normally changes with loading level and loading block. In addition, Miner's rule ignores the interactions between each loading block and the influence of transient behavior of materials such as strain hardening (or softening).

Considering that the critical threshold damage D_C normally does not equal to 1, the relative Miner's rule was developed. A large amount of fatigue tests have demonstrated that the critical threshold damage D_C ranges from 0.3 to 3. For example, many fatigue tests under two-level loadings have a critical threshold damage of 0.7. Some researchers also pointed out that the critical threshold damage should be different under different situations. In such a situation, some researchers pointed out that the critical threshold damage should be determined by using the average accumulation damage. As pointed out by the relative Miner's rule, to eliminate the influence of loading sequence, loading level, materials, and specimen geometries, specimens with a similar loading spectrum and similar specimen geometries hold the same critical threshold damage D_f, which is expressed as:

$$\sum_{i=1}^{m} \frac{n_i}{N_i} = D_f \tag{2.10}$$

Equation (2.10) provides an efficient strategy for the definition of critical threshold damage, however, engineering components are normally designed into various shapes and bear complex loading spectra. Therefore, the application of the relative Miner's rule is limited.

Miner's rule is the most widely utilized damage accumulation rule in engineering, but the differences between the calculated and the experimental lives can be as much as 5–10 times. The relative Miner's rule integrates the influence load sequence effect and material dispersion into the standard spectrum experiment, greatly eliminating the bias in life prediction using Miner's rule.

2.4.2.2 Double Linear Fatigue Damage Accumulation Rule

The double linear fatigue damage accumulation rule was developed by Manson by considering two types of fatigue damage, i.e. crack initiation and propagation. Specifically, fatigue damage accumulates under two stages according to different linear fatigue damage accumulation rules, and the mathematical formula for the transition "inflection point" is established. However, due to the fact that the mathematical formula is established by using tested data from only one type of material, its validity needs to be further verified. In addition, the physical interpretation of the two stages of double linear fatigue damage accumulation rule is not reasonable. Firstly, the crack length at the inflection point is uncertain and unmeasurable, and secondly, the inflection point changes with different combinations of applied loads. In view of this, Manson [9] further treats the double linear fatigue damage accumulation rule as a simplification of the damage curve model and views the inflection point as the transition point after the simplification of the damage curve. The two stages determined by the inflection point were renamed as stage 1 and stage 2. In such a way, the aforementioned two problems can be avoided with simple numerical analysis, as shown in Figure 2.10 [10].

As shown in Figure 2.10, fatigue life N_1 corresponding to stage 1 and fatigue life N_2 corresponding to stage 2 can be expressed as: $N_1 = N_f \exp\left(ZN_f\right)^{\Phi}$ and $N_2 = N_f - N_1$, where

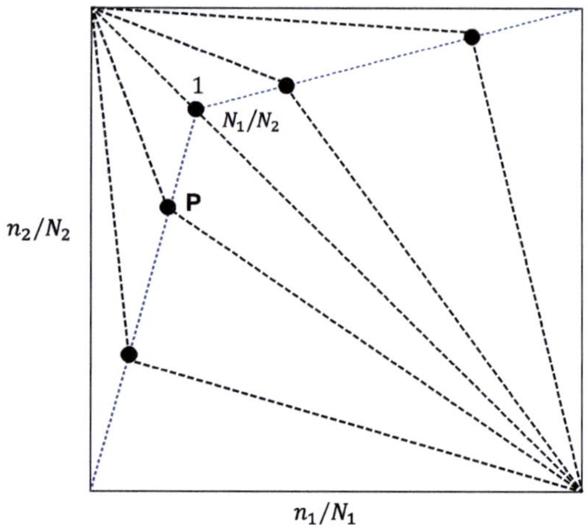

FIGURE 2.10 Illustration of the double linear fatigue damage accumulation rule. (The figure is created based on the work of Ref. [10].)

$$\begin{cases} \Phi = \left[\dfrac{1}{\ln\left(N_{f1}/N_{f2}\right)} \right] \ln\left[\dfrac{\ln(0.35Q)}{\ln(1-0.65Q)} \right] \\ Z = \dfrac{\ln(0.35Q)}{N_{f1}^{\Phi}} \\ Q = \left(\dfrac{N_{f1}}{N_{f2}} \right)^{0.25} \end{cases} \tag{2.11}$$

Specifically, the mathematical formula for the inflection point can be expressed as:

$$\begin{cases} \dfrac{n_1}{N_{f1}} = 0.35 \left(\dfrac{N_{f1}}{N_{f2}} \right)^{0.25} \\ \dfrac{n_2}{N_{f2}} = 0.65 \left(\dfrac{N_{f1}}{N_{f2}} \right)^{0.25} \end{cases} \tag{2.12}$$

where N_{f1} and N_{f2} denote the maximum and minimum fatigue lives of loading blocks which causes fatigue damage, respectively. Then, the Miner's rule is applied for these two stages, only when both the ratio of the cyclic loading for stage 1 and stage 2 reach 1 will the fatigue failure occur. Specifically, the accumulated fatigue damage is:

$$\begin{cases} \sum \dfrac{n_i}{N_{1,i}} = \sum \dfrac{n_i}{N_f \exp\left(ZN_f\right)^{\Phi}} = 1 \\ \sum \dfrac{n_i}{N_{2,i}} = \sum \dfrac{n_i}{N_f - N_1} = 1 \end{cases} \tag{2.13}$$

Based on the comparison of several types of damage accumulation rules, it is concluded that double linear damage accumulation rule yields better estimations than the Miner's rule. The double linear fatigue damage accumulation rule is developed based on the assumption that the crack propagation is the main reason for the fatigue damage, which has been widely utilized on steel components. Later, some researchers developed the three-stage damage accumulation rules by using characteristics of changes in the microstructure and properties of metal materials and quantification of the loading sequence effect, namely the cyclic hardening/softening stages, nucleation and propagation of short/long crack propagation and final fracture. Although these rules take the load sequence effect into account, they still can not conquer the shortcoming of linear damage rule.

2.4.3 Nonlinear Fatigue Damage Accumulation Rule

It's assumed by the linear fatigue damage accumulation rules that damage caused by each loading cycle is the same for the whole life regime. However, damage is not linear accumulated during the fatigue process of the material. In addition, note that during tension-compression and rotation-bending fatigue tests, affected by exercise low stress effect, the critical threshold damage in the low-to-high loads order is often greater than 1, while the critical threshold is often less than 1 in the high-to-low loads. To explain such phenomena, various nonlinear fatigue damage accumulation rules have been established, e.g. Corten-Dolan theory and damage curve approach, as shown in Table 2.1, which will be introduced herein.

2.4.3.1 Damage Curve Approach

The damage curve approach was by treating the damage as proportional to the cycle ratio

$$D \propto \left(\frac{n_i}{N_i}\right)^C \tag{2.14}$$

where C is a constant which is larger than 1. Specifically, with the increase of the loading level, C is more closed to 1. The relation between cycle ratio and damage is presented in Figure 2.11 [11].

As shown in Figure 2.11, for the low-to-high loads, the process of fatigue damage is expressed as OBAP, where the critical threshold damage D_C can be expressed as:

$$\sum_{i=1}^{m} \frac{n_i}{N_i} = \frac{n_2}{N_2} + \left(1 - \frac{n_1}{N_1}\right) > 1 \tag{2.15}$$

As for the high-to-low loadings, the process of fatigue damage is expressed as OABP, where the critical threshold damage D_C can be expressed as:

$$\sum_{i=1}^{m} \frac{n_i}{N_i} = \frac{n_1}{N_1} + \left(1 - \frac{n_2}{N_2}\right) < 1 \tag{2.16}$$

As shown in Figure 2.11, the damage curve approach provides a good explanation to the loading sequence effect under two-level loading. In addition, it's assumed that the damage rate is proportional to the loading cycle, which is in line with the actual

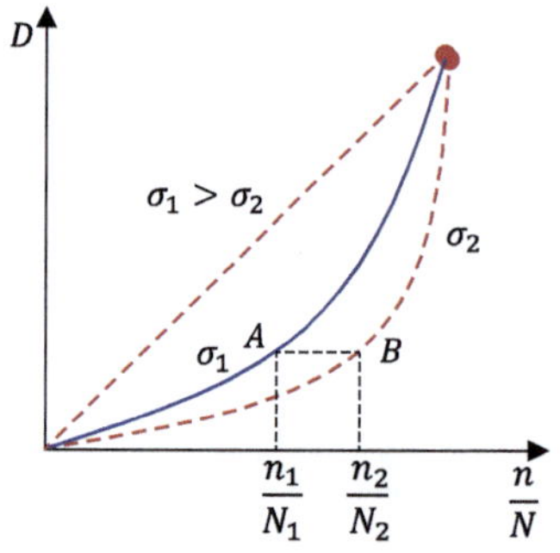

FIGURE 2.11 The relation between the cycle ratio and the damage. (The figure is created based on the work of Ref. [11].)

process of fatigue failure. However, the damage curve approach does not separate the two stages of damage evolution. Namely, the damage curve approach ignores that the damage accumulation rules in the two stages are diverse. In addition, the damage rate corresponding to higher stress levels in the second half of the fatigue process in this model is actually lower, which is not in line with the actual situation. In addition, it only provides a qualitative explanation without providing a specific formula for the fatigue damage, and cannot be used for actual calculation of fatigue life.

2.4.3.2 Corten-Dolan Theory

The Corten-Dolan theory assumes that damage is imagined as accumulation and combination of the micro cracks, and is related to the number of cracks and crack growth rate. More specifically, it's assumed that damage is affected by the maximum variable loading in the loading sequence, and cracks grow at all load levels. Specifically, Corten-Dolan theory assumes that fatigue damage is expressed as:

$$D = m^c r^d \tag{2.17}$$

where m represents the number of the damage core, which increases with loading level, r denotes damage evolution rate, which is related with loading level, c and d are material constants. In addition, fatigue damage under n loading cycles for the CAF and VAF loadings are calculated by:

$$\begin{cases} D = nm^c r^d \\ D = \sum_{i=1}^{p} n_i m_i^c r_i^d \end{cases} \tag{2.18}$$

where n_i denotes the loading cycles of the ith loading level. The critical threshold damage D_C can be expressed as:

$$D_C = N_1 m_1^c r_1^d \tag{2.19}$$

For the CAF loading, N_1 denotes fatigue life under CAF loadings. For the VAF loading, N_1 denotes the fatigue life under the largest loading level

$$D = \sum_{i=1}^{p} n_i m_i^c r_i^d = N_1 m_1^c r_1^d \tag{2.20}$$

Considering that the accumulated fatigue damage continues to increase with the increase of the loading cycle, hence $m_i = m_1$, Eq. (2.20) can be expressed as:

$$\sum_{i=1}^{p} n_i r_i^d = r_1^d N_1 \tag{2.21}$$

Considering the relationship between loading level with the damage evolution rate, therefore:

$$1 = \sum_{i=1}^{p} \frac{n_i}{N_1\left(\frac{S_1}{S_i}\right)^d} \tag{2.22}$$

where S_1 denotes the largest loading level among all load levels before the current loading cycle. About the calculation of the material constant d, Corten and Dolan have also developed a relevant procedure based on the two-level loading test, which is explained as: (1) The stress amplitude of the first-level load is applied as the largest loading level; (2) The stress amplitude of the second-level load of all specimens gradually decrease; (3) The cycle ratio of the first-level load to the second-level load equals to the frequency ratio on the loading frequency curve. Specifically, based on linear fitting of the tested results on the $\lg\sigma_2 - \lg N_2'$ coordinate, the S-N curve under two-level loading test can be plotted, as shown in Figure 2.12 [12].

Note from Figure 2.12, σ_2 denotes the stress amplitude of the second-level load, N_2' denotes the fatigue life at the second-level load, which is calculated as:

$$N_2' = N_2 + n_1\left(\frac{\sigma_1}{\sigma_2}\right)^d \tag{2.23}$$

It's worth mentioning that the horizontal ordinate represents the equivalent fatigue life N_2' rather than the tested fatigue life N_2. The reason for this is that this curve reflects the influence of high stress on fatigue performance under low stress, and should not include damage caused by high stress itself. In fact, during two-level loading test, higher stress not only affects the fatigue lives under lower stress but also causes damage itself. To avoid repeated calculations of damage caused by high stress, the fatigue lives under high stress σ_1 should be converted into the fatigue lives under low stress σ_2. In addition, the parameter d is expressed as the negative reciprocal of slope of S-N curve, see Figure 2.12, which can be calculated as:

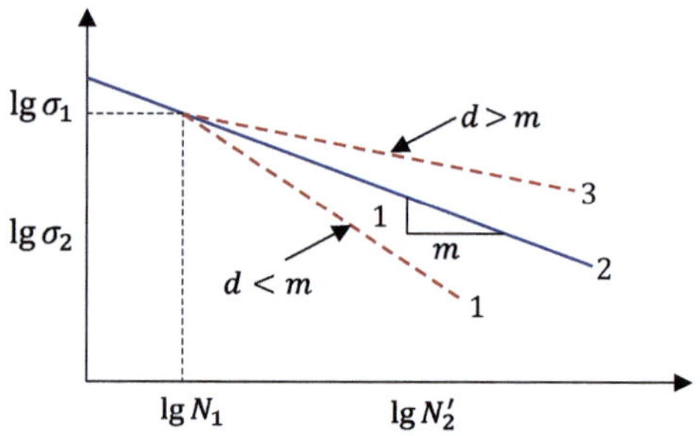

FIGURE 2.12 S-N curve under the two-level loading test [12].

$$d = \frac{\lg(\alpha_2 N) - \lg(N_1 - \alpha_1 N)}{\lg\left(\frac{\sigma_1}{\sigma_2}\right)} \tag{2.24}$$

where N denotes the total fatigue life during the two-level loading test, N_1 denotes the fatigue life at first-level load, α_1 denotes the ratio of the cycles under the first-level load to the total cycle, α_2 denotes the ratio of cycles in second-level load to total cycle. Specifically, Corten and Dolan suggested that

$$d = \begin{cases} 4.8 & \text{steels} \\ 5.8 & \text{other} \end{cases} \tag{2.25}$$

Although the Corten-Dolan theory considers the nonlinearity of damage development, it also shows several drawbacks: (1) It suggests that the number of the damage core m is only affected by the loading level, but large amounts of experiments have shown that m is also related to the number of stress actions; (2) There is still a significant error in estimating fatigue lives under random loads using the d value determined by the two-level loading test. The d value is not material constant but is affected by the stress state. Hence, the Corten-Dolan theory still needs further development.

A series of experiments have shown the loading sequence effect during fatigue damage evolution cannot be ignored. Compared with the linear fatigue damage accumulation rule, the nonlinear fatigue damage accumulation rule is more in line with the physical mechanism of the damage evolution process. The linear damage theory of variable damage not only reflects the influence of loading sequence, but also includes the loading history, load interaction effect, and average stress effect. Until now, various nonlinear fatigue damage accumulation rules have been developed, see Table 2.1. The idea of establishing a fatigue damage model according to the changes of S-N curve during the damage process in the life curve correction model is creative, which can reflect the interaction between the loading process and the load level. The Corter-Dolan theory explains the loading sequence effect and the influence of high stress on total damage value. However, the exponent d needs to be determined through two-level loading test. Some researchers found that the value d is related to both the component and loading level, and there is a significant difference in damage development under spectrum load and two-level loading test, Therefore, it is difficult and random to determine d. Specifically, considering that the value d is strongly related to the loading spectra, Zhu et al. [13] developed a new model for derivation and calculation of it. Specifically, the value d is treated as a function related to the loading level. Results indicated that the proposed method is more effective in fatigue damage assessment than traditional methods.

2.4.4 Probabilistic Fatigue Damage Accumulation Rule

The abovementioned damage accumulation rules are developed based on "certainty". Until now, probabilistic fatigue damage accumulation rules have made significant progress to meet the requirements of fatigue design under uncertainty. The probabilistic fatigue damage accumulation rule was developed on the basis of understanding of fatigue damage evolution process. Specifically, the nature of fatigue failure is irreversible randomness of damage evolution. From the micro aspect, the irreversible process of damage evolution process is considered as the continuous increase of the micro defects. From the macro aspect, irreversible process of damage evolution process can be viewed as that the accumulation fatigue damage is a generalized monotone increasing function of the number of loading cycles. In addition, the randomness of fatigue damage evolution is caused by two reasons. For one thing, the randomness of fatigue damage is related to the dispersion of material properties and the uncertainties during its measurement and statistics due to insufficient database and error in operation (including test environment and operation process), internal initial defect distribution, crack interference, differences in processing of the parent materials (different batches), material's inherent inhomogeneity, anisotropy, and grain orientation. For another, it is related to the service environment, service time, installation and commissioning, maintenance and repair differences, load level and frequency, random variation of external loads (different working loading spectrum compilation), difference of bearing capacity of each unit of the structure, connection mode, difference of stress distribution caused by assembly, difference in boundary conditions. In view of the unavoidable uncertainty issues, a probabilistic fatigue damage accumulation rule is more in line with the process of fatigue failure under multi-source uncertainties. In this section, the commonly used probabilistic fatigue damage accumulation rules include probabilistic Miner's rule and dynamic statistical model.

2.4.4.1 Probabilistic Miner's Rule

Many experimental statistical facts have indicated that Miner's rule is effective for fatigue life predictions of engineering structures under random loads. Although a lot of fatigue accumulation damage theories were developed, Miner's rule is still a widely adopted engineering fatigue resistance design criterion. However, the Miner's rule can not take the material and load dispersion into consideration during fatigue damage quantification. In view of this, the probabilistic Miner's rule emerged, specifically, the fatigue damage is expressed as:

$$D = \frac{1}{N} \tag{2.26}$$

where N denotes fatigue life corresponding to the current loading level, which is a random variable herein. After n loading cycles, the fatigue damage is called as instantaneous accumulation damage, which is a combination of internal and external dispersions. If the internal dispersion is represented as random variable $D^{(1)}$, and external one is described as random variable $D^{(2)}$, then the distribution of instantaneous accumulation damage $D(n)$ is:

$$D(n) = D^{(1)} \cup D^{(2)} \tag{2.27}$$

For the internal dispersion such as the material property, its distribution can only be obtained through a series of tests. If fatigue life under the CAF loadings S is denoted as N. Therefore:

$$D^{(1)} = \frac{n}{N} \tag{2.28}$$

where n represents the number of loading cycles. If probability density function of fatigue life N is denoted as $f_N(x)$, and the distribution function is $F_N(x)$, so distribution function of fatigue damage $D^{(1)}$ is:

$$\begin{cases} F_D(y) = P\left\{D^{(1)} \le y\right\} = 1 - F_N\left(\dfrac{n}{y}\right) = \displaystyle\int_{n/y}^{\infty} f_N(x)\,dx \\ f_D(y) = \dfrac{dF_D(y)}{dy} = \dfrac{n}{y^2} f_N\left(\dfrac{n}{y}\right) \end{cases} \tag{2.29}$$

Normally, the lognormal distribution can be applied for the description of the fatigue life, as:

$$f_N(x) = \begin{cases} \dfrac{1}{\sqrt{2\pi}\sigma_N x} \exp\left[-\dfrac{(\ln x - \mu_N)^2}{2\sigma_N^2}\right] & x > 0 \\ 0 & x \le 0 \end{cases} \tag{2.30}$$

Combination of Eqs. (2.29) and (2.30) yields the distribution function of the internal damage $D^{(1)}$:

$$f_D(y) = \begin{cases} \dfrac{1}{\sqrt{2\pi}\sigma_D y} \exp\left[-\dfrac{(\ln y - \mu_D)^2}{2\sigma_D^2}\right] & y > 0 \\ 0 & y \le 0 \end{cases} \tag{2.31}$$

where $\mu_D = \ln n - \mu_N$, which is the logarithmic mean value of instantaneous internal damage $D^{(1)}$. $\sigma_D = \sigma_N$, which is the logarithmic standard deviation of $D^{(1)}$.

For the external dispersion such as the loading and environment, its distribution can not be obtained through a series of tests. The external dispersion of the instantaneous accumulation damage $D(n)$ mainly depends on the randomness of loads. Specifically, the randomness of loads contains two aspects: (1) random arrangement of the order of external loads; (2) stochastic distribution of the load level at the given time. Normally, the fatigue loading spectrum is given in the form of a block spectrum, and if the number of blocks in the spectrum is appropriately increased, the randomness of the second aspect can be basically avoided. Therefore, it can be considered that the randomness of external loads mainly comes from the random arrangement of loading sequence.

For a given loading spectrum, the influence of the loading sequence can be considered with the modified linear fatigue damage accumulation rule. For a random loading issue, the determination of the distribution function of the external damage $D^{(2)}$ is difficult. One efficient strategy to solve this issue is using the numerical method, where the procedure is described as: (1) adopt the random sampling technique for the loading spectrum to generate a random loading series; (2) apply the modified Miner's rule to calculate the instantaneous accumulation damage $D^{(2)}(n)$ at different time; (3) sampling and calculating for m times and m series of $D^{(2)}(n)$ can be obtained for describing its distribution information.

As for the critical threshold damage D_C, mainly two opinions exist. For the first, researchers assume that the critical threshold damage D_C is a constant, where the dispersion of fatigue damage has been considered during the modeling of the instantaneous accumulation damage $D(n)$. For another, researchers assume that the critical threshold damage D_C can be treated as material property, and as random variable. Therefore, in engineering practice, whether the critical threshold damage D_C should be treated as constant or random variable should be further investigated.

2.4.4.2 Dynamic Statistical Model

The process of damage accumulation is an irreversible stochastic process both macroscopically and microscopically, which is expressed as:

$$D(n) = F\left(\{D_0\}, \left\{D^{(1)}\right\}, \left\{D^{(2)}\right\}\right) \tag{2.32}$$

where $\{D_0\}$ denotes random variables describing initial defects inside materials, $\left\{D^{(1)}\right\}$ denotes a family of random variables describing internal dispersion of the damage accumulation, $\left\{D^{(2)}\right\}$ denotes a family of random variables describing external dispersion of damage accumulation. In view of the crack initiation process, some researchers have developed a dynamic statistical model for modeling of damage accumulation. Specifically, the fatigue damage induced by n loading cycles is expressed as:

$$D(n) = \sum_{i=1}^{n} \Delta D_i \tag{2.33}$$

It's assumed by the dynamic statistical model that if n is big enough, the distribution of $D(n)$ can be expressed by using the central limit theorem, as:

$$F_D(n) = \Phi\left[\frac{D(n) - \mu_{D(n)}}{\sigma_{D(n)}}\right] \tag{2.34}$$

As for the critical threshold damage D_C, it's a random variable with the mean value of 1. In addition, its variation coefficient equals to coefficient of variation of the fatigue life.

2.5 FATIGUE LIFE PREDICTION UNDER VAF LOADINGS

Recently, the issue of fatigue failure can be found in several applications, such as marine structures, aircraft components, automotive, etc. Therefore, it is of vital importance to investigate the main origins of fatigue failure and assess the fatigue behaviors of these main components. In view of this, researchers have developed a lot of methods to understand fatigue failure mechanism and predict fatigue lives of these main components. However, these methods are normally developed and validated based on tested fatigue data from the laboratory, which can not reflect the fatigue behaviors of real service environment with so many variables to be considered. In practice, fatigue loading is normally random, i.e. VAF loading. For the cases of VAF loading conditions, it is more complicated to estimate the fatigue lives accurately since fatigue damage is highly related to real service loading during the whole service period.

Normally, fatigue life prediction of main components is conducted based on material properties obtained from fatigue tests and fatigue damage accumulation rules. To develop a reasonable fatigue life prediction model, the influencing factors should be considered. The evolution of damage is a complicated process and reasonable methods/theories for fatigue damage modeling are the foundations of fatigue life predictions. Until now, a lot of life prediction models under VAF loadings have been developed from various aspects. Among them, stress-based methods, strain-based methods, energy-based methods, and fracture mechanics-based methods are commonly utilized for life predictions, which are systematically introduced in this section.

2.5.1 Stress-Based Method

The stress-based method is a commonly utilized method for life predictions of main components, which was developed using S-N curves (see Figure 2.13) to relate the fatigue lifetime N_f with applied stress S_a. The procedure for fatigue tests is explained as follows: (1) a set of specimens (e.g. smooth specimens, notched specimens, structures, etc.) are prepared; (2) cyclic variable amplitude loading is applied; (3) record fatigue lives of the specimen and reapply the test in an abovementioned test condition for 4–6 times; (4) reapply the tests under several different stress levels and plot all recorded points in the log-log space; (5) Then, the fitted curve in Figure 2.13 is

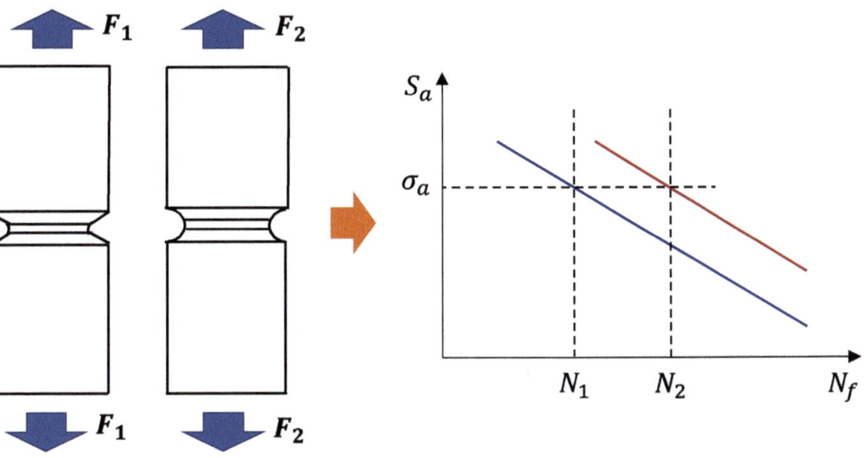

FIGURE 2.13 Graphically illustration of the S-N curve-based method.

namely S-N curve. Normally, the relation between fatigue lives and the true stress amplitudes is expressed by using the Basquin law, as:

$$\sigma_a = \frac{E\Delta\varepsilon_e}{2} = \sigma'_f (2N)^b \tag{2.35}$$

where E represents the elastic modulus, $\Delta\varepsilon_e$ represents the elastic strain range, σ_a is the loading amplitude, N represents the number of loading cycles, σ'_f represents the fatigue strength coefficient, and b represents the fatigue strength exponent.

The S-N curve illustrated in Figure 2.13 is obtained under constant amplitude loadings. A graphic illustration of mean stress effect is presented in Figure 2.14a [14], which can be described by the modified Goodman and Gerber equations:

$$\frac{\sigma_a}{S_e} + \frac{\sigma_m}{S_u} = 1 \tag{2.36}$$

$$\frac{\sigma_a}{S_e} + \left(\frac{\sigma_m}{S_u}\right)^2 = 1 \tag{2.37}$$

where S_e is the loading amplitude at zero mean stress, σ_m denotes the mean stress, S_u denotes the ultimate tensile strength. Specifically, modified Goodman and Gerber equations are graphically illustrated in Figure 2.14b [14]. For the VAF loading, the impact of mean stress on fatigue performance is non-negligible.

Moreover, to consider the localized stress related to loading interaction effect during VAF loadings, the relation between stress σ and strain ε is expressed by using the Ramberg-Osgood equation:

$$\varepsilon = \frac{\sigma}{E} + \left(\frac{\sigma}{K}\right)^{1/n} \tag{2.38}$$

where K denotes the strain hardening coefficient, n denotes the strain hardening exponent. Due to the simplicity of stress-based method, it has been widely applied for fatigue life predictions.

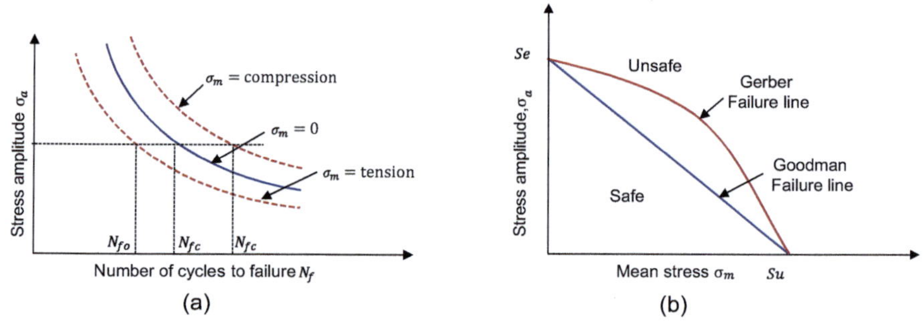

FIGURE 2.14 Illustration of the influence of the mean stress on the fatigue life. (a) Fatigue life under different mean stress; (b) Goodman and Gerber correction. (The figure is created based on the work of Ref. [14].)

2.5.2 Strain-Based Method

Different from the stress-based method which accounts for the relation between life and stress, the strain-based method is developed based on the general strain-life model. Strain-life model for life predictions needs to describe the mechanical response to the cyclic elastic-plastic strain. Early in 1910, Basquin assumed that the relation between endurance and loading level was linear. Later, Coffin and Manson obtained the relation between stress/strain amplitude and fatigue life, as:

$$\sigma_a = \sigma'_f \left(2N_f\right)^b \tag{2.39}$$

$$\varepsilon_{a,p} = \varepsilon'_f \left(2N_f\right)^c \tag{2.40}$$

where ε'_f denotes the fatigue ductility coefficient, c denotes the fatigue ductility exponent. Through combining the Coffin-Manson equation with the Basquin equation, the total strain amplitude ε_a is:

$$\varepsilon_a = \varepsilon_{a,e} + \varepsilon_{a,p} = \frac{\sigma'_f}{E}\left(2N_f\right)^b + \varepsilon'_f\left(2N_f\right)^c \tag{2.41}$$

Particularly, the two terms in Eq. (2.41) denote elastic strain $\varepsilon_{a,e}$ (high-cycle region) and plastic strain $\varepsilon_{a,p}$ (low-cycle region), respectively. The relationship between elastic strain and plastic strain is defined by the stable hysteresis loop, as shown in Figure 2.15. An advantage of strain-based method is that fatigue life prediction can be performed under oscillatory random loadings, as presented in Figure 2.15.

Although Coffin-Manson equation presents strong curve-fitting ability, a total of five material properties need to be calculated. Therefore, Manson further simplified it as [15]:

$$\Delta\varepsilon = 3.5\frac{S_u}{E}\left(N_f\right)^{-0.1\varepsilon} + \varepsilon_f^{0.6}\left(N_f\right)^{-0.6} \tag{2.42}$$

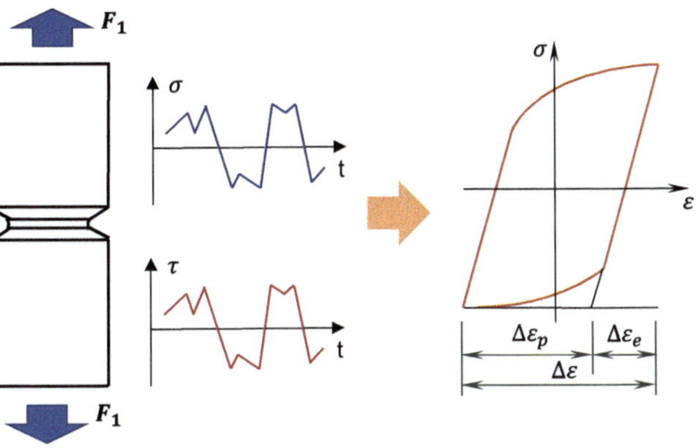

FIGURE 2.15 Graphically illustration of the strain-based method.

where the parameters S_u, E and ε_f can be obtained based on the monotonic tensile test. Note from Eq. (2.42), it is assumed that the fatigue behaviors of all materials are controlled by only the three parameters S_u, E and ε_f. Later, Eq. (2.42) was further modified by Muralidharan and Manson, as [16]:

$$\Delta\varepsilon = 0.0266D^{0.115}\left(\frac{S_u}{E}\right)^{-0.53} N_f^{-0.56} + 1.17\left(\frac{S_u}{E}\right)^{0.832} N_f^{-0.09} \tag{2.43}$$

Results indicated that a good correlation between predicted life and tested life was found.

Based on investigation of the relationship between monotonic tensile and fatigue tests, the following model was developed by Roessle and Fatemi [17]:

$$\frac{\Delta\varepsilon}{2} = \frac{4.25(\mathrm{HB}) + 225}{E}\left(2N_f\right)^{-0.009} + \frac{0.32(\mathrm{HB})^2 - 487(\mathrm{HB}) + 191{,}000}{E}\left(2N_f\right)^{-0.56} \tag{2.44}$$

where HB denotes the hardness. Note from Eq. (2.44), only the hardness and the elastic modulus are adopted for the construction of the strain-life model, which greatly improves the applicability of it. To consider the mean stress effect, the parameter σ'_f can be replaced with $\sigma'_f - \sigma_m$

$$\frac{\Delta\varepsilon}{2} = \frac{\left(\sigma'_f - \sigma_m\right)\left(2N_f\right)^b}{E} + \varepsilon'_f\left(2N_f\right)^c \tag{2.45}$$

where σ_m is positive for tensile loading. Another method to consider the mean stress effect was developed by Smith et al., as:

$$\sigma_{\max}\varepsilon_a E = \left(\sigma'_f\right)^2\left(2N_f\right)^{2b} + \sigma'_f\varepsilon'_f E\left(2N_f\right)^{b+c} \tag{2.46}$$

where $\sigma_{\max} = \sigma_m + \sigma_a$. Recently, many strain-life models were proposed for life predictions. However, more efficient models which balance the calculation efficiency and prediction accuracy should be developed.

2.5.3 Energy-Based Method

Different from stress/strain-based methods, the energy-based method takes the energy as the driving factor for the fatigue fracture. The fatigue failure is mainly controlled by the distribution and evolution of the microscopic defects, where this internal state can only be changed when energy is provided. Inglis first developed the connection between hysteresis energy and fatigue failure process. Specifically, the existence of Bausching effect and hysteresis phenomenon confirmed that energy dissipation is along with the fatigue failure process. Therefore, the process of energy dissipation can be used to characterize the damage process of the material. Specifically, the physical meaning of this characterization is clearer due to the close relationship between material microstructure evolution and energy dissipation, which better describes the irreversible thermodynamic essence of fatigue process. In addition, the scalar properties of energy also make it easier to calculate. Based on these advantages, the energy criterion has become the main method for life predictions.

The energy-based method was developed by Morrow, who assumed plastic work as the main reason for plastic deformation in materials leading to fatigue failure of components. Specifically, it is assumed that the relationship between the plastic work of each cycle W^p and the fatigue life can be described as:

$$N_f = AW_p^m \tag{2.47}$$

$$W^p = \int \sigma d\varepsilon^p + \tau d\gamma^p \tag{2.48}$$

Specifically, for tension-compression loadings, the plastic strain energy ΔW^p is:

$$\Delta W^p = \Delta\sigma\Delta\varepsilon^p\left(\frac{1-n'}{1+n'}\right) + \Delta\tau\Delta\gamma^p\left(\frac{1-n'}{1+n'}\right) \tag{2.49}$$

where $\Delta\sigma$ represents the equivalent stress amplitude, $\Delta\varepsilon^p$ denotes the plastic equivalent strain amplitude, n' denotes the cyclic strain hardening exponent.

Based on the model developed by Morrow, Wang et al. [18] developed a generalized methodology for predicting both the high temperature low cycle fatigue and creep-fatigue lives based on the strain energy density rate concept. Specifically, the strain energy density ΔW^t consists of the positive elastic strain energy density ΔW^{e+} and plastic strain energy density ΔW^p (see Figure 2.16 [19]), as:

$$\begin{cases} \Delta W^t = \Delta W^p + \Delta W^{e+} \\ \Delta W^p = \int_{\text{cycle}} \sigma_{ij}\, d\varepsilon_i^e \\ \Delta W^{e+} = \int_t^{t+T} H(\sigma_i) H\left(d\varepsilon_i^e\right)\sigma_i\, d\varepsilon_i^e \end{cases} \tag{2.50}$$

where $i, j = 1,2,3$, H is Heaviside function, where:

$$\begin{cases} H(x) = 1 & (x \geq 0) \\ H(x) = 0 & (x < 0) \end{cases} \tag{2.51}$$

By introducing the multiaxial constraint factor $\bar{\rho}$, the tested fatigue data under different loading path were fitted into energy-life curve by Ellyin [20], as:

$$\Delta W^{e+} + \frac{\Delta W^p}{\bar{\rho}} = k_u\left(2N_f\right)^a + C_u \tag{2.52}$$

where k_u, a and C_u are material constants. However, due to the fact that the calculation of the multiaxial constraint factor $\bar{\rho}$ is very complex, its application on fatigue

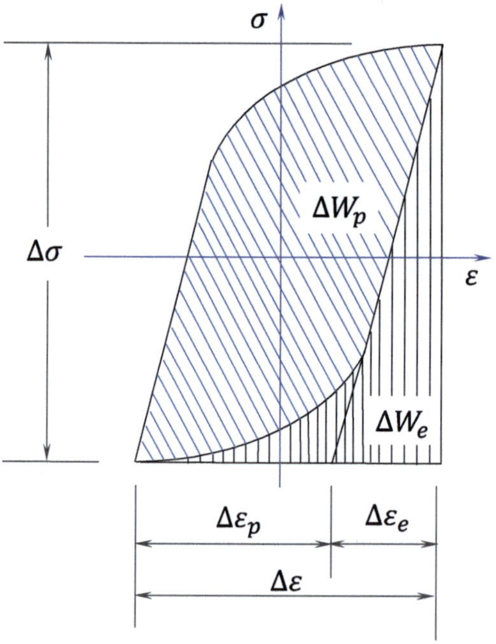

FIGURE 2.16 Schematic diagram of strain energy density. (Reprinted with permission from Ref. [19], Elsevier.)

life predictions under VAF loadings is limited. In view of this, Liao and Zhu [21] developed the concept of modified positive total strain energy ΔW^{t+} under multiaxial loading conditions. Specifically, ΔW^{t+} is composed of total axial strain energy density ΔW_a^{t+} and total shear strain energy density ΔW_s^{t+}, as:

$$\Delta W^{t+} = \Delta W_a^{t+} + \Delta W_s^{t+} = p\left(2N_f\right)^q \tag{2.53}$$

where p and q represent material parameters. In recent years, energy-based life prediction models have attracted extensive attention, since the fatigue damage process shows a good correlation with the strain energy density. In addition, cyclic hardening behavior is quantified in the energy-based fatigue damage parameters. Recently, Branco et al. [22] compared the prediction accuracy of fatigue models on the notched structures based on various criteria, and the results indicated that the energy-based fatigue model have shown the best prediction ability.

2.5.4 Fracture Mechanics-Based Method

Another efficient method often utilized for fatigue life prediction is the fracture mechanics-based method which focuses on fatigue crack information. Due to the flaws distributed in the material owing to manufacturing or fabricating process, fatigue crack initiation or propagation occurs, and eventually leads to fatigue failure.

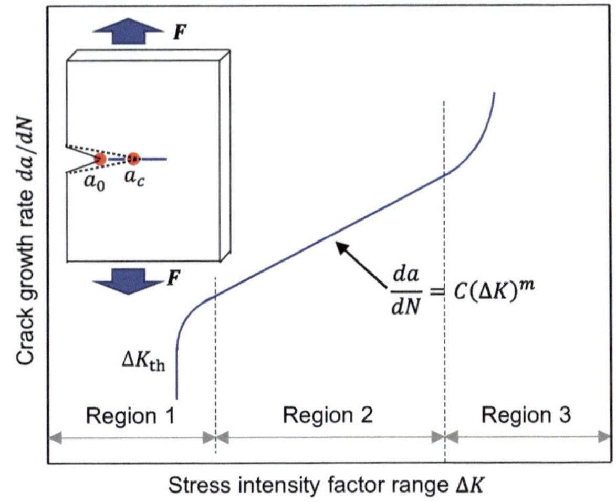

FIGURE 2.17 Graphically illustration of crack growth and the Paris equation.

Specifically, stress intensity factor range ΔK is the basis of fracture mechanics-based fatigue analysis, as follows:

$$\Delta K = YS_A\sqrt{\pi a} \tag{2.54}$$

where the geometrical correction factor Y is related to applied far-field stress, crack size, and structural geometry. S_A denotes the applied stress range and a denotes crack length. In particular, it has been demonstrated that the relation between crack growth rate da/dN and ΔK can be expressed as (see Figure 2.17):

$$\frac{da}{dN} = C(\Delta K)^{\theta} \tag{2.55}$$

where C and θ denote empirical constants and N denotes loading cycles.

By combining Eq. (2.55) from a_0 to a_c (fatigue lifetime from 0 to N), the fatigue life N_f is expressed as:

$$N_f = \int_0^N dn = \int_{a_0}^{a_c} \frac{da}{C\left(YS_R\sqrt{\pi a}\right)^{\alpha}} \tag{2.56}$$

Later, researchers found that the crack growth rate curve should be classified into three stages, as shown in Figure 2.17. Note from Figure 2.17, for Region 1, the crack growth rate is close to zero with the threshold value ΔK_{th}. Namely, no crack growth can be found below the threshold value ΔK_{th}. Specifically, the crack growth rule in Region 1 is described as:

$$\frac{da}{dN} = C(\Delta K - \Delta K_{\text{th}})^{\theta} \tag{2.57}$$

As for Region 2, the crack growth rule obeys the Paris law, see Eq. (2.55). As for Region 3, the crack growth shows a rapidly increasing growth rate, which can be described as:

$$\frac{da}{dN} = \frac{C(\Delta K)^{\theta}}{(1-R)K_c - \Delta K} \tag{2.58}$$

where K_c denotes the fracture toughness. Specifically, at high and low ΔK, the relations combining the departures from power-law behavior also exist, which can be described by the model developed by McEvily and Groeger:

$$\frac{da}{dN} = A(\Delta K - \Delta K_{\text{th}})^2 \left(1 + \frac{\Delta K}{K_c - K_{\max}}\right) \tag{2.59}$$

To consider the load ratio effect on crack growth rate, effective cyclic stress intensity factor ΔK_{th} was introduced by Elber, as:

$$\Delta K_{\text{eff}} = K_{\max} - K_{\text{op}} \tag{2.60}$$

where $K_{\max}$ represents the stress intensity factor obtained at the maximum loading, K_{op} represents the stress intensity factor obtained under the crack-opening loading. Later, a crack closure model was developed to improve the traditional model, as:

$$\Delta K_{\text{effM}} = K_{\max} - K_{\text{op}}\left[1 + \left(\frac{2}{\pi} - 1\right)g\right] \tag{2.61}$$

where the parameter g can be expressed as:

$$g = \exp\left[-\left(\frac{K_{\max}}{K_{\max\text{TH}}}\right) - 1\right] \tag{2.62}$$

where $K_{\max\text{TH}}$ denotes the maximum stress intensity factor at given load ratio. Later, it was found that Eq. (2.62) can not take the stress ratio effect into account. In view of this, the following parameters were defined:

$$\Delta K^{*} = (K_{\max})^{\alpha}\left(\Delta K^{+}\right)^{1-\alpha} \tag{2.63}$$

where ΔK^{+} is the positive portion of ΔK.

2.5.5 Recent Advances

As abovementioned, the mainly adopted fatigue life prediction methods are classified into stress-based methods, strain-based methods, energy-based methods, and fracture mechanics-based methods. These methods have been developed for a long time and great development as well as some new methods have emerged. In this section, to catch developing status and trends and enhance engineering applications of

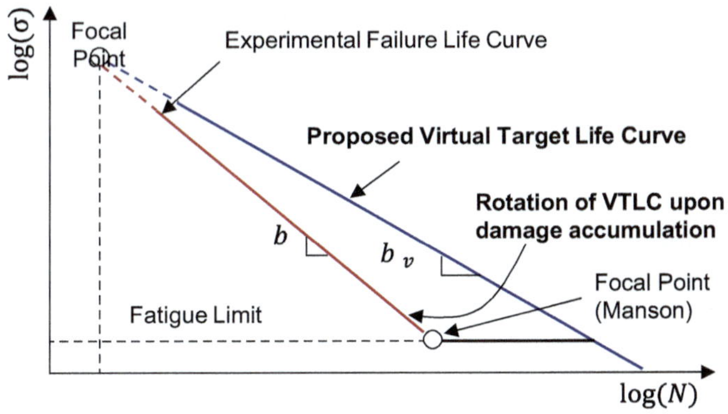

FIGURE 2.18 Comparison between virtual target life curve and tested curve. (Reprinted with permission from Ref. [23], Elsevier.)

fatigue life prediction methods under VAF loadings, recent advances on fatigue life predictions are explored and discussed.

2.5.5.1 Stress-Based Method

As the most commonly adopted method for fatigue life predictions under VAF, it has been widely applied in various aspects. By adding the time dimension to the S-N curve, a novel stress-life model was developed by Aghoury et al. [23]. Specifically, the concept of "Virtual Target Life Curves" was developed by assuming that a virtual expected life exists which is bigger than tested lives, as shown in Figure 2.18 [23]. Note from Figure 2.18, comparison between virtual target life curves and other methods indicates that various methods show different methods of establishing the S-N curve.

To study the fatigue performance of welded joints with VAF loadings, a new advanced probabilistic approach using the S-N curve was developed by D'Angelo et al. [24]. Specifically, the new method is developed by combining Maximum Likelihood approach with Monto-Carlo Simulation method for predicting the S-N curves under VAF loading. To consider the load sequence and interaction effects, a novel nonlinear damage accumulation rule was developed by Zhu et al. [25] based on a newly developed damage function. The developed model was verified based on tested data of three types of metals. Results indicated that the developed method shows higher precision than the other four existing methods. To conquer the problem that applications of nonlinear fatigue damage accumulation rules require modifications of the S-N curve, Aeran et al. [26] developed a new damage model based on commonly available S-N curve parameters. Specifically, a novel damage index D_i is proposed on the basis of fatigue damage evolution under VAF loading, as:

$$D_i = 1 - \left(1 - \frac{n_i}{N_i}\right)^{\delta_i} \tag{2.64}$$

where n_i denotes loading cycles at the loading level σ_i, N_i denotes fatigue life, model parameter δ_i is calculated based on the S-N curve. Model validation based on two types of materials indicated that the proposed model provides better estimations of fatigue lives than other models.

To sum up, the stress-based method has been widely applied for fatigue damage modeling due to its simplicity. However, this method normally shows unacceptable estimation accuracy of fatigue life within the low-cycle fatigue regime. In view of this, the strain-based method is an efficient strategy.

2.5.5.2 Strain-Based Method

Different from stress-based method, the strain-based method is developed by using the strain as fatigue damage driving factor. In recent years, a lot of new life prediction methods under VAF loading by using the strain were proposed. Huffman and Beckman [27] developed a life model under VAF loading by using tested data from CAF loading. Specifically, the fatigue damage is modeled by using the tested data from CAF loading.

Based on fatigue tests on TC4 alloy, Wu et al. [28] found that the shear strain-based critical plane method (i.e. Wu-Hu-Song model) provides promising life predictions accuracy than the Smith-Watson-Topper (SWT) equation. Specifically, the relation between maximum shear strain amplitudes $\Delta\gamma_{\max}/2$ and fatigue lives is:

$$\frac{\Delta\gamma_{\max}}{2}+k\left(\frac{\sigma_{n,\max}\Delta\varepsilon_n}{E}\right)^{0.5}=\frac{\pi'_f}{G}\left(2N_f\right)^{b_0}+\gamma'_f\left(2N_f\right)^{c_0} \quad (2.65)$$

where $\sigma_{n,\max}$ denotes maximum normal stress at the maximum shear plane, $\Delta\varepsilon_n$ denotes normal strain range at the maximum shear plane, k is a model parameter. Later, by combining the Wu-Hu-Song model, rain-flow cycle counting, and Miner's rule, Wu et al. [29] proposed a new model for the fatigue assessment of metallic materials. Specifically, the framework for life predictions under VAF loadings is illustrated in Figure 2.19 [29]. Results indicated that Wu's model presents great ability for life predictions of TC4 alloy.

By coupling critical plane criteria with Miner's rule, a framework for fatigue life predictions under VAF loading was developed by Tao et al. [30]. Specifically, it shows acceptable results for fatigue life predictions of five types of metallic materials under VAF loading. To sum up, the strain-based method has been widely applied for fatigue damage modeling due to its applicability for low-cycle fatigue. However, this method normally concentrates on fatigue damage assessment while ignores the essence of fatigue damage accumulation. In view of this, fracture mechanics-based method is an efficient strategy.

2.5.5.3 Energy-Based Method

Considering that the energy-based method could be applied in both low and high-cycle fatigue regimes, it was commonly used for life predictions under VAF loadings. By combining the energy concept with the critical plane approach, Xue et al. [31] developed a new fatigue damage parameter without any weight coefficient, as:

$$\left(\frac{\tau_{\max}}{\tau'_f}+\frac{\sigma_{n,\max}}{\sqrt{3}\sigma'_f}\right)\sqrt{3\varepsilon_n^{*2}+\left(\frac{\Delta\gamma_{\max}}{2}\right)^2}=\frac{\tau'_f}{G}\left(2N_f\right)^{2b'}+\gamma'_f\left(2N_f\right)^{2b'+c'} \quad (2.66)$$

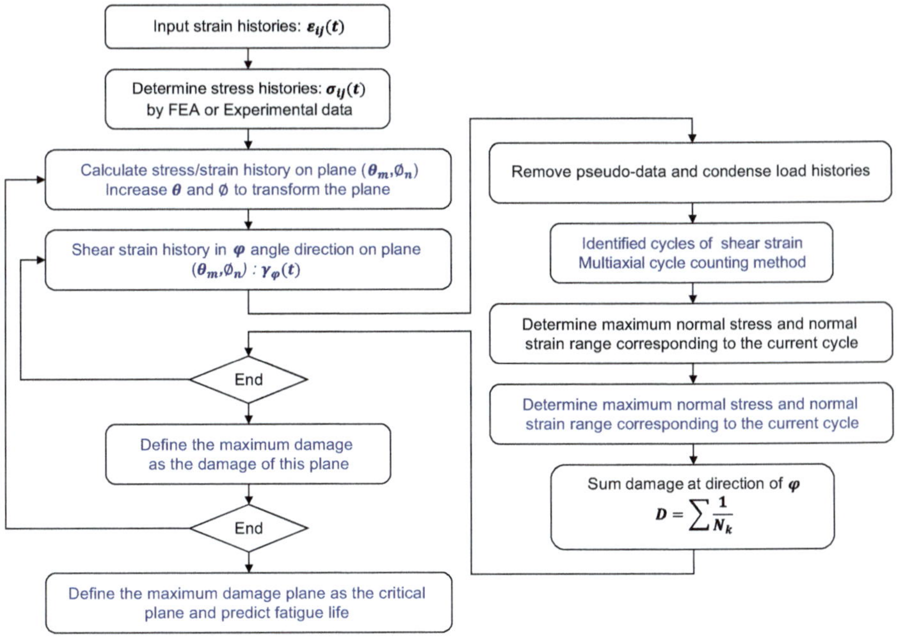

FIGURE 2.19 Procedure for Wu's model for VAF life prediction. (The figure is created based on the work of Ref. [29].)

Specifically, with the simplified form of the shear strain energy, both non-proportional additional hardening and mean stress effect are considered. By combining the maximum shear stress range with the maximum normal stress on the critical plane, a new path-independent damage parameter was developed by Tao et al. [32] to investigate the multiaxial VAF behavior. Specifically, the framework for fatigue life predictions under VAF loading is illustrated in Figure 2.20 [32]. According to tested data of En15R steel and 7075-T7451 aluminum alloy, the developed life predictions framework is validated. Results indicated that the predicted lives agree with tested lives under VAF loading.

Recently, a review about the multiaxial fatigue issues under VAF loading was published by Deng et al. [33], where the commonly applied energy-based critical plane methods were presented. For the VAF loading, damage plane of the material varies with the time. Until now, various methods have been developed for accurately defining the critical plane. Among these methods, the maximum damage method works as a reasonable method which is able to solve the fatigue damage in all planes. However, the accuracy and reliability of these new models need further validation.

2.5.5.4 Fracture Mechanics-Based Method

As mentioned above, stress/strain/energy-based methods have been commonly applied for damage modeling owing to its simplicity. However, these methods normally show unacceptable estimation accuracy in the low-cycle fatigue regime. In view of this, fracture mechanics-based method is an efficient strategy. Different from the

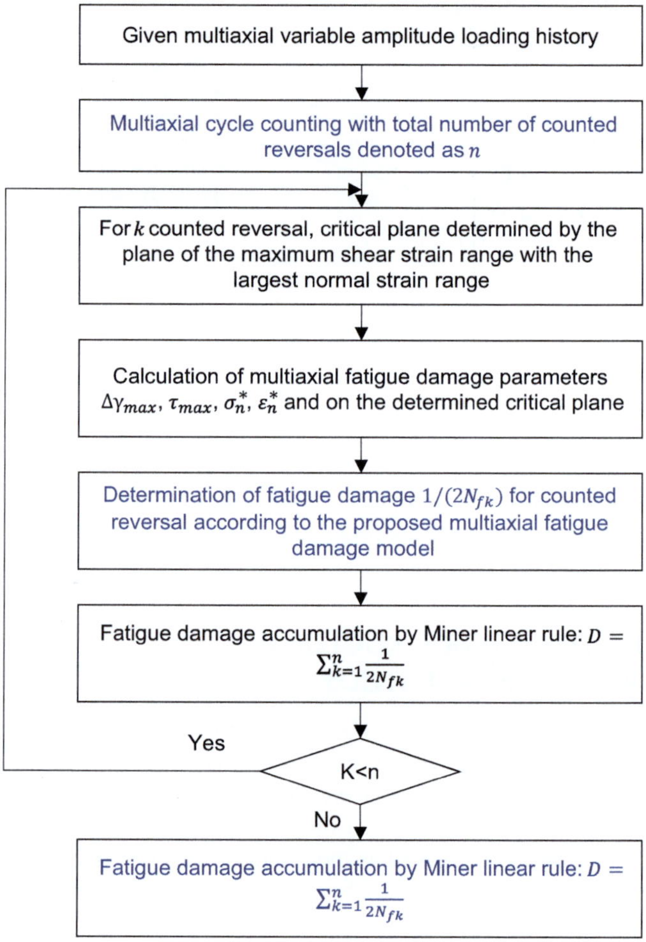

FIGURE 2.20 Fatigue life predictions under VAF loading by using Tao's model [32].

crack propagation behavior under CAF loadings, the crack growth state in current loading cycle is highly related to the previous loading history. Among published research, the influence of overload on fatigue performance is widely studied, as shown in Figure 2.21 [34].

Based on experimental data from VAF tests, Correia et al. [35] investigated the crack propagation behavior of metallic material. Specifically, the crack closure effect was analyzed based on the electrical strain gauges. Results indicated that Paris's law in conjunction with linear damage accumulation rule yields the best estimations of fatigue behavior. Zhang et al. [36] developed a crack growth model under VAF loading by using the back-propagation neural network. Specifically, it was found that the proposed model shows good predictive ability in assessing the crack growth life. However, the application of the artificial neural network in engineering practice needs to be further studied since the structural shape and loading spectrum of real components is complex. To assess the fatigue lives of concretes under VAF loading,

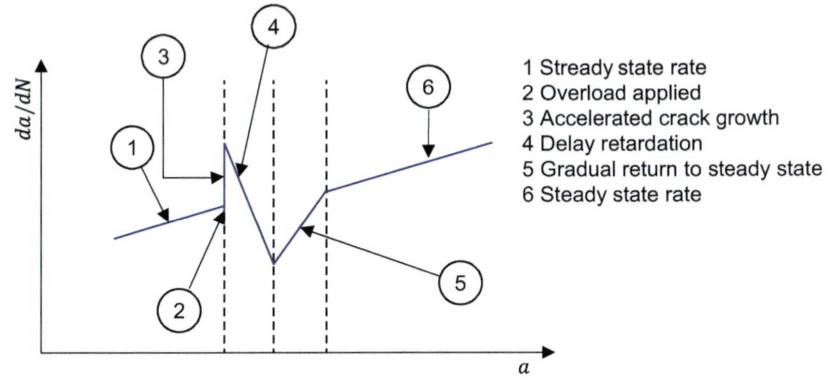

FIGURE 2.21 Influence of overload on the fatigue crack growth rate. (Reprinted with permission from Ref. [34], Elsevier.)

the analytical approach was developed by Ray et al. [37]. Specifically, the quantifiable equation for modeling the crack propagation was developed by using intermediate asymptotic and self-similarity.

2.6 APPLICATIONS FOR STRUCTURAL COMPONENTS: CASE STUDY

Efficiently assessing the fatigue lives of main structures is a vital task in engineering practice. As mentioned in Section 2.5, various fatigue life prediction methods under VAF loadings have been developed. However, the validity of these models needs to be further investigated. In this section, several newly proposed models by the authors are presented and compared, as well as their applications for structural components, in order to provide some references for future studies in fatigue life prediction under VAF loadings.

2.6.1 Case 1: Stress-Based Fatigue Life Prediction

Considering that fatigue life predictions of main structures under VAF loading are significant for maintaining their structural integrity, based on Manson-Halford theory, Yuan et al. [38] developed a modified nonlinear damage accumulation rule. Specifically, two sets of attested data were used for method verification and results indicated that the developed method shows good agreement between tested and predicted data. Recently, Zhu et al. [39] systematically compared five types of nonlinear fatigue damage accumulation rules by using tested lives of P355NL1 steel. Specifically, to quantify the influence of load sequence on fatigue behavior, a novel nonlinear damage accumulation rule was developed. Specifically, the load interaction effect is described by:

$$\frac{n_2}{N_{f2}} = \left(1 - \frac{n_1}{n_{f1}}\right)\left(\frac{\ln N_{f1}}{\ln N_{f2}}\right)^{(\sigma_1/\sigma_2)} \tag{2.67}$$

where σ_1 and σ_2 are two stresses expressing two-block loadings. Therefore, damage accumulation with VAF loadings is expressed as:

$$\frac{n_i}{N_{fi}} = \left(1 - \frac{n_1}{n_{f1}} - \frac{n_2}{n_{f2}} - \cdots - \frac{n_{i-1}}{n_{fi-1}}\right)\left(\frac{\ln N_{f1}}{\ln N_{fi}}\right)^{(\sigma_{i-1}/\sigma_i)} \tag{2.68}$$

Based on Eq. (2.68), the nonlinear fatigue damage evolution is modeled for further fatigue life predictions under VAF loading.

To validate the developed model by Zhu et al. (shorted as Model A herein), tested lives of P355NL1 steel under H-L and L-H block loadings (see Tables 2.2 and 2.3) were applied [39]. To further validate Model A, a comparison between the proposed model with Lv's model, Gao's model, Peng's model, Mesmacque's model, and Zuo's model is presented in Figure 2.22 [39]. Note, from Figure 2.22, that for both the H-L and L-H block loadings, Model A provides acceptable estimations of the fatigue lives.

TABLE 2.2
Tested Lives of P355NL1 Steel under H-L Block Loadings

Loading Sequence	Experimental Data					
H-L	$\Delta\varepsilon_H$ (%)	$\Delta\varepsilon_L$ (%)	n_H (Cycles)	n_H/N_H	n_L (Cycles)	n_L/N_L
1	1	0.5	337	0.25	3062	0.24
2	1	0.5	673	0.5	2727	0.22
3	1	0.5	1010	0.75	1597	0.13
4	1.5	0.75	121	0.251	1464	0.472
5	1.5	0.75	141	0.5	1374	0.443
6	1.5	0.75	362	0.751	1015	0.327
7	1.5	0.75	60	0.124	2430	0.783
8	1.5	0.75	181	0.376	1400	0.451
9	1.5	0.75	301	0.624	499	0.161
10	1.5	0.75	421	0.873	940	0.303

TABLE 2.3
Tested Lives of P355NL1 Steel under L-H Block Loadings

Number	Experimental Data					
L-H	$\Delta\varepsilon_L$ (%)	$\Delta\varepsilon_H$ (%)	n_L (Cycles)	n_L/N_L	n_H (Cycles)	n_H/N_H
1	0.5	1	3620	0.29	1007	0.745
2	0.5	1	7240	0.58	1275.5	0.95
3	0.5	1	9377	0.75	1610	1.2
4	0.75	1.5	776	0.25	585	1.214
5	0.75	1.5	1551	0.5	472.5	0.887
6	0.75	1.5	2327	0.75	259.5	0.538

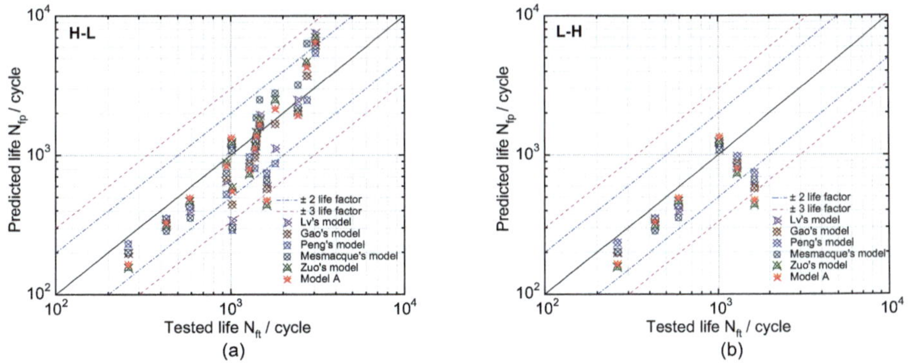

FIGURE 2.22 Model comparison with tested data of P355NL1 steel. (a) H-L loadings and (b) L-H loadings [39].

Specifically, almost all predicted lives fall within the two-factor bands. In addition, it can be found that for the other five models, conservative estimations of the fatigue lives normally exist, which proves that Model A can simultaneously consider the load sequence and load interaction effects.

2.6.2 Case 2: Strain Energy Density-Based Fatigue Life Prediction

To quantify the mean stress effect and loading history on the fatigue behavior, a strain energy density-based damage accumulation model was proposed by Zhu et al. [40]. Specifically, the equivalent elastic strain energy is calculated by

$$W_{\text{eq}}^{e} = \frac{1}{2}\varepsilon_{a,\text{eq}}^{e}\frac{\Delta\sigma}{2} = \frac{1}{2}\frac{\Delta\varepsilon^{e}\sigma_{\max}\sigma_{a}}{2\sigma_f'} = \frac{\sigma_f'^{2}}{2E}\left(2N_f\right)^{3b} \tag{2.69}$$

where W_{eq}^{e} denotes the equivalent elastic strain energy, $\varepsilon_{a,\text{eq}}^{e}$ denotes the equivalent elastic strain amplitude, $\sigma_{\max}$ denotes the maximum stress amplitude. In this work, to quantify load interaction effect, the following parameter was used

$$W_{\text{equiv}} = \frac{W_u - W_{i+1}}{W_u - W_i}\alpha_{i-1,i}\left(W_{i(\text{ed})} - W_i\right) + W_{i+1} \tag{2.70}$$

where W_i denotes the energy at ith level of the loading, and the parameter $\alpha_{i-1,i} = \dfrac{\sigma_{i+1}}{\sigma_i}$.

To validate the developed method for fatigue life predictions under VAF loading, four types of materials including the Al 2024 T42, 300CVM steel, 30NiCrMoV12 steel, and the 45 steel were utilized for model validation [40]. Specifically, a comparison between the proposed model (shorted as Model B herein) with the Miner's rule, damaged stress model (DSM), and damaged energy model (DEM). Comparisons between predicted and tested lives are shown in Figure 2.23 [40]. Note from Figure 2.23, for the four types of metallic materials, Model B provides better estimations of fatigue lives than the other three models.

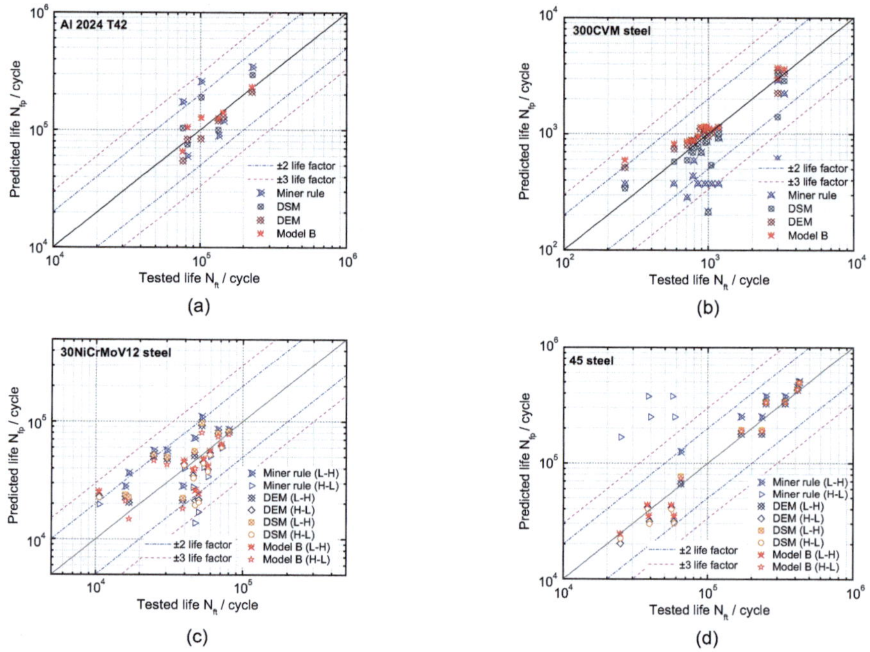

FIGURE 2.23 Comparisons between tested and the predicted fatigue lives. (a) Al 2024 T42; (b) 300CVM steel; (c) 30NiCrMoV12 steel; (d) 45 steel [40].

2.6.3 Case 3: Isodamage Curve-Based Fatigue Life Prediction

To conquer the shortcomings of linear fatigue damage accumulation rules, a novel nonlinear damage rule was developed by Zhu et al. [25] according to the isodamage curve. Specifically, a damage function corresponding to remaining life and isodamage curve was introduced. Considering that the lives under VAF loadings are affected by loading history, the isodamage curve exponential should be described as function to the loading history. Specifically, based on isodamage curve, a new nonlinear damage accumulation rule was developed by redefining the damage exponent. Therefore, the accumulated fatigue damage in the *i*th loading step is:

$$D_i = \left(\frac{\log N_e - \log N_i}{\log N_e - \log n_i} \right)^{(l \log n_i - s \log \varepsilon_{a,i})} \tag{2.71}$$

where $\varepsilon_{a,i}$ denotes strain amplitude at *i*th loading step, l and s respectively denote load sequence weighted coefficient and load weighted coefficient. Specifically, the accumulated damage under multi-blocks loading is able to be assessed by using Eq. (2.71).

To validate the isodamage curve-based nonlinear damage model (shorted as Model C herein), tested lives of P355NL1 steel under diverse types of loadings (including two blocks, multi-blocks, and spectrum loadings) were utilized [25]. Specifically, comparisons between predicted and tested lives under two blocks, multi-blocks, and spectrum loadings are presented in Figures 2.24–2.26 [25], respectively.

Note from Figure 2.24, for the predicted lives under H-L and L-H loadings, almost all predicted fatigue lives fall into two-factor bands, which demonstrate that Model C can be applied for life predictions under two-block loadings. Specifically, Model C provides better estimation of the fatigue lives than Subramanyan's model, Rege's model, Mesmacque's model, and double linear rule. Note from Figure 2.25, for the predicted lives under multi-blocks loading, predicted lives based on the three models agree well with the tested ones, which demonstrates that Model C can be utilized for life predictions under multi-block loadings. Note from Figure 2.26, for the predicted lives under spectrum loading, Model C provides good estimations of the fatigue lives under H-L loading spectrum, L-H loading spectrum, L-H-L loading spectrum, and random spectrum.

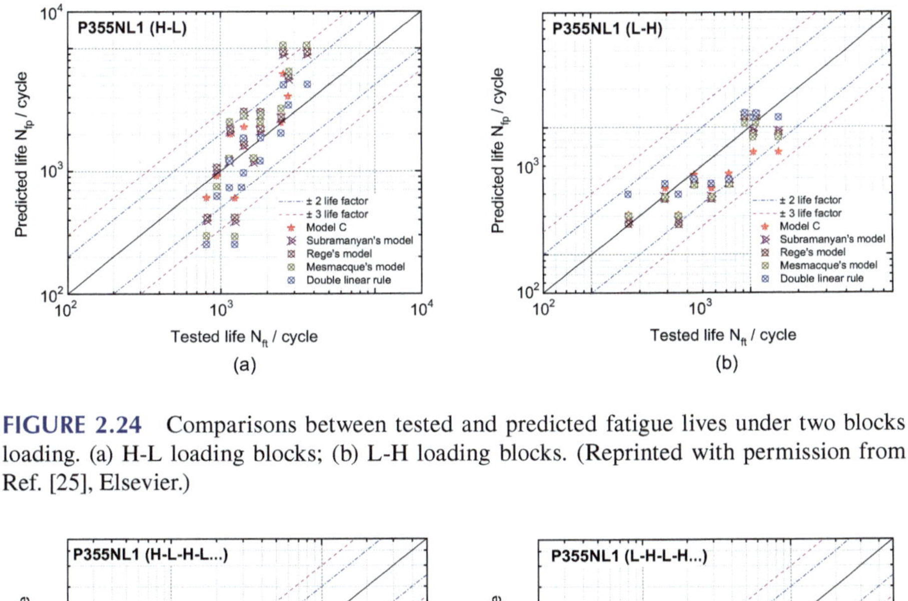

FIGURE 2.24 Comparisons between tested and predicted fatigue lives under two blocks loading. (a) H-L loading blocks; (b) L-H loading blocks. (Reprinted with permission from Ref. [25], Elsevier.)

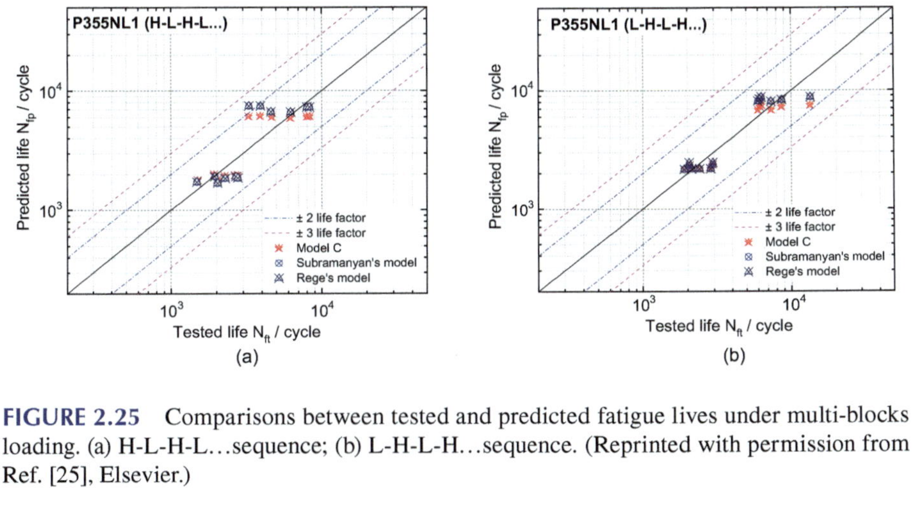

FIGURE 2.25 Comparisons between tested and predicted fatigue lives under multi-blocks loading. (a) H-L-H-L...sequence; (b) L-H-L-H...sequence. (Reprinted with permission from Ref. [25], Elsevier.)

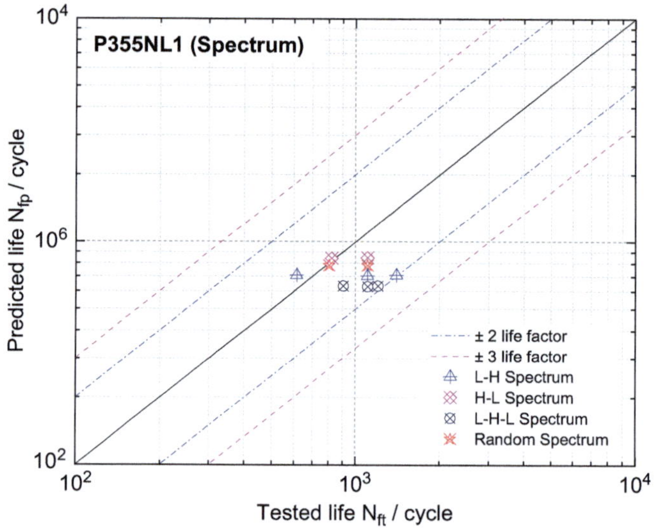

FIGURE 2.26 Comparison between tested and predicted fatigue lives under spectrum loading. (Reprinted with permission from Ref. [25], Elsevier.)

2.6.4 Case 4: Application for Turbine Blades

Considering that the combined low and high-cycle fatigue (CCF) is one of the main failure modes of the aircraft gas turbine, Zhu et al. [41] investigated the CCF behavior of the turbine blades and developed a new damage accumulation rule based on the Miner's rule. Specifically, the accumulated damage is obtained by using the life fraction model, as:

$$D = N\left(\frac{n}{N_{\mathrm{HCF}}} + \frac{1}{N_{\mathrm{LCF}}} + \frac{1}{(n+1)\lg(N_{\mathrm{HCF}})^{\alpha}}\right) \tag{2.72}$$

Then, the number of combined cycle blocs N_B at critical fatigue damage is:

$$N_B = \left[\left(\frac{n}{N_{\mathrm{HCF}}} + \frac{1}{N_{\mathrm{LCF}}} + \frac{1}{(n+1)\lg(N_{\mathrm{HCF}})^{\alpha}}\right)\right]^{-1} \tag{2.73}$$

Based on Eq. (2.73), the fatigue life is calculated by $N_f = (n+1)N_B$.

To validate the developed method for CCF life prediction (shorted as Model D herein) of the turbine blades under VAF loading, four types of materials (containing TC11, Ti-6Al-4V, Al 2024-T3, and DZ22 alloys) were applied for model validation and comparison [41]. In addition, a comparison of the prediction accuracy between Model D, Miner's rule, Manson-Halford model, and Trufyakov-Kovalchuk modes was conducted, see Figure 2.27 [41]. Note, from Figure 2.27, that the predictions based on Miner's rule tend to be conservative. Compared with linear fatigue damage

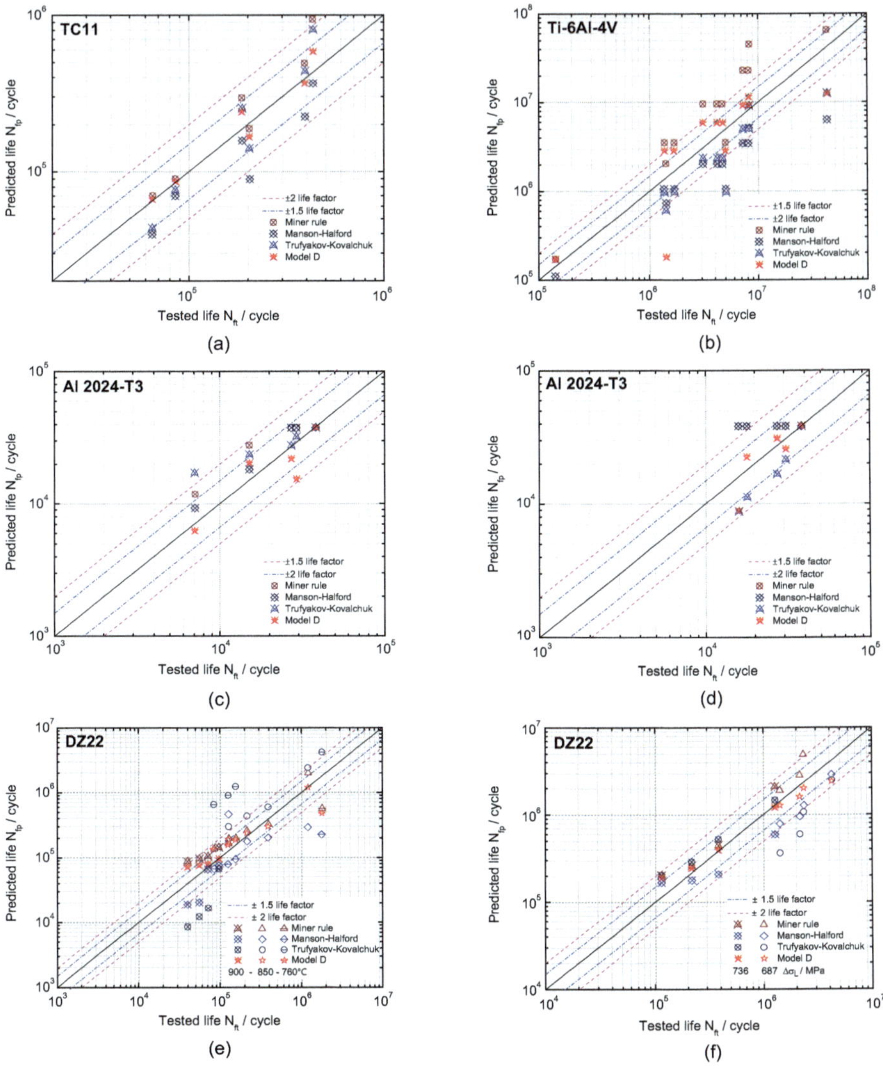

FIGURE 2.27 Comparisons between tested and predicted fatigue lives. (a) TC11 alloy; (b) Ti-6Al-4V alloy; (c) Al 2024-T3 for different α; (d) Al 2024-T3 for different n; (e) DZ22 alloy for different temperatures; (f) DZ22 alloy for different $\Delta\sigma_L$ [41].

rule, the Manson-Halford model and Trufyakov-Kovalchuk models provide better estimations of the CCF life. Specifically, it is found that Model D yields the most accurate estimations of the fatigue life among the four models.

Note from Figure 2.27, Model D yields acceptable estimations of the fatigue life for four types of turbine blade alloys. To further validate Model D for application of the fatigue life prediction of real components under CCF loading, CCF tests on turbine blades as well as the fatigue life predictions were conducted, as shown in Figure 2.28 [41].

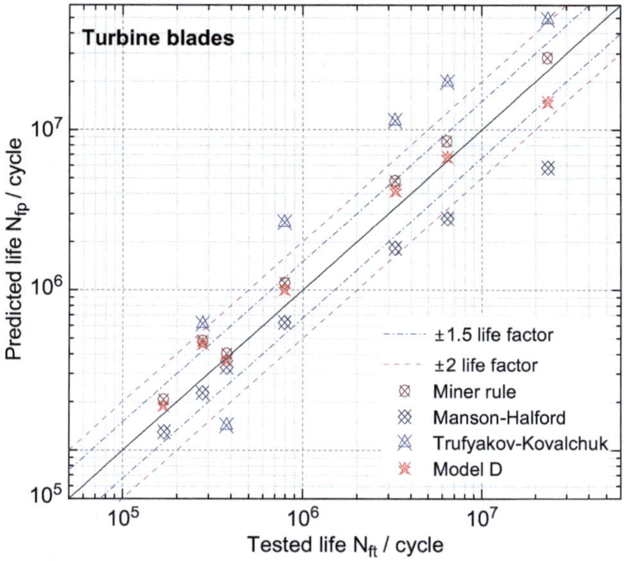

FIGURE 2.28 CCF life prediction of the turbine blades [41].

Note from Figure 2.28, fatigue life predictions by using the Manson-Halford model tend to be underestimated. For Miner's rule and the proposed model, more accurate estimations of the fatigue life are obtained since almost all predicted lives fall within the 1.5 factor bands. However, the Trufyakov-Kovalchuk model provides an estimation of fatigue life with unacceptable scatter, since an extra fitting parameter is introduced which restrained its applicability in limited data conditions.

2.7 SUMMARY

To provide a comprehensive understanding of life prediction challenges under VAF loadings, this chapter introduces key aspects commonly addressed in VAF studies. Firstly, commonly applied methods to describe and determine the loading spectrum are introduced. Then, fatigue test criteria under VAF loadings as well as commonly adopted methods for modeling the fatigue damage accumulation are clarified. Finally, commonly applied approaches for life predictions under VAF loadings are presented as well as four case studies.

ABBREVIATIONS

CAF Constant amplitude fatigue
CCF Combined low and high-cycle fatigue
DEM Damaged energy model
DSM Damaged stress model
SWT Smith-Watson-Topper
VAF Variable amplitude fatigue

REFERENCES

1. Marquis, G.B. (2011). Fatigue assessment methods for VAF loading of welded structures. In K.A. MacDonald (ed.), *Fracture and Fatigue of Welded Joints and Structures*. Cambridge: Woodhead Publishing, pp. 208–238.
2. ASTM, E. (2005). Standard practices for cycle counting in fatigue analysis. *Annual Book of ASTM Standards*, 3(1): 614–620.
3. Xiao, J.Q. (2009). *Theoretical and Experimental Investigation on Fatigue Properties of Rock under Cyclic Loading*. Changsha: Central South University.
4. Gassner, E. (1939). Strength experiments under cyclic loading in aircraft structures. *Luftwissen*, 6: 61–64.
5. Xu, Z.G., Guo, X., & Deng, C.G. (2019). *Brief Introduction of VAF Test*. Sichuan Building Science, China.
6. Zhu, S.P., Huang, H.Z., Li, Y., Liu, Y., & Yang, Y. (2015). Probabilistic modeling of damage accumulation for time-dependent fatigue reliability analysis of railway axle steels. *Proceedings of the Institution of Mechanical Engineers, Part F: Journal of Rail and Rapid Transit*, 229(1): 23–33.
7. Yao, W.X. (2003). *Fatigue Life Prediction of Structures*. Beijing: Defense Industry Press.
8. Zhu, S.P., Huang, H.Z., & Wang, Z.L. (2011). Fatigue life estimation considering damaging and strengthening of low amplitude loads under different load sequences using fuzzy sets approach. *International Journal of Damage Mechanics*, 20(6): 876–899.
9. Manson, S.S., & Halford, G.R. (1981). Practical implementation of the double linear damage rule and damage curve approach for treating accumulation fatigue damage. *International Journal of Fracture*, 17: 169–192.
10. Ji, Y.F., Yao, W.X., & Xia, T.X. (2015). An applicability assessment of linear accumulation fatigue damage rules. *Mechanics in Engineering*, 37(6): 674–682.
11. Li, R., Qiu, H.X., & Chun, Q. (2005). Research review of fatigue accumulative damage rule. *Journal of Jinling Institute of Technology*, 21(3): 23–27.
12. Zhao, S.B. (2015). *Anti-Fatigue Design Manual* (2nd ed). Beijing: China Machine Press.
13. Zhu, S.P., Huang, H.Z., Liu, Y., He, L.P., & Liao, Q. (2012). A practical method for determining the Corten-Dolan exponent and its application to fatigue life prediction. *International Journal of Turbo & Jet-Engines*, 29: 79–87.
14. Zakaria, K.A., Abdullah, S., & Ghazali, M.J. (2016) A review of the loading sequence effects on the fatigue life behaviour of metallic materials. *Journal of Engineering Science & Technology Review*, 9(5): 189–200.
15. Manson, S.S. (1965). Fatigue: A complex subject-some simple approximations. *Experimental Mechanics*, 5(7): 193–226.
16. Muralidharan, U., & Manson, S.S. (1988). A modified universal slopes equation for estimation of fatigue characteristics of metals. *Journal of Engineering Materials and Technology*, 110: 55–58.
17. Roessle, M.L., & Fatemi, A. (2000). Strain-controlled fatigue properties of steels and some simple approximations. *International Journal of Fatigue*, 22(6): 495–511.
18. Wang, Q., Xu, Z., & Wang, X. (2020). An efficient fatigue and creep-fatigue life prediction method by using the hysteresis energy density rate concept. *Fatigue & Fracture of Engineering Materials & Structures*, 43(7): 1529–1540.
19. Li, X.K., Chen, S., Zhu, S.P., Liao, D., & Gao, J.W. (2021). Probabilistic fatigue life prediction of notched components using strain energy density approach. *Engineering Failure Analysis*, 124: 105375.
20. Ellyin, F. (2012). *Fatigue Damage, Crack Growth and Life Prediction*. Dordrecht: Springer Science & Business Media.

21. Liao, D., & Zhu, S.P. (2019). Energy field intensity approach for notch fatigue analysis. *International Journal of Fatigue*, 127: 190–202.
22. Branco, R., Costa, J.D., Borrego, L.P., Berto, F., Razavi, S.M.J., & Macek, W. (2021). Comparison of different one-parameter damage laws and local stress-strain approaches in multiaxial fatigue life assessment of notched components. *International Journal of Fatigue*, 151: 106405.
23. El Aghoury, I., & Galal, K. (2013). A fatigue stress-life damage accumulation model for VAF loading based on virtual target life. *Engineering Structures*, 52: 621–628.
24. D'Angelo, L., & Nussbaumer, A. (2017). Estimation of fatigue SN curves of welded joints using advanced probabilistic approach. *International Journal of Fatigue*, 97: 98–113.
25. Zhu, S.P., Liao, D., Liu, Q., Correia, J.A.F.O., & de Jesus, A.M.P. (2019). Nonlinear fatigue damage accumulation: Isodamage curve-based model and life prediction aspects. *International Journal of Fatigue*, 128: 105185.
26. Aeran, A., Siriwardane, S.C., Mikkelsen, O., & Langen, I. (2017). A new nonlinear fatigue damage model based only on SN curve parameters. *International Journal of Fatigue*, 103: 327–341.
27. Huffman, P.J., & Beckman, S.P. (2013). A non-linear damage accumulation fatigue model for predicting strain life at VAF loadings based on CAF fatigue data. *International Journal of Fatigue*, 48: 165–169.
28. Wu, Z.R., Hu, X.T., & Song, Y.D. (2014). Multiaxial fatigue life prediction for titanium alloy TC4 under proportional and nonproportional loading. *International Journal of Fatigue*, 59: 170–175.
29. Wu, Z.R., Hu, X.T., Li, Z.X., & Song, Y.D. (2015). Evaluation of fatigue life for titanium alloy TC4 under variable amplitude multiaxial loading. *Fatigue & Fracture of Engineering Materials & Structures*, 38(4): 402–409.
30. Tao, Z.Q., Shang, D.G., Liu, H., & Chen, H. (2016). Life prediction based on weight-averaged maximum shear strain range plane under multiaxial VAF loading. *Fatigue & Fracture of Engineering Materials & Structures*, 39(7): 907–920.
31. Xue, L., Shang, D.G., Li, D.H., Li, L.J., Liu, X.D., & Chen, H. (2020). Equivalent energy-based critical plane fatigue damage parameter for multiaxial LCF under VAF loading. *International Journal of Fatigue*, 131: 105350.
32. Tao, Z.Q., Qian, G., Sun, J., Zhang, Z.L., & Hong, Y. (2022). Multiaxial fatigue life prediction by equivalent energy-based critical plane damage parameter under VAF loading. *Fatigue & Fracture of Engineering Materials & Structures*, 45(12): 3640–3657.
33. Deng, Q.Y., Zhu, S.P., He, J.C., Li, X.K., & Carpinteri, A. (2022). Multiaxial fatigue under VAF loadings: Review and solutions. *International Journal of Structural Integrity*, 13(3): 349–393.
34. Yuen, B.K.C., & Taheri, F. (2006). Proposed modifications to the Wheeler retardation model for multiple overloading fatigue life prediction. *International Journal of Fatigue*, 28(12): 1803–1819.
35. Correia, J., Carvalho, H., Lesiuk, G., Mourão, A., Grilo, L.F., de Jesus, A.M.P., & Calçada, R. (2020). Fatigue crack growth modelling of Fão Bridge puddle iron under VAF loading. *International Journal of Fatigue*, 136: 105588.
36. Zhang, L., & Wei, X. (2022). Prediction of fatigue crack growth under VAF loading by artificial neural network-based Lagrange interpolation. *Mechanics of Materials*, 171: 104309.
37. Ray, S., Jeshna, C.C., & Gupta, N. (2020). Fatigue life assessment of concrete members under VAF loading: An analytical approach. *Engineering Fracture Mechanics*, 223: 106735.

38. Yuan, R., Li, H., Huang, H. Z., Zhu, S. P., & Gao, H. (2015). A nonlinear fatigue damage accumulation model considering strength degradation and its applications to fatigue reliability analysis. *International Journal of Damage Mechanics*, 24(5): 646–662.
39. Zhu, S.P., Hao, Y.Z., Correia, J.A.F.O., Lesiuk, G., & de Jesus, A.M.P. (2019). Nonlinear fatigue damage accumulation and life prediction of metals: A comparative study. *Fatigue & Fracture of Engineering Materials & Structures*, 42(6): 1271–1282.
40. Yue, P., Lei, Q., Zhang, C.L., Zhu, S.P., & Huang, H.Z. (2017). A modified cumulative damage model for fatigue life prediction under VAF loadings. *Applied Mechanics and Materials*, 853: 62–66.
41. Zhu, S.P., Yue, P., Yu, Z.Y., & Wang, Q. (2017). A combined high and low cycle fatigue model for life prediction of turbine blades. *Materials*, 10(7): 698.

3 Multiaxial Fatigue

Ding Liao, Shun-Peng Zhu, and Qingyuan Wang

3.1 INTRODUCTION

Since 1829, when Albert finished the first research paper with regard to metal fatigue, it has been about 200 years of history. Nevertheless, up to now, fatigue fracture accidents are still out of well control. One of the prime reasons is that much attention was paid to the simplest uniaxial fatigue study. The critical parts of aero & space vehicles, pressure vessels, nuclear power plants, and traffic conveyances are subjected to not simple uniaxial but complex multiaxial cyclic loadings.

To date, for uniaxial fatigue theories widely used in structural anti-fatigue design and life assessment, rich material performance data and practical experiences were accumulated. However, they cannot serve the strength and life design of the aforementioned critical components. Consequently, driven by the practical demand, multiaxial fatigue research has attracted increasing attention in fatigue studies in recent decades. Multiaxial fatigue is the fatigue damage behaviour under stress/strain-controlled loadings along multiple directions. To be more specific, there should be two or three stress/strain components that change cyclically.

Since 1982, when the First Symposium on Multiaxial Fatigue initiated by the American Society for Testing and Materials (ASTM) was held, this conference series has been held 13 times, which is now titled "International Conference on Multiaxial Fatigue and Fracture". Particularly, in the recent International Fatigue Congresses, multiaxial fatigue was frequently set as a special topic. Note, from the papers published on these two conference series, that the focus of the multiaxial fatigue study has been transferred from material fatigue behaviour under proportional or non-proportional multiaxial loadings to fatigue damage evaluation and fatigue life assessment of practical engineering details subjected to multiaxial loadings during operation. To conclude, perfecting multiaxial fatigue theory and improving life prediction accuracy are critical challenges we face now, which deserve further efforts by researchers and designers.

In the early stages of multiaxial fatigue research, due to limited analytical devices, it is hard to know the damage mechanism of multiaxial fatigue. On the question of which parameter to choose as the damage parameter, one can only rely on speculation. As a result, the early multiaxial fatigue failure criteria were developed based on the static strength theory. In the selection of damage parameters, equivalent stress or strain is used as a parameter to convert the multiaxial stress/strain state into the equivalent uniaxial one; after that, the fatigue life can be estimated conveniently.

In this period, multiaxial fatigue assessment methods were mainly built based on the following three criteria: the first is the maximum principal stress/strain criterion, which believes that even under multiaxial stress conditions, material damage is

DOI: 10.1201/9781003405122-3

dominated by the maximum principal stress (strain) and has nothing to do with other factors; the second is the von Mises equivalent stress (strain) criterion, which suggests that under multiaxial stress–strain conditions, material damage is dominated by von Mises equivalent stress–strain; the third is Tresca's maximum shear stress/strain criterion, which attributes the damage process to the maximum shear stress (strain) only. For proportional loaded situations, these guidelines are valid, simple, and practical. However, for the cases of non-proportional loadings, especially when hydrostatic stress affects fatigue life, none of the above criteria can provide a satisfactory prediction.

Morrow proposed the energy method in 1965, believing that the accumulation of plastic work makes the dominant contribution to irreversible material damage, which leads to final fatigue failure. The core idea of this method is that in each cycle, a part or a component absorbs the energy delivered by the external loading, thereby causing irreversible damage to it. The extent of damage is directly proportional to the amount of energy absorbed. With gradual damage accumulation, it fails due to fatigue once a critical value is reached. Garud perfected the plastic work theory and further applied it to multiaxial fatigue scenes.

In 1973, after analyzing sufficient multiaxial low-cycle fatigue data, Brown and Miller summarized that using one parameter to describe multiaxial fatigue is no longer effective and, therefore, used two strain parameters to model the fatigue process. They thought that cyclic shear and normal strain components on the plane of the maximum shear strain must be considered together (cyclic shear strain contributes to crack initiation, while the normal one promotes crack growth), which is the embryonic form of the critical plane approach (CPA). Later, CPAs of various definitions were proposed by later researchers, which not only consider the stress and strain magnitudes but also consider their directions when selecting parameters. In general, the damage parameters are getting more meaningful, making the CPA closer to reality and more accurate in fatigue strength assessment of engineering components.

Research on multiaxial fatigue damage behaviour and life prediction method is related to the service safety of structures subjected to complex loadings. Therefore, it always receives much attention from science and engineering fields. The following sections will introduce multiaxial fatigue tests, multiaxial fatigue damage mechanisms, multiaxial fatigue criteria, and two case studies for reference.

3.2 MULTIAXIAL FATIGUE TESTS

In recent decades, due to the invention of electro-hydraulic servo fatigue testing machines, it has become feasible to simulate/apply the complex loading histories of engineering components during practical service. In 1935, Gough and Pollard first used a classic mechanically driven fatigue testing machine, applying alternating bending and torsional stresses of the same frequency to smooth cylindrical specimens [1]. Kinds of testing machines provide effective tools for multiaxial fatigue study, which help us to intensively study the multiaxial fatigue damage mechanism so as to find a way to improve multiaxial fatigue resistance, conduct accurate fatigue life prediction, and promote multiaxial fatigue study.

FIGURE 3.1 MTS® series 809 axial/torsional test system.

To date, there are three kinds of typical multiaxial fatigue testing facilities, including tension–torsion, bending–torsion, and planar biaxial ones. Axial–torsional fatigue testing is the most widely used one. Most of the commercial multiaxial fatigue testing apparatus use this loading method (see Figure 3.1) [2]. For such kind of test, thin-walled tubular smooth and annular notched bar specimens are suggested (see Figure 3.2). When using the stress-controlled mode, the loads are measured by force sensor integrated into the fatigue testing apparatus. While for the strain-controlled case, a multiaxial (tension–torsion) extensometer is usually utilized for real-time monitoring (see Figure 3.3 [3]). An illustration of proportional (or in-phase) and non-proportional (or 90° out-of-phase) loadings is shown in Figure 3.4. In addition to proportional loaded cases, fatigue damage evolution under various non-proportional multiaxial loaded scenarios (see Figure 3.5) was intensively studied.

In the axial–torsional multiaxial fatigue test, the stress–strain state of the body can be expressed with strain vector $\boldsymbol{\varepsilon}$ in strain space $(\varepsilon, \gamma/\sqrt{3})$ and stress vector $\boldsymbol{\sigma}$ in stress space $(\sigma, \sqrt{3}\tau)$:

$$\boldsymbol{\sigma} = \sigma \boldsymbol{n}_1 + \sqrt{3}\tau \boldsymbol{n}_3 \tag{3.1}$$

$$\boldsymbol{\varepsilon} = \varepsilon \boldsymbol{n}_1 + \gamma/\sqrt{3}\boldsymbol{n}_3 \tag{3.2}$$

where $\boldsymbol{n}_1$ and $\boldsymbol{n}_3$ are unit vectors (orthogonal basis vectors in two-dimensional vector space), σ and ε are respectively axial stress and strain, τ and γ are the shear ones.

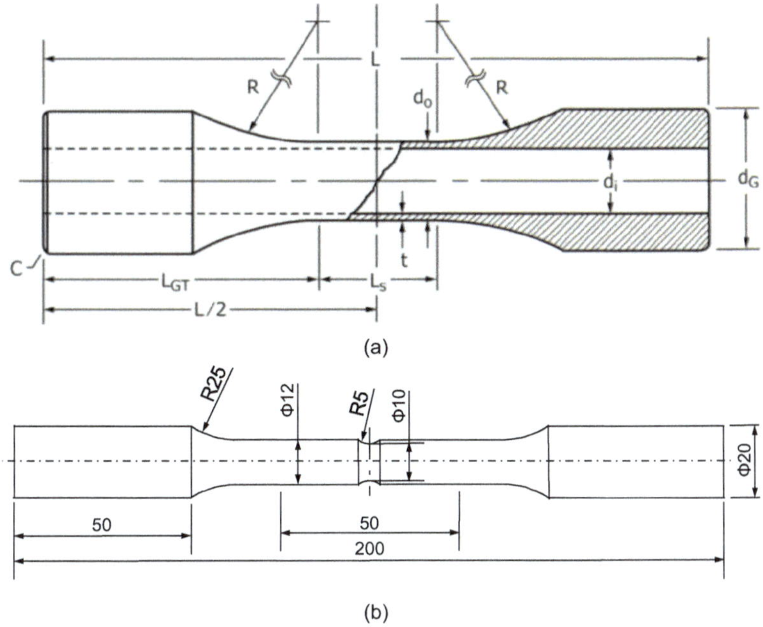

FIGURE 3.2 Multiaxial fatigue specimens: (a) thin-walled tubular specimen [2]; (b) biaxial cruciform specimen. (Refer to the original texts for the definitions of the symbols.)

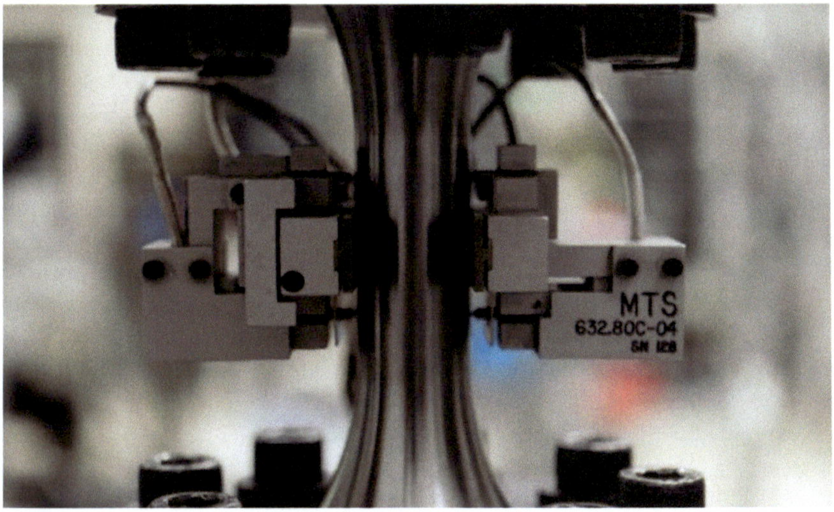

FIGURE 3.3 MTS® 632.80C-04 multiaxial (tension–torsion) extensometer [3].

The values of σ and ε are respectively equal to those of von Mises equivalent stress and strain, which are defined as follows:

$$\sigma_{eq} = \sqrt{\sigma^2 + 3\tau^2} \tag{3.3}$$

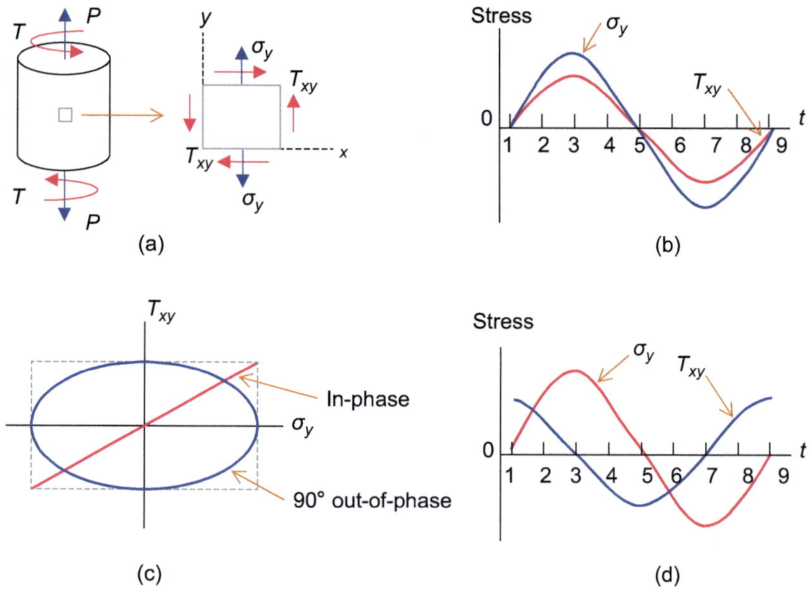

FIGURE 3.4 Illustration of proportional (or in-phase) and non-proportional (or 90° out-of-phase) loadings. (a) Stress element in axial–torsion loading; (b) proportional histories; (c) stress paths for proportional and 90° non-proportional loadings; (d) 90° non-proportional histories.

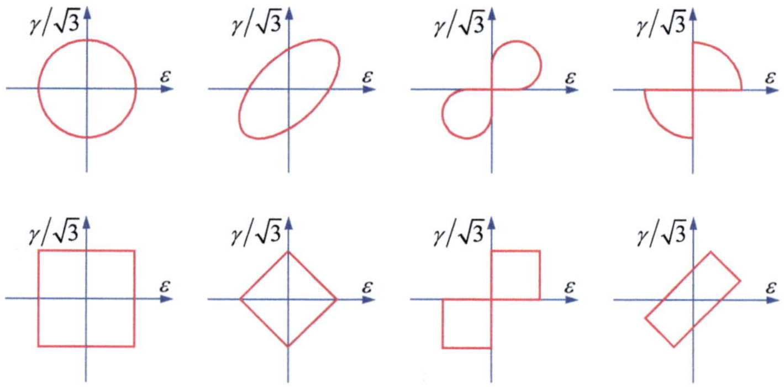

FIGURE 3.5 Non-proportional multiaxial loading paths.

$$\varepsilon_{\text{eq}} = \sqrt{\varepsilon^2 + \gamma^2/3} \tag{3.4}$$

Under the stress-controlled mode, it follows:

$$\sigma = \frac{4F}{\pi(d_o + d_i)(d_o - d_i)} \tag{3.5}$$

$$\tau = \frac{16T}{\pi(d_o + d_i)^2(d_o - d_i)} \tag{3.6}$$

where σ and τ are respectively the axial and shear stresses, F is the axial force, T is the torque, d_o and d_i are respectively the external and internal diameters.

Under strain-controlled mode, it follows:

$$\Delta l = l - l_0 = l_0\left(e^{\varepsilon} - 1\right) \tag{3.7}$$

$$\Delta\theta = \gamma\frac{4\times l\times 180^{\circ}}{\pi(d_o + d_i)} \tag{3.8}$$

where Δl is the axial displacement, l is the gauge length of the extensometer, l_0 is the original gauge length of the extensometer, and $\Delta\theta$ is the torsion angle.

In 1985, Downing and Galliart reported a two-channel test rig that was able to apply torque and bending moment synchronously on a round-notched specimen [4] (see Figure 3.6 [5]). The test system consists of a customized set of hardware for the dimensions and geometry of the Society of Automotive Engineers (SAE) Fatigue

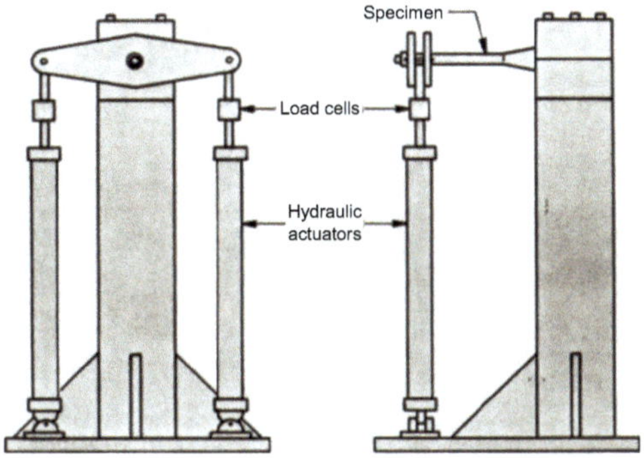

(a)

(b)

FIGURE 3.6 Combined bending and torsion fatigue testing system developed in support of the SAE [5]. (a) Diagrammatic sketch; (b) picture of the real product.

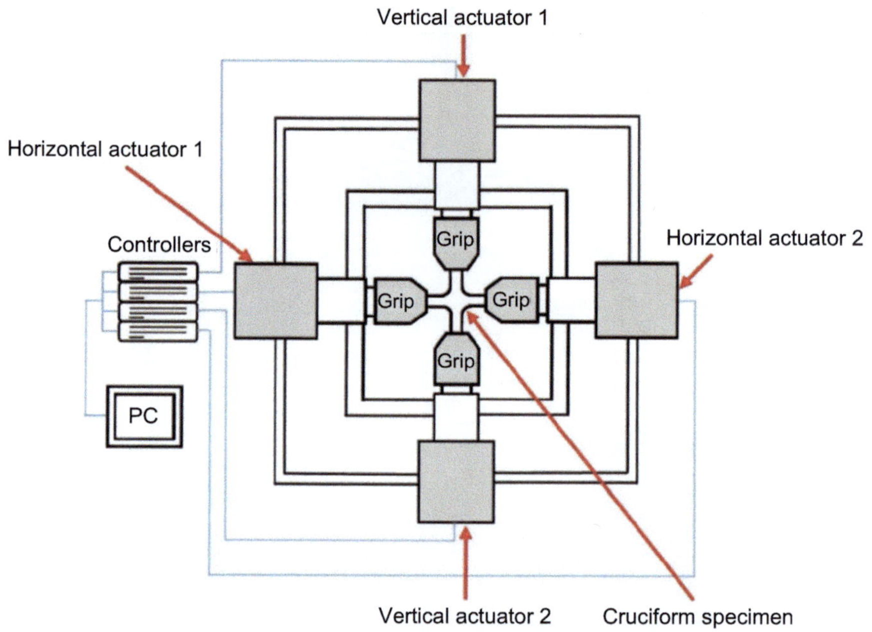

FIGURE 3.7 Schematic of planar biaxial fatigue testing apparatus. (With the permission from Ref. [6], Elsevier.)

Design and Evaluation Committee's bar-shaped specimen. In particular, the system should be capable of providing any combination of proportional and non-proportional bending and torsional loadings.

Under actual service conditions, metallic platy structures will be subjected to planar multiaxial fatigue load, and the non-proportional additional hardening effect will reduce their fatigue lives. Accordingly, planar biaxial fatigue testing machines were developed using biaxial cruciform multiaxial fatigue specimens. The schematic of the planar biaxial fatigue testing apparatus is shown in Figure 3.7 [6], and an MTS® planar biaxial fatigue testing system is shown in Figure 3.8. The advantages of such kind of fatigue testing include a measurable homogeneous strain field on the specimen, adjustable loading stress ratio, strain rate, and the phase difference between the two loading axes. Now, it has become a common tool for multiaxial performance study of planar metal plates, including both low- and high-cycle fatigue campaigns. As for designing biaxial cruciform fatigue specimens (see Figure 3.9 [7]), a detailed introduction was made by the Chinese standard GB/Z 40387—2021 [7].

In addition to material-level multiaxial tests, some research institutes exerted their efforts on developing component-level multiaxial fatigue testing facilities in view of the difficulty of achieving the smooth transformation between material fatigue properties and fatigue strength of full-scale engineering components.

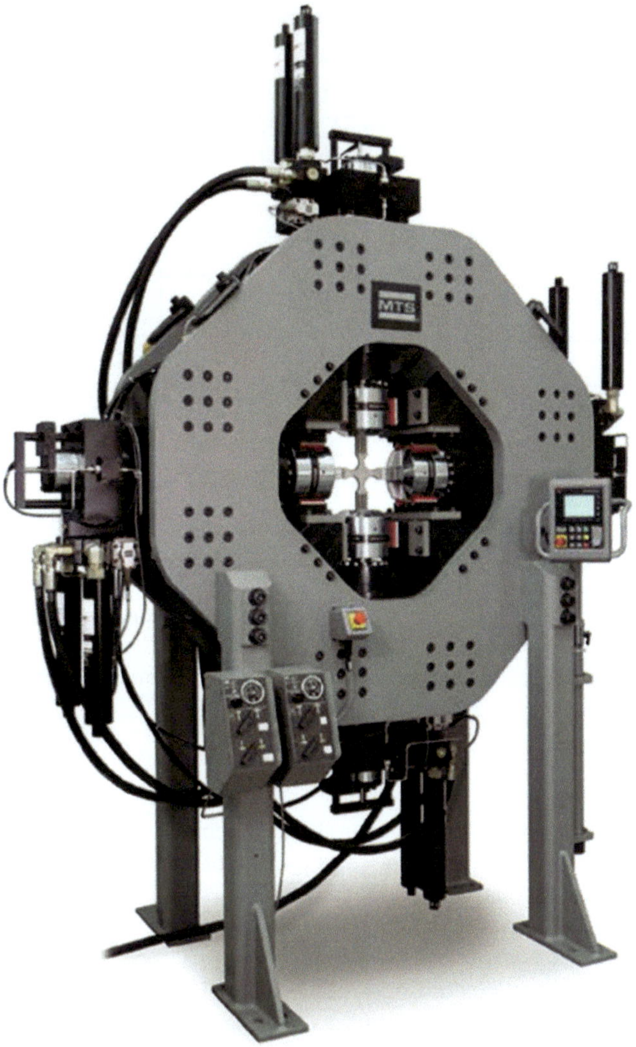

FIGURE 3.8 MTS® planar biaxial fatigue testing system.

Empirically, providing multiaxial stress for test components is a hard task in view of the challenging stress measurement under multiaxial loading and the prohibitive costs. Nowadays, only very limited universities and institutes have the capability to perform accurate multiaxial stress tests (see Figure 3.10).

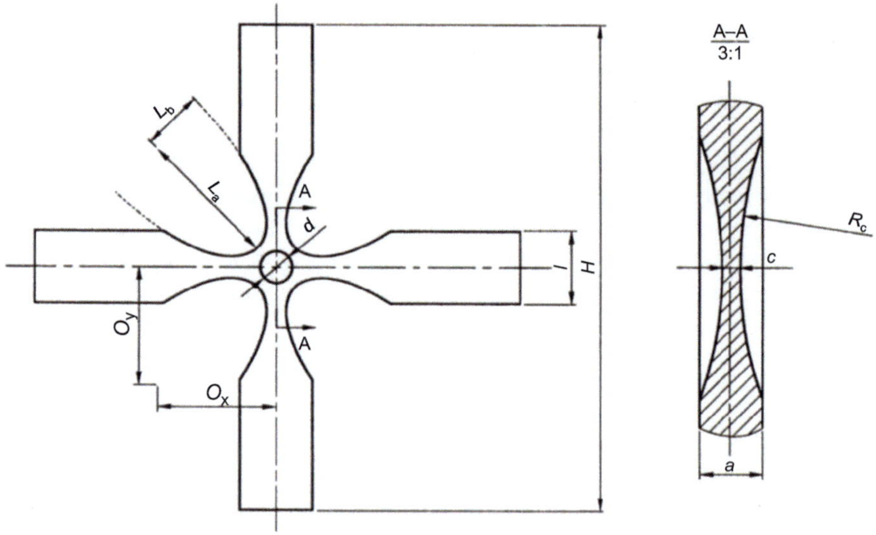

FIGURE 3.9 Biaxial cruciform multiaxial fatigue specimen [7]. (Refer to the original texts for the definitions of the symbols.)

FIGURE 3.10 Component-level multiaxial fatigue testing apparatus in the TU Delft.

3.3 MULTIAXIAL FATIGUE DAMAGE MECHANISMS

A complete fatigue damage process includes crack initiation, coalescence, and stable growth, ultimately causing yielding or brittle fracture at the net section. This section briefly reviews the mechanisms of crack formation and growth. Familiarity with basic fatigue and fracture mechanics terminology is assumed. Two excellent books are available to readers interested in additional discussion of many topics [8,9].

In 1903, Ewing and Humfrey [10], inspired by the research of Wöhler and Baushinger, finished their representative work "Fracture of metals under repeated alterations of stress." This paper reported experimental data on high-quality annealed Swedish iron flat specimens. In particular, optical microscopy was utilized to observe the fractography of the same specimen region at different fatigue damage periods.

They describe the breakdown process as follows: The first examination carried out after a few stress reversals, found slip lines on some grains (crystals), which are very similar in appearance to those that emerge when applying tensile stress over the yield limit. Later, after sufficient reversals, they evolved into comparatively wide bands with rather hazily defined edges. With an increasing number of reversals, they continued to widen, and some regions of the surface were full of dark markings. At this stage, cracks can be captured on some crystals, which occurred along widened slip bands. In some cases, they appeared on a single grain (crystal) first but soon joined from crystal to crystal. Eventually, a long continuous crack formed, across the surface of the specimen.

Additionally, in case an incipient crack begins to develop across a certain set of crystals, the influence of further reversals is mainly limited to the crack vicinity. The basic cause of fatigue damage was established by the turn of the century. Later work with electron microscopes, X-rays, and other powerful facilities proved that fatigue crack formation results from alternating shear stresses and strains.

3.3.1 Crack Nucleation and Early Growth

A schematic illustration of Ewing and Humfrey's observations on how a crack forms on the surface of a ductile metal is illustrated in Figure 3.11. In all of these examples, the tensile stress is applied vertically, which results in 45° shear stress. Eventually, cyclic plastic shear strain contributes to the formation of the slip band, which is shown as one of the fine lines in the diagrammatic sketch. In particular, the grains, whose crystallographic slip planes and directions are favourably oriented with the applied cyclic shear stress, will be the first to develop a slip band.

Every grain actually has its preferred slip plane, and the applied shear stress can be resolved on this plane. All grains (crystals) have the same critical resolved shear stress, i.e., the shear stress at which plastic deformation initiates. However, because of the different orientations, they will yield and show plastic deformation at different stress levels. At low stress and strain levels, only limited grains have favourable orientations, and therefore, only a few slip bands form; while at high stress levels, the critical resolved shear stress is exceeded in almost every grain, and a great deal of slip bands form. Under continuous cyclic loading, these slip bands gradually grow and coalesce into a sole dominant fatigue crack.

The nucleation process of polycrystalline metals is relatively complex, which is affected by grain boundaries, precipitates, impurities, and inclusions. The intrusion–extrusion model in Figure 3.12 [11], in spite of its simplicity, is valid for modeling the nucleation process. This figure shows a cross-section view of a deforming grain in the material. Slip bands are evidence of dislocation movement, which can move only along a slip plane determined by the crystal structure within individual grains. Cyclic shear stress will drive dislocation movement. Finally, plastic deformation

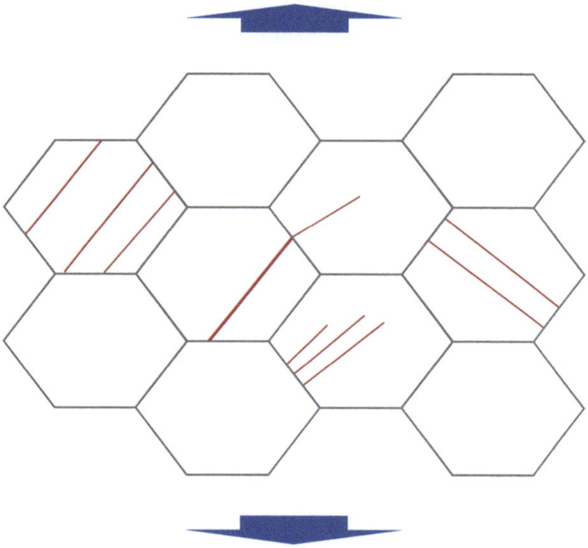

FIGURE 3.11 Crack nucleation within grains.

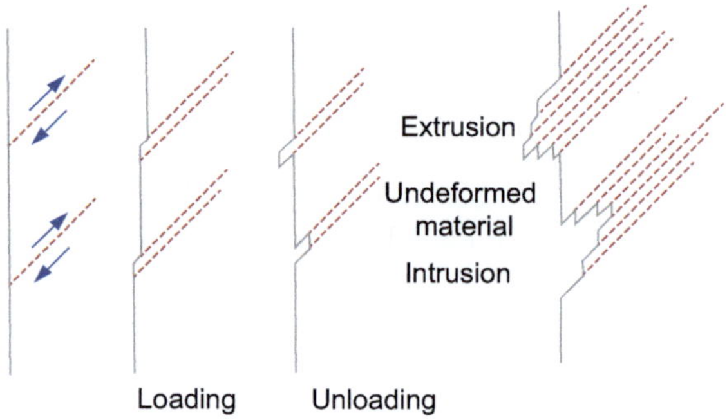

FIGURE 3.12 Slip band formation [11].

occurs as these stresses get higher than the critical resolved shear stress, resulting in a permanent offset in adjacent atomic planes.

Cyclic plastic deformation finally results in the development of a slip band. Some slip bands, called extrusions, come out of the material surface. Intrusions go into the surface. Deformation is localized in the slip bands, with relatively undeformed material between them. The dependence of crack nucleation on the amount of plastic strain explains the strength dependence of the fatigue limit. Smaller grains also make the movement of dislocations more difficult and result in a material with a higher fatigue strength. Similarly, inclusions provide a local stress concentration that can nucleate a slip band at a lower applied stress level. Repeated plastic straining of a slip band eventually causes a crack to form. Essentially, fatigue crack nucleation is a

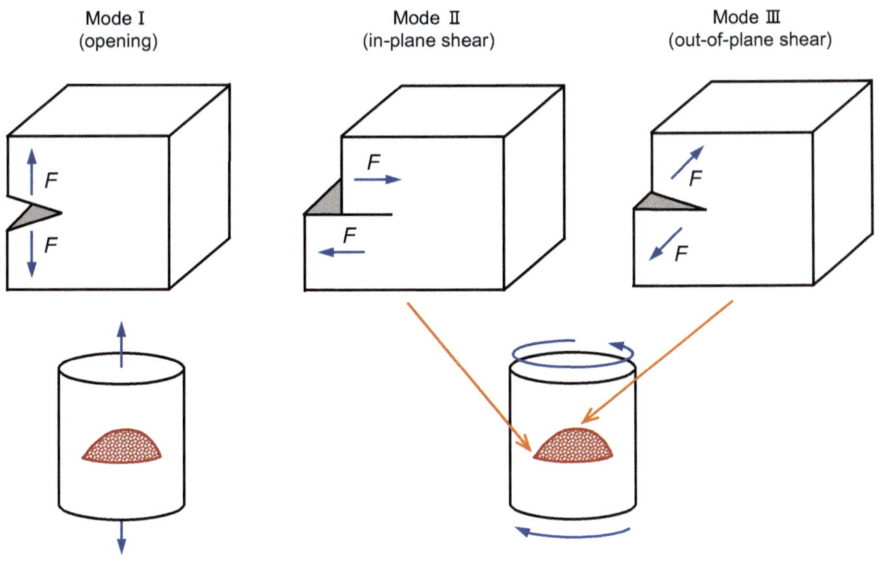

FIGURE 3.13 Surface crack subjected to mode I, mode II, and mode III loads [11].

surface phenomenon. Cracks rarely nucleate on the subsurface. When they do, it is usually at an inclusion, porosity, or interface, even these may be considered internal surfaces.

Fracture mechanics terminology is utilized to characterize how a crack is loaded. Figure 3.13 [11] shows an edge crack and a semi-elliptical surface crack in a round bar or shaft. The crack's fracture mechanics stress intensity factor is related to the crack shape, crack size, and loading mode. Tension loadings will produce Mode I or opening loadings. Such kind of loading is the commonest and is generally treated as Mode I unless otherwise noted. Torsion loading of a surface crack introduces both Mode II and Mode III. In particular, Mode II is an in-plane shear loading and can be found on the surface, whereas Mode III is an out-of-plane tearing loading appearing at the deepest region of the crack. Mode III loading drives the crack growing into the surface, whereas Mode II promotes the crack growing along the surface in torsion loading. For other points along the crack front, they undergo mixed Mode II and Mode III loading.

Knowledge of how cracks form on the surface is only the first aspect of fatigue. In reality, the length of the surface crack is not as important as the crack depth. Understanding how cracks grow into the material surface is important for a complete description of the process. A cross-section of a polycrystalline material loaded in cyclic tension is illustrated in Figure 3.14. A crack starts in a single surface grain and then goes into the neighbouring one.

Note that in the diagrammatic sketch, the crack orientation in the second grain is mostly slightly different than in the first, given the random grain orientations within a material. Overall, the growth path is in accord with the plane with the maximum

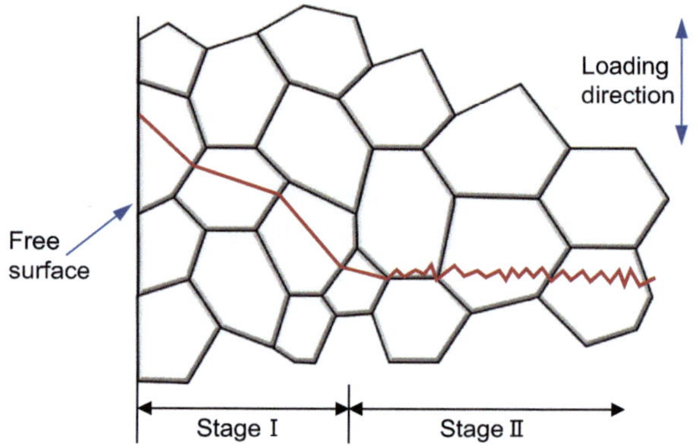

FIGURE 3.14 Stage I and stage II growth processes.

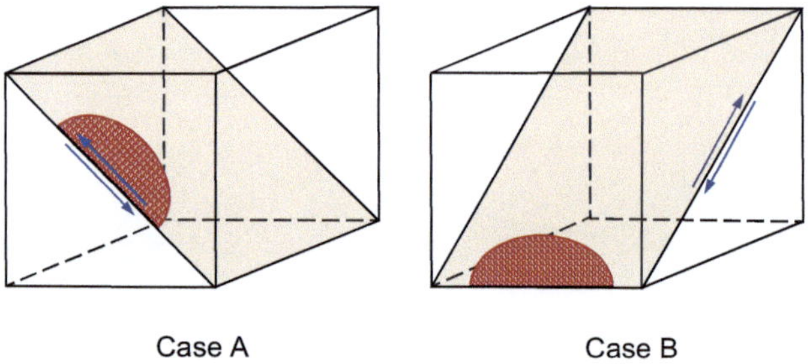

FIGURE 3.15 Case A and Case B crack growth.

shear stress, Mode II in fracture mechanics terminology, with deviations within individual grains. Eventually, as the crack becomes long enough, it will turn to grow perpendicular to the tensile stress direction. At that stage, the crack is large enough to produce plasticity and will grow planarly through the grains, named Mode I crack in fracture mechanics terminology.

This transition is determined by the material microstructure and the applied loading. Compression loadings favour Stage I shear growth, the same goes for torsion. Particularly, materials with low ductility have predominantly Stage II growth. The environment also influences the transition, particularly in aluminium alloys.

As given by Case A and Case B in Figure 3.15, shear cracks can grow in one of two ways. For a Case A crack, the in-plane shear stress acts on the free surface parallel to the crack length. In particular, no shear stress acts perpendicular to the free surface along the crack depth. Consequently, they tend to be shallow with low aspect ratios. Case B cracks are the result of out-of-plane shear and are characterized by the intrusion–extrusion model. For tension loading, it has the same shear stress for both Case A and Case B and can display either crack growth mode. In particular, torsion or combined tension–torsion loading always has Case A cracks.

3.3.2 Tensile Mechanisms – Mode I Growth

As cracks grow in length, they become less sensitive to microstructural variables than the damage process zone traversing several grains. The process is driven by intense shear bands formed at the crack tip. In a crystallographic view, even if the crack tip undergoes Mode I opening displacements, the crack grows in shear.

Macroscopically, the crack grows perpendicular to the applied tensile stress. Microscopically, the process is governed by the local shear stresses generated at the crack tip. The slip process is irreversible, and a small increment of crack extension occurs during each cycle. Figure 3.16 illustrates the duplex shear band model for Stage II crack propagation [11]. This mechanism results in the formation of striations on the crack surface. Figure 3.17 provides a typical electron micrograph of a fracture

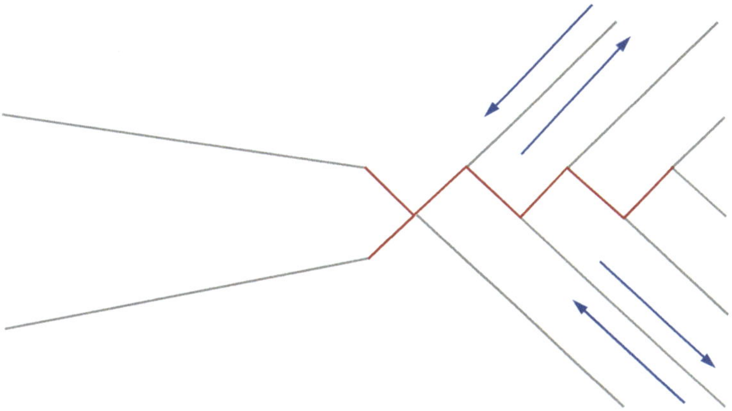

FIGURE 3.16 Shear band model for crack growth [11].

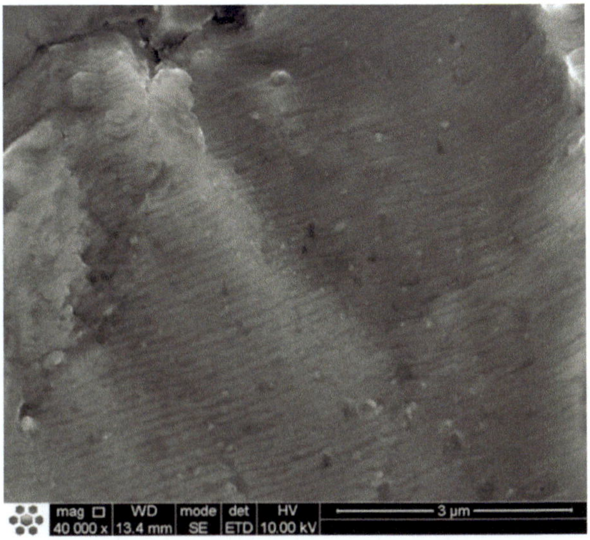

FIGURE 3.17 Striations on mode I fracture surface (Al 7075-T6511 alloy).

surface containing striations. Note the well-defined surface topography where the striations are perpendicular to the crack growth direction.

Each striation represents an increment of crack growth during each cycle. Many mechanisms, in addition to the one described in Figure 3.16, are possible, and fatigue striations can take many forms. Sometimes, they are difficult to find in high-strength steels where crack closure forces damage the fracture surfaces.

Scholars have put forward numerous multiaxial fatigue models for different materials or loading paths. These models can be roughly divided into three categories: equivalent stress/strain models, energy criteria, and critical plane-based approaches. Among them, equivalent stress/strain models and energy models cannot explain the micro failure mechanism of multiaxial fatigue, while the CPA has been widely spread and developed due to its ability to explain better the generation and growth mechanisms of fatigue fracture modes or cracks. Therefore, the following will mainly review the multiaxial models based on the critical plane.

Crack propagation is influenced not only by the conditions ahead of the crack tip but also by residual stresses behind the crack tip. The material in front of a growing crack is stretched during tensile loading, resulting in the stress distribution in Figure 3.18 [11].

Upon unloading, this material is forced to return to zero strain by the surrounding elastic material. This results in residual compressive stresses in the plastic zone and behind the crack in the plastic wake. On reloading, these residual stresses must be overcome before tensile stresses can advance the crack tip. Crack closure can be observed even if the applied stress cycle is tensile.

Additional sources of crack closure include oxide films, surface roughness, and trapped fluids. Crack closure is most pronounced during low ΔK and low R ratio fatigue and reduces the effectiveness of the applied stress intensity range. Crack closure can result from several sources, as shown in Figure 3.19 [11].

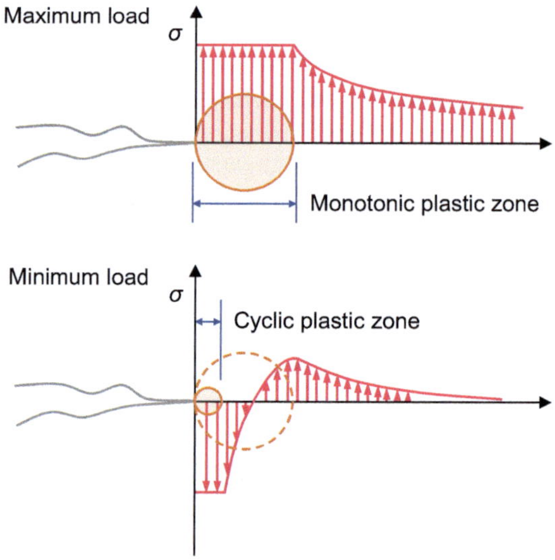

FIGURE 3.18 Stress fields during loading and unloading [11].

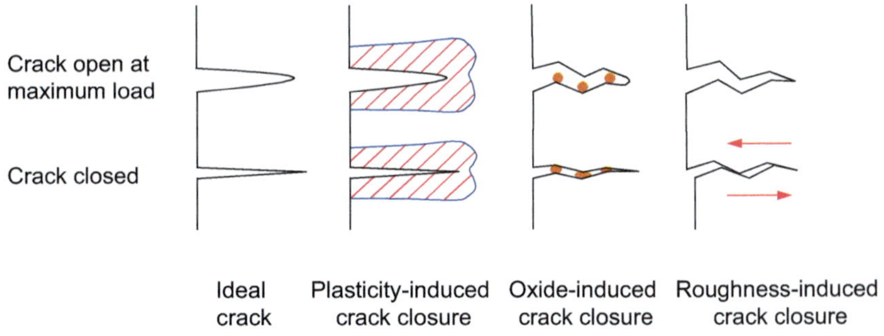

FIGURE 3.19 Crack closure mechanisms [11].

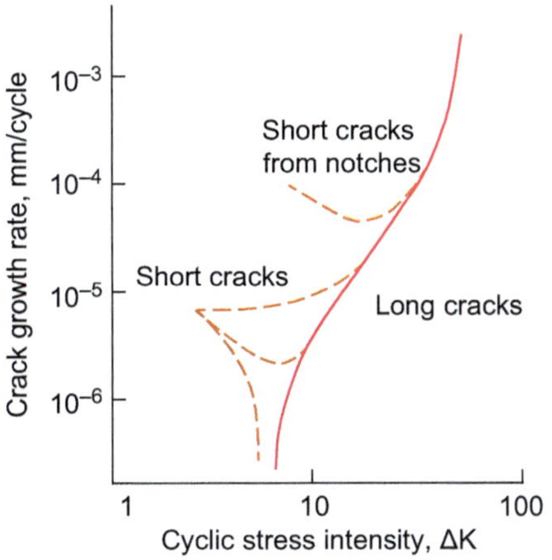

FIGURE 3.20 Small crack growth [11].

Small cracks propagate faster than the longer ones when applying the same cyclic stress intensity and at stress intensities much lower than the long crack threshold. There are various definitions of small cracks. A microstructurally small crack is of the same magnitude as the material microstructural features, e.g. grain size and interparticle spacing. Mechanically small flaws are comparable in size to the crack tip plastic zone. They represent cracks growing in the plastic zones of a notch. Physically small cracks usually are those less than a millimeter in length and are comparable to the crack size that can be detected with non-destructive examination.

Figure 3.20 shows that short cracks propagate much faster than long cracks [11]. Any factors contribute to the difference between the growth of short and long fatigue cracks. Small cracks display several kinds of behaviour. They may decelerate with increasing crack length as they approach some barrier to crack growth. At high stress levels, they may accelerate with increasing crack length. Sometimes, they initially decelerate and then accelerate, Figure 3.20 shows all of these behaviours.

Mechanisms of small crack growth are different than those for long cracks. Fracture mechanics theory suggests the stress–strain fields surrounding a crack are

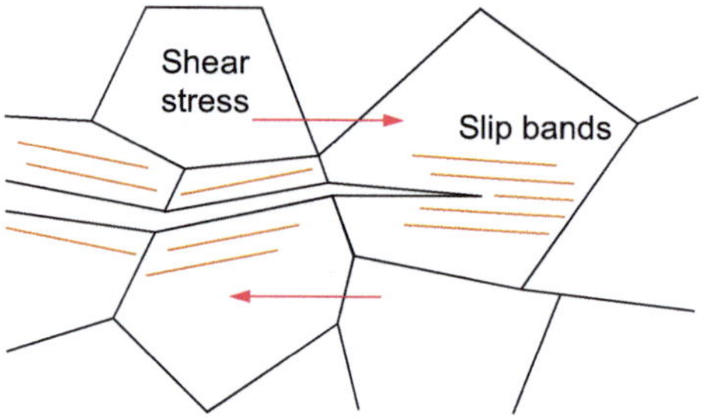

FIGURE 3.21 Single slip mechanism for a shear crack [11].

determined only by the stress intensity factor. Note that for small cracks in a notch, the stress–strain fields are related to both the notch and crack stress–strain fields. As a polycrystalline material cannot be regarded as an ideal homogeneous solid on this scale, continuum mechanics assumptions no longer serve microstructural cracks. Additionally, closure conditions are different for small cracks because of the small crack wake.

3.3.3 Shear Mechanisms – Mode II Growth

In contrast to the duplex shear growth mechanism for Mode I cracks, shown in Figure 3.16, Mode II crack growth is characterized by single shear. This is illustrated in Figure 3.21 [11], which shows a crack extension parallel to the direction of shear stress. Pure Mode II loading results in significant rubbing of the crack faces because the faces are always in contact. This leads to slower crack growth, crack arrest, or crack branching as cracks tend to grow in a more favourable Mode I direction.

Mode II loading produces closure at the crack surfaces, similar to the roughness-induced closure previously discussed. These closure forces will reduce the driving force at the crack tip and lead to lower crack growth rates. The closure forces are high enough to leave wear marks on the fracture surface, similar to a pin-on-disc wear test.

The fracture surface is relatively featureless, and extensive rubbing is apparent on the Mode II fracture surface. Compare this photo with the one in Figure 3.17 for Mode I loading. A superimposed Mode I crack opening load can greatly accelerate Mode II crack propagation by reducing the degree of crack face interference and rubbing. Similarly, a compressive Mode I load will retard crack growth by increasing the friction forces.

3.4 MULTIAXIAL FATIGUE CRITERIA

Multiaxial fatigue criteria can be divided into classical and critical plane-based approaches. Established in the late 19th to mid-20th century, classical strength theories' hypotheses generally don't reflect the damage mechanism. According to the damage parameter, they can be further classified into stress-, strain- and energy-based criteria. Among them, using von Mises as the energy distortion criterion is the commonest one. However, under diverse ratios, it can neither correlate in-phase and 90° out-of-phase experimental data nor axial–torsional fatigue data. Therefore, these models show considerable error when applying them to non-proportional biaxial loaded cases. Besides, they are insensitive to deformation memory effects.

To overcome the shortcomings of classical fatigue criteria under non-proportional multiaxial loadings, over the last few decades, critical plane-based approaches, which pay attention to the specific plane(s) with the maximum damage, were developed and continuously perfected. CPAs are applicable for both proportional and non-proportional loaded scenarios, which consider localized crack initiation phenomenon and can predict the angle of crack initiation within the material.

In addition to the classification mentioned above, some researchers categorized multiaxial fatigue criteria according to fatigue regime [12], e.g. low- and high-cycle fatigue, which helps choose an appropriate method in engineering practice.

3.4.1 Stress-Based Criteria

The stress parameter model based on the critical surface applies to the high-cycle fatigue (HCF) area. The stress amplitude on the critical plane mainly determines the damage parameters. Findley [13] determines the critical surface for the first time by the maximum value of the linear combination of the maximum normal stress and the shear stress amplitude in all material planes, as shown below:

$$\max\left\{\tau_{s,a} + k\sigma_{n,\max}\right\} = f = a\left(N_f\right)^b \tag{3.9}$$

where k and f were determined by the axial tensile and torsional fatigue tests, and this model was used to predict the multiaxial fatigue life under shear failure mode and gave excellent results [14].

McDiarmid [15] took the maximum shear stress as an important parameter to determine the critical plane analysis and developed a new fatigue model:

$$\frac{\left(\tau_{s,a}\right)_{\max}}{2\tau_{\text{af}}} + \frac{\sigma_{n,\max}}{2\sigma_{\text{af}}} = f \tag{3.10}$$

where τ_{af} is the shear fatigue strength, σ_{af} is the ultimate tensile strength, and Matake [16] presented a similar model based on the critical plane of the maximum shear stress.

For multiaxial random loads, Macha et al. [17,18] used the maximum variance of the shear stress to determine the critical plane and proposed a general model:

$$\max_t\left\{A\tau_s(t) + B\sigma_n(t)\right\} = f \tag{3.11}$$

where A, B, and f are material constants, depending on the selection of different failure criteria.

The Carpinteri–Spagnoli–Vantadori (CSV) criterion [19] defines the included angle α between the averaged maximum principal stress directions and the critical plane orientation as:

$$\alpha = 45^\circ \frac{3}{2}\left[1 - \left(\frac{\tau_{\mathrm{af}}}{\sigma_{\mathrm{af}}}\right)^2\right] \tag{3.12}$$

Carpinteri et al. [20,21] used different weight functions to derive the direction of the averaged maximum principal stress. Later, by combining with the equivalent stress $\sigma_{\mathrm{eq},a}$, the scope of application of the CSV criterion [22] was extended to random loading scenarios. Particularly, the relationship between fatigue damage criterion and material parameters was proposed:

$$\frac{\Delta\sigma_{\mathrm{eq}}}{2} = \sqrt{\sigma_{n,\max}^2 + \left(\frac{\sigma_{\mathrm{af}}}{\tau_{\mathrm{af}}}\right)^2 \tau_{s,a}^2} = f \tag{3.13}$$

3.4.2 Strain-Based Criteria

For low-cycle fatigue failure, the strain-based models work better than the stress-based models, which mainly use strain parameters as the main damage parameters.

Kandil, Brown, and Miller [23] took the maximum shear strain plane as the critical plane and obtained the Kandil–Brown–Miller (KBM) model by combining normal strain and shear strain with the strain life equation. Wang and Brown [24] furtherly considered the influence of average stress, resulting in the more widely used Wang–Brown (WB) model:

$$(\gamma_s)_{\max} + k\Delta\varepsilon_n = A\frac{\sigma'_f - 2\sigma_{n,\mathrm{mean}}}{E}(2N_f)^b + B\varepsilon'_f(2N_f)^c \tag{3.14}$$

where A, B, and k are material constants. Society [25] also observed that the average normal stress presents a significant impact on crack initiation and propagation, and proposed an improved model:

$$(\gamma_s)_{\max} + \varepsilon_{n,a} + \frac{\sigma_{n,\mathrm{mean}}}{E} = f \tag{3.15}$$

To consider the influence of normal stress components on material crack propagation, Fatemi and Socie [26] combined the shear form strain life equation and further modified the above equation to obtain the classic Fatemi–Socie (FS) model:

$$\gamma_s\left(1 + k\frac{\sigma_{n,\max}}{\sigma_y}\right) = \frac{\tau'_f}{G}(2N_f)^{b_0} + \gamma'_f(2N_f)^{c_0} \tag{3.16}$$

In particular, the normal stress sensitivity parameter k is usually fitted from additional experimental data, which introduces inconvenience for practice especially under limited testing data conditions. In this regard, Xu et al. [27] put forward a simple critical plane-based damage parameter with no additional material constants, which provides a robust method for multiaxial fatigue analysis of turbine disk alloys.

Macha et al. [28] developed a generalized fatigue criterion for multiaxial randomly loaded scenarios, in terms of the maximum shear strain $\gamma_{ns}(t)$ and the normal strain $\varepsilon_n(t)$ in the critical plane for multiaxial random loading, as follows:

$$\max_t\left\{A\gamma_{ns}(t)+B\varepsilon_n(t)\right\}=f \tag{3.17}$$

where A, B , and f are constants relevant to the specific model. The critical plane direction is taken as the mean direction of the maximum shear strain. Particularly, the maximum shear strain rotates and has variable amplitude.

Shang et al. [29] established a novel multiaxial fatigue model based on the maximum shear strain $\gamma_{ns}(t)$ plane:

$$\frac{\Delta\varepsilon_{\text{eq}}}{2}=\sqrt{\left(\Delta\varepsilon_n^*\right)^2+\frac{1}{3}\left(\frac{\Delta\gamma_{\max}}{2}\right)^2}=f \tag{3.18}$$

3.4.3 Energy-Based Criteria

The energy parameter model based on the critical surface considers the strain energy physical quantity constituted by the elastic–plastic stress/strain hysteresis loop. This energy parameter will cause irreversible damage on the microscopic scale of the structure until the microcrack initiates and then expands to the critical crack size, which is considered to explain the main causes of fatigue failure; these models have, therefore, been adequately developed in past studies. Compared with stress- and strain-based criteria, energy-based criteria have shown superior ability to unify microscopic and macroscopic experimental evidence and formulate multiaxial life prediction models [30].

Smith, Watson and Topper [31] took the plane of the maximum normal strain amplitude plane as the damage critical plane, and the strain energy obtained by the product of normal stress and strain on the critical plane was used for life prediction based on the tensile failure mode model (the Smith–Watson–Topper (SWT) model):

$$\sigma_{n,\max}\frac{\Delta\varepsilon_n}{2}=\frac{\left(\sigma_f'\right)^2}{E}\left(2N_f\right)^{2b}+\sigma_f'\varepsilon_f'\left(2N_f\right)^{b+c} \tag{3.19}$$

This model can provide better fatigue life prediction under uniaxial tension and compression loading, but the fatigue prediction effect for the shear failure mode is not good. Based on the SWT model to consider the defect of insufficient failure mode, Jiang [32] introduced the shear strain energy density and defined the maximum damage parameter plane as the critical plane, and the expression of its modification (the modified Smith–Watson–Topper (MSWT) model) is as follows:

$$\left(2a\Delta\varepsilon_n\sigma_{n,\max}+\frac{1-a}{2}\Delta\tau_s\Delta\gamma_s\right)_{\max}=4a\frac{\left(\sigma_f'\right)^2}{E}\left(2N_f\right)^{2b}+4a\sigma_f'\varepsilon_f'\left(2N_f\right)^{b+c} \tag{3.20}$$

where a is an additional material constant whose value depends on different material failure modes.

Liu [33] formulated an energy critical plane model based on the virtual strain energy, which is linked to diverse fatigue cracking modes. For Mode I fracture, ΔW_{I} is the sum of the maximum normal strain energy density $(\Delta\sigma_n\Delta\varepsilon_n)_{\max}$ and the shear strain energy density $(\Delta\tau\Delta\gamma)$ on the plane of $(\Delta\sigma_n\Delta\varepsilon_n)_{\max}$:

$$\Delta W_{\mathrm{I}} = (\Delta\sigma_n\Delta\varepsilon_n)_{\max} + (\Delta\tau\Delta\gamma) = f \tag{3.21}$$

For Mode II fracture, ΔW_{II} is computed in the plane of $(\Delta\tau\Delta\gamma)_{\max}$:

$$\Delta W_{\mathrm{II}} = (\Delta\tau\Delta\gamma)_{\max} + (\Delta\sigma_n\Delta\varepsilon_n) = f \tag{3.22}$$

Glinka et al. [34] put forward an energy criterion combining the elastic and plastic energy densities in the plane of the maximum shear strain range:

$$\Delta W_a = \frac{(\Delta\gamma_s)_{\max}}{2}\frac{\Delta\tau_s}{2} + \frac{\Delta\sigma_n}{2}\frac{\Delta\varepsilon_n}{2} = f \tag{3.23}$$

Then, Glinka et al. [35] improved it by including the mean stress terms:

$$W_a = \frac{(\Delta\gamma_s)_{\max}}{2}\frac{\Delta\tau_s}{2}\left[\frac{1}{1-\tau_{s,\max}/\tau'_f} + \frac{1}{1-\sigma_{n,\max}/\sigma'_f}\right] = f \tag{3.24}$$

Karolczuk and Macha [36], and Walat et al. [37] raised a model for multiaxial random loaded scenarios:

$$W_a = \frac{(\Delta\gamma_s)_{\max}}{2}\frac{\Delta\tau_s}{2}\left[\frac{1}{1-\tau_{s,\max}/\tau'_f} + \frac{1}{1-\sigma_{n,\max}/\sigma'_f}\right] = f \tag{3.25}$$

where A, B, and f are constants depending on the specific model.

Considering the tensile behaviour, Xu et al. [38] employed the maximum principal stress to amend the virtual shear strain energy in the plane of the maximum shear strain range:

$$\Delta W = \Delta\tau(\Delta\gamma)_{\max}\left(1 + \frac{\sigma_{n,\max}}{\sigma'_f}\right) = f \tag{3.26}$$

Under purely torsional loading, the maximum normal stress $\sigma_{n,\max}$ is zero, which is more reasonable.

Yu et al. [39] considered that the contribution of shear behaviour on the maximum shear strain plane to failure differs from the normal behaviour, and used multitudes of correction methods to characterize materials failure:

$$\Delta W = \frac{\tau_{s,a}}{\tau'_f}\frac{(\Delta\gamma_s)_{\max}}{2} + \frac{2\sigma_{n,\max}}{\sigma_y + \sigma'_f}\frac{\Delta\varepsilon_n}{2} = f \tag{3.27}$$

TABLE 3.1
A Collection of SED-Based Equations with Multiaxial Constraint Factor [40]

Year	Parameter	Equation	Study
1981	ξ	$N_f = A\left[\int_{\text{cycle}} \left(\sigma_x d\varepsilon_x^p + \xi \cdot \tau_{xy} d\gamma_{xy}^p\right)\right]^\alpha$	[41,42]
1996	$\bar{\rho}$	$\frac{\Delta W^p}{\bar{\rho}} + \Delta W^{e+} = \kappa_u (2N_f)^\alpha + C_u$	[43]
2019	f_{NP}	$N_f = \frac{\left[1+(K-1)\left(\frac{2}{\pi}\arctan\lambda^b\right)\right]E_{C,A}}{\left(1+mf_{\text{NP}}^n\right)d_{1,\text{cycle}}}$	[44]
2020	λ_{NP}	$\Delta W_{\text{DP}} = \Delta\sigma\Delta\varepsilon^p\left(\frac{1-h_c}{1+h_c}\right) + \Delta\tau\Delta\gamma^p\left(\frac{1-h_c'}{1+h_c'}\right) = D(\lambda_{\text{NP}})(N_f)^{m(\lambda_{\text{NP}})}$	[45]

Notice *All equations are presented in their original forms, and all parameters have nothing to do with those in the main text. Readers of interest may find what they want to know by reading the original publications.*

Among the strain energy density-based fatigue models, to improve their predicting accuracy under various loading conditions, it is common to introduce multiaxial constraint factors (or non-proportional factors), see Table 3.1 [40] Nevertheless, despite their complex form, to date, none of the life models built by integrating strain– and stress–life expressions is clear and widely accepted.

Recently, referring to axial and shear strain range–fatigue life relationships in ASTM [2], Liao et al. [40] developed a normative form of total strain energy density-based fatigue life equations and a standardized procedure to derive the relevant material fatigue parameters.

$$\Delta W_a^t = \Delta W_a^{e+} + \Delta W_a^p = E'_{e+,a}(2N_f)^{b_{e+,a}} + E'_{p,a}(2N_f)^{c_{p,a}} \tag{3.28}$$

$$\Delta W_s^t = \Delta W_s^{e+} + \Delta W_s^p = E'_{e+,s}(2N_f)^{b_{e+,s}} + E'_{p,s}(2N_f)^{c_{p,s}} \tag{3.29}$$

$$\Delta W^t = \Delta W_a^t + \Delta W_s^t \tag{3.30}$$

where ΔW^t is the total strain energy density, ΔW_a^t is the total axial strain energy density, ΔW_s^t is the total shear strain energy density, ΔW_a^{e+} is the positive axial elastic strain energy density, ΔW_a^p is the axial plastic strain energy density, ΔW_s^{e+} is the positive shear elastic strain energy density, ΔW_s^p is the shear plastic strain energy density, $E'_{e+,a}$ is the energy-based axial fatigue strength coefficient, $b_{e+,a}$ is the energy-based axial fatigue strength exponent, $E'_{p,a}$ is the energy-based axial fatigue ductility coefficient, $c_{p,a}$ is the energy-based axial fatigue ductility exponent, $E'_{e+,s}$ is the energy-based shear fatigue strength coefficient, $b_{e+,s}$ is the energy-based shear fatigue strength exponent, $E'_{p,s}$ is the energy-based shear fatigue ductility coefficient, and $c_{p,s}$ is the energy-based shear fatigue ductility exponent.

With axial fatigue data of specimens made from a virgin Al 7075-T6511 alloy and an in-service puddle iron, Eq. (3.28) was found to correlate experimental results better than the axial strain range–fatigue life relationship. It can be taken as a prototype to establish more effective energy-based multiaxial fatigue models.

3.4.4 Critical Plane-Based Criteria

CPAs are generally based on the maximum principal or shear planes. Like classical multiaxial fatigue criteria, they can also be classified into three types based on damage parameters. The first type is stress-based criteria, suitable for the HCF regime with negligible plastic deformation (Table 3.2 [18]). Findley and McDiarmid models are two famous stress-based CPAs for materials subjected to shear failure mode. The second type is the strain-based criteria, such as the KBM model for materials that fail on the maximum shear plane, see Table 3.3 [18]. These models are suitable when plastic deformation is present; nevertheless, they cannot consider

TABLE 3.2
Critical Plane Criteria Based on Stresses [18]

No.	Criterion	Critical Plane Orientation Based on Stress Crack Mode			Successfully Applied under Loading	Mean Stress
		I	II/III	Mixed		
1	Maximum normal stress (Rankine)	+	–	–	Cyclic, proportional	–
2	Maximum shear stress (Tresca)	–	+	–	Cyclic, proportional	–
3	Octahedral shear stress (HMH)	–	–	+	Cyclic, proportional	–
4	Findley (1956)	–	+	+	Cyclic, proportional	+
5	McDiarmid (1972)	–	+	–	Cyclic, proportional	–
	McDiarmid (1990)	–	+	–	Cyclic, non-proportional	+
6	Dietman and Issler (1974)	–	–	+	Cyclic, non-proportional	+
7	Simbürger and Crubisic (1976)	+	+	+	Cyclic, non-proportional	+
8	Matake (1977)	–	+	–	Cyclic, proportional	–
9	Macha (1979)	+	+	+	Random, non-proportional	-
10	Dang Van (1982)	–	+	–	Cyclic, non-proportional	+
11	Robert (1992)	–	+	+	Random, proportional	+
12	Papadopoulos (1993)	–	+	–	Cyclic, non-proportional	–
13	Zolochevski et al. (2000)	+	+	+	Cyclic, uniaxial	+
14	Carpinteri et al. (2001)	+	+	+	Cyclic, non-proportional	+
	Carpinteri et al. (2003)	+	+	+	Random, proportional	+
15	Küppers and Sonsino (2003)	–	–	+	Cyclic, non-proportional	–

TABLE 3.3
Critical Plane Criteria Based on Strains [18]

No.	Criterion	Critical Plane Orientation Based on Crack Mode: I	II/III	Mixed	Successfully Applied under Loading	Mean Stress
1	Maximum normal strain (de Saint-Venant)	+	–	–	Cyclic, proportional	–
2	Maximum shear strain (Coulomb-Tresca-Guest)	–	+	–	Cyclic, proportional	–
3	Octahedral shear strain (HMH)	–	–	+	Cyclic, proportional	–
4	Brown et al. (1973)	–	+	–	Cyclic, proportional	–
	Brown et al. (1982)	–	+	–	Cyclic, proportional	–
	Brown et al. (1993)	–	+	–	Random, non-proportional	–
5	Lohr and Ellison (1980)	–	+	–	Cyclic, proportional	–
6	Socie et al. (1985)	–	+	–	Cyclic, non-proportional	+
	Socie et al. (1988)	–	+	–	Cyclic, non-proportional	–
7	Macha (1988)	+	+	+	Random, non-proportional	–
8	Shang et al. (2007)	+	+	+	Cyclic, non-proportional	+
9	Wu, Hu and Song (2014)	+	+	+	Random, non-proportional	+
10	Yu et al. (2017)	+	+	+	Cyclic, non-proportional	+

the material's constitutive behaviour like non-proportional additional hardening and deformation memory effects.

Models including both stress and strain terms are the third group, such as the SWT and the FS models proposed respectively for materials with tensile and shear failure modes. These models can reflect the mean stress effect and non-proportional additional hardening. It was reported that the FS model provided excellent predictions for diverse materials and loading paths over the past 20 years. Additionally, energy-based critical plane models put forward by Liu, Chu et al., and Glinka et al., which incorporate stress–strain terms, are also applicable for such loading types, see Table 3.4 [18].

In recent decades, given the exclusive advantage of critical plane-based approaches for predicting both fatigue life and cracking direction, various fatigue criteria of different damage parameters were put forward. Karolczuk and Macha conducted a comprehensive review of the orientations of the critical plane in multiaxial fatigue of metallic materials, in which some representative models are tabulated in three tables, respectively, according to the type of damage parameter [18]. Some recently proposed models have also been included.

When applying a CPA, the most critical point is to locate the critical plane. The body subjected to complicated loadings results in 3D time-varying stress–strain states at the internal reference point O, which is the benchmark to establish a local coordinate system, O_{xyz}. The stress–strain states at that point can be fully characterized by the following stress–strain tensor:

TABLE 3.4
Critical Plane Criteria Based on Strain Energy Densities [18]

No.	Criterion	Critical Plane Orientation Based on Stress Crack Mode: I	II/III	Mixed	Successfully Applied under Loading	Mean Stress
1	Smith et al. (SWT, 1970)	+	–	–	Cyclic, uniaxial	+
	Socie (1987)	+	–	–	Cyclic, non-proportional	+
2	Nitta et al. (1988)	+	+	–	Cyclic, non-proportional	–
3	Liu (1993)	+	+	–	Cyclic, non-proportional	–
4	Glinka et al. (1995)	–	+	–	Cyclic, proportional	–
	Glinka et al. (1995)	–	+	–	Cyclic, non-proportional	+
	Pan et al. (1999)	–	+	–	Cyclic, non-proportional	–
5	Lagoda and Macha (1998)	+	+	+	Random, non-proportional	–
6	Rolovic and Tipton (1999)	–	–	+	Cyclic, non-proportional	+
7	Chen et al. (1999)	+	+	+	Cyclic, non-proportional	–
8	Hoffmeyer et al. (2001)	+	+	–	Cyclic, non-proportional	–
9	Varvani–Farahani (2002)	–	+	–	Cyclic, non-proportional	+
10	Lee et al. (2003)	–	–	+	Cyclic, non-proportional	+
11	Zhu et al. (2018)	+	+	+	Cyclic, non-proportional	+

$$[\sigma(t)] = \begin{bmatrix} \sigma_x(t) & \tau_{xy}(t) & \tau_{xz}(t) \\ \tau_{xy}(t) & \sigma_y(t) & \tau_{yz}(t) \\ \tau_{xz}(t) & \tau_{yz}(t) & \sigma_z(t) \end{bmatrix} \tag{3.31}$$

$$[\varepsilon(t)] = \begin{bmatrix} \varepsilon_x(t) & \frac{\gamma_{xy}(t)}{2} & \frac{\gamma_{xz}(t)}{2} \\ \frac{\gamma_{xy}(t)}{2} & \varepsilon_y(t) & \frac{\gamma_{yz}(t)}{2} \\ \frac{\gamma_{xz}(t)}{2} & \frac{\gamma_{yz}(t)}{2} & \varepsilon_z(t) \end{bmatrix} \tag{3.32}$$

where $t \in T$, T represents the time period of a loading cycle; $\sigma_x(t), \sigma_y(t), \sigma_z(t), \varepsilon_x(t)$, $\varepsilon_y(t)$ and $\varepsilon_z(t)$ are the normal stress–strain components, whereas $\tau_{xy}(t)$, $\tau_{xz}(t)$, $\tau_{yz}(t)$, $\gamma_{xy}(t)$, $\gamma_{xz}(t)$ and $\gamma_{yz}(t)$ are the shear ones.

A generic material plane Δ represented by its unit normal vector n presented in Figure 3.22 can be defined by the angles φ and θ. Referring to the current schematization in Figure 3.22, φ is the angle between the projection of the unit vector n on the $x-y$ plane and the x-axis, and θ is the angle between the unit vector n and the z-axis. For the second reference system, O_{anb}, it can be defined by three unit vectors determined by the angles θ and φ as well:

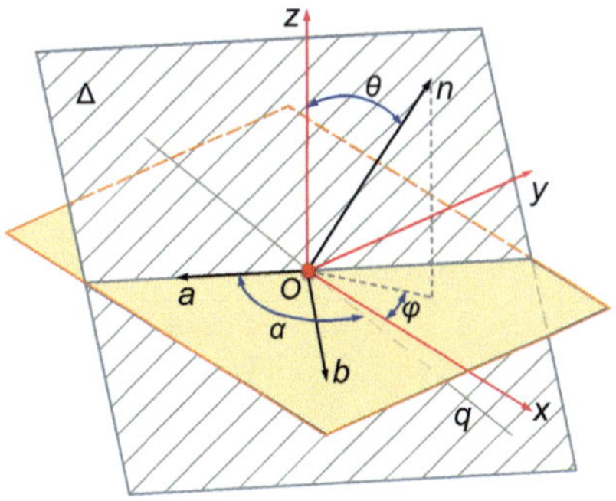

FIGURE 3.22 Determination of critical plane under multiaxial stress–strain state.

$$n = \begin{bmatrix} n_x \\ n_y \\ n_z \end{bmatrix} = \begin{bmatrix} \sin\theta\cos\varphi \\ \cos\theta\sin\varphi \\ \cos\theta \end{bmatrix} \tag{3.33}$$

$$a = \begin{bmatrix} a_x \\ a_y \\ a_z \end{bmatrix} = \begin{bmatrix} \sin\varphi \\ -\cos\varphi \\ 0 \end{bmatrix} \tag{3.34}$$

$$b = \begin{bmatrix} b_x \\ b_y \\ b_z \end{bmatrix} = \begin{bmatrix} \cos\theta\cos\varphi \\ \cos\theta\sin\varphi \\ -\sin\theta \end{bmatrix} \tag{3.35}$$

The unit vector q denotes a generic direction on the Δ plane passing through the point O is:

$$q = \begin{bmatrix} q_x \\ q_y \\ q_z \end{bmatrix} = \begin{bmatrix} \cos\alpha\sin\varphi + \sin\alpha\cos\theta\cos\varphi \\ -\cos\alpha\cos\varphi + \sin\alpha\cos\theta\sin\varphi \\ -\sin\alpha\sin\theta \end{bmatrix} \tag{3.36}$$

where α is the angle between the q and the a-axis. The instantaneous values of the normal stress and strain of the Δ plane are:

$$\sigma_n(t) = n^T\left[\sigma(t)\right]n \tag{3.37}$$

$$\varepsilon_n(t) = n^T\left[\varepsilon(t)\right]n \tag{3.38}$$

The shear stress $\tau_q(t)$ and shear strain $\gamma_q(t)$ along the direction q are:

$$\tau_q(t) = q^T\left[\sigma(t)\right]n \tag{3.39}$$

$$\frac{\gamma_q(t)}{2} = q^T\left[\varepsilon(t)\right]n \tag{3.40}$$

Based on the abovementioned equations, after extracting the studied object's stabilized stress–strain response, the critical plane can be obtained with regard to its definition by periodically scanning with an appropriate angular interval. After that, the damage parameter can be determined, and the fatigue life can be predicted. Additionally, given the low efficiency of periodical scanning, some researchers put forward some optimized search method [46,47].

Finally, let's talk about the application of the CPAs. After decades of development, countless criteria are now available. For the scholars or engineers dedicated to fatigue study, they can select the suitable ones from them according to corresponding material types and failure modes. Nevertheless, for designers who have not received specialized training in this field, it can be hard. In this regard, some review works were conducted [48–50]. Zhu et al. [49] evaluated and compared the performance of some commonly-used critical plane criteria in multiaxial fatigue analysis of ductile and brittle materials, and proposed a modified generalized strain energy criterion that yields better predictions for both ductile and brittle materials. Besides, critical plane-based criteria have been found of highly flexibility, which can be coupled with other models for fatigue strength evaluation of engineering structures under complex working conditions [51]. Given that practical components are of complex geometrical configurations, stress concentration is unavoidable. Notch effect will significantly reduce the fatigue life/strength, which was not considered by models developed from multiaxial fatigue data of smooth material-level specimens. Recently, Liao et al. [52] studied the possibility of combining the CPA and the theory of critical distance (TCD) for multiaxial notch fatigue life assessment. Taking the FS model for instance, different coupling sequences of the CPA and the TCD were discussed, see Figure 3.23 [52]. Finally, it was found the coupling sequence that applies the TCD on the determined critical plane yields better prediction.

3.5 APPLICATIONS – CASE STUDIES OF AEROENGINE COMPONENTS

This subsection presents two case studies of aeroengine components, including a high-pressure turbine disc–blade attachment and a compressor turbine assembly. Specifically, two multiaxial fatigue models developed by the authors were used.

3.5.1 A High-Pressure Turbine Disc–Blade Attachment

High-pressure turbine services under harsh cyclic loadings, which is vulnerable to fatigue damage and generally determines the service life of the aeroengine. In this case, a high-pressure turbine disc–blade attachment is studied [53]. The disc is made by GH4169, whose material properties are shown in Table 3.5. To obtain the stress–strain distribution of the turbine blade, finite element analysis (FEA) on the high-pressure turbine disc–blade attachment was performed under actual service loadings, and the mesh is shown in Figure 3.24.

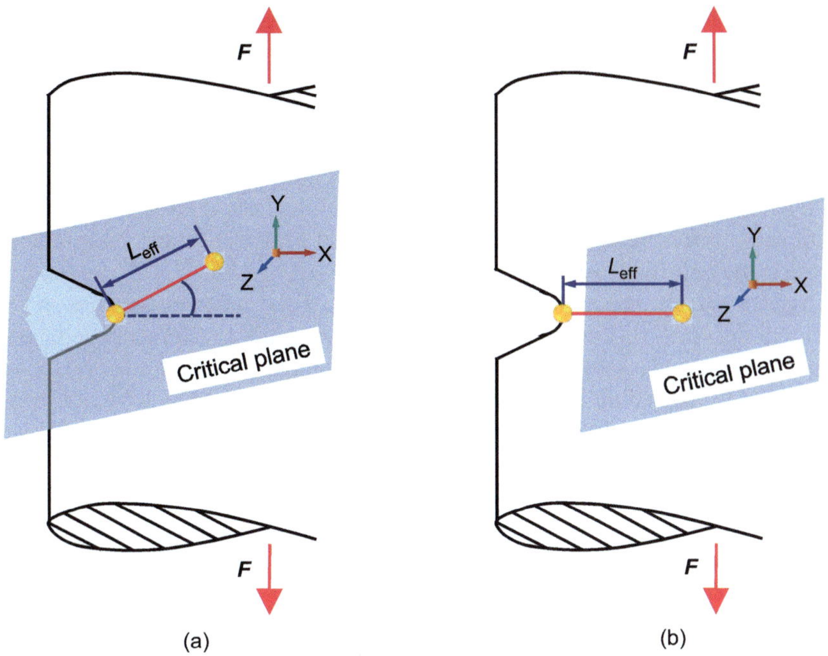

FIGURE 3.23 Schematic diagram of two strategies for combined critical plane–critical distance approaches analysis. (With the permission from Ref. [52], Elsevier.) (a) Sequence 1 (CPA → TCD); (b) Sequence 2 (TCD → CPA).

TABLE 3.5
Material Properties of GH4169 Nickel-Base Alloy

T (°C)	E (GPa)	σ_y (MPa)	σ_f' (MPa)	ε_f'	b	c	K' (MPa)	n'
650	182	626.4	1476	0.162	−0.086	−0.58	1933	0.1483

Twenty-node higher-order hexahedral elements (SOLID186 in ANSYS®) were used for meshing the turbine disc and fir-tree root of the blade, while ten-node tetrahedral elements (TETRA10 in ANSYS®) were adopted for the blade body. Surface-to-surface contact elements were used at the contact region. In particular, at the fir-tree root, a mesh convergence study was performed to get reliable and precise stress–strain response for fatigue strength assessment. The load spectrums were applied in accordance with the actual service conditions of the turbine blade, which includes three typical loading cycles: 0–maximum–0 (0–450 r/s–0), idle–maximum continue–idle (230 r/s–450 r/s–230 r/s), cruise–maximum continue–cruise (431 r/s–450 r/s–431 r/s).

The FEA result of the turbine blade at the 0–maximum–0 load spectrum is given in Figure 3.23. It is visible that the fir-tree region is the critical region, in accord with the conclusion of Sinclair et al. [54]. Plastic deformation happens at the first tooth of the fir-tree tenon structure owing to the compression stress attributed to the geometrical volume expansion under centrifugal loadings from the blade.

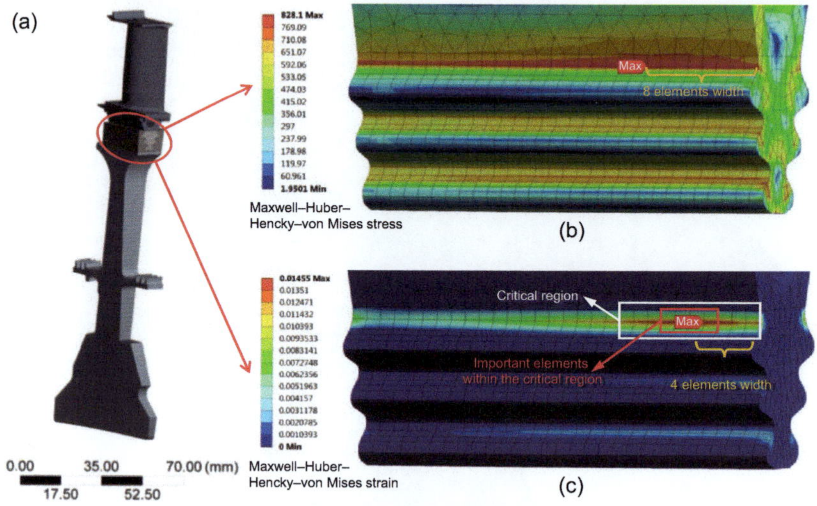

FIGURE 3.24 FEA of the turbine disc–blade attachment. (a) Mesh; (b) von Mises equivalent stress field; (c) equivalent plastic strain field of turbine blade under the 0–maximum speed–0 cycle.

TABLE 3.6
Comparison of Model Predicted Blade Life

	Evaluated Life (Cycles)		
Model	**0–450 r/s–0**	**230r/s–450 r/s–230 r/s**	**431 r/s–450 r/s–431 r/s**
WB	993,346	59,560,715	>1E+12
MWBDP	13,481,520	113,213,394	>1E+12
FS	360,185	63,404,561	>1E+12
MDPFS	253,996	53,007,514	>1E+12
SWT	71,540	676,766	>1E+12
ECP	85,932	936,842	>1E+12

Notably, the maximum von Mises stress and plastic strain appear at diverse locations. The location of the maximum von Mises stress is eight elements length away from the fir-tree root edge, while the other one is only four elements length away.

Moreover, the maximum shear strain and damage parameter don't appear at the location of the maximum von Mises stress but close to the maximum equivalent plastic strain, which to some extent meets the conclusion of Maktouf et al. [55]. Finally, four elements near the maximum equivalent plastic strain point are identified as critical elements, whose stabilized stress–strain responses were extracted. Finally, according to the definitions of different critical planes, the elements of most significant damage were utilized for fatigue life computation.

Table 3.6 provides the predicted fatigue lives of the turbine blade under diverse service conditions. Note that the predicted fatigue lives of the three models with

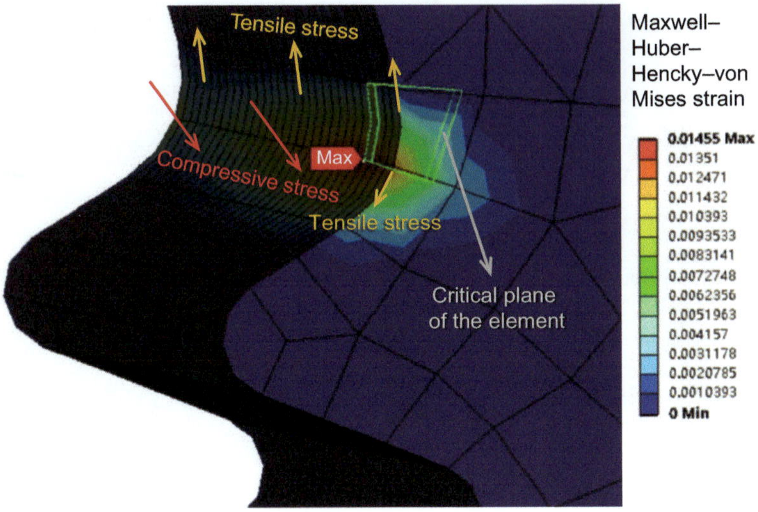

FIGURE 3.25 Stress analysis of the element with the critical plane.

diverse critical plane definitions differ from each other. Although they are valid under both uniaxial and multiaxial loaded scenarios, in engineering practice, especially for the components subjected to complicated failure modes, e.g. turbine blades, the stress–strain repose must be preliminarily checked to select an appropriate model. Note from Figure 3.25 that the critical element undergoes the tensile stress exerted by the upper elements. Besides, the lower elements suffer the compressive stress of the fir-tree tenon of the turbine disc, resulting in diagonal tensile stress. Tension stresses from diverse directions lead to shear stress. Therefore, at the critical region, it's a mixed failure mode involving tension and shear failure.

However, the shear strain originates from the tension stress in the asymmetrical direction. To be specific, it's a tensile-dominant mixed failure mode. The energy critical plane (ECP) damage parameter yields an excellent correlation of the fatigue life with the SWT model at the flight missions 0–maximum–0 and idle–maximum continue–idle, which confirms the rationality of the earlier analysis on the failure mode. However, given the empirical constants k (the FS model) and S (the WB model) are related to fatigue life, their predicted results are longer than the SWT and the ECP models. Note that the predicted fatigue life using the maximum WB damage parameter (MWBDP) is longer than that of the WB model, which indicates that for the WB model, it's unreasonable to define the critical plane according to the maximum damage parameter, as it ignores the mean stress effect [53]. In contrast, for the FS and the SWT models, under complex loading, it's possible to define the critical plane using the maximum damage parameter, see Table 3.6.

3.5.2 A Compressor Turbine Assembly

This subsection introduces a casting TC4 compressor turbine disc, which undergoes complicated multiaxial loading at the highest operating temperature of 100°C [56]. During elasto-plastic FEA, centrifugal loading and temperature field were applied in

(a)

(b)

(c)

Performance indicator	
Maximum diameter of the rotor	1200 mm
Rotation speed range	02500 rpm
Maximum temperature	800±10°C
Rotation speed spectrum	Trapezoidal, triangular, random waveforms

(d)

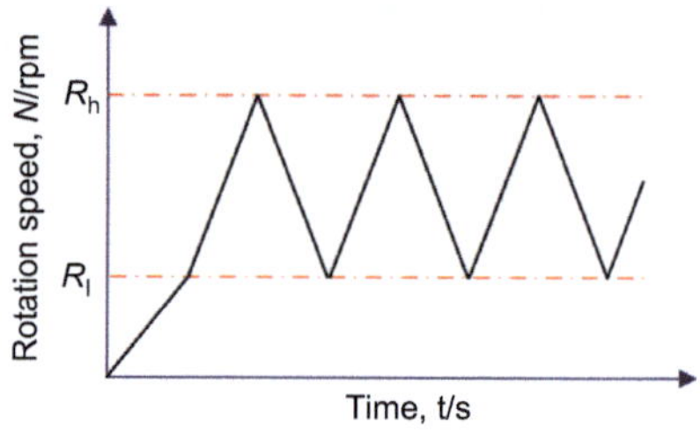

FIGURE 3.26 The vertical disc test facility. (a) Vacuum case; (b) power system; (c) performance indicator; (d) rotation speed spectrum.

accordance with the service condition. There are three typical load spectrums with regard to the actual flight missions of the aero engine [56]: 0–maximum continue–0 (0–450 r/s–0), idle–maximum continue–idle (230 r/s–450 r/s–230 r/s), cruise–maximum continue–cruise (431 r/s–450 r/s–431 r/s). Accordingly, full-scale low-cycle fatigue (LCF) tests at elevated temperatures were performed at the vertical disc test facility of the China Gas Turbine Establishment. To be more specific, the spinning facility includes a vacuum case, data acquisition & control systems as well as a main power system, see Figure 3.26a and b. Its performance indicators are tabulated in Figure 3.26c. Three sets of tests were conducted under the aforementioned rotation speeds given in Figure 3.26d. Particularly, the service life of the disc was regarded as the number of loading cycles when an engineering crack at the critical regions of the disc reaches 0.78 mm, which can be located at the assembly holes, the hub zone, and the dovetail-rim area.

Given the cyclic symmetry of the turbine disc, a 1/48 model was created. The critical region was identified by combining preliminary computations and filed data of the engine. In that region, locally refined hexahedral mesh was utilized, see Figure 3.27a.

Due to the mass of the disc and blades, together with the constraint of the bolt along the circumferential direction, stress concentration appears at the assembly hole edges. Maktouf et al. [55] stated that designers shouldn't focus only on the region with the maximum von Mises stress. Actually, fatigue damage is highly relevant to cyclic plastic deformation, particularly for strain- and energy-based models. It was noticed that the location of the maximum damage parameter of the strain-based model is not the same as that of the maximum von Mises stress but is near the maximum equivalent plastic strain. The equivalent plastic strain field near the critical region is presented in Figure 3.27b.

By taking the stabilized cyclic stress–strain responses of the elements within the critical region as the inputs, the minimum values of the computed fatigue lives of

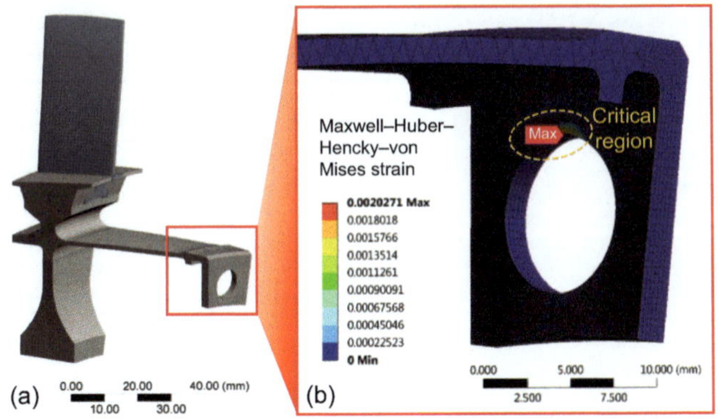

FIGURE 3.27 FEA of the compressor turbine assembly. (a) Mesh; (b) equivalent plastic strain distribution in the critical region under 0–maximum continue–0.

TABLE 3.7
Comparison of Model Predicted Life of the Compressor Turbine Disc

Model	Evaluated Life (Cycles) 0–450 r/s–0	230 r/s–450 r/s–230 r/s	431 r/s–450 r/s–431 r/s
Liu	166,755	3,486,815	>1E+12
Modified Ince–Glinka	251,427	4,916,421	>1E+12
Proposed	308,620	3,826,588	>1E+12

these elements by diverse criteria were taken as the predicted lives of the turbine disc, see Table 3.7. Note that two flight missions, including 0–maximum continue–0 and idle–maximum continue–idle, contribute mostly to the fatigue damage of the disc. Accordingly, the fatigue damage under these two load spectrums should be carefully considered in fatigue strength assessment. Compared with tested lives in the full-scale experimental campaign, the model provides a promising fatigue life correlation with the other two criteria [56], which gives a reference for structural fatigue assessment.

3.5.3 Prospects on Multiaxial Fatigue Study

Understanding multiaxial fatigue is crucial in numerous engineering applications where components are subjected to varying and often unpredictable loading conditions. Comprehensively assessing the fatigue behaviour under multiaxial loading is critical to enhancing critical structures' reliability, safety, and longevity [57].

Physics-based criteria can reflect fatigue damage mechanisms, including tensile versus shear failure modes. Generally, shear stress–strain cause cracks in ductile materials, while tensile ones typically initiate crack initiation in brittle materials. Furthermore, physics-based multiaxial fatigue criteria can also reveal the mechanism that tension loading opens cracks and intensifies damage. In contrast, compressive loading causes crack closure and lowers the driving force of fatigue damage.

Multiaxial fatigue criteria should consider plastic deformation in LCF and reflect the material's constitutive behaviour, including deformation history and non-proportional hardening effect. As a competent constitutive model, yield surface expansion (isotropic hardening) and translation (kinematic hardening) should be considered during material yielding. Another critical consideration is non-proportional additional hardening, which raises stress and shortens fatigue life. This character is essential for improving the accuracy of stress analysis and fatigue life prediction in non-proportional multiaxial loaded scenarios.

Multiaxial fatigue criteria should show that normal mean stress influences tension and torsion fatigue, while shear mean stress impacts neither tension nor torsion fatigue before yielding. In addition, multiaxial fatigue criteria consider the reality that tensile mean stress reduces fatigue life, whereas the compressive one is helpful.

Multiaxial fatigue criteria should be suitable for various stress–strain states. Under bending–torsion loading, principal stresses are in the 2nd or 4th quadrants of two-dimensional stress space. Nevertheless, principal stresses are in the 1st and 3rd quadrants for tension–tension loadings. Multiaxial fatigue criteria are not prepared for a particular loading scenario and should serve all stress quadrants and loading situations. Besides, they should be applicable under variable amplitude loadings.

Under multiaxial non-proportional loading, material type, loading path, and loading level have different effects on the additional hardening and life of low- and high-cycle fatigue [58]. The additional hardening mechanisms are different, and material ductility significantly affects the multiaxial HCF behaviour. Lately, Deng et al. [59] performed load path sensitivity analysis and multiaxial fatigue life prediction of metals under non-proportional loadings. A new definition of non-proportional degree on generalized plane for non-proportional loading was given and a novel fatigue model without additional material constants was put forward, whose prediction accuracy was verified by experimental data of three materials under different loading paths.

3.6 SUMMARY

Unlike uniaxial fatigue with a predominant stress component, multiaxial fatigue acknowledges the complex nature of real-world loading conditions experienced by engineering components. Multiaxial fatigue criteria should be robust, related to the physics of failure, sensitive to material deformation behaviour, and suitable for complex variable/stochastic amplitude-loaded scenarios. Among multiaxial fatigue life criteria, CPAs and strain energy density-based methods are widely used because they can provide fatigue damage processes with better explanations. Moreover, fatigue models whose damage parameters in the form of energy are suitable for both low- and high-cycle fatigue.

ABBREVIATIONS

ASTM American Society for Testing and Materials
CPA Critical plane approach
CSV Carpinteri–Spagnoli–Vantadori
ECP Energy critical plane

FEA	Finite element analysis
FS	Fatemi–Socie
HCF	High-cycle fatigue
KBM	Kandil–Brown–Miller
LCF	Low-cycle fatigue
MSWT	Modified SWT
MWBDP	Maximum WB damage parameter
SAE	Society of Automotive Engineers
SWT	Smith–Watson–Topper
TCD	Theory of critical distance
WB	Wang–Brown

REFERENCES

1. Gough HJ, Pollard HV. The strength of metals under combined alternating stresses. *Proc Inst Mech Eng* 1935, 131: 3–103. https://doi.org/10.1243/PIME_PROC_1935_131_008_02
2. American Society for Testing and Materials. ASTM E2207-15(2021) Standard Practice for Strain-Controlled Axial-Torsional Fatigue Testing with Thin-Walled Tubular Specimens. America, 2021. https://doi.org/10.1520/E2207-15R21
3. Wächter M, Linn A, Wuthenow R, Esderts A, Gaier C, Kraft J, Fällgren C, Vormwald M. On scaled normal stresses in multiaxial fatigue and their exemplary application to ductile cast iron. *Appl Mech* 2022, 3: 259–95. https://doi.org/10.3390/applmech3010018
4. Downing SD, Galliart DR. *A Fatigue Test System for a Notched Shaft in Combined Bending and Torsion.* ASTM International, West Conshohocken, PA, 1985. https://doi.org/10.1520/STP36215S
5. Shawki GSA. A review of fatigue testing machines. *Eng J Qatar Univ* 1990; 3: 55–69.
6. Zhang XS, Ma YE, Yang M, Zhou CT, Fu NK, Huang W, et al. A review of in-plane biaxial fatigue behavior of metallic materials. *Theor Appl Fract Mech* 2023, 123: 103726. https://doi.org/10.1016/j.tafmec.2022.103726.
7. State Administration for Market Regulation. Standardization Administration of the People's Republic of China. GB/Z 40387—2021 Metallic materials — Principles and designs for multiaxial fatigue testing. People's Republic of China, 2021.
8. Suresh S. *Fatigue of Materials.* 2nd ed. Cambridge University Press, Cambridge, MA, 1998. https://doi.org/10.1017/CBO9780511806575
9. Hertzberg RW, Vinci RP, Hertzberg JL. *Deformation and Fracture Mechanics of Engineering Materials.* 5th ed. Wiley, Hoboken, NJ, 2012.
10. Ewing JA, Humfrey JCW. VI. The fracture of metals under repeated alternations of stress. *Philos Trans R Soc London Ser A Contain Pap a Math or Phys Character* 1903, 200: 321–330. https://doi.org/10.1098/rsta.1903.0006
11. Socie DF, Marquis GB. *Multiaxial Fatigue.* Society of Automotive Engineers, Inc., Warrendale, PA, 2000.
12. Sun GQ, Shang DG, Wang Y. Research progress on fatigue behavior and life prediction under multiaxial loading for metals (in Chinese). *J Mech Eng* 2021, 57: 153–172. https://doi.org/10.3901/JME.2021.16.153
13. Findley WN. A theory for the effect of mean stress on fatigue of metals under combined torsion and axial load or bending. *J Eng Ind* 1959, 81: 301–305. https://doi.org/10.1115/1.4008327

14. Bäckström M, Marquis GB. A review of multiaxial fatigue of weldments: experimental results, design code and critical plane approaches. *Fatigue Fract Eng Mater Struct* 2001, 24: 279–291. https://doi.org/10.1046/j.1460-2695.2001.00284.x
15. McDiarmid DL. A general criterion for high cycle multiaxial fatigue failure. *Fatigue Fract Eng Mater Struct* 1991, 14: 429–453. https://doi.org/10.1111/j.1460-2695.1991.tb00673.x
16. Matake T, Imai Y. Fatigue strength of notched specimen under combined stress. *Bull JSME* 1980, 23: 623–629. https://doi.org/10.1299/jsme1958.23.623
17. Macha E. Generalization of fatigue fracture criteria for multiaxial sinusoidal loadings in the range of random loadings. In: Brown MW, Miller KJ, editors. *Biaxial and Multiaxial Fatigue*, London: Mechanical Engineering Publications Limited; 1989, pp. 425–436.
18. Karolczuk A, Macha E. A review of critical plane orientations in multiaxial fatigue failure criteria of metallic materials. *Int J Fract* 2005, 134: 267–304. https://doi.org/10.1007/s10704-005-1088-2
19. Carpinteri A, Spagnoli A, Vantadori S. Multiaxial fatigue assessment using a simplified critical plane-based criterion. *Int J Fatigue* 2011, 33: 969–976. https://doi.org/10.1016/j.ijfatigue.2011.01.004
20. Carpinteri A, Macha E, Brighenti R, Spagnoli A. Expected principal stress directions under multiaxial random loading. Part I: Theoretical aspects of the weight function method. *Int J Fatigue* 1999, 21: 89–96. https://doi.org/10.1016/S0142-1123(98)00046-2
21. Carpinteri A, Brighenti R, Macha E, Spagnoli A. Expected principal stress directions under multiaxial random loading. Part II: Numerical simulation and experimental assessment through the weight function method. *Int J Fatigue* 1999, 21: 89–96. https://doi.org/10.1016/S0142-1123(98)00047-4
22. Carpinteri A, Spagnoli A, Vantadori S. A multiaxial fatigue criterion for random loading. *Fatigue Fract Eng Mater Struct* 2003, 26: 515–522. https://doi.org/10.1046/j.1460-2695.2003.00620.x
23. Brown MW, Miller KJ. A theory for fatigue failure under multiaxial stress–strain conditions. *Proc Inst Mech Eng* 1973, 187: 745–755. https://doi.org/10.1243/PIME_PROC_1973_187_161_02
24. Wang CH, Brown MW. A path-independent parameter for fatigue under proportional and non-proportional loading. *Fatigue Fract Eng Mater Struct* 1993, 16:1285–1297. https://doi.org/10.1111/j.1460-2695.1993.tb00739.x
25. Socie DF, Waill LA, Dittmer DF. *Biaxial Fatigue of Inconel 718 Including Mean Stress Effects.* ASTM Internationa, West Conshohocken, PA, 1985. https://doi.org/10.1520/STP36238S
26. Fatemi A, Socie DF. A critical plane approach to multiaxial fatigue damage including out-phase loading. *Fatigue Fract Eng Mater Struct* 1988, 11: 149–165. https://doi.org/10.1111/j.1460-2695.1988.tb01169.x
27. Xu S, Zhu SP, Hao YZ, Liao D. Critical plane-based multiaxial fatigue life prediction of turbine disk alloys by refining normal stress sensitivity. *J Strain Anal Eng Des* 2018, 53: 719–729. https://doi.org/10.1177/0309324718779922
28. Macha E. Generalized fatigue criterion of maximum shear and normal strains on the fracture plane for materials under multiaxial random loadings. *Materwiss Werksttech* 1991, 22: 203–210. https://doi.org/10.1002/mawe.19910220605
29. Shang DG, Wang DJ. A new multiaxial fatigue damage model based on the critical plane approach. *Int J Fatigue* 1998, 20: 241–245. https://doi.org/10.1016/S0142-1123(97)00123-0
30. Liao D, Zhu SP. Energy field intensity approach for notch fatigue analysis. *Int J Fatigue* 2019, 127: 190–202. https://doi.org/10.1016/j.ijfatigue.2019.06.010

31. Calhoun CA, Wheeler R, Baxevanis T, Lagoudas DC. Actuation fatigue life prediction of shape memory alloys under the constant-stress loading condition. *Scr Mater* 2015, 95: 58–61. https://doi.org/10.1016/j.scriptamat.2014.10.005
32. Jiang YY. Fatigue criterion for general multiaxial loading. *Fatigue Fract Eng Mater Struct* 2000, 23: 19–32. https://doi.org/10.1046/j.1460-2695.2000.00247.x
33. Liu KC. A method based on virtual strain-energy parameters for multiaxial fatigue life prediction. In: McDowell DL, Eills R, editors. *STP 1191 Adv. Multiaxial Fatigue*, Philadelphia, PA: ASTM International; 1993, pp. 67–84. https://doi.org/10.1520/stp24796s
34. Glinka G, Shen G, Plumtree A. A multiaxial fatigue strain energy density parameter related to the critical fracture plane. *Fatigue Fract Eng Mater Struct* 1995, 18: 37–46. https://doi.org/10.1111/j.1460-2695.1995.tb00140.x
35. Glinka G, Wang G, Plumtree A. Mean stress effects in multiaxial fatigue. *Fatigue Fract Eng Mater Struct* 1995, 18: 755–764. https://doi.org/10.1111/j.1460-2695.1995.tb00901.x
36. Karolczuk A, Macha E. Selection of the critical plane orientation in two-parameter multiaxial fatigue failure criterion under combined bending and torsion. *Eng Fract Mech* 2008, 75: 389–403. https://doi.org/10.1016/j.engfracmech.2007.01.021
37. Walat K, Kurek M, Ogonowski P, Łagoda T. The multiaxial random fatigue criteria based on strain and energy damage parameters on the critical plane for the low-cycle range. *Int J Fatigue* 2012, 37: 100–111. https://doi.org/10.1016/j.ijfatigue.2011.09.013
38. Xu S, Zhu SP, Hao YZ, Liao D, Qian GA. A new critical plane-energy model for multiaxial fatigue life prediction of turbine disc alloys. *Eng Fail Anal* 2018, 93: 55–63. https://doi.org/10.1016/j.engfailanal.2018.07.001
39. Yu ZY, Zhu SP, Liu Q, Liu YH. Multiaxial fatigue damage parameter and life prediction without any additional material constants. *Materials (Basel)* 2017, 10: 923. https://doi.org/10.3390/ma10080923
40. Liao D, Zhu SP, Gao JW, Correia JAFDO, Calçada RAB, Lesiuk G. Generalized strain energy density-based fatigue indicator parameter. *Int J Mech Sci* 2023, 254: 108427. https://doi.org/10.1016/j.ijmecsci.2023.108427
41. Garud YS. A new approach to the evaluation of fatigue under multiaxial loadings. *J Eng Mater Technol* 1981, 103: 118–125. https://doi.org/10.1115/1.3224982
42. Nourian-Avval A, Khonsari MM. A new model for fatigue life prediction under multiaxial loadings based on energy dissipation. *Int J Fatigue* 2021, 151: 106255. https://doi.org/10.1016/j.ijfatigue.2021.106255
43. Ellyin F. *Fatigue Damage, Crack Growth and Life Prediction.* Chapman & Hall, London, 1996. https://doi.org/10.1007/978-94-009-1509-1
44. Feng ES, Wang XG, Jiang C. A new multiaxial fatigue model for life prediction based on energy dissipation evaluation. *Int J Fatigue* 2019, 122: 1–8. https://doi.org/10.1016/j.ijfatigue.2019.01.003
45. Gan L, Wu H, Zhong Z. Multiaxial fatigue life prediction based on a simplified energy-based model. *Int J Fatigue* 2021, 144: 106036. https://doi.org/10.1016/j.ijfatigue.2020.106036
46. Svärd H. A branch and bound algorithm for evaluation of the Findley fatigue criterion. *Int J Fatigue* 2015, 73: 27–38. https://doi.org/10.1016/j.ijfatigue.2014.11.008
47. Sunde SL, Berto F, Haugen B. Efficient implementation of critical plane for 3D stress histories using triangular elements. *Int J Fatigue* 2020, 134: 105448. https://doi.org/10.1016/j.ijfatigue.2019.105448
48. Zhu SP, Huang HZ, Ontiveros V, He LP, Modarres M. Probabilistic low cycle fatigue life prediction using an energy-based damage parameter and accounting for model uncertainty. *Int J Damage Mech* 2012, 21: 1128–1153. https://doi.org/10.1177/1056789511429836

49. Zhu SP, Yu ZY, Correia JAFDO, De Jesus AMP, Berto F. Evaluation and comparison of critical plane criteria for multiaxial fatigue analysis of ductile and brittle materials. *Int J Fatigue* 2018, 112: 279–288. https://doi.org/10.1016/j.ijfatigue.2018.03.028
50. Arora P, Gupta SK, Samal MK, Chattopadhyay J. Comparing fatigue life prediction capability of critical plane models using multiaxial test database on 17 materials. *Fatigue Fract Eng Mater Struct* 2023, 46: 1330–1356. https://doi.org/10.1111/ffe.13928
51. Leidermark D, Moverare J, Simonsson K, Sjöström S. A combined critical plane and critical distance approach for predicting fatigue crack initiation in notched single-crystal superalloy components. *Int J Fatigue* 2011, 33: 1351–1359. https://doi.org/10.1016/j.ijfatigue.2011.05.009
52. Liao D, Zhu SP, Qian GA. Multiaxial fatigue analysis of notched components using combined critical plane and critical distance approach. *Int J Mech Sci* 2019, 160: 38–50. https://doi.org/10.1016/j.ijmecsci.2019.06.027
53. Yu ZY, Zhu SP, Liu Q, Liu YH. A new energy-critical plane damage parameter for multiaxial fatigue life prediction of turbine blades. *Materials (Basel)* 2017, 10: 513. https://doi.org/10.3390/ma10050513
54. Sinclair GB, Comier NG, Griffin JH, Meda G. Contact stresses in dovetail attachments: finite element modeling. *Proceedings of the ASME 1999 International Gas Turbine and. Aeroengine Congress and Exposition*, Indianapolis: 1999, pp. 1–11. https://doi.org/10.1115/99-GT-387
55. Maktouf W, Ammar K, Ben Naceur I, Saï K. Multiaxial high-cycle fatigue criteria and life prediction: application to gas turbine blade. *Int J Fatigue* 2016, 92: 25–35. https://doi.org/10.1016/j.ijfatigue.2016.06.024
56. Zhu SP, Yu ZY, Liu Q, Ince A. Strain energy-based multiaxial fatigue life prediction under normal/shear stress interaction. *Int J Damage Mech* 2019, 28: 708–739. https://doi.org/10.1177/1056789518786031
57. Dantas RGR, Correia JAFDO, Lesiuk G, Rozumek D, Zhu SP, De Jesus AMP, Susmel L, Berto F. Evaluation of multiaxial high-cycle fatigue criteria under proportional loading for S355 steel. *Eng Fail Anal* 2021, 120: 105037. https://doi.org/10.1016/j.engfailanal.2020.105037
58. Xu L, Wang RZ, Wang J, He L, Itoh T, Miura H, Zhang XC, Tu ST. On multiaxial creep–fatigue considering the non-proportional loading effect: constitutive modeling, deformation mechanism, and life prediction. *Int J Plast* 2022, 155: 103337. https://doi.org/10.1016/j.ijplas.2022.103337
59. Deng QY, Zhu SP, Niu XP, Lesiuk G, Macek W, Wang QY. Load path sensitivity and multiaxial fatigue life prediction of metals under non-proportional loadings. *Int J Fatigue* 2023, 166: 107281. https://doi.org/10.1016/j.ijfatigue.2022.107281

4 Fatigue Crack Growth

Grzegorz Lesiuk, Mieczysław Szata, Kayode Olaleye, Dariusz Rozumek, Abílio M.P. De Jesus, and José A.F.O. Correia

4.1 INTRODUCTION

Metallic structure fatigue failures are a well-known technical issue. The first laboratory studies were conducted, and multiple significant fatigue failures were observed as early as the 1800s. August Wöhler conducted noteworthy studies on weariness and realized that a single application of load that was well below a structure's static strength did not do any harm to the structure. However, repeatedly applying the same stress could result in total failure. Because fatigue damage was invisible in the 19th century, fatigue was believed to be a mystery occurrence in the material. It appears that there was no prior notice before the failure happened. It has been discovered in the 20th century that repetitive load applications can initiate a fatigue process in the material, which may result in the nucleation of a small crack, further growth of the crack, and eventual catastrophic damage/fracture. There have been many fatigue failures of machinery, moving vehicles, welded structures, aircraft, etc., throughout the history of engineering structures/components. Occasionally, these malfunctions have resulted in disastrous incidents such as a pressure vessel explosion, a bridge collapse, or an additional major structure failure. Although many fatigue issues remained off the front pages of newspapers, non-catastrophic fatigue failures have had a huge economic impact. Structure fatigue is now widely acknowledged as a serious issue [1–3]. Walter Schütz analyzed the history of fatigue from 1837 to 1994 in a lengthy study [4]. The historical milestone papers were gathered by Sanford [5] and Hanewinckel and Zenner [6]. Since then, there have been many papers on fatigue; in 2000, there may have been as many as 100,000. Thankfully, computerized literature retrieval technologies have made it possible to consult the literature on particular themes. Numerous studies and real-world experience have led to a great deal of knowledge being acquired regarding the material's fatigue mechanism and the fatigue of structures. According to the author's survey research [7], qualitatively, our understanding of fatigue problems has advanced significantly in the 20th century. The results of the laboratory study have taught us a lot. However, accident investigations have also considerably contributed to the current state of the art. Service-related tiredness failures can offer valuable insights and compelling proof that fatigue could be a significant issue. Failure analysis frequently identifies a number of flaws that contribute to a structure's inadequate fatigue resistance [8].

Following a fatigue failure, the two most frequent queries are: (1) Was the material's fatigue resistance too low? or (2) Was the stress level at the failure spot too high? But there are more questions on the list. For example, (3) What makes the load

 DOI: 10.1201/9781003405122-4

spectrum of a failed motorcycle item so unique, and did the motorcycle industry know about and consider the fatigue load spectrum? (4) What was the condition of the material's surface where the fatigue cracks first appeared? This inquiry relates to production. Each question's answers may provide a distinct hint for a solution. It is essential to design and build a structure so that unwanted fatigue failures do not happen while the structure is still under design. There seems to be an issue known as "designing against fatigue". The following sections will address the different design solutions that can guarantee adequate life, safety, and economy during fatigue. They are associated with several structural ideas, including more meticulous detail design, materials that are less susceptible to fatigue, better material surface treatments, different kinds of joints, and reduced design stress levels. Less visible strategies can also be considered, such as design for damage tolerance (fail-safe), damage avoidance (e.g., corrosion protection), and reducing dynamic stresses in operation. The wide range of variables that might impact a structure's fatigue behaviour means that there are many options. Questions about the cost-effectiveness of design initiatives to enhance a structure's fatigue quality also impact fatigue-resistant design scenarios. Employees in an industry's design office typically use standard operating procedures (SOPs) to calculate fatigue strength, fatigue life, crack growth, and residual strength. Though they have their uses, standardized processes have the potential to be excessively or unnecessarily conservative. These computation processes begin with some generalized conditions, which don't have to match the conditions of the operational structure. One needs knowledge, expertise, and engineering judgement to assess the relevance of computed results (Figure 4.1). There could be a limit to the predictions' dependability and accuracy. Supporting fatigue tests should be run to help with uncertainty regarding computed predictions. For others, an experiment is far more useful than theoretical calculations. It's well known that experiments never lie. Regretfully, the results of an experiment are only useful within the study's parameters. Is the state of the tests an accurate depiction of the state of the service?

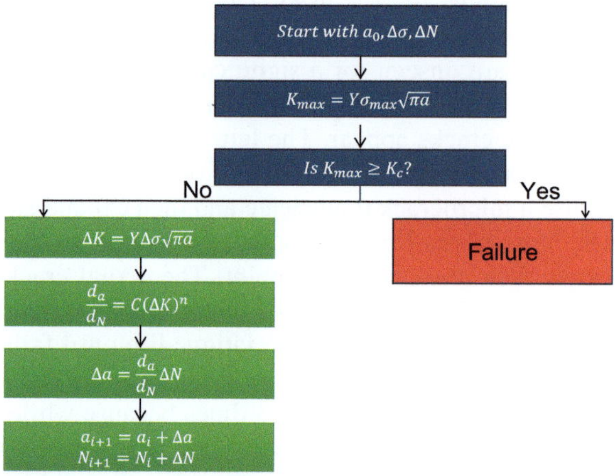

FIGURE 4.1 Standard operating procedures (SOPs) to calculate fatigue strength, fatigue life, crack growth, and residual strength.

Additionally, comprehension, judgement, and experience are requested in this question. Stated differently, designing against fatigue, be it through analysis, computations, or experiments, necessitates a deep understanding of the fatigue phenomena in materials and structures and the wide range of variables that can impact fatigue.

4.2 PHYSICS OF FCG – FUNDAMENTALS

When a structural member undergoes cyclic stress, a fatigue crack nucleus may form on the microscopic level, broaden to a macroscopic size, and ultimately fail the specimen in the ultimate cycle of its fatigue life. The fatigue phenomena will be covered in this section as a mechanism that affects metallic materials or materials in general, initially on a microscale and then on a macroscale. Knowing the fatigue mechanism is crucial for taking into account the many technical factors, such as material surface quality, residual stress, and environmental effects, that affect fatigue life and fatigue crack propagation. The examination of an engineering structure's fatigue qualities requires this knowledge. Only when fatigue is viewed as a fracture initiation mechanism preceding a crack growth period can fatigue prediction methods be assessed. Because of this, most of the book's chapters require reading this one first.

A *crack growth* phase and a *crack initiation* period typically comprise the *fatigue life*. Although some microcrack growth is expected during the beginning phase, the fatigue cracks are still too tiny to be noticeable. The crack widens throughout the second phase and eventually fails. Because various real-world factors significantly impact the fracture initiation period but have little to no impact on the crack development period, it is technically significant to examine the crack initiation and growth periods independently. Microscopic studies [9] revealed that the microcracks in slipping bands that precede fatigue crack nuclei are undetectable. Following the release of additional microscopic data about the development of tiny cracks, it was discovered that microcrack nucleation typically happens relatively early in the fatigue life. There are signs that it could happen immediately if a cyclic stress that exceeds the fatigue limit is used. The cycle stress level below which a fatigue failure does not occur is known as the *fatigue limit*. Even with early fracture nucleation, microcracks are unseen for a significant portion of the fatigue life. A laboratory specimen's residual fatigue life is often only a small portion of its complete life after noticeable cracks appear. The latter percentage might be far higher for buildings like ships, airplanes, etc. The actions of microstructures, such as grain boundaries, can cause a microcrack to nucleate and then expand slowly and irregularly. Nevertheless, more normal growth is shown once some microcrack growth has taken place away from the nucleation point. The actual crack-growing phase is about to begin. Figure 4.2 shows the many steps in the fatigue life. It's crucial to remember that there are two phases in the fatigue life until failure: the fracture growth phase and the crack initiation phase.

It is crucial to distinguish between the two times because while various surface conditions do impact the crack growth period, they have very little effect on the initiation period.

The fact that the two periods' fatigue prediction techniques differ should already be mentioned. The key variable for crack initiation predictions is the stress

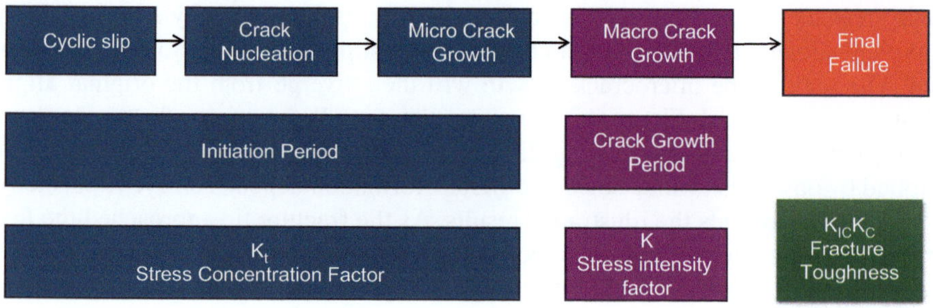

FIGURE 4.2 Steps in the fatigue life.

concentration factor, or K_t. The crack growth forecasts are based on the stress intensity factor K.

Cyclic slide leads to the initiation and propagation of fatigue cracks. It suggests dislocation activity or cyclic plastic deformation. At stress magnitudes lower than the yield stress, fatigue sets in. Plastic deformation at this low-stress level is restricted to many of the material's grains. Due to the reduced constraint on slip, this microplasticity preferentially occurs in grains at the material's surface. A substance's free surface is the only place where the surrounding material is present. The environment is the opposing side, typically a liquid (like ocean water) or a gaseous form (e.g. air). Because of this, plastic deformation in surface grains is possible at a lower stress level and is less restricted by nearby grains than in subsurface grains. A cyclic shear stress is necessary for a cyclic slide. The shear stress in the material doesn't spread uniformly on a microscale. According to the material's size, shape, crystallographic orientation, and elastic anisotropy, the shear stress on crystallographic slip lines varies from grain to grain. These circumstances are more conducive to cyclic slide in some material surface grains than in others.

A slip step will form at the material's surface if a slide happens in the grain. A slip step suggests that there will be a fresh material rim exposed to the elements. In most conditions, at least for most structural materials, an oxide layer will form immediately on top of the fresh surface material. These monolayers stick firmly to the surface of the substance and are difficult to remove. Slipping during an increase in load also suggests some strain hardening in the slip band, another critical factor. Consequently, a greater shear force in the opposite direction will be seen on the same slip band following unloading. It cannot be entirely reversible for the two reasons we have already discussed. Firstly, it is not possible to remove the oxide monolayer from the slip stage. Second, there is some irreversibility in the strain hardening that occurs in the slip band. Consequently, even in the same slip band, a reversed slip will happen on adjacent parallel slip planes.

The microcrack is evident in an elastically anisotropic material with a crystalline structure and multiple distinct slip systems as long as its size is still on the order of a single grain. With a stress concentration near the microcrack's tip, the microcrack contributes to an uneven stress distribution on a microscale. Consequently, many slip systems can be engaged. Furthermore, the constraint on slip displacements will increase if the crack develops into the material in some nearby grains because of their presence. Similarly, accommodating the slip displacements by slip on a single slip plane alone will get harder and harder. More slip planes ought to have that happen. The path of the microcrack growth will then diverge from the original alignment of the slip band. Figure 4.3 illustrates the overall propensity for development to be perpendicular to the loading direction. A threshold for crack growth may be implied by obstacles to slide since microcrack formation depends on cyclic plasticity.

Figure 4.4 presents the illustrative results. As the fracture tip approached the first grain boundary, the crack development rate—calculated as the crack length increment per cycle—began to drop. The crack development rate increased as it grew into the next grain after breaking through the first one, but it decreased once more as it got closer to the second grain boundary. Following the passage of that grain limit,

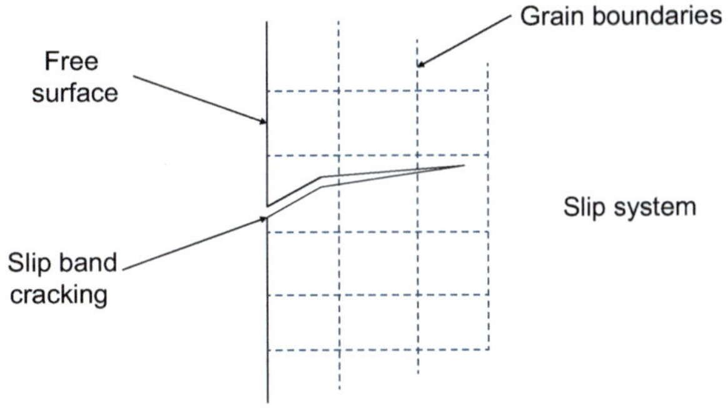

FIGURE 4.3 Microcrack cross section.

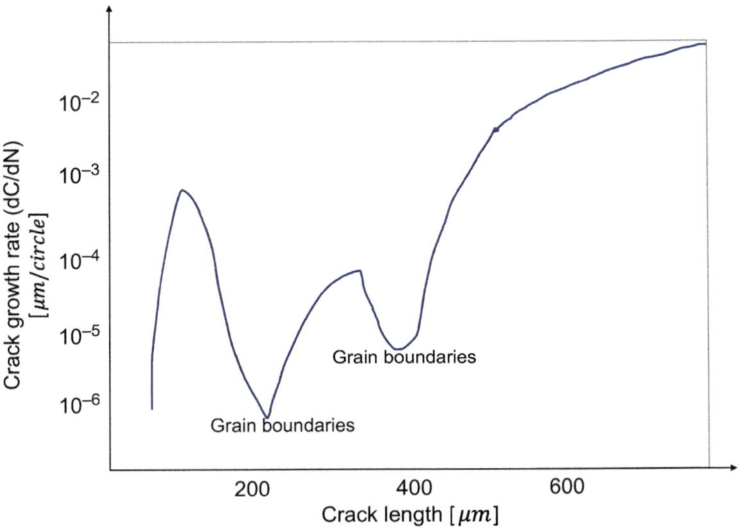

FIGURE 4.4 Grain boundary effect on crack growth in an Al-alloy. (The figure is created based on the work of Ref. [13].), the crack length c was measured along the material surface.

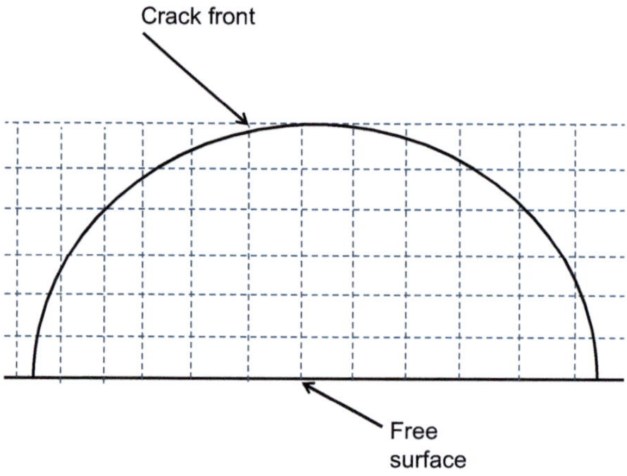

FIGURE 4.5 Crack tip view showing the crack front through several grains.

the microcrack grew at a rate that increased gradually. As the crack is enough large, fatigue crack growth can be described by power-law function [10,11, 12].

Numerous instances of initially inhomogeneous microcrack formation—which begins with a relatively high crack growth rate and after that slows down or even stops—are documented in the literature. These observations are thought to be caused by material structural barriers. However, as graphically shown in Figure 4.5, if the crack front, after some fracture propagation, passes through a significant number of grains, the picture changes. The crack cannot expand in any direction or growth rate apart from the crack growth in the neighboring grains since the crack front needs to stay cohesive crack front.

A fatigue crack with a macroscale propagates gradually once it reaches the fatigue crack initiation stage until it reaches its critical size, at which point it causes structural failure. Fracture mechanics theory's-controlled failure assumptions have been used to explain this consistent crack propagation. Linear elastic fracture mechanics, or LEFM, is applicable when the materials show a linear elastic behaviour or a low degree of plasticity around the crack tip. According to LEFM, the stress intensity factor, K, should represent the stress condition at the fracture tip. If the stress intensity factor range, ΔK, is assumed, LEFM is still appropriate for fatigue fracture propagation analysis under cyclic loadings.

4.3 DESCRIPTION OF FCG PROCESS BASED ON LEFM

The study of fatigue crack growth kinetics generally involves establishing a relationship between the change in crack size and the time or number N of loading cycles. It follows from fracture mechanics that the external load applied to the body, the geometry of the body, and the size of the crack in determining the rate of fatigue crack growth can be characterized by a well-known parameter – the stress intensity factor (SIF), which describes the stress-strain field in front of the crack tip. For brittle materials, the crack growth rate should depend primarily on the stress intensity factor

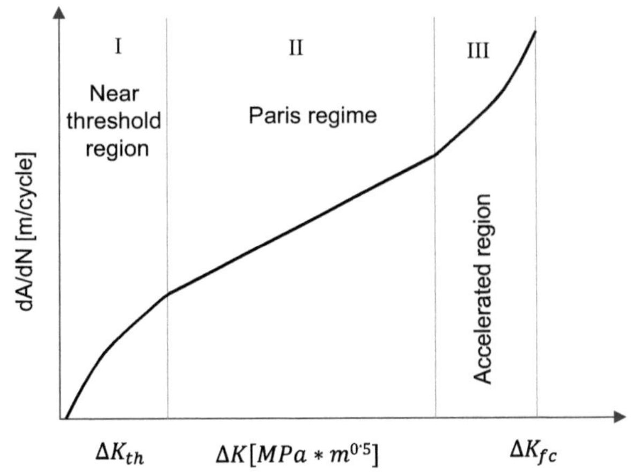

FIGURE 4.6 Schematic of a typical kinetic fatigue fracture diagram (KFFD). (The figure is created based on the work of Ref. [37].)

(SIF), the main factor determining the fracture process. Schematic, typical crack growth rate V dependence against SIF-K is shown in Figure 4.6.

These graphs show the dependence of the crack growth rate $V = da/dN$ (da - the increase in crack length during dN load cycles) on the range of change of SIF during the cycle $\Delta K = K_{\text{I max}} - K_{\text{I min}}$ or its maximum value. In a double logarithmic system, this graph has a characteristic sigmoidal shape, bounded by the two asymptotes (Figure 4.6). The threshold SIF value K_{th} denotes the magnitude below which the crack does not grow, K_{fc} the critical SIF (cyclic fracture toughness) value at which failure of the specimen occurs.

Paris [10, 11, 12] developed the most widespread empirical law of fatigue crack growth. According to Paris's pioneering work [12] the kinetic fatigue fracture equation (in II stage of KFFD – Figure 4.6) can be written as:

$$\frac{da}{dN} = C(\Delta K)^m \tag{4.1}$$

where C and m are material constants determined by experimental data. The generalized stress intensity factor range, $\Delta K = K_{max} - K_{min}$ defined for a cycle and related to the stress range, $\Delta\sigma$ and crack length, a, is given by Eq. (4.2).

$$\Delta K = Y\Delta\sigma\sqrt{\pi a} \tag{4.2}$$

In which Y represents a geometric constraint parameter. Aside from regime II, Eq. (4.1) might not be appropriate to describe the propagation rate when ΔK is low enough. Given these circumstances, regime I, where a threshold value for ΔK may be observed, is recognized as the regime where crack propagation occurs. This indicates that no crack propagation occurs if ΔK is less than the range of the threshold stress intensity factor, ΔK_{th}.

As it was mentioned, this Paris law formula is one of the most recognized laws of crack growth. As an example, many mathematical concepts of subcritical fracture propagation behaviour (regime II) is based on power-law function description [13–21]. A good example of extension Paris' concept is the relationship covering near threshold region (4.3)

$$\frac{da}{dN} = A_{th}\left(\Delta K - \Delta K_{th}\right)^{p_{th}} \tag{4.3}$$

where A_{th} and p_{th} represent the respective regressors which can also be determined experimentally.

However, a rapid increase of crack propagation is noticed at propagation regime III, which leads to the collapse quickly when the maximum stress intensity factor value approximates the critical stress intensity factor, K_c, or the fracture toughness, K_{IC}. Since mean stress plays a significant role in fatigue life approaches, it stands to reason that this variable also affects the growth of fatigue cracks. In both regimes, a rise in the mean stress tends to increase the crack propagation rate, *da/dN*, although less so in regime II. Significant changes in the crack propagation rate in regime III were made based on the material's fracture toughness. Regime I, on the other hand, is most affected, indicating that mean stress significantly impacts threshold stress intensity factors above and beyond material dependency. Based on Eq. (4.1), several models that consider the properties of fracture materials and the mean stress effect have been presented. Modified Eq. (4.1) to account for various stress ratios, R_σ, in regime II [20] such that:

$$\frac{da}{dN} = C_{w,\text{II}}\left(\frac{\Delta K}{(1-R_\sigma)^{1-\gamma}}\right)^{m_{w,\text{II}}} = C_{w,\text{II}}\left(\overline{\Delta K}\right)^{m_{w,\text{II}}} \tag{4.4}$$

where the exponent $m_{w,\text{II}}$ and the coefficient $C_{w,\text{II}}$ are fitted to experimental data, and ΔK indicates an equivalent value of the stress intensity factor range. For most metals, γ varies between 0.3 and 1, with typical values around 0.5 [20]. The γ is material-dependent and directly relates to the stress ratio effect; for lower values, it indicates a greater impact of R_σ on fatigue crack propagation behaviour. Walker's parameter can be calculated using the following equation [20] based on the material's ultimate tensile strength, σ_{uts}, in the absence of experimental data:

$$\gamma = -0.0002\sigma_{uts} + 0.8818 \tag{4.5}$$

Forman integrated the mean stress effect with regimes II and III [20] in addition to the Walker model, producing (Eq. 4.6), which:

$$\frac{da}{dN} = \frac{C_f\left(\Delta K\right)^{m_f}}{(1-R_\sigma)K_c - \Delta K'} \tag{4.6}$$

where the empirical parameters C_f and m_f are also fitted by the available experimental data. Walker incorporated both regimes I and II of crack propagation in addition to the Walker model's broad applicability (Eq. 4.4) and the significance of regime I. This led to the following outcomes:

$$\frac{da}{dN} = C_{w,\mathrm{I,II}}\left(\overline{\Delta K}\right)^{m_{w,\mathrm{I,II}}} = C_{w,\mathrm{I,II}}\left(\frac{\Delta K}{\left(1-R_\sigma\right)^{1-\gamma}} - \Delta K_{\mathrm{th}}\right)^{m_{w,\mathrm{I,II}}} \tag{4.7}$$

where K_{th} is a function of gamma, γ, such that

$$\Delta K_{\mathrm{th}} = \Delta K_{\mathrm{th}R0=0}\left(1-R_\sigma\right)^{1-\gamma} \tag{4.8}$$

This leads to a new equation by Walker that does not depend on ΔK_{th} as described by Eq. (4.9) below:

$$\frac{da}{dN} = C_{w,\mathrm{I,II}}\left(\frac{\Delta K}{\left(1-R_\sigma\right)^{1-\gamma}} - \Delta K_{\mathrm{th}(R=0.0)}\left(1-R_\sigma\right)^{1-\gamma}\right)^{m_{w,\mathrm{I,II}}} \tag{4.9}$$

where $C_{w,\mathrm{I,II}}$ and $m_{w,\mathrm{I,II}}$ are the only factors that need to be calculated. It is important to note that ΔK_{th} ($R_\sigma = 0$) in Eq. (4.9), which was derived for different stress ratios, is a known value that was previously determined using Eq. (4.8) using different ΔK_{th} data.

Elber [21], who demonstrated that a fatigue crack may be closed even under tensile strain, highlighted the significance of crack closure effect. The reduction in crack driving force, which results in the contact of fracture faces before the minimum loading is attained, is caused by residual compressive deformation surrounding the crack tip. Crack closure is commonly expressed as a closure ratio, U, which denotes the loading fraction while the crack is open. This can be expressed as follows:

$$U = \frac{\Delta K_{\mathrm{eff}}}{\Delta K} = \frac{1-\left(\dfrac{K_{\mathrm{op}}}{K_{\mathrm{max}}}\right)}{1-R_\sigma} \tag{4.10}$$

where $U = 1$ indicates the absence of a closure effect, while $U << 1$ indicates a significant closure impact. ΔK_{eff}, which is $K_{\mathrm{max}} - K_{\mathrm{op}}$ in Eq. (4.10), can be interpreted as the effective driving force of the crack, less than or equal to the nominal driving force of the crack tip, ΔK. The stress intensity factor in the opening is called K_{op}, and it can be either slightly more or equal to K_{cl}, which is the closing stress intensity factor.

According to the literature, regime I of fatigue crack propagation may be more affected by the fracture closure effect. Numerous developments in the theory of fracture closure have been put forth, including roughness-, oxide-, and plasticity-induced closure [22, 23, 24]. The cyclic plastic zone at the fracture tip, as well as a wake of plasticity in deformed material along crack faces, are taken into account by the plasticity-induced closure theory. The mathematical expression of this model is represented by Newman's closure Equation (4.11), which separates the value of ΔK_{eff} from Eq. (4.10) [24].

$$\Delta K_{\text{eff}} = \frac{1-f}{1-R_\sigma}\Delta K \tag{4.11}$$

The models discussed above mainly describe empirical relationships built based on experimental results. There are also known attempts to create models - based on theoretical considerations - appropriate models, several of which will be discussed.

Wu et al. [25] developed a *K*-driving force fatigue crack growth model based on the Dugdale model and damage accumulation theory:

$$V = \frac{\pi}{2\sigma_{\text{pl}}^2}\frac{(\Delta K)^{2\alpha+2} f_1(R)}{[2(1-R)K_c]^{2\alpha} - (\Delta K)^{2\alpha}(1-R)^2} I(\alpha, R) \tag{4.12}$$

where σ_{pl} is the yield point of the material, α is constant,

$$f_1(R) = 0.09286 + 0.02556R - 0.35166R^2 + 0.2342R^3 \tag{4.13}$$

$$I(\alpha, R) = \int_0^{\xi_0} \frac{[g(\xi\omega^*/\omega) - A_0 - A_2(\xi\omega^*/\omega)^2 -}{A_3(\xi\omega^*/\omega)^3 - (1-R)^2\Delta K_0^2/\Delta K^2]^\alpha} d\xi \tag{4.14}$$

where ω is the length of static (monotonic) plastic zone, ω^* is the length of cyclic plastic zone, $\omega^*/\omega = f_1(R)\,K_{fc}$ is resistance of material to fracture.

$$A_0 = f_2(R) = 0.8562 - 0.0145R + 0.41592R^2 - 0.25241R^3 \tag{4.15}$$

$$A_2 = \frac{-(\omega^*/\omega)g'(\omega^*/\omega) - 3\left[f_2(R) - g(\omega^*/\omega)\right]}{(\omega^*/\omega)^2} \tag{4.16}$$

$$A_3 = \frac{-2\left[g(\omega^*/\omega) - f_2(R)\right] + (\omega^*/\omega)g'(\omega^*/\omega)}{(\omega^*/\omega)^3} \tag{4.17}$$

$$g(\xi) = \sqrt{1-\xi} - \frac{1}{2}\xi\log\left|\frac{1+\sqrt{1-\xi}}{1-\sqrt{1-\xi}}\right| \tag{4.18}$$

$$g'(\xi) = \frac{dg(\xi)}{d\xi} = -\frac{1}{2}\log\left|\frac{1+\sqrt{1-\xi}}{1-\sqrt{1-\xi}}\right| \tag{4.19}$$

The threshold range of change of SIF is of the form $\Delta K_{\text{th}} = \sqrt{2}\Delta K_0$, and ξ_0 is the solution of the equation,

$$g(\xi\omega^*/\omega) - A_0 - A_2(\xi\omega^*/\omega)^2 - A_3(\xi\omega^*/\omega)^3 - (1-R)^2\Delta K_0^2/\Delta K^2 = 0 \tag{4.20}$$

One of the method's assumptions is the hypothesis that a fatigue crack develops through the accumulation of damage in the cyclic plastic zone occurring in ahead of the crack front.

The prediction of crack growth rate under complex loading conditions, considering the influence of fracture geometry and material properties, was developed based on the strain energy density factor ΔS [26]. It was assumed that the crack growth in the direction requires the lowest energy ΔS and can be described by a formula with a classical structure:

$$V = C(\Delta S_{\min})^{n} \tag{4.21}$$

An additional correction factor was introduced, including the stress ratio factor R, yield stress, and ultimate tensile stress. The quantities appearing in the resulting complex formulae for calculating fracture velocity were determined numerically. Raju [27], using a mathematical description of the function, $\delta = f(\varepsilon)$ performed an approximate analysis of the energy of plastic deformation at the crack front. From the energy balance equations, it was obtained the following expression [27]:

$$V = A_R \frac{(1-R)^{4-m} - K_{\max}^4}{2\left(K_c^2 - K_{\max}^2\right)} \tag{4.22}$$

in which K_c and $K_{\max}$ are derived from the size of the plastic deformation zone, and stress ratio $R = \sigma_{\min}/\sigma_{\max}$.

The feasibility of using such a description of fatigue crack growth in different steels, aluminium alloys, and titanium alloys is presented in the work of Badaliance [28]. Cherepanov [29] studied the quasi-static development of fracture in elastic-plastic bodies under varying loading conditions, assuming that the energy dispersion due to crack growth, related to a unit of newly formed area, is a material constant. He denoted the dissipated energy by γ and called his assumption the γ concept. Cherepanov showed that the crack growth rate depends on the largest K value and smallest K value - stress intensity factor per cycle, the number of cycles N, the energy γ and the material constants E, R_e (flow stress), v (Poisson's ratio). As a result of the dimensional analysis, he obtained the equation [29]:

$$\frac{dl}{dN} = \frac{K_{\mathrm{I}\max}^2}{R_e^2}\psi\left(\frac{K_{\mathrm{I}\max}^2}{E\gamma}, N, \frac{R_e}{E}, v\right) \tag{4.23}$$

in which ψ is a specific dimensionless function. Assuming that there is no change in the length of the crack when relieved during one cycle from $K_{\mathrm{I}\max}$ to $K_{\mathrm{I}\min}$ leads to the following formula:

$$\frac{dl}{dN} = -\beta\left(\frac{K_{\mathrm{I}\max}^2 - K_{\mathrm{I}\min}^2}{K_C^2} + \ln\frac{K_C^2 - K_{\mathrm{I}\max}^2}{K_C^2 - K_{\mathrm{I}\min}^2}\right) \tag{4.24}$$

In Eq. (4.24), β denotes the material constant of the length dimension (comprising the constants E, R_e, K_C, and v), which is determined experimentally. It describes the increase in crack length, and its magnitude corresponds to the increase in crack length when K_I changes from zero to the value K_C. The value of K_C is approximately equal to the fracture toughness K_{IC}, although there are certain limitations (for example, $K_C \neq K_{IC}$ for cracking at a small number of cycles). The constant β can be calculated similarly to the size of the plastic deformation zone in front of the crack front.

Expanding the right-hand part of Eq. (4.24) into a Taylor series and omitting parts with higher powers leads to a Paris-type formula if $K_{min} = 0$. For $K_{max}/K_c \leq 0.5$ the difference does not exceed 15%. The theoretical derivation of the crack growth relation by P.G. Cherepanov [29] thus largely coincides with the basic experimental formula.

An intensively developing area of fracture mechanics is fatigue crack growth research under an elastoplastic regime. This issue will be discussed in the next section by introducing a new energy predictor of fatigue crack growth rate. On the other hand, various parts of machines and structures work under multiaxial (especially biaxial) fatigue loading. This is also a domain of application of several models based on stress intensity factors as well as energy approach (discussed in a further part of this chapter).

4.4 ENERGY DESCRIPTION OF FCG PROCESS

An energy balance for an infinite plate with a central through thickness crack will be performed for energy consideration. Let us consider an infinite disc with an internal flat defect (crack) with a surface S_0 defined by the smooth contour L_o (Figure 4.7). The external stress varies cyclically and is perpendicular to the plane of the crack [30, 31, 32]:

$$\sigma_{zw} = \frac{1}{2}(\sigma_{zw\,max} - \sigma_{zw\,min})\sin \omega t \tag{4.25}$$

Determining the period N_s of subcritical crack growth (according to Szata's concept [30,31]) requires the application of the first law of thermodynamics [30]:

$$A^* + Q^* = W^* + K_e^* + \Gamma^* \tag{4.26}$$

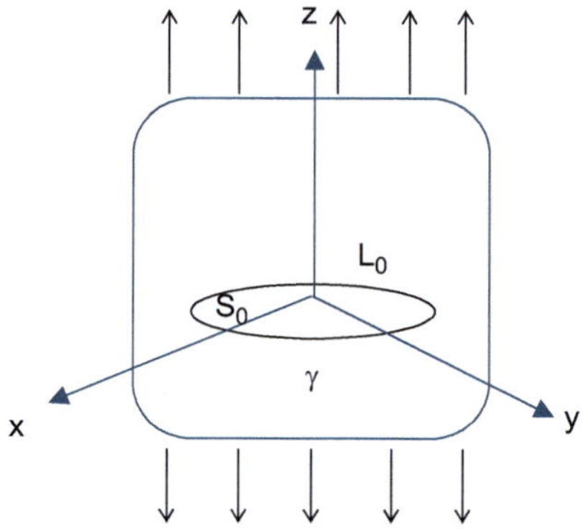

FIGURE 4.7 Scheme of an infinite plate with a crack and its coordinate system.

where A^* is external stress work σ_{zw} after N load cycles, Q^* is heat delivered to the body during loading, W^* is Strain energy after N load cycles, K_e^* is kinetic energy of the body, Γ^* is the energy of destruction associated with the change in fracture area from S_0 to S.

Assuming, the quantities A, Q, W, K_e, Γ per unit thickness of the body B, i.e. [30]:

$$A^* = BA, Q^* = BQ, W^* = BW, K_e^* = BK_e, \Gamma^* = B\Gamma \tag{4.27}$$

Equation (4.26) can be written in the following form:

$$A + Q = W + K_e + \Gamma \tag{4.28}$$

Using the notation $t = NT$, in which T is the period of the cyclic loading, it is possible to switch from the derivative after time to the differentiation after the number of cycles:

$$\frac{d}{dt} = T\frac{\partial}{\partial N} \tag{4.29}$$

After differentiating Eq. (4.26) by the number of cycles (after time), with the assumption that the slow increase in crack length is not accompanied by thermal processes and neglecting small changes in kinetic energy (for low cyclic loading frequencies), it can be written as [30]:

$$\frac{\partial A}{\partial N} = \frac{\partial W}{\partial N} + \frac{\partial \Gamma}{\partial N} \tag{4.30}$$

Equation (4.30) describes the condition in which slow crack growth occurs. Considering also the situation for such values of the external load σ_{zw}, which turn out to be too small to start the crack propagation. Equation (4.30) can then be written in another form [30]:

$$\frac{\partial (A - W)}{\partial N} = \frac{\partial W_p^\circ}{\partial N} \tag{4.31}$$

where W_p° is the energy dissipated for plastic deformation.

Szata [30,31] makes the following assumptions in the proposed criterion: After ΔN cycles, an elementary act of fracture occurs, resulting in an increase in the fracture area by ΔS (a change in the fracture area from S_0 to $S = S_0 + \Delta S$). Once the plastic strain energy W_p° exceeds a critical value equal to the failure energy Γ, new surface formation and crack propagation are initiated. If $W_p^\circ < \Gamma$, the fracture process does not occur.

It is assumed that:

$$W_p^\circ = W_c^\circ + W_s^\circ \tag{4.32}$$

whereby W_c° is the energy of cyclic plastic deformation, W_s° is the static (monotonic) component of energy, corresponding to $\sigma_{zw\ max}$, which changes with the development of the fatigue crack.

If $W_p^\circ = W_p = \Gamma$, there will be an elementary increase in the crack area by ΔS. Then:

$$W_p = \Gamma = W_c + W_s \tag{4.33}$$

The two components W_s and W_c depend on the value of the applied cyclic external load σ_{zw}, so their sum (at a fixed value of Γ), determined by Eq. (4.33), can be reached with different combinations of $\sigma_{zw\ max}$ and $\Delta\sigma_{zw} = \sigma_{zw\ max} - \sigma_{zw\ min}$. The magnitude of $\sigma_{zw\ max}$ determines W_s, while W_c depends on $\Delta\sigma_{zw}$.

The value Γ is defined as the maximum value of the dispersion of the static component of energy per plastic deformation ($\Gamma = W_{s\ max}$), triggering the fracture process without the contribution of the energy of cyclic strain changes ($W_c = 0$).

The static energy component for plastic deformation W_s occurs only once in Eq. (4.33). The energy dispersion for cyclic plastic deformation W_c can instead be represented as the sum of the energy for cyclic plastic deformation for all ΔN loading cycles [30]:

$$W_c = \frac{\partial W_c}{\partial N} \Delta N \tag{4.34}$$

Assuming a constant value of $W_c^{(1)} = \dfrac{\partial W_c}{\partial N}$ in each cycle of a load block of ΔN cycles.

Once a crack has occurred, the plastic zone ahead of the crack front will increase in length from part of its length, and the remainder, together with a new batch of material in front of the crack front, will form a new plastic zone for which the elementary act of cracking described above will be repeated. This means that the subsequent first loading cycle will induce a new static plastic zone in front of the crack front, just enlarged by ΔS. In the next step, due to the cyclic nature of the external load σ_{zw}, there will again be alternating loading ΔN cycles, forming an area of cyclic plastic zone in front of the crack front, until the next elementary act of cracking and the displacement of the plastic zone in front of the enlarged crack front [30].

Equation (4.33) can be transformed to the form [30,31,32]:

$$\Gamma - W_s = W_c \tag{4.35}$$

which differentiate by number of cycles (after time):

$$\frac{\partial(\Gamma - W_s)}{\partial N} = \frac{\partial W_c}{\partial N} \tag{4.36}$$

Equation (4.36) describes the elementary act of cracking related to one cycle. The quantities Γ and W_s in their entirety are assigned to the elementary act of cracking only once, so it is more convenient to change the differentiation by introducing a derivative after the increase of the crack area ∂S [30,31,32]:

$$\frac{\partial(\Gamma - W_s)}{\partial N} = \frac{\partial(\Gamma - W_s)}{\partial S} \cdot \frac{\partial S}{\partial N} \tag{4.37}$$

Taking into account Eq. (4.37), Eq. (4.36) is obtained in the form of:

$$\frac{\partial(\Gamma - W_s)}{\partial S} \cdot \frac{\partial S}{\partial N} = \frac{\partial W_c}{\partial N} \tag{4.38}$$

which, after transformations, gives us:

$$\frac{\partial S}{\partial N} = \frac{\partial W_c}{\partial N} / \frac{\partial(\Gamma - W_s)}{\partial S} \tag{4.39}$$

where $\frac{\partial S}{\partial N}$ is the rate of increase of the surface area S of the crack.

The next step is determining the existing energy components in Eq. (4.39). It should be recalled here that the following loading model for a material with a fracture is assumed based on experimental studies, as presented in Wu et al. [25], among others. For the small size of the plastic zone in front of the crack front (this is the most common case), the plastic deformation zone is surrounded by a much larger volumetric elastic deformation material (depending on the value of the applied load σ_{zw}). The elastic deformation field controls the plastic deformation zone near a crack tip, inducing cyclic deformation. The loading scheme of the specimen is shown in Figure 4.8. Two stages can be distinguished here: loading the specimen up to the value $P_{zw\ max}$ and then cyclic loading between the values and $P_{zw\ max} P_{zw\ min}$ by ΔN cycles, schematically marked by hysteresis loops. The plastic zone ahead of a crack tip can be identified as cyclic and static [33-36]. As noted earlier, analytical function for calculation of displacements ahead of a crack tip according to the Dugdale-Panasiuk model [33-36], has been adopted for the calculations, the state is shown in Figure 4.9. This means that the elastic-perfectly plastic body model was adopted in the zone in front of the crack front.

Let us consider the situation shown in Figure 4.9. As mentioned before, and known from the literature [33–36], under cyclic loading, two zones of plastic deformation are formed ahead of the crack front - the first one of length l_s (static) associated with

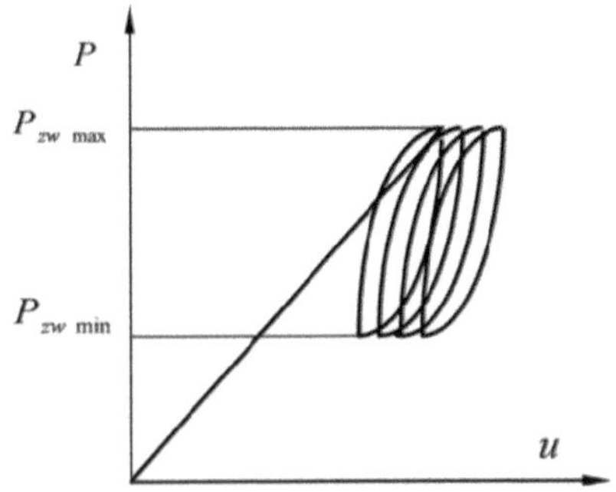

FIGURE 4.8 Schematic of sample loading.

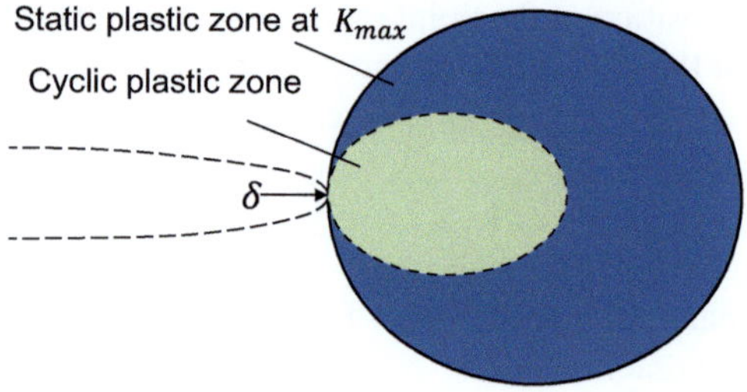

FIGURE 4.9 Areas - static and cyclic - zones in front of the crack front. (The figure is created based on the work of Ref. [37].)

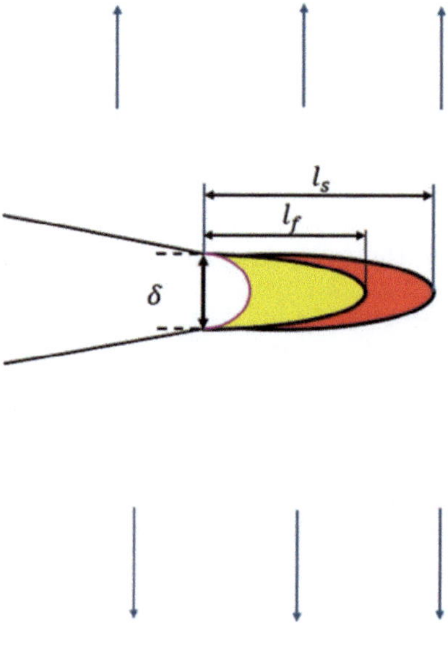

FIGURE 4.10 Model diagram of the full load cycle for the material in front of the crack front.

static loading, the second one related to cyclic loading of length l_f. The shape of these zones around the crack is defined by the contours L_s and L_f, respectively. Area I in Figure 4.10 corresponds to stage I of loading of the specimen and the physical quantities associated with this area have the subscript '*s*' (indicating static loading). Area II in Figure 4.10 refers to the second stage - cyclic (fatigue) loading, hence the subscript '*f*' (fatigue) next to the physical quantities appearing.

The area under the stress-strain curve $\sigma - \varepsilon$ in the range $0 - \varepsilon_{\max}$ represents the volumetric energy density of static strain, which can be denoted $\frac{\partial W_s}{\partial S}$. In contrast, the area under the curve in the $\sigma - \varepsilon$ and $\varepsilon_{\min} - \varepsilon_{\max}$ range relates to the volumetric energy density of cyclic strain after one loading cycle, which will be denoted as $\frac{\partial^2 W_c}{\partial S \, \partial N}$.

Depending on the position of the point in the zone in front of the crack front, there are different values of the deformations and ε_{max} - ε_{min}, corresponding to the external stresses and $\sigma_{zw\ max}$ -$\sigma_{zw\ min}$. In zone II, this relationship can be written:

$$\varepsilon_{\max}^{(f)} = \varepsilon_{\max}^{(f)}(t, x_i), \quad \varepsilon_{\min}^{(f)} = \varepsilon_{\min}^{(f)}(t, x_i), \tag{4.40}$$

where t is the coordinate along the contour L_o (Figure 4.7), x_i is the coordinate along the zone in front of the crack front.

The magnitude $\frac{\partial^2 W_c}{\partial S\, \partial N}$ determines the (volumetric) energy dispersion density of cyclic plastic deformation in front of the crack front in one loading cycle (the area between and ε_{min} -ε_{max} in Figure 4.10).

The volumetric density of static strain energy $\frac{\partial W_s}{\partial S}$ is determined by the change in plastic strain in area 1, and the linear density of cyclic strain energy after one loading cycle $\frac{\partial W_c}{\partial N}$ is determined by the change in plastic strain in area 2. The strains in both areas are determined from the Dugdale-Panasiuk model, shown in Figure 4.11, where the applied uniform stresses σ_{plf} are marked.

The magnitude of the static stress σ_{pl} is determined from the static strength characteristics of the material. However, the cyclic characteristic may be written in the form of:

$$\sigma_{plf} = \eta \sigma_{pl} \tag{4.41}$$

η with >1. The η factor is determined experimentally for each material by comparing the diagrams for cyclic and static tension. As the sizes of l_f and l_s areas ahead of the crack front are much smaller than the size of the crack itself, the radius of the crack contour in polar coordinates, r, ϕ (Figure 4.7).

According to Figure 4.10, the amount of cyclic plastic deformation energy dissipation in front of the crack front per load cycle is equal to [30,31,32]:

$$\frac{\partial W_c}{\partial N} = \int_L \int_0^{l_f} \frac{\partial^2 W_c}{\partial S\, \partial N} dF \tag{4.42}$$

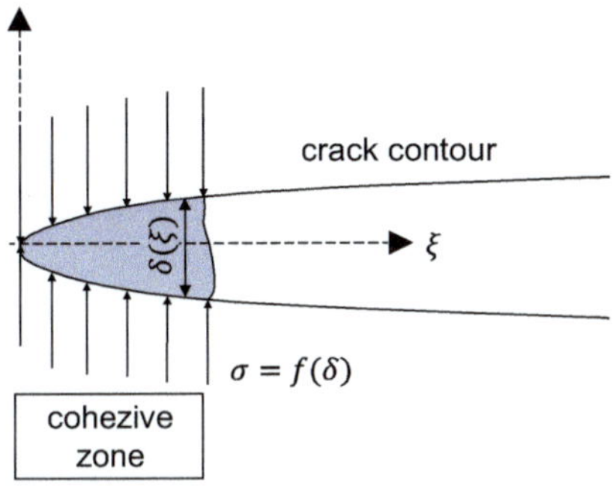

FIGURE 4.11 Dugdale-Panasyuk model of the zone ahead of a crack. (The figure is created based on the work of Ref. [32].)

where $dF = dx_i dt$ is the surface element of the cyclic plastic zone ahead of a crack tip.

The density (volumetric) $\frac{\partial^2 W_c}{\partial S \partial N}$ of the energy dispersion of cyclic plastic deformation ahead of a crack tip during one loading cycle is represented by the expression [30]:

$$\frac{\partial^2 W_c}{\partial S \partial N} = \sigma_{\text{plf}} \left[\varepsilon_{\max}^{(f)}(t, x_i) - \varepsilon_{\min}^{(f)}(t, x_i) \right] \tag{4.43}$$

After using the relationship between the magnitude of the deformation ε in front of the crack tip and the crack opening δ in the Dugdale-Panasyuk model [33] $\varepsilon = \varepsilon_{fc}\delta / \delta_{fc}$, one obtains [30]:

$$\begin{aligned} \frac{\partial W_c}{\partial N} &= \int_L \int_0^{l_f} \sigma_{\text{plf}} \left[\varepsilon_{\max}^{(f)}(t, x_i) - \varepsilon_{\min}^{(f)}(t, x_i) \right] dx_i \, dt \\ &= \frac{\sigma_{\text{plf}} \varepsilon_{fc}}{\delta_{fc}} \int_L \int_0^{l_f} \left[\delta_{\max}^{(f)}(t, x_i) - \delta_{\min}^{(f)}(t, x_i) \right] dx_i \, dt \end{aligned} \tag{4.44}$$

In Eq. (4.44), dt denotes the change in coordinate length along the L contour (Figure 4.7), dx_i is the change in coordinate length along area, $\delta_{\max}^{(f)}(t, x_i)$ - $\delta_{\min}^{(f)}(t, x_i)$ is the maximum and minimum crack opening displacement, respectively, in the area in front of the crack front; δ_{fc} is critical value, corresponding to the critical strain ε_{fc} under cyclic loading. The critical value corresponds to the critical load at which the crack development ceases to be stable and the process of unstable crack development begins. The magnitude of $\delta_{\min}^{(f)}(t, x_i)$ depends not only on the minimum value of the external stress $\sigma_{zw\ \min}$, but also on the profile of the joint surfaces and their mutual contact (top and bottom surfaces) [30]:

$$\delta_{\min}^{(f)}(t, x_i) = \max \left\{ \delta_{\min}^{(fp)}(t, x_i); \delta^{(fk)}(t, x_i) \right\} \tag{4.45}$$

where $\delta^{(fk)}(t, x_i)$ is the crack opening displacement, resulting from contact between uneven surfaces, which also depends on the load history, $\delta_{\min}^{(fp)}(t, x_i)$ is the minimum crack opening displacement due to external stress $\sigma_{zw} = \sigma_{zw\ \min}$, which is of the form:

$$\delta_{\min}^{(fp)}(t, x_i) = \delta_{\max}^{(f)}(t, x_i) \left[1 - \frac{1}{2}(1 - R)^2 \right] \tag{4.46}$$

Equation (4.46) will be derived after analyzing the subsequent states, which are the elements of the final state superposition in the form of a cyclic load. The stress ratio R is equal obviously:

$$R = \frac{K_{\text{I}\min}}{K_{\text{I}\max}} = \frac{\sigma_{zw\ \min}}{\sigma_{zw\ \max}} \tag{4.47}$$

By analogy with Eq. (4.44), it is possible to write the static component W_s of the dispersion of plastic strain energy per unit area in the zone ΔF ahead of the crack front:

$$\frac{\partial W_s}{\partial S} = \frac{\sigma_{\text{plf}}\varepsilon_{fc}}{\delta_{fc}\Delta F}\int_L \int_0^{l_s} \delta_{\max}^{(s)}(t, x_i)\,dx_i dt \tag{4.48}$$

where l_s is the length of the plastic zone in front of the crack tip, ΔF is the area of this zone (with length l_s).

The critical value of the static component W_s of the plastic strain energy per unit area of the crack is equal to the destruction energy per unit area and has the form [30]:

$$\frac{\partial \Gamma}{\partial S} = \frac{\sigma_{\text{plf}}\varepsilon_{fc}}{\delta_{fc}\Delta F^*}\int_L \int_0^{l_s^*} \delta_{\max}^{(s)}(t, x_i)\,dx_i\, dt \tag{4.49}$$

where l_s^* is critical length of the plastic zone in front of the fracture tip, ΔF^* is critical area of this zone (with length l_s^*). The explicit form of the criterion Eq. (4.39), expressing the rate of increase of the fatigue gap area, requires the insertion of the corresponding formulae (4.44), (4.48), and (4.49).

A superposition of two solutions of the classical Dugdale crack model for mode I - opening (Figure 4.12) and mode II – in plane shearing (Figure 4.13) is used to obtain the solution for the dynamic load. The solutions for both states mode I (Figure 4.12) and mode II (Figure 4.13) are obtained from the classical approach shown in Figure 4.14. The final result, i.e. the solution for state III, is obtained by the superposition of states I and II. The external load σ_{zw} of a body with a gap (Figure 4.14) is described by the known relations:

$$K_{\text{I}} = \sigma_{zw} f(a) \tag{4.50}$$

$$\delta = \frac{K_{\text{I}}^{\;2}}{E\sigma_{\text{plf}}} = \frac{\sigma_{zw}^2 f^2(a)}{E\sigma_{\text{plf}}} \tag{4.51}$$

$$l_s = \frac{\pi}{8}\frac{K_{\text{I}}^{\;2}}{\sigma_{\text{plf}}^2} = \frac{\pi}{8}\frac{\sigma_{zw}^2 f^2(a)}{\sigma_{\text{plf}}^2} \tag{4.52}$$

where a is crack length.

As a result of substituting $\sigma_{zw} = \sigma_{zw\,\max}$ in Eqs. (4.50)–(4.52), relations corresponding to this case are obtained:

$$K_{\text{I}\max} = \sigma_{zw\,\max} \tag{4.53}$$

$$\delta_{\max} = \frac{K_{\text{I}\max}^{\;\;2}}{E\sigma_{\text{plf}}} = \frac{\sigma_{zw\,\max}^2 f^2(a)}{E\sigma_{\text{plf}}} \tag{4.54}$$

$$l_s = \frac{\pi}{8}\frac{K_{\text{I}\max}^{\;\;2}}{\sigma_{\text{plf}}^2} = \frac{\pi}{8}\frac{\sigma_{zw\,\max}^2 f^2(a)}{\sigma_{\text{plf}}^2} \tag{4.55}$$

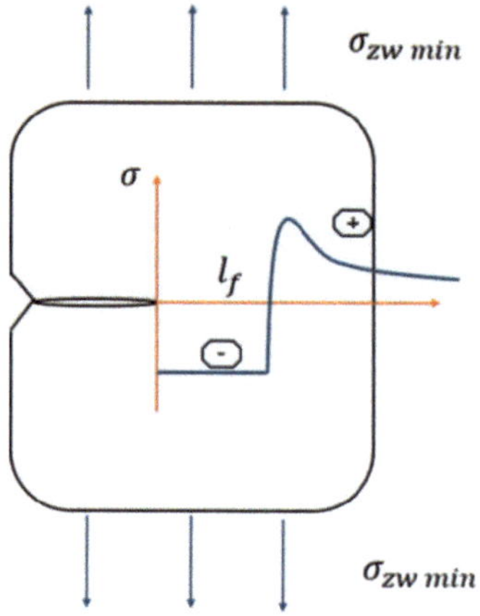

FIGURE 4.12 A body with a crack loaded with external stress $\sigma_{zw\ \max}$.

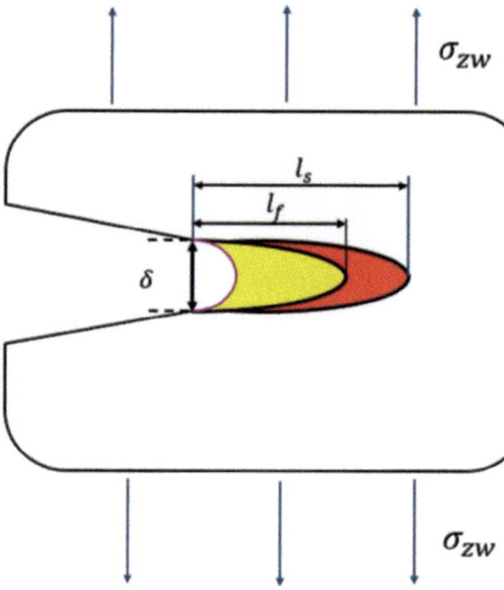

FIGURE 4.13 A body with a gap loaded with external stress $\sigma_{zw\ \min}$.

The analysis of the cyclic area of the zone in front of the crack tip, resulting from the cyclic external load with an amplitude between $\sigma_{zw\ \max}$ and $\sigma_{zw\ \min}$, requires a superposition of cases I and II (Figure 4.13), realizing case III. Body loading with an amplitude equal to $\sigma_{zw} = \sigma_{zw\ \max} - \sigma_{zw\ \min}$ (Figure 4.15).

Based on the above, the following relationships are obtained:

$$K_{\mathrm{I}} = \left(\sigma_{zw\ \max} - \sigma_{zw\ \min}\right) f(a) \tag{4.56}$$

$$\delta^{(f)} = \delta_{\max} - \delta_{\min} \tag{4.57}$$

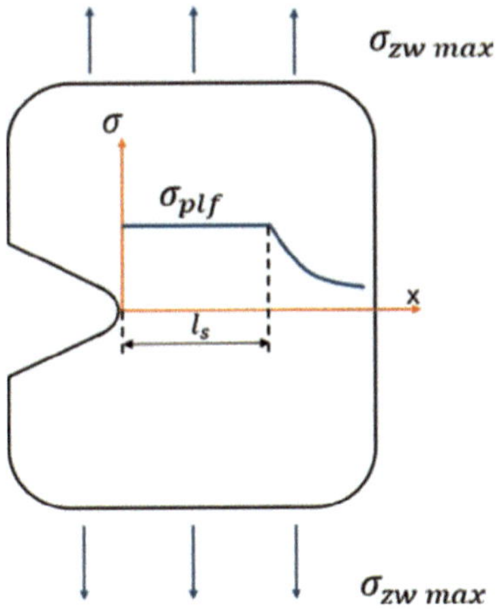

FIGURE 4.14 A body with a crack loaded with external stress σ_{zw}.

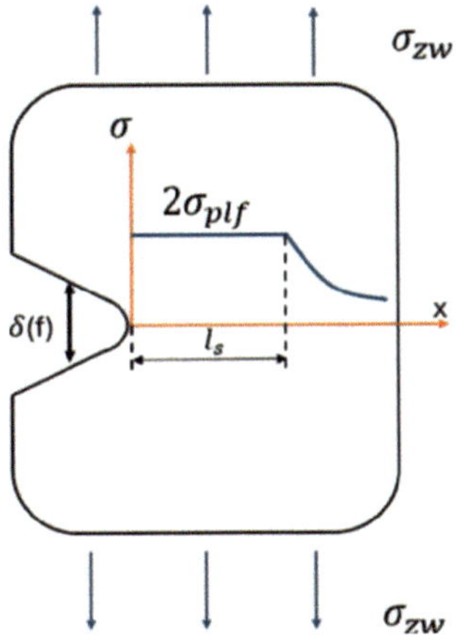

FIGURE 4.15 A body with a gap loaded with external stress $\sigma_{zw} = \sigma_{zw\ max} - \sigma_{zw\ min}$.

$$\delta^{(f)} = \frac{K_I^2}{E2\sigma_{plf}} = \frac{\left(\sigma_{zw\ max} - \sigma_{zw\ min}\right)^2 f^2(a)}{E2\sigma_{plf}} \tag{4.58}$$

$$l_f = \frac{\pi}{8}\frac{K_I^2}{4\sigma_{plf}^2} = \frac{\pi}{8}\frac{\left(\sigma_{zw\ max} - \sigma_{zw\ min}\right)^2 f^2(a)}{4\sigma_{plf}^2} \tag{4.59}$$

After substituting $\sigma_{zw} = \sigma_{zw\,\max}$ and $\sigma_{zw} = \sigma_{zw\,\min}$ into Eq. (4.50), one obtains:

$$K_{\mathrm{I\,max}} = \sigma_{zw\,\max} f(a), \quad K_{\mathrm{I\,min}} = \sigma_{zw\,\min} f(a) \tag{4.60}$$

Taking into account Eq. (4.60), Relation (4.59) can be transformed to the form:

$$l_f = \frac{\pi}{8}\frac{(K_{\mathrm{I\,max}} - K_{\mathrm{I\,min}})^2}{4\sigma_{\mathrm{plf}}^2} \tag{4.61}$$

By substituting $K_{\mathrm{I\,min}} = RK_{\mathrm{I\,max}}$ (Eq. 4.47) into Eq. (4.61), one obtains:

$$l_f = \frac{\pi}{8}\frac{K_{\mathrm{I\,max}}^2 (1-R)^2}{4\sigma_{\mathrm{plf}}^2} \tag{4.62}$$

A comparison of Eqs. (4.55) and (4.62) yields the expression:

$$l_f = \beta\, l_s (1-R)^2 \tag{4.63}$$

in which $\beta = 0.25$.

In an analogous way, the expression for the vertex gap opening is obtained. After inserting the formulae for $\sigma_{zw\,\max}$ and $\sigma_{zw\,\min}$ determined from (4.60) into Eq. (4.58), one obtains the relation:

$$\delta^{(f)} = \frac{(K_{\mathrm{I\,max}} - K_{\mathrm{I\,min}})^2}{E2\sigma_{\mathrm{plf}}} \tag{4.64}$$

which, after using formula (4.47), can be written as:

$$\delta^{(f)} = \frac{K_{\mathrm{I\,max}}^2 (1-R)^2}{2E\sigma_{\mathrm{plf}}} \tag{4.65}$$

After substituting (4.54) into Eq. (4.65) and taking into account Eq. (4.57), one obtains:

$$\delta^{(f)} = \delta_{\max}\frac{1}{2}(1-R)^2 \tag{4.66}$$

$$\delta_{\max} - \delta_{\min} = \delta_{\max}\frac{1}{2}(1-R)^2 \tag{4.67}$$

Finally, the relationship shown by Eq. (4.46) has the form:

$$\delta_{\min} = \delta_{\max}\left(1 - \frac{1}{2}(1-R)^2\right) \tag{4.68}$$

The opening of $\delta^{(s)}(t,x_i)$ is described by equation [30]:

$$\delta^{(s)}(t,x_i)=\frac{8\sigma_{\text{plf}}\left(1-\nu^2\right)}{\pi E}l_s(t)\left[\sqrt{1-\xi}-\xi\ln\frac{1+\sqrt{1-\xi}}{\sqrt{\xi}}\right] \tag{4.69}$$

where $\xi=\dfrac{x_i}{l_s}$, ν, E denote elastic material constants.

The next step is to derive the expressions, $\dfrac{\partial W_c}{\partial N}\dfrac{\partial W_s}{\partial S}\dfrac{\partial \Gamma}{\partial S}$ enabling the kinetic growth equation of the fatigue crack area (4.39) to be represented in a detailed form.

The starting point for the expression $\dfrac{\partial W_c}{\partial N}$ is Eq. (4.44);

$$\frac{\partial W_c}{\partial N}=\frac{\sigma_{\text{plf}}\varepsilon_{fc}}{\delta_{fc}}\int_L\int_0^{l_f}\left[\delta_{\max}^{(f)}(t,x_i)-\delta_{\min}^{(f)}(t,x_i)\right]dx_i\,dt \tag{4.70}$$

After inserting after and $\delta_{\max}^{(f)}(t,x_i)\delta_{\min}^{(f)}(t,x_i)$, it can be concluded that [30]:

$$\frac{\partial W_c}{\partial N}=\frac{\sigma_{\text{plf}}\varepsilon_{fc}\alpha}{\delta_{fc}}\int_L\left(\delta_{\max}^{(f)}(t,0)-\delta_{\min}^{(f)}(t,0)\right)\cdot l_f(t)\,dt \cdot\int_0^1\left[\sqrt{1-\xi}-\xi\ln\frac{1+\sqrt{1-\xi}}{\sqrt{\xi}}\right]d\xi \tag{4.71}$$

where $\xi=\dfrac{x}{l_f}$, integral $A2=\displaystyle\int_0^1\left[\sqrt{1-\xi}-\xi\ln\frac{1+\sqrt{1-\xi}}{\sqrt{\xi}}\right]d\xi=0.33$, hence

$$\frac{\partial W_c}{\partial N}=\frac{0.33\cdot\sigma_{\text{plf}}\varepsilon_{fc}}{\delta_{fc}}\int_L\left(\delta_{\max}^{(f)}(t,0)-\delta_{\min}^{(f)}(t,0)\right)\cdot l_f(t)\,dt \tag{4.72}$$

The magnitude of $\dfrac{\partial W_s}{\partial S}$ is determined from Eq. (4.48),

Finally, it can be expanded as:

$$\frac{\partial W_s}{\partial S}=\frac{\sigma_{\text{plf}}\varepsilon_{fc}}{\delta_{fc}\Delta F}\int_L\delta^{(s)}(t,0)l_s(t)\,dt\int_0^1\left[\sqrt{1-\xi}-\xi\ln\frac{1+\sqrt{1-\xi}}{\sqrt{\xi}}\right]d\xi \tag{4.73}$$

$$\frac{\partial W_s}{\partial S}=\frac{0.33\sigma_{\text{plf}}\varepsilon_{fc}}{\delta_{fc}\Delta F}\int_L\delta^{(s)}(t,0)l_s(t)\,dt \tag{4.74}$$

The relationship $\dfrac{\partial \Gamma}{\partial S}$ above is determined from equation:

$$\frac{\partial \Gamma}{\partial S}=\frac{\sigma_{\text{plf}}\varepsilon_{fc}}{\delta_{fc}\Delta F^*}\int_L\int_0^{l_s^*}\delta_{\max}^{(s)}(t,x_i)\,dx_i\,dt \tag{4.75}$$

After inserting after $\delta_{\max}^{(s)}(t,x_i)$:

$$\frac{\partial\Gamma}{\partial S}=\frac{\sigma_{\text{plf}}\varepsilon_{fc}}{\delta_{fc}\Delta F}\int_L \delta_c^{(f)}(0)l_s\,dt\int_0^1\left[\sqrt{1-\xi}-\xi\ln\frac{1+\sqrt{1-\xi}}{\sqrt{\xi}}\right]d\xi=0.33\sigma_{\text{plf}}\varepsilon_{fc} \quad (4.76)$$

The determined quantities, $\frac{\partial W_c}{\partial N}\frac{\partial W_s}{\partial S}\frac{\partial\Gamma}{\partial S}$ enabling the kinetic growth equation of the fatigue crack area (4.39) to be represented in detailed form can be written as Eqs. (4.72), (4.74), and (4.76).

The size of the area ΔF is expressed by the formula:

$$\Delta F=\int_L l_s(t)\,dt \quad (4.77)$$

Expressing the magnitude $\delta_{\max}^{(f)}(t,0)$, $l_s(t)$ by the stress intensity factors $K_{\text{I}\max}$ assuming $\delta_{\min}^{(fP)}(t,x_i)>\delta^{(fK)}(t,x_i)$, allows Eqs. (4.72), (4.74) and (4.77) to be written as a function of $K_{\text{I}\max}$. The next step is to derive these relationships. When formulae (4.46) and (4.70) are inserted into (4.72), the following is obtained [30]:

$$\frac{\partial W_c}{\partial N}=\frac{0.33\sigma_{\text{plf}}\cdot\varepsilon_{fc}\alpha}{\delta_{fc}}\left[1-\left(1-\frac{1}{2}(1-R)^2\right)^2\right]\int_L \delta_{\max}l_f\,dt \quad (4.78)$$

$$\frac{\partial W_c}{\partial N}=\frac{0.33\sigma_{\text{plf}}\cdot\varepsilon_{fc}\alpha}{\delta_{fc}}\left[\frac{1}{2}(1-R)^2\right]\int_L \delta_{\max}\beta l_s(1-R)^2\,dt \quad (4.79)$$

Because, $\beta=\frac{1}{4}$, $\delta_{fc}=\frac{K_{fc}^2}{E\sigma_{\text{plf}}}$, $l_s=\frac{\pi}{8}\frac{K_{\text{I}\max}^2}{\sigma_{\text{plf}}^2}$, i, $\delta_{\max}=\frac{K_{\text{I}\max}^2}{E\sigma_{\text{plf}}}$,

Inserting the above parameters into Eq. (4.79), then we have:

$$\begin{aligned}\frac{\partial W_c}{\partial N}&=\frac{0.33\sigma_{\text{plf}}^2\cdot\varepsilon_{fc}\cdot E\alpha\beta}{K_{fc}^2}\cdot\frac{1}{2}(1-R)^4\int_L\frac{K_{\text{I}\max}^2}{E\sigma_{\text{plf}}}\cdot\frac{\pi}{8}\cdot\frac{K_{\text{I}\max}^2}{\sigma_{\text{plf}}^2}dt\\&=\frac{0.33\pi}{16}\alpha\beta\varepsilon_{fc}(1-R)^4K_{fc}^{-2}\sigma_{\text{plf}}^{-1}\int_L K_{\text{I}\max}^4(t)\,dt\end{aligned} \quad (4.80)$$

For $\beta=\frac{1}{4}$ after simplification, we have:

$$\begin{aligned}\frac{\partial W_c}{\partial N}&=\frac{0.33\pi}{16\cdot 4}\alpha\varepsilon_{fc}(1-R)^4K_{fc}^{-2}\sigma_{\text{plf}}^{-1}\int_L K_{\text{I}\max}^4(t)\,dt\\&=0.0162\cdot\alpha\cdot\varepsilon_{fc}(1-R)^4K_{fc}^{-2}\sigma_{\text{plf}}^{-1}\int_L K_{\text{I}\max}^4(t)\,dt\end{aligned} \quad (4.81)$$

K_{fc} is cyclic critical stress intensity factor.

The starting point for deriving the relationship $\frac{\partial W_s}{\partial S}$ is given below:

$$\frac{\partial W_s}{\partial S} = \frac{0.33\sigma_{\text{plf}}\varepsilon_{fc}}{\delta_{fc}\Delta F}\int_L \delta^{(s)}(t,0)l_s(t)\,dt \tag{4.82}$$

Because, $\delta_{fc} = \frac{K_{fc}^2}{E\sigma_{\text{plf}}}$, $l_s = \frac{\pi}{8}\frac{K_{\text{I}\max}^2}{\sigma_{\text{plf}}^2}$, i, $\delta^{(s)}(t,0) = \delta_{\max} = \delta\frac{K_{\text{I}\max}^2}{E\sigma_{\text{plf}}}$,

By inserting the above into Eq. (4.79), we then have:

$$\frac{\partial W_s}{\partial S} = \frac{0.33\sigma_{\text{plf}}\varepsilon_{fc}E\sigma_{\text{plf}}}{K_{fc}^2\cdot\Delta F}\int_L \frac{K_{\text{I}\min}^2}{E\sigma_{\text{plf}}}\cdot\frac{\pi}{8}\cdot\frac{K_{\text{I}\min}^2}{\sigma_{\text{pl}}^2}\,dt = \frac{0.33\cdot\pi}{8}\varepsilon_{fc}K_{fc}^{-2}\sigma_{\text{plf}}^{-1}(\Delta F)^{-1}\int_L K_{\text{I}\max}^4\,dt$$

$$= 0.129\varepsilon_{fc}K_{fc}^{-2}\sigma_{\text{plf}}^{-1}(\Delta F)^{-1}\int_L K_{\text{I}\max}^4\,dt \tag{4.83}$$

As a result, the kinetics of fatigue crack growth (surface expansion) as a function of cycles is obtained in the form of:

$$\frac{\partial S}{\partial N} = 0.049\alpha(1-R)^4 K_{fc}^{-2}\sigma_{\text{plf}}^{-2}\int_L K_{\text{I}\max}^4(t)\,dt$$
$$\left[1 - 0.9959K_{fc}^{-2}\left(\int_L K_{\text{I}\max}^2(t)\,dt\right)^{-1}\int_L K_{\text{I}\max}^4(t)\,dt\right]^{-1} \tag{4.84}$$

taking into account the initial conditions:

$$N = 0,\quad S = S_0 \tag{4.85}$$

and final conditions:

$$N = N_g,\quad S = S_* \tag{4.86}$$

S_* denotes here the critical size of the fracture surface, after which fracture development ceases to be stable and rapid enlargement of the fracture occurs. This area can be determined from Irwin's criterion:

$$K_{\text{I}\max}(\ldots,S_*,\ldots) = K_{fc} \tag{4.87}$$

However, the solution of an exact equation can be problematic, due to its theoretical nature. Based on [30,31,37], the simplified form of energy equation can be presented as below:

$$\frac{\partial \Gamma}{\partial S} \cong \sigma_{\text{plf}}\varepsilon_{fc} \tag{4.88}$$

$$\frac{\partial W_S}{\partial S} \cong \sigma_{\text{plf}}\varepsilon_{\text{fmax}} \tag{4.89}$$

Finally, we have

$$\frac{\partial(\Gamma - W_S)}{\partial S} = \sigma_{\text{plf}}\varepsilon_{fc} - \sigma_{\text{plf}}\varepsilon_{\text{fmax}} = \sigma_{\text{plf}}\varepsilon_{fc}\left(1 - \frac{\varepsilon_{\text{fmax}}}{\varepsilon_{fc}}\right) = \sigma_{\text{plf}}\varepsilon_{fc}\left(1 - \frac{K_{\text{I max}}^2}{K_{fc}^2}\right) \tag{4.90}$$

Based on the above, it can be rewritten that the expansion rate of crack surface can be expressed as:

$$\frac{dS}{dN} = \frac{\alpha W_c^{(1)}}{\sigma_{\text{plf}}\varepsilon_{fc}\left(1 - K_{\text{I max}}^2 / K_{fc}^2\right)} \tag{4.91}$$

Finally, coming back to the crack length dimension a, we obtain:

$$\frac{da}{dN} = \frac{A W_c^{(1)}}{B\sigma_{\text{plf}}\varepsilon_{fc}\left(1 - K_{\text{I max}}^2 / K_{fc}^2\right)} \tag{4.92}$$

where A – experimentally determined constant, B – specimen thickness. $W_c^{(1)}$ in Eqs. (4.91) and (4.92) represents energy irrevocably dissipated during each loading cycle. The new energy parameter can be defined as:

$$\Delta S = \frac{W_c^{(1)}}{1 - \dfrac{K_{\text{Imax}}^2}{K_{fc}^2}} \tag{4.93}$$

As a main advantage of this energy-based approach, it should be noted that fatigue crack growth rate diagrams can be independent of stress ratio R. This was experimentally confirmed in works [30,31,32] – see Figures 4.16 and 4.17. As can be noticed, new energy-based kinetic fatigue fracture diagrams describe the fatigue crack growth process in a robust form independent from stress ratios R.

However, Lesiuk [37] extended the energy model by physically interpreting dissipated energy. This concept was based on applying the cyclic – J-integral parameters, initially used by Dowling and Begley [38], using elastic and plastic strain energy density components based on the J-integral (Figure 4.18). Defining two-parametric energy-based crack driving force, it can be represented by:

$$\Delta S^* = \sqrt{\Delta J^+ \cdot J_{\text{max}}} \tag{4.94}$$

Similarly, like for ΔS, ΔS^* describes synonymously fatigue crack growth rate in the elastoplastic range without the influence of stress ratio. Exemplary results are presented for 18G2A (S355 grade) steel in [37] – Figure 4.19.

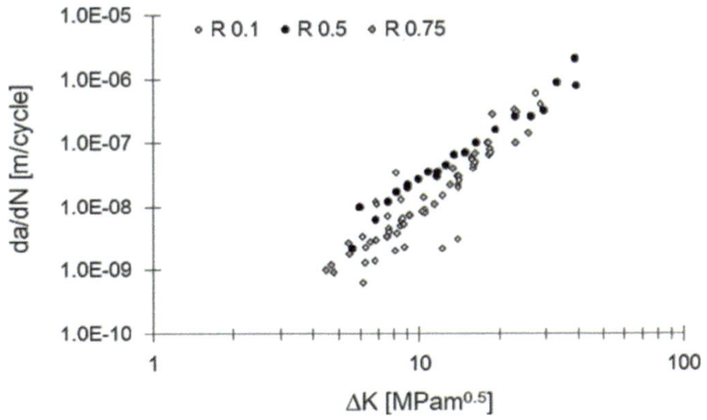

FIGURE 4.16 Fatigue crack growth rate data for S355 steel grade for different R (K-approach). (Data from Refs. [30,32,37].)

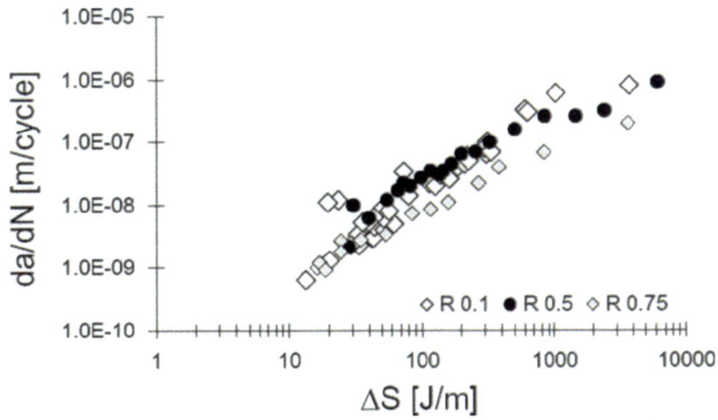

FIGURE 4.17 Fatigue crack growth rate data for S355 steel grade for different R (ΔS - energy-based approach). (Data from Refs. [30,32,37].)

4.5 MIXED MODE FATIGUE CRACK GROWTH

Material fatigue, particularly the formation of cracks and their development, is often focused on load cases causing cracks according to the mode I loading [39]. However, fatigue crack growth is more often observed in conditions under mixed modes loading [40,41], such as modes I and II, I and III or II and III, which lead to damage or failure of the material. When analyzing crack growth under multiaxial loads, we must consider two main factors that determine the failure of the element (specimen). One is the fatigue crack growth rate, and the other is the growth direction. Various criteria have been developed for determining the direction of crack growth under multiaxial loads, some of which will be presented in this paper. Different specimen geometries are used since crack growth tests

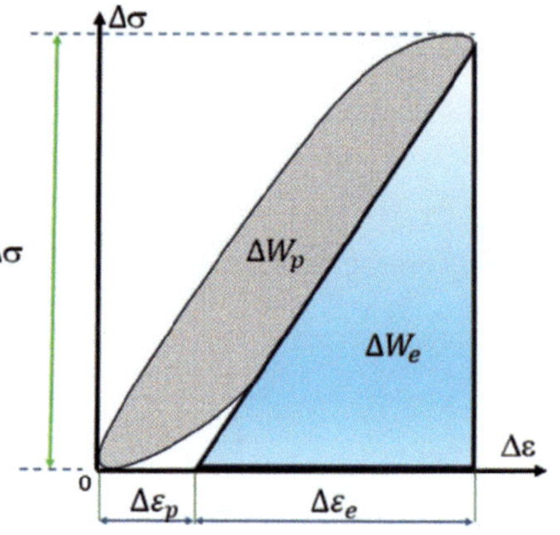

FIGURE 4.18 Decomposition of the hysteresis loop on the energy components.

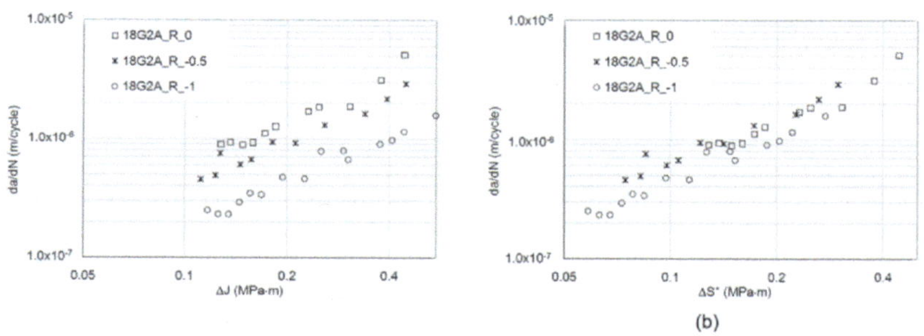

FIGURE 4.19 Fatigue crack growth curves for 18G2A steel (a) based on the ΔJ crack driving force, (b) based on the new, averaged ΔS^* crack driving force, [37].

under mixed modes loading are not standardized. Locally, mixed modes cracks can be observed in machine elements, caused by an external load and a crack direction different from the three typical modes crack. This means that material loads induce asymmetric stress fields near the crack front. In practice, such cracks can occur, for example, in bends, branches, notches, or welded joints. The loads that induce them can be static, dynamic, or thermal. Examples of mixed modes loading are shown in Figure 4.20. The specimen shown in Figure 4.20a was used to analyze fatigue crack growth with mixed modes I and II loading obtained by initiating the crack at an angle but with a uniaxial force acting on the specimen [42]. Similarly, Figure 4.20b shows crack growth in mixed modes I and II loading of a CT-type specimen, but with the load acting diagonally on the specimen [43]. Four-point bending and shearing of the specimen (modes II and III loading) for a semi-elliptical edge crack are shown in Figure 4.20c and described in Tohgo et al. [44]. The twisted circular notched specimen subjected to tension, shown in Figure 4.20d, is used for testing modes I and III loading [45].

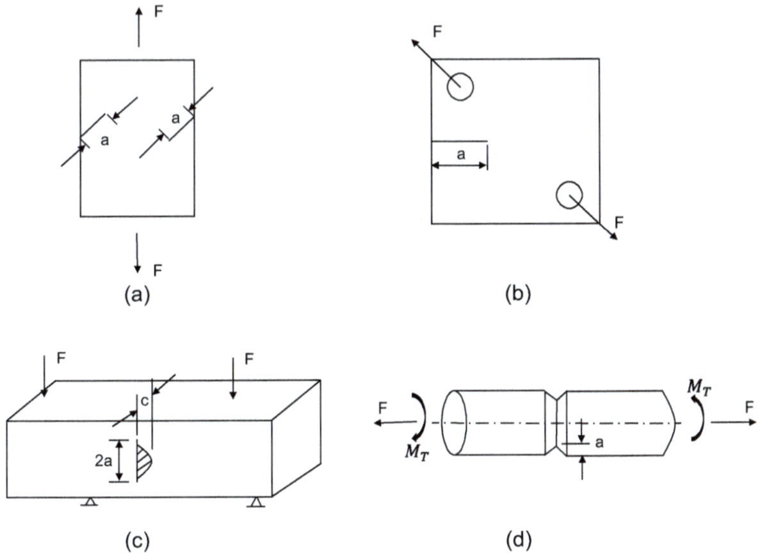

FIGURE 4.20 Examples of specimens for mixed modes loading tests. (Reprinted with permission from Ref. [40], Elsevier.)

The criteria related to mixed modes apply to different types of materials and the ranges they describe, and so they can be stress criteria related to the linear range, strain criteria, or crack tip opening displacement (CTOD) criteria up to the yield strength, and energy criteria covering the entire scope of research [39,40].

Under uniaxial or multiaxial loading, crack growth occurs when the stress intensity factor (SIF) K for mode I or equivalent load near the crack front reaches a critical value K_{Ic}. Then the fracture criterion can be written as:

$$K_I(\Delta K_{eq}) = K_{Ic} \tag{4.95}$$

The criteria presented in this paper can be based on the threshold value K_{th} or relate to the critical value K_c. The criteria based on the critical value are more often presented, and this paper also concerns them because they are concerned with the description of the entire range of fatigue cracking.

One of the first and well-known works on the criteria is the work of Erdogan and Sih [46], referring to the criterion of maximum shear stresses. In the linear elastic range, Pook proposed a criterion for three modes (I + II + III) loading in the form of Pook [47,48]:

$$K_{(I,II,III)eq} = \frac{K_{(I,II)eq}(1+2\nu)+\sqrt{K^2_{(I,II)eq}(1-2\nu)^2+4K^2_{III}}}{2} \le K_{Ic} \tag{4.96}$$

Based on the above equation, Pook determined the relationships between the critical stress intensity factors (SIF): $K_{IIc} = 0.87\ K_{Ic}$ and $K_{IIIc} = K_{Ic}$.

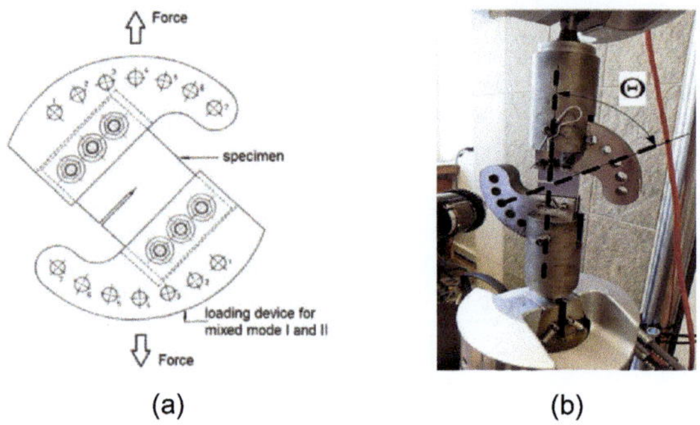

FIGURE 4.21 (a) Specimen scheme with a holder for mixed-mode I + II, (b) the experimental setup for fatigue fracture test for mixed-mode I + II.

Richard [41] proposed a generalized fracture criterion for three modes of loading as:

$$K_{eq} = \frac{K_I}{2} + \frac{1}{2}\sqrt{K_I^2 + 4(\beta_1 K_{II})^2 + 4(\beta_2 K_{III})^2} \le K_{Ic} \quad (4.97)$$

where the coefficient values are $\beta_1 = K_{Ic}/K_{IIc} = 1.155$, $\beta_2 = K_{Ic}/K_{IIIc} = 1.0$.

Richard and his team's significant achievement was developing a specimen holder for mixed modes crack testing. An example of such a holder together with a specimen for mixed modes I + II tests is shown in Figure 4.21. Figure 4.21a shows a diagram of the specimen together with the holder, and Figure 4.21b shows the holder mounted on a fatigue machine for uniaxial force (F) tests. Still, thanks to the holder, we have a biaxial test mode [49].

Tanaka [50] formulated the following expression for the effective range of the SIF for three cracking modes:

$$\Delta K_{eq,eff} = \left(\Delta K_I^4 + 8\Delta K_{II}^4 + \frac{8\Delta K_{III}^4}{(1-\nu)} \right)^{0.25} \quad (4.98)$$

where ΔK is range of the stress intensity factor for modes I, II, III, ν is Poisson's ratio.

This equation assumes that the fatigue crack growth occurs when displacements in the plastic zone reach the critical value. Moreover, for the mixed modes I, II, and III, it is necessary to introduce an assumption that there is no interaction between deformations coming from those crack modes.

The literature has many stress criteria for mixed modes because they are easy to verify during tests. This parameter is especially helpful for brittle and elastic-plastic materials in the linear range or short cracks. The criteria based on crack tip opening displacement (CTOD) are related to multiaxial fatigue cracking and can only be applied up to the yield strength. Panasyuk and Andreykiv [35] proposed the cracking criterion for multiaxial loads, taking into account the crack tip displacements for three modes and the critical values of the crack opening. He formulated the following conditions for crack development:

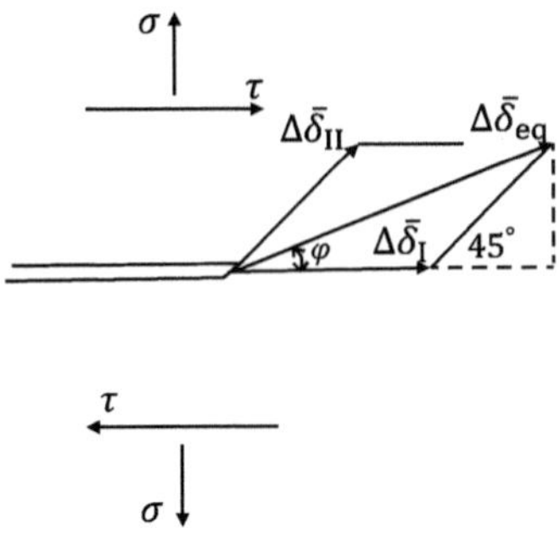

FIGURE 4.22 Vector CTOD for mixed mode I+II. (Reprinted with permission from Ref. [39], Springer.)

$$\left(\frac{\delta_{\mathrm{I}}}{\delta_{\mathrm{I}c}}\right)^{n_1}+\left(\frac{\delta_{\mathrm{II}}}{\delta_{\mathrm{II}c}}\right)^{n_2}+\left(\frac{\delta_{\mathrm{III}}}{\delta_{\mathrm{III}c}}\right)^{n_3}=1 \tag{4.99}$$

where n_1, n_2, and n_3 are material constants determined during experiments, δ_i is the crack tip opening displacement for three modes $i=\mathrm{I}$, II, and III.

For description of the mixed mode I and II, Li [51] applied the criterion based on the CTOD range ($\Delta\delta_{eq}$). The crack vector displacement was defined at the angle φ deflected from mode I according to Figure 4.22. The author states in Li [51] that at the initial stage of the tests the crack developed according to mode I, and after a certain number of cycles the crack changed direction and growth according to the mixed mode I+II.

The author proposed [51] a criterion for the mixed mode I+II:

$$\Delta\delta_{eq}=\frac{4}{\pi E\sigma_Y}\sqrt{\left(\Delta K_{\mathrm{I}}^2+3\Delta K_{\mathrm{II}}^2\right)\left(\Delta K_{\mathrm{I}}^2+2\Delta K_{\mathrm{II}}^2+2\Delta K_{\mathrm{I}}\Delta K_{\mathrm{II}}\right)} \tag{4.100}$$

where E - Young's modulus, σ_Y - yield strength under tension, and ΔK_{I} and ΔK_{II} are ranges of the SIF for modes I and II. Multiaxial fatigue crack growth criteria based on CTOD are not widely applicable because of measurement problems during mixed mode crack tip displacement.

The third group of criteria related to fatigue cracks in different directions are energy criteria, which are based on the strain energy density W or the J parameter. They are starting to be widely applied because they take into account the entire range of tests and can be related to both the linear elastic and elastic-plastic ranges.

The equivalent fatigue crack growth range ΔJ_{eq} (previously in the linear range ΔK_{eq}) in mixed mode loading cases is calculated in the limits from the threshold to the critical value:

$$\Delta J_{\mathrm{I,th}}(\Delta K_{\mathrm{I,th}})<\Delta J_{eq}(\Delta K_{eq})<J_{\mathrm{I}c}(K_{\mathrm{I}c}) \tag{4.101}$$

For the entire range of elastic-plastic materials subjected to multiaxial fatigue loading, in the work [39], the following energy criterion for cracking was proposed, using the range of the parameter J:

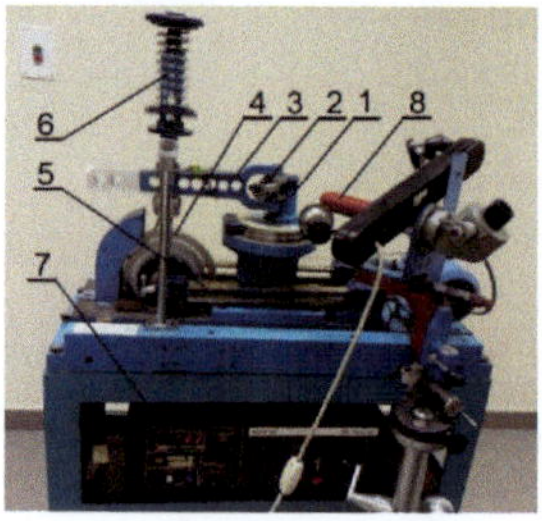

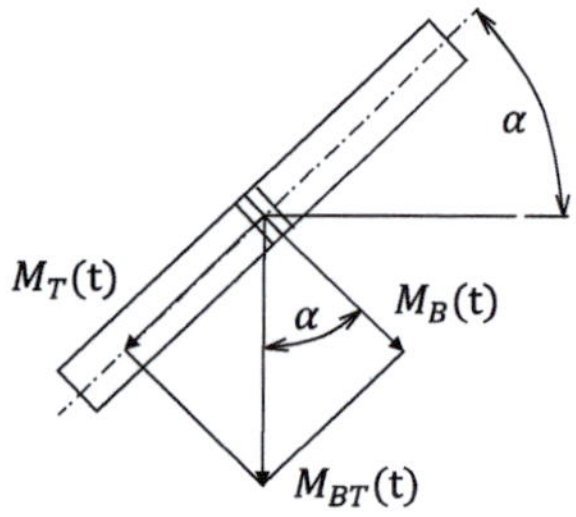

FIGURE 4.23 Fatigue test stand MZGS-100 (a) 1 – rotational head with a clamp, 2 – holder, 3 – lever (length = 0.2 m), 4 – rotating disk, 5 – flat spring, 6 – spring, 7 – control box, 8 – microscope, (b) description of the angle α on the specimen for mixed mode I + III.

$$\left(\frac{\Delta J_{\mathrm{I}}}{J_{\mathrm{I}c}}\right)^2+\left(\frac{\Delta J_{\mathrm{II}}}{J_{\mathrm{II}c}}\right)^2+\left(\frac{\Delta J_{\mathrm{III}}}{J_{\mathrm{III}c}}\right)^2=1 \tag{4.102}$$

where $J_{\mathrm{I}c}$, $J_{\mathrm{II}c}$, and $J_{\mathrm{III}c}$ are the critical values of the J parameter for mixed modes I, II, and III.

The criterion shown in Eq. (4.102) was verified while tests of aluminium alloy and steel, among others, for proportional bending and torsion (mixed mode I + III) on the MZGS-100 machine (Figure 4.23) with a sinusoidal waveform [52]. The device enables testing of bending, torsion, and proportional bending with torsion. The existing shear force on the lever has small values less than 2% of the maximum bending moment. The tests on the device can be performed with cyclic loading of the specimen and static loading by the spring pressure. During the tests, the ratio of the torsional moment to the bending moment was $M_T/M_B = \tan\alpha$. The definition of the angle α under bending with torsion on a specimen is shown in Figure 4.23b. The tests were carried out with force control (with controlled total torque amplitude) at a load frequency of 28.4 Hz. The development of fatigue cracks was observed with a light microscope and measured with a micrometric optical measurement system. During the tests, the number of cycles N was also recorded. For bending and torsion, the crack length "a" is defined as the increments in length and angle measured on the lateral surface of the specimen.

In several Authors' papers [49, 52, 53, 54] different combination of mixity levels (I+II, I+III etc.) were tested in elastic-plastic materials. The criteria were verified on three different steels: an aluminium alloy and a titanium alloy. The test results were described with the range of the parameter ΔJ. In modes I, II, and III, the range of the equivalent parameter ΔJ_{eq} is equal to the sum of the ranges of these parameters ΔJ for the individual loading modes:

$$\Delta J_{\mathrm{eq}} = \Delta J_{\mathrm{I}} + \Delta J_{\mathrm{II}} + \Delta J_{\mathrm{III}} \tag{4.103}$$

The ΔJ parameter ranges for particular modes include both elastic and plastic parts.

The ΔJ parameter ranges for modes I, II, and III were calculated from the dependence [53]:

$$\Delta J_{\mathrm{I}} = \left(1 - \nu^2\right)\frac{\Delta K_{\mathrm{I}}^2}{E} + \pi Y_{\mathrm{I}}^2\left(\frac{\Delta\sigma\Delta\varepsilon_p}{\sqrt{n'}}\right)a \tag{4.104}$$

$$\Delta J_{\mathrm{II}} = \left(1 - \nu^2\right)\frac{\Delta K_{\mathrm{II}}^2}{E} + \pi Y_{\mathrm{II}}^2\left(\frac{\Delta\tau\Delta\gamma_p}{\sqrt{n'}}\right)a \tag{4.105}$$

$$\Delta J_{\mathrm{III}} = (1 + \nu)\frac{\Delta K_{\mathrm{III}}^2}{E} + \pi Y_{\mathrm{III}}^2\left(\frac{\Delta\tau\Delta\gamma_p}{\sqrt{n'}}\right)a \tag{4.106}$$

where a is length of the crack, E is Young's modulus, ν is Poisson's ratio, $\Delta\sigma$, $\Delta\tau$ is stress ranges for modes I, II, and III near the crack (notch) tip, $\Delta\varepsilon_p$, $\Delta\gamma_p$ is ranges of plastic strains for modes I, II, and III near the crack (notch) tip, $Y_{\mathrm{I,II,III}}$ is correction coefficients for modes I, II, and III [52,53] n' is cyclic strain hardening exponent. We can present a graphical diagram of the ranges in which a crack develops using the three fracture modes. Figure 4.24 illustrates the stable fatigue crack growth in mixed mode domains for the energy approach. It occurs between the threshold boundary surface ΔJ_{th} and the critical boundary curves (described by the critical values J_{Ic}, J_{IIc}, and J_{IIIc}), forming the crack interface. The tests were performed in mixed modes I+II and I+III (grey areas in the diagram), and these ranges are described schematically in Figure 4.24.

Examples of cracks observed in the mixed mode I+II and I+III experiments are shown in Figure 4.25 [49]. In Figure 4.25a the crack grew at an angle of $\theta = 30°$, while in Figure 4.25b for an angle of $\alpha = 60°$ (Figure 4.25) the fatigue crack growth was at an angle of $\alpha_1 = 40°$.

Criteria describing the growth of fatigue cracks in mixed mode are often used in the description and various analyses. Conducting experiments is expensive and

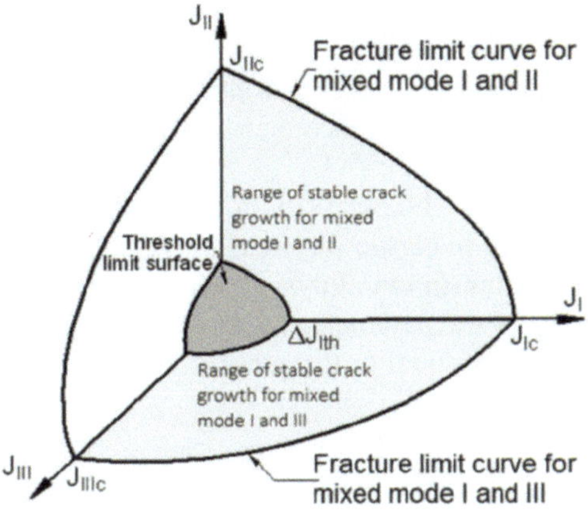

FIGURE 4.24 Diagram for stable fatigue crack growth for mixed modes energy, including fracture limits.

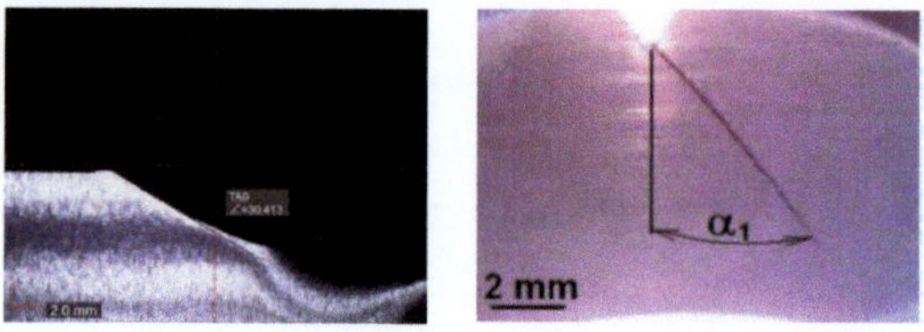

FIGURE 4.25 Fatigue crack paths for (a) mode I+II and $\theta=30°$, (b) mode I+III and $\alpha=60°$ angle.

laborious; therefore, new computational techniques are increasingly being used, which, at the initial stage, seem to be a good approach. Many authors believe the stress approach does not describe all phenomena occurring during fatigue crack growth, especially in elastic-plastic materials. Therefore, the energy approach is used, fully explaining the phenomena occurring during research on the material.

Experimentally J-integral can be easily interpreted as:

$$J_{eq} = -\frac{1}{2\pi L}\left(\frac{\partial U_i}{\partial a}\right)_{u,\alpha} \tag{4.107}$$

where L represents ligament length (non-fractured specimen length measured along the crack path), u is displacement, α is mixing parameter, and U_i is strain energy density for a specific combination of fracture modes. For mixed-mode loading, a possible extension of the energy-based crack driving force ΔS^*_{eq} (Eq. 4.108) can be represented as follows [54]:

$$\Delta S^*_{eq} = \sqrt{\Delta J^+_{eq} J^{max}_{eq}} \tag{4.108}$$

where ΔJ^+_{eq} is an equivalent positive range of the J-integral obtained for mixed-mode loading conditions, $J^{(max)}_{eq}$ is an equivalent maximal value of the J-integral obtained for mixed-mode loading conditions.

Exemplary kinetic fatigue fracture diagrams for P355NL1 steel obtained for multiaxial loading (I+II) are presented in Figures 4.26 and 4.27.

As it can be concluded, based on the presented KFFDs, the energy approach is an efficient way to describe FCGR under mixed-mode loading conditions with respect to the mean stress effect.

4.6 FCG PREDICTIONS AND LIFETIME CALCULATIONS: CASE STUDY

Lug joints are widely used in various fields of engineering, such as aircraft design (the connection between fuselage and wing), automotive (as a part of suspension system), civil engineering (bridges), and many more. The main advantage of lug connectors is the simplicity of assembling and disassembling parts or mechanical systems. Conversely, lack of adequate lubrication conditions, the existence of

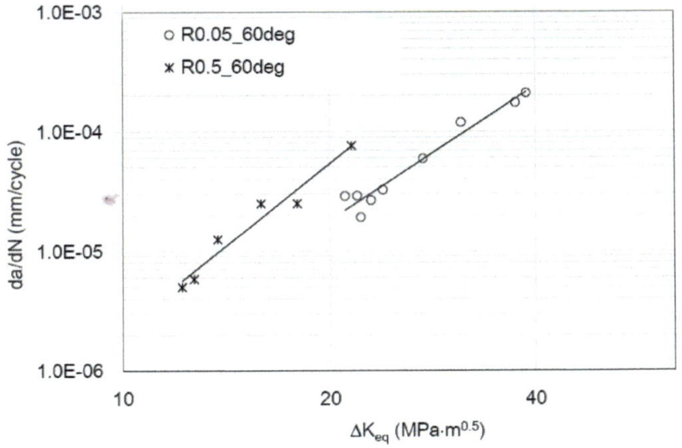

FIGURE 4.26 Kinetic fatigue fracture diagram for P355NL1 steel (R=0.05, 0.5, load angle=60°) based on ΔK approach. (Reprinted with permission from Ref. [54], Elsevier.)

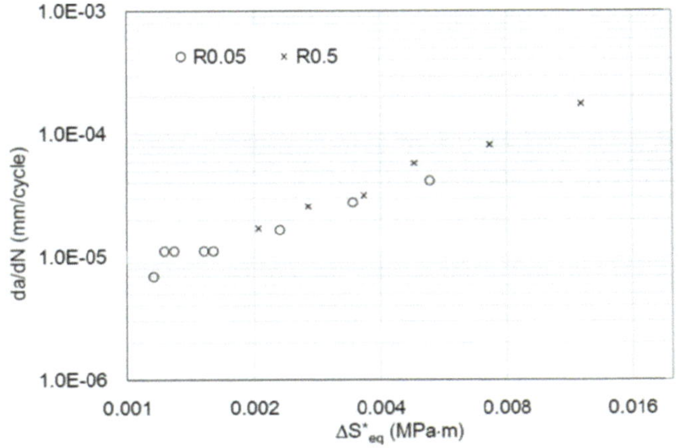

FIGURE 4.27 Kinetic fatigue fracture diagram for P355NL1 steel (R=0.05, 0.5, load angle=60°) based on ΔS^* approach. (Reprinted with permission from Ref. [54], Elsevier.)

transversal clearance, exposure to a corrosive environment and high surface stress rates may lead to crack initiation and propagation. Owing to the above facts, the usage of numerical methods with fracture mechanics tools as its basis is completely justified.

In this subsection, the procedure of stress intensity factors evaluation was presented, as well as the modelling method and assumed boundary conditions. Boundary conditions and load distribution in lug connector are presented in Figure 4.28 [55]. All the numerical computations were performed in *Abaqus/CAE* 6.14. Results were compared with analytical closed solutions [55]. Finally, the experimental fatigue lifetime obtained in works [56] for Al7075-T6 alloy [57] will be compared with the

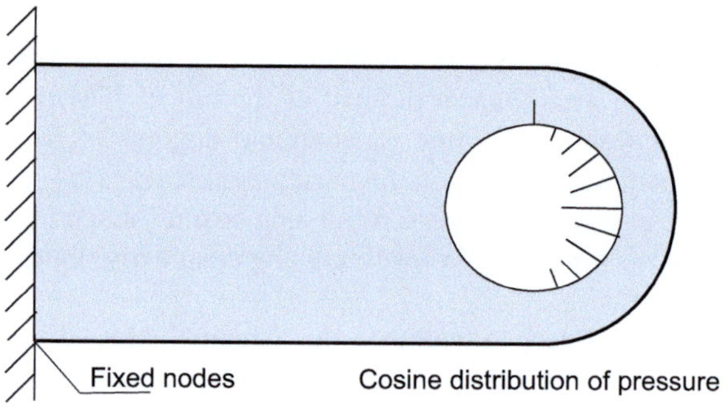

FIGURE 4.28 Boundary Conditions and load distribution in lug connector. (The figure is created based on the work of Ref. [55].)

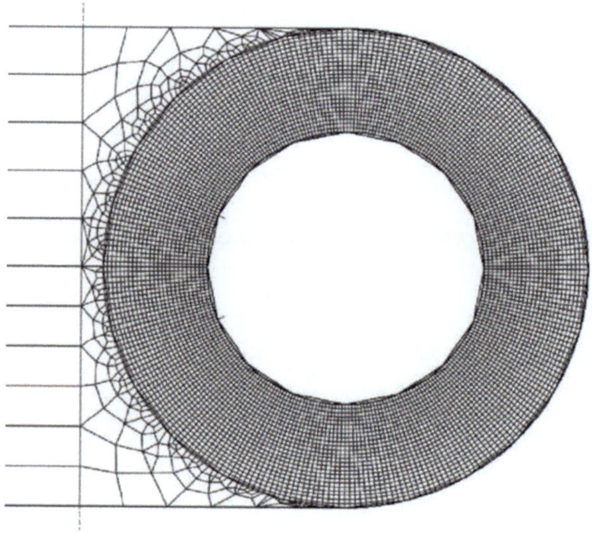

FIGURE 4.29 Elements pattern in lug connector discrete model. The existence of finer mesh in the region of potential crack propagation [55].

proposed numerical tool FM-TOOL as a fatigue crack growth solver. Initially, the FEM-xFEM method was used for the validation of existing solutions of SIFs and stress-strain field analysis in the lug connector.

The lug connector was modelled as a three-dimensional solid body. The stationary crack approach was used for simplification with a constant 0.75 mm increment of crack growth for each simulation (values between 0.5 and 8.75 mm). Crack was modelled in regard to xFEM methodology with through-the-thickness geometry. The discrete model has 180000 8-node linear elements (HEXA8). The choice was coerced by the eXtended finite element method being limited to that order of elements. The mesh pattern is presented in Figure 4.29 [55].

A number of elements and their quality (aspect ratio, distortion) are crucial to getting accurate results in the vicinity of the crack tip due to the fact the mesh is much finer in the propagation area than in the rest of the model. The planar end surface of the lug connector was fixed (three translational degrees of freedom were constrained). To avoid contact problems in the enrichment region (lug–pin pair), traction force between these elements was performed as a cosine–shaped pressure distribution. The load was distributed by pin loading using cosine distribution of pressure in the Abaqus environment.

The crack was located perpendicular to the principal stress direction. Thus, criteria were assumed. Stress intensity factors were obtained for each crack increment and stress and displacement distribution. SIFs for each crack growth increment were estimated and gathered in the form of a graph shown in Figure 4.31. Results obtained from the simulation were compared to other analytical or empirical solutions presented. It should be noted that K_I magnitudes differ for each model. It is worth mentioning that the behaviour of crack propagation is more complex for crack lengths larger than 8.5 mm due to the relatively large yield region near the crack tip. In literature [58], some formulae describe generalized equations for SIF in different configurations of lug geometry:

$$K_I = G_w G_L G f_b \sigma_n \sqrt{\pi a} = M_m \sigma_n \sqrt{\pi a} \tag{4.109}$$

where σ_n represents the normal, gross stress in cross section and G, G_L, G_w, and f_b are defined as:

$$G_w = \sqrt{\sec\left(0.5\pi \frac{2r+a}{W-a}\right)} \tag{4.110}$$

$$G_l = \sqrt{\sec\left(\frac{\pi d}{2W}\right)} \tag{4.111}$$

$$G = 0.5 + \frac{W}{\pi(2r+a)}\sqrt{\frac{r}{r+a}} \tag{4.112}$$

$$f_b = 0.707 - 0.18z + 6.55z^2 - 10.54z^3 + 6.84z^4 \tag{4.113}$$

The variable z represents the non-dimensional crack length function: $z = 1/(1+2a/r)$. Results of the comparison of SIF solutions are presented in Figure 4.30. An example plot of a nondimensionalized function M_m is presented in Figure 4.31.

In the following steps, a crack growth simulation was performed in an FM-TOOL numerical environment delivered by Nobo Solutions Company (Poland). For simulation, the same geometry of the lug connector was used as in the papers [56]. The scheme of the lug connector and dimensions are shown in Figure 4.32 (thickness $B = 12.7$ mm, $d = 38.1$ mm, $W = 85.7$ mm). The load type was modelled as pin-loaded central force F_n, which caused stress in a rectangular cross section on level 41.38 MPa

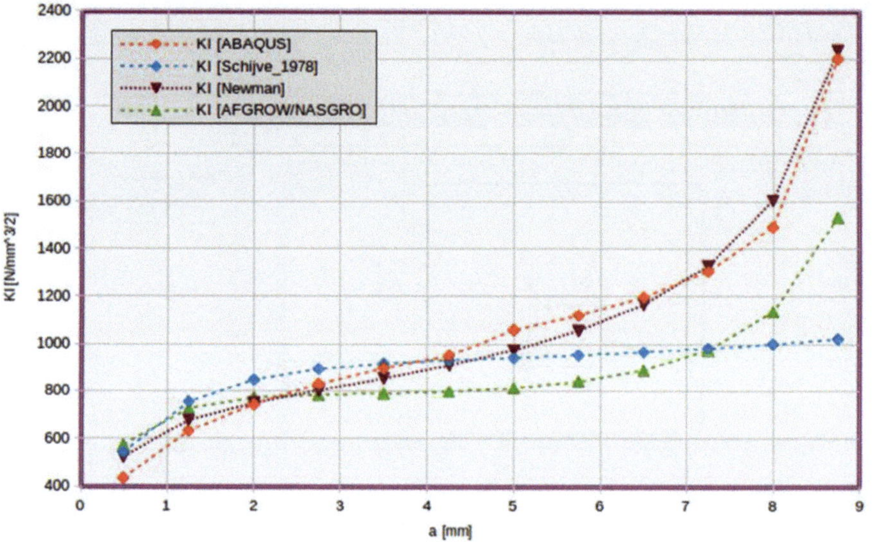

FIGURE 4.30 Comparison between results obtained from Abaqus and other analytical solutions [55].

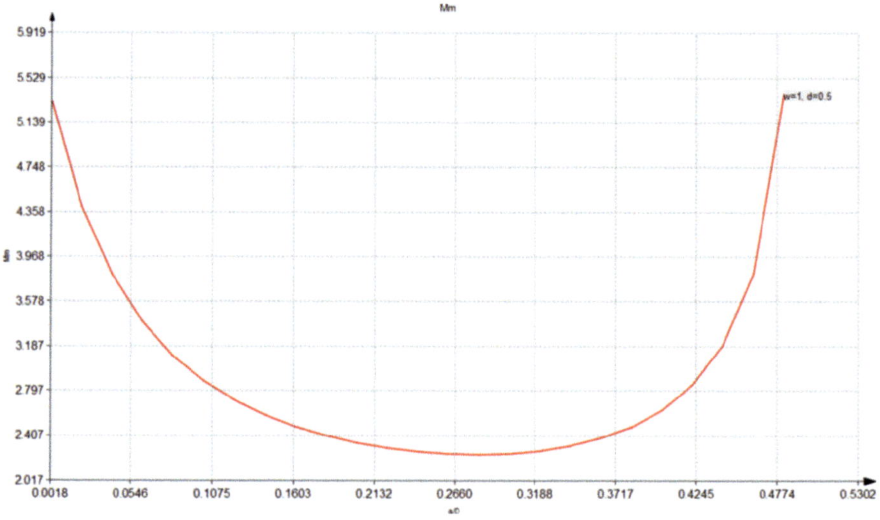

FIGURE 4.31 Nondimensionalized M_m function (Eq. 4.109) for normalized crack length in lug connector.

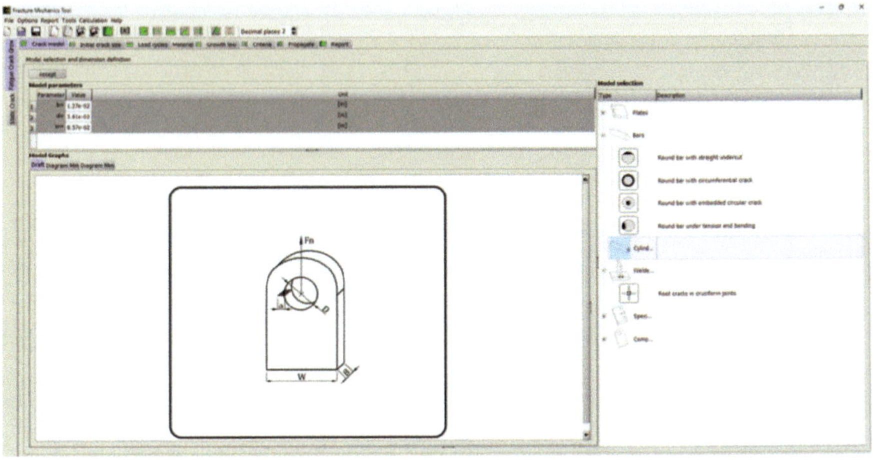

FIGURE 4.32 Geometry window and boundary conditions view in FM-TOOL environment.

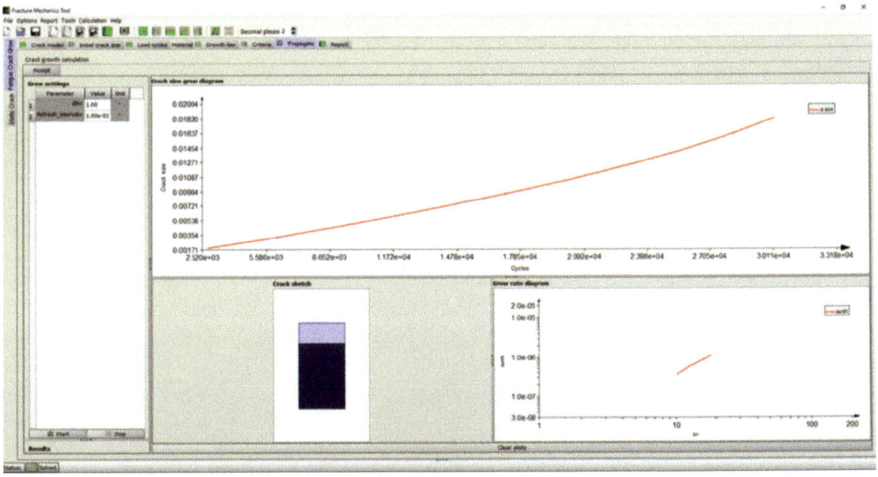

FIGURE 4.33 The output of simulation results in FM-TOOL environment.

(similar to reference experimental works [56]) as material data for stress ratio $R=0.1$ was used AL7075-T6 expressed in Paris law's constants $C=4.33\times10^{-9}$, $m=1.91$ as reported in paper [57].

The view window of the FM-TOOL system with numerical analysis results is presented in Figure 4.33. As a stop criterion, the same crack length was published in experimental works [56].

Predicted fatigue lifetime is compared with experimental data for $R=0.1$ taken from Bahloul et al. [56]. A comparison is presented in Figure 4.34.

As it can be concluded, prediction results are satisfactory even for complicated structure joints like lug connectors. It is worth noting that in the case of analytical solvers, computation time is much more efficient than in FEM results. However, a

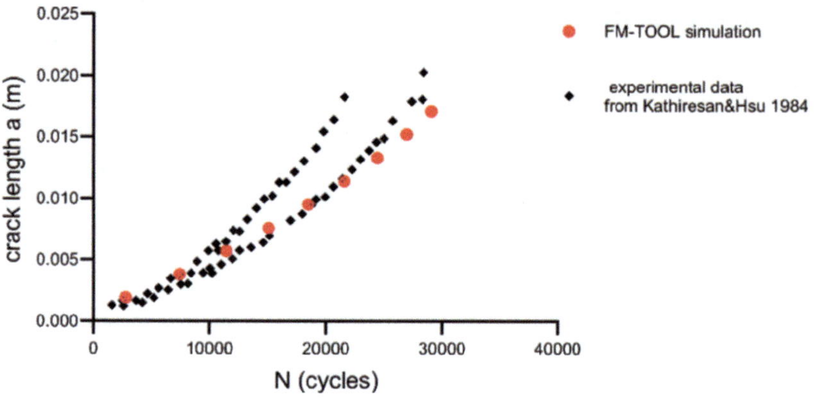

FIGURE 4.34 Comparison of fatigue lifetime predicted by FM-TOOL solution with experimental. (Data from Ref. [56].)

problem can be specific geometry finding proper SIF solutions, and in this case, a hybrid approach is required by building its own analytical formulae of SIF and inputting them into numerical fatigue calculators.

4.7 SUMMARY

The principles underlying the mechanics of fatigue fracture propagation have been covered in this chapter. Additionally, a summary of the basic development rules based on linear fracture mechanics was given. Detailed equations are used to discuss the applicability of the fracture propagation laws. Also described was the energy foundation for fracture growth models. It was demonstrated that strain energy, independent of stress ratio R (mean stress effect), enables efficient fatigue crack growth modelling. It was demonstrated that crack propagation in the elastic-plastic region may be described by relationships between the wasted energy and the cyclic value of the J-integral. The issue of fatigue fracture of structural parts under intricate stress scenarios was then covered.

ABBREVIATIONS

COD	Crack Opening Displacement
CTOD	Crack Tip Opening Displacement
FEM	Finite Element Method
FM	Fracture Mechanics
FCG	Fatigue Crack Growth
FCGR	Fatigue Crack Growth Rate
KFFD	Kinetic Fatigue Fracture Diagram
LEFM	Linear Elastic Fracture Mechanics
SIF	Stress Intensity Factor
SOP	Standard Operating Procedures
xFEM	Extended Finite Element Method

ACKNOWLEDGEMENTS

The authors would like to thank the following institutions and programmes thanks to which this work has been created, namely the Wroclaw University of Science and Technology through the *Academia Professorum Iuniorum* project and the Visegrad Fund grant ID #62320197 (Fatigue Crack Growth rate description in structural steels for Civil Engineering Structures).

REFERENCES

1. Zamanzadeh M, Larkin E, Mirshams R (2015) Fatigue failure analysis case studies. *J Failure Anal Prev, 15*:803–809.
2. Slattery JC Cizmas PG (2018) Macro-scale fatigue fracture analysis of multiphase bodies, aircraft design, and catastrophic failure: Two aircraft accidents. *Eng Fract Mech*, 199:274–279.
3. Gomes VMG, Lesiuk G, Correia JAFO, De Jesus AMP (2024) Fatigue crack propagation of 51CrV4 steels for leaf spring suspensions of railway freight wagons. *Materials* 17:1831.
4. Schütz W (1996) A history of fatigue. *Eng Fract Mech* 54:263–300.
5. Sanford RJ (1997) *Selected Papers on Foundations of Linear Elastic Fracture Mechanics.* SPIE Milestone Series 137. Bethel: SPIE Optical Engineering Press.
6. Hanewinckel D, Zenner H (1989) *Fatigue Strength, a Collection of Facsimile, Historical Papers until 1950.* Clausthal: Technical University.
7. Schijve J (2003) Fatigue of structures and materials in the 20th century and the state of the art. *Int J Fatigue* 25:679–702.
8. Li B, Shen Y, Hu W (2011) Casting defects induced fatigue damage in aircraft frames of ZL205A aluminum alloy–A failure analysis. *Mater Des*, 32:2570–2582.
9. Heywood RB (1962) *Designing against Fatigue.* London: Chapman and Hall.
10. Paris PC Sih GC (1965) Stress analysis of cracks. *ASTM stp, 381*:30–81.
11. Paris P, Erdogan F (1963) A critical analysis of crack propagation laws. *J Basic Eng* 85(4):528.
12. Paris CP (1961) A rational analytic theory of fatigue. *Trends Eng* 13:9–14.
13. Blom AF, Hedlund A, Zhao W (1986) Short fatigue crack growth behaviour in Al 2024 and Al 7475. In: KJ Miller, ER de los Rios (eds.), *The Behaviour of Short Fatigue Cracks*, Mechanical Engineering Publications, pp. 37–66.
14. Melander A, Larsson M (1993) The effect of stress amplitude on the cause of fatigue crack initiation in a spring steel. *Int J Fatigue* 15:119–131.
15. Ceyhanli UT, Bozca M (2020) Experimental and numerical analysis of the static strength and fatigue life reliability of parabolic leaf springs in heavy commercial trucks. *Adv Mech Eng* 12:1687814020941956.
16. Wu SC, Li CH, Luo Y, Zhang HO, Kang GZ (2020) A uniaxial tensile behavior based fatigue crack growth model. *Int J Fatigue* 131:105324.
17. Lee CS, Lee KA, Li DM, Yoo SJ, Nam WJ (1998) Microstructural influence on fatigue properties of a high-strength spring steel. *Mater Sci Eng A* 241:30–37.
18. Dowling NE, Iyyer NS (1987) Fatigue crack growth and closure at high cyclic strains. *Mater Sci Eng, 96*: 99–107.
19. Dowling NE, Kampe SL, Kral MV (1999) *Mechanical Behavior of Materials: Engineering Methods for Deformation, Fracture, and Fatigue.* Upper Saddle River: Prentice Hall.
20. Forman RG, Kearney VE, Engle RM (1967) Numerical analysis of crack propagation in cyclic-loaded structures. *J Fluids Eng* 89:459.

21. Elber W (1997) Fatigue crack closure under cyclic tension. *Eng Fract Mech* 2, 37–44.
22. Suresh S (1998) *Fatigue of Materials*. Cambridge: Cambridge University Press.
23. Maierhofer J, Gänser H-P, Pippan R (2015) Modified Kitagawa–Takahashi diagram accounting for finite notch depths. *Int J Fatigue* 70:503–509.
24. Newman Jr JC (1984) A crack opening stress equation for fatigue crack growth. *Int J Fract* 24: R131–R135.
25. Wu S-X, Mai Y-W, Cotterell B (1992) A model of fatigue crack growth based on Dugdale model and damage accumulation. *Int J Fract* 57:253–267.
26. Sih GC, Barthelemy BM (1980) Mixed mode fatigue crack growth predictions. *Eng Fract Mech* 13:439–451.
27. Raju KN (1972) An energy balance criterion for crack growth under fatigue loading from considerations of energy of plastic deformation. *Int J Fract Mech* 8:1–14.
28. Badaliance R (1980) Application of strain energy density factor to fatigue crack growth analysis. *Eng Fract Mech* 13:657–666.
29. Cherepanov, G. P. (1967). Crack propagation in a continuum. Prikl. Mat. Mekh.;(USSR), 31(3).
30. Szata M (2002) *Modeling of Fatigue Crack Growth Using Energy Method*. Wroclaw: Publishing House of Wroclaw University of Technology.
31. Szata M, Lesiuk G (2009) Algorithms for the estimation of fatigue crack growth using energy method. *Arch Civil Mech Eng* 9:119–134.
32 Lesiuk G, Szata M, Rozumek D, Marciniak Z, Correia J, De Jesus A (2018) Energy response of S355 and 41Cr4 steel during fatigue crack growth process. J Strain Anal Eng Des 53:663–675
33. Panasyuk VV, Andreikiv OE, Darchuk OI (1995) Calculation model of fatigue crack growth in elastic-plastic materials under mixed mode loading. In: *7th International Conference on the Mechanical Behaviour of Materials* (May–June, 1995, Hague). pp 119–120.
34. Panasyuk VV, Andreikiv OE, Darchuk OI, Kun PS (1994) Analysis of short and long fatigue cracks growth kinetics under nonregular loading. *Struct Integr Exp Models Appl* 2:1271–1276.
35. Panasyuk VV, Andreykiv Oye (1997) Fracture mechanics of materials and methods of assessment of structures service life-time: success and prospects. In: BL Karihaloo (ed.), *Advances in Fracture Research*, Pergamon, pp. 105–116.
36. Panasyuk VV, Savruk MP, Datsyshyn AP (1977) A general method of solution of two-dimensional problems in the theory of cracks. *Eng Fract Mech* 9:481–497.
37. Lesiuk, G (2019) Application of a new, energy-based ΔS^* crack driving force for fatigue crack growth rate description. *Materials*; 12:518.
38. Dowling NE, Begley JA (1976) Fatigue crack growth during gross plasticity and the J-integral. In: JR Rice, PC Paris (eds.), *Mechanics of Crack Growth*. ASTM International. pp. 82–103.
39. Rozumek D, Macha E (2009) A survey of failure criteria and parameters in mixed-mode fatigue crack growth. *Mater Sci* 45:190–210.
40. Qian J, Fatemi A (1996) Mixed mode fatigue crack growth: a literature survey. *Eng Fract Mech* 55:969–990.
41. Richard HA (2003) Theoretical crack path determination. In: *International Conference on Fatigue Crack Paths* (18–20 September, 2003, Parma). Citeseer.
42. Lal D (1970) Fatigue crack propagation and fracture in compressive members. MS Thesis, Syracuse University.
43. Mahanty DK, Maiti SK (1990) Experimental and finite element studies on mode I and mixed mode (I and II) stable crack growth—I. Experimental. *Eng Fract Mech* 37:1237–1250.

44. Tohgo K, Otsuka A, Yoshida M (1990) Fatigue behavior of a surface crack under mixed mode loading. In: *Proceedings of the Fourth International Conference on Fatigue and Fatigue Thresholds* (15–20 July, 1990, Honolulu), MCPE. pp. 567–572.
45. Kimachi H, Tanaka K, Akiniwa Y, Yu H (2003) Elastic-plastic fatigue crack propagation under mixed-mode (I+III) cyclic torsion and axial loading. In: *International Conference on Fatigue Crack Paths* (18–20 September, 2003, Parma).
46. Erdogan F, Sih GC (1963) On the crack extension in plates under plane loading and transverse shear. *J Basic Eng* 85:519–525.
47. Pook LP (1985) The fatigue crack direction and threshold behaviour of mild steel under mixed mode I and III loading. *Int J Fatigue* 7:21–30.
48. Pook LP (1971) The effect of crack angle on fracture toughness. *Eng Fract Mech* 3:205–218.
49. Rozumek D, Marciniak Z, Lesiuk G, Correia JA, de Jesus AMP (2018) Experimental and numerical investigation of mixed mode I+II and I+III fatigue crack growth in S355J0 steel. *Int J Fatigue* 113:160–170.
50. Tanaka K (1974) Fatigue crack propagation from a crack inclined to the cyclic tensile axis. *Eng Fract Mech* 6:493–507.
51. Li C (1989) Vector CTD criterion applied to mixed mode fatigue crack growth. *Fatigue Fract Eng Mater Struct* 12:59–65.
52. Rozumek D, Macha E (2006) Elastic–plastic fatigue crack growth in 18G2A steel under proportional bending with torsion loading. *Fatigue Fract Eng Mater Struct* 29:135–144.
53. Rozumek D (2009) Influence of the slot inclination angle in FeP04 steel on fatigue crack growth under tension. *Mater Des* 30:1859–1865.
54. Lesiuk G (2019) Mixed mode (I+II, I+III) fatigue crack growth rate description in P355NL1 and 18G2A steel using new energy parameter based on J-integral approach. *Eng Fail Anal* 99:263–272.
55. Kucharski P, Lesiuk G, Czapliński T, Fratczak R, Maciejewski Ł (2016) Numerical estimation of stress intenisty factors and crack propagation in lug connector with existing flaw. XXVI Sympozjum *Zmęczenie i Mechanika Pękania*, May 17–20, 2016, *Bydgoszcz* – Fojutowo, In: *AIP Conference Proceedings*. AIP Publishing.
56. Bahloul A, Ahmed AB, Bouraoui C (2018) An engineering predictive approach of fatigue crack growth behavior: the case of the lug-type joint. *Comptes Rendus Mécanique* 346:1–12.
57. Shukla SS, Murthy K (2023) A study on the effect of different Paris constants in mixed mode (I/II) fatigue life prediction in Al 7075-T6 alloy. *Int J Fatigue* 176:107895.
58. Schijve J, Hoeymakers AHW (1979) Fatigue crack growth in lugs. *Fatigue Fract Eng Mater Struct* 1:185–201.

5 Ultra-Low-Cycle Fatigue

João Carlos Pereira and Abílio M.P. De Jesus

5.1 INTRODUCTION

Traditionally, fatigue of materials has been investigated in two primary fatigue regimes: high-cycle fatigue (HCF) and low-cycle fatigue (LCF). High-cycle fatigue is characterized by elastic deformation (elastic fatigue), where fatigue damage develops with minimal macroscopic plasticity. In this fatigue regime, plasticity or other non-linear material behaviors are typically confined to localized features, such as notches or the material's microstructure. This was the initial focus of fatigue research, often limited to fatigue lives around 10^7 cycles, corresponding to the material's true or conventional fatigue limit.

However, with advancements in testing technologies and the growing industrial demand for components with exceptionally high fatigue lives, ranging from 10^7 to 10^9~10^{10} cycles, the interest in new fatigue regime emerged. This regime is known as ultra high-cycle fatigue (UHCF), also referred to as giga-cycle fatigue (GCF) or very high-cycle fatigue (VHCF) [1]. Although UHCF is considered a subdomain of HCF, it warrants distinct attention due to differing damage mechanisms. In UHCF, damage mechanisms can change from typical surface cracking to internal cracking originating from material defects. In this regime, surface defects alone may not be responsible for fatigue damage; instead, internal material defects can lead to failure after an exceptionally high number of cycles. Testing technologies resorting to high-frequency systems makes frequency and dissipation dependencies worth to be investigated in this fatigue regime. Both HCF and VHCF are relevant for applications where significant plastic deformation is not permissible during the design and operation of equipment.

After the works of Coffin-Manson [2,3] about the effects of cyclic plasticity on fatigue damage, a new fatigue regime was established, named as low-cycle fatigue regime or plastic fatigue. In this fatigue regime, plasticity controls fatigue crack initiation and propagation and leads to shorter fatigue lives, which for many engineering metallic materials could fit between 10^4 and 10^5 cycles. However, new advanced alloys may show higher resistance to cyclic plastic deformations leading to higher fatigue lives than the ones typically observed in more traditional engineering materials.

More recently (last two decades), some research was performed to explore the ELCF regimes, which are characterized by intense cyclic plastic deformations leading to failure after an exceptionally low number of cycles, i.e., between 10^0 and 10^2 cycles. This fatigue regime is seen in the practice as the transition between the ductile (plastic) monotonic failure and the LCF, being seen as a transition material damage progress, where there is a mix of damage mechanisms between the ductile damage and the LCF damage. This fatigue regime is very often called as ultra low-cycle

DOI: 10.1201/9781003405122-5

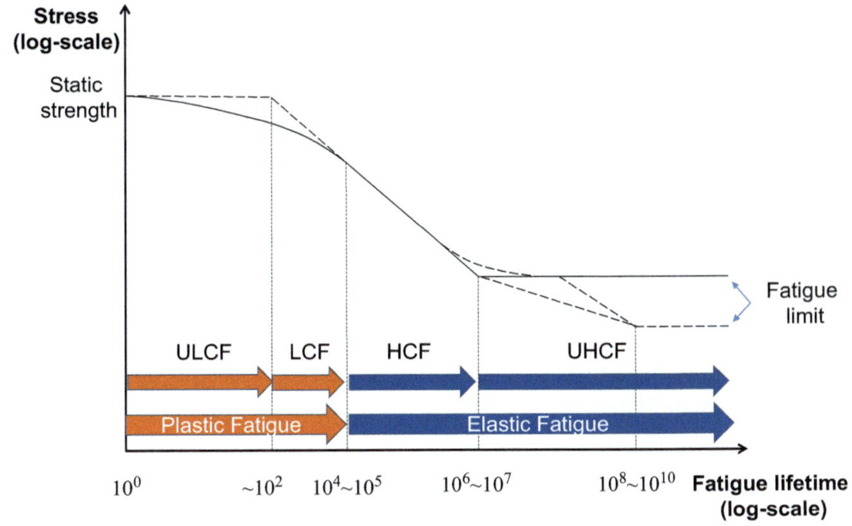

FIGURE 5.1 Fatigue regimes.

fatigue (ULCF) regime or extremely low-cycle fatigue regime (ELCF), appearing related to the material and structural damage under extreme cyclic loading conditions as those raised by earthquakes or other catastrophic dynamic events. Critical metallic structures and infrastructures (e.g., buildings, cranes, bridges, and pipelines) may be exposed to these extreme conditions and must be designed to resist without catastrophic failures [4].

Figure 5.1 illustrates the different fatigue regimes in the S-N curve representation with indicative fatigue lives. There are some attempts in the literature to find a unique model to encompass all fatigue regimes [5]. Nevertheless, those models are of phenomenological nature, as they do not account for the specificities of different damage micro-mechanisms involved in each fatigue regime.

This chapter presents an overview of the investigation about the ULCF, and some developments resulted from the ULCF European project [6–13], oriented for pipelines, will be presented, namely about a new model for ULCF, which was applied to pipelines.

5.2 OVERVIEW OF ULTRA LOW-CYCLE FATIGUE MODELS

The ultra low-cycle fatigue has been characterized in the literature as a fatigue damage regime characterized by intensive and widespread cyclic plastic deformations, which typically may lead to material failure at a maximum of around 10^2 cycles, but failures may appear just after a few cycles being applied. Therefore, this fatigue regime has been modeled according to two different perspectives: (1) continuum damage mechanics (CDM) coupled models; (2) uncoupled models.

For continuum damage mechanics coupled models, damage and plasticity are coupled. These models, which are very common to predict ductile failure, have been extended to ULCF, accounting for the specificities of the cyclic plasticity, but assuming

ULCF damage of the same nature of the monotonic ductile. These models lead to high computational costs and demands for robust numerical integration schemes, but as the number of cycles to failure is limited, its applicability is still feasible. These models can model fatigue crack initiation (damage onset) and fatigue crack propagation (damage spread). An illustration of such formulations was put forward by Lemaitre [14]. Subsequently, Leblond et al. [15] introduced an adaptation of the GTN model to account for cyclic loading. This adaptation incorporated kinematic hardening into the porous plasticity model, resulting in the development of the GTN-LPD model [16].

Phenomenological/empirical fatigue damage models that rely on independent cyclic plasticity analysis and fatigue damage computation. The state-of-the-art cyclic plasticity models can be applied to model the stress/strain histories. Then, a fatigue damage model is applied to account for fatigue cracking. The applicability of these models for fatigue crack propagation is limited, as the stress/strain histories previously estimated will be significantly changed by the presence of a macroscopic crack. Nevertheless, it is expected that fatigue crack propagation in the extreme conditions of ULCF will be relatively short. The efficiency of this approach is significantly higher, and damage models can more easily include the stress/strain damage parameters physically relevant for ULCF fatigue damage.

Several uncoupled ULCF models have been proposed in the literature, each based on different physical assumptions. According to Komotori and Shimizu [17], damage mechanisms during fatigue loading can vary depending on whether small or large plastic strain ranges are developed. As a result, fatigue endurance in the ULCF regime is significantly influenced by the material ductility. Intense plastic deformations can account for a substantial portion of the monotonic fracture strain, potentially triggering damage mechanisms more commonly associated with monotonic ductile damage, such as the nucleation and growth of voids [18]. Building on these observations, Kuroda [19] developed a fatigue model that categorizes ULCF damage into three distinct components: (1) damage from tensile deformation, (2) damage from ductility exhaustion due to cyclic loading, and (3) damage from crack progression. Alternatively, Tateishi and Hanji [20] proposed a model that defines ULCF damage as the linear sum of tensile ductile and cyclic loading damages.

Xue [21] introduced a damage rule for predicting fatigue life in the ULCF regime, addressing the overestimation issue associated with the classical Coffin-Manson approach [22]. Kanvinde and Deierlein [23] also proposed a fatigue model based on the cyclic behavior of micro-voids. This model suggests that material degradation due to micro-void growth is influenced by plastic strain and weighted by a triaxiality function. A few more ULCF models have been proposed in the literature [24]. However, due to higher popularity, some models are briefly described in the following subsections.

5.2.1 Coffin-Manson Foundational Model

The first empirical relation proposed to correlate LCF data was proposed by Coffin and Manson [2,3]:

$$\frac{\Delta\varepsilon^P}{2} = \varepsilon'_f\,(2N_i)^c \tag{5.1}$$

This relation firstly correlated the plastic strain range ($\Delta\varepsilon^P$) with the fatigue crack initiation life ($2N_i$) in the form of a power relation, involving two parameters (ε'_f and c). This relation has been proposed to cover LCF, including extreme cases of ULCF, due to its simplicity [4]. For example, it has been adopted by several design codes of practice such as the ASME B&PVC, Section VIII, Div. 2 [25]. Despite the original Coffin-Manson relation being proposed assuming uniaxial loading conditions, multiaxial stress-strain conditions can be also considered adopting a proper equivalent multiaxial strain. Some authors [19–22] verified that Coffin-Manson relation does not provide an acceptable portrayal of the ULCF regime, for some tested metallic alloys, resulting in a fatigue life overestimation in that fatigue regimes. Therefore, other models have been proposed in the literature to overcome this limitation.

5.2.2 Tateishi Empirical Model

ULCF has attracted the attention of Japanese researchers, motivated by the earthquake high risks of Japan and the need to design resilient metallic structures to that extreme loading conditions. In this line, an empirical fatigue model was proposed by Tateishi. This model is an update of the Coffin-Manson relation, which assumes that damage generated by ULCF loading is due to monotonic ductile damage and cyclic damage which accumulates linearly [20]. The model assumes the following form:

$$\Delta\varepsilon^P \cdot N_i^k = C \cdot C_m \tag{5.2}$$

where C_m is defined as:

$$C_m = \begin{cases} \left(\dfrac{\overline{\varepsilon_f} - \varepsilon_{\max}}{\overline{\varepsilon_f} - \varepsilon_{\text{pD}}} \right)^k & \text{if } \varepsilon_{\max} > \varepsilon_{\text{pD}} \\ 1 & \text{if } \varepsilon_{\max} \le \varepsilon_{\text{pD}} \end{cases} \tag{5.3}$$

In the previous equations, C and k are the original Coffin-Manson constants. $\varepsilon_{\max}$ is the maximum strain of the cycle, ε_{pD} is the accumulated plastic strain damage threshold, and $\overline{\varepsilon_f}$ is the critical ductile failure strain. This model clearly states that if the maximum strain of a cycle exceeds the ductile damage threshold, there is a need for a correction of the Coffin-Manson model, as there is a ductile damage component, which will be superimposed to the fatigue damage, leading to a reduction in the expected number of cycles to failure.

5.2.3 Kuroda Empirical Model

An empirical damage model was proposed by Kuroda [19] for the ULCF regimes including three components: (1) damage due to tensile loading; (2) damage due to ductility exhaustion under cyclic loading and (3) damage due to small cracks progression. Assuming a linear damage accumulation rule, the damage relation can be written as:

$$D = \underbrace{D_{\text{tensile}} + D_{\text{cyclic}}}_{\text{ductility exaustion}} + D_{\text{crack}} \tag{5.4}$$

Damage due to the tensile loading part of the stress cycle is related to the maximum strain level reached during the extreme loading conditions and their proximity to the ductile material limit. The cyclic loading due to its intensity may reduce the material ductility limit, which is a typical phenomenon of the ULCF regime. Finally, there is fatigue damage dominated by small fatigue crack growth, common in LCF. The three damage components can be linearly combined to compute the failure cycles:

$$D = \frac{\varepsilon_{P\max}}{\overline{\varepsilon_f}} + 4N_f\left(\frac{\Delta\varepsilon^P}{2\overline{\varepsilon_f}}\right)^{a'} + \frac{\Delta\varepsilon^P N_f^{\alpha'}}{C'} = 1 \tag{5.5}$$

where $\overline{\varepsilon_f}$, a', α', and $C'$are material constants; $\varepsilon_{P\max}$ is the applied maximum plastic strain. When the cyclic loading is less severe, then $\varepsilon_{P\max}$ and $\Delta\varepsilon^P$ can be neglected with respect to the ductility limit of the material, $\overline{\varepsilon_f}$, and Eq. (5.5) results in a simple power relation, common in LCF.

5.2.4 Xue Empirical Model

Xue [21] verified that neither the power law nor the exponential damage rules fit the whole range of life cycles from both ULCF and LCF. To account for fatigue lives in the range of 1/2~10⁴ cycles, a damage potential function for the full rage fatigue regimes (ULCF+LCF) was proposed:

$$\psi\left(\frac{\varepsilon_d}{\overline{\varepsilon_f}}\right) = \frac{e^{\lambda\left(\frac{\varepsilon_d}{\overline{\varepsilon_f}}\right)^m} - 1}{e^{\lambda} - 1} \tag{5.6}$$

Differentiating the damage potential, one can obtain the following differential relation:

$$dD = \frac{m\lambda\left(\frac{\varepsilon_d}{\overline{\varepsilon_f}}\right)^{m-1} e^{\lambda\left(\frac{\varepsilon_d}{\overline{\varepsilon_f}}\right)^m}}{e^{\lambda} - 1}\frac{d\varepsilon_p}{\overline{\varepsilon_f}} \tag{5.7}$$

Integrating the relation along the cyclic loading history, assuming a constant amplitude cyclic plastic loading, one retrieves a relation for the crack initiation cycles:

$$N_i = \frac{1}{2}\frac{e^{\lambda} - 1}{e^{\lambda\left(\frac{\varepsilon_d}{\overline{\varepsilon_f}}\right)^m} - 1} \tag{5.8}$$

where m and λ are material constants; ε_d is the equivalent plastic distortion, which can be computed using the three principal strains from the plastic strain tensor:

$$\varepsilon_d = \sqrt{\frac{2}{3}}\sqrt{\varepsilon_1^2 + \varepsilon_2^2 + \varepsilon_3^2} \tag{5.9}$$

5.2.5 Kanvinde and Deierlein Model

Kanvinde and Deierlein [23] developed a model for ULCF as an extension of a ductile model also proposed by the authors, assuming that materials have internal defects which will propagate under the plastic deformation action. Based on Rice and Tracey's work [26], the growth rate for a spherical void in an infinite continuum solid as:

$$\frac{dR}{R} = Ce^{1.5\eta} d\varepsilon_p \tag{5.10}$$

where R is the average void radius, C is a material constant, η is the stress triaxiality, and $d\varepsilon_p = \sqrt{2/3\varepsilon^P : \varepsilon^P}$ is the incremental equivalent plastic strain; and integrating the previous relation, between an initial defect R_0 size and R, results:

$$\ln\left(\frac{R}{R_0}\right) = \int_0^{\varepsilon_p} Ce^{1.5\eta}\, d\varepsilon_p \tag{5.11}$$

For monotonic ductile damage, the failure conditions can be established as:

$$\ln\left(\frac{R}{R_0}\right)_{\text{monotonic}}^{\text{critical}} = \int_0^{\varepsilon_p^{\text{critical}}} Ce^{1.5\eta}\, d\varepsilon_p \tag{5.12}$$

or

$$\frac{\ln(R/R_0)_{\text{monotonic}}}{C} = \text{VGI}_{\text{monotonic}} = \int_0^{\varepsilon_p} e^{1.5\eta}\, d\varepsilon_p \le \text{VGI}_{\text{monotonic}}^{\text{critical}} = \int_0^{\varepsilon_p^{\text{critical}}} e^{1.5\eta}\, d\varepsilon_p \tag{5.13}$$

where $\text{VGI}_{\text{monotonic}}$ stands for the void growth index under monotonic loading and $\text{VGI}_{\text{monotonic}}^{\text{critical}}$ corresponds to its critical value at failure, considered a material property.

The monotonic ductile model was generalized to cyclic loading under the following premises:

i. Cyclic loading induces stress triaxiality histories changing from positive to negative values at components critical locations.
ii. Voids will expand under positive triaxialities and will shrink under negative triaxialities.

The sign of the triaxiality can be used to differentiate tensile from compressive loading, governing the voids expansion or contraction, while the intensity of the stress triaxiality and the equivalent plastic strain rule the rate of voids expansion/contraction [23].

For ULCF loading, the monotonic model proposed originally by Kanvinde and Deierlein [23] was rewritten as

$$\ln\left(\frac{R}{R_0}\right)_{\text{cyclic}} = \sum_{\text{tensile cycles}} C_1 \int_{\varepsilon_1^P}^{\varepsilon_2^P} e^{\lfloor 1.5\eta \rfloor}\, d\varepsilon_p - \sum_{\text{tensile cycles}} C_2 \int_{\varepsilon_1^P}^{\varepsilon_2^P} e^{\lfloor 1.5\eta \rfloor}\, d\varepsilon_p \tag{5.14}$$

Kanvinde and Deierlein considered $C_1 = C_2 = C$ resulting:

$$\text{VGI}_{\text{cyclic}} = \frac{\ln(R/R_0)_{\text{cyclic}}}{C} = \sum_{\text{tensile cycles}} \int_{\varepsilon_1^P}^{\varepsilon_2^P} e^{\lfloor 1.5\eta \rfloor} d\varepsilon_p - \sum_{\text{tensile cycles}} \int_{\varepsilon_1^P}^{\varepsilon_2^P} e^{\lfloor 1.5\eta \rfloor} d\varepsilon_p \leq \text{VGI}_{\text{cyclic}}^{\text{critical}} \tag{5.15}$$

where $\text{VGI}_{\text{cyclic}} \geq 0$ and

$$\text{VGI}_{\text{cyclic}}^{\text{critical}} = \text{VGI}_{\text{monotonic}}^{\text{critical}} e^{-\lambda \varepsilon_P^{\text{accumulated}}} \tag{5.16}$$

where λ is a constant.

This model has been updated and reinterpreted by some authors and applied to different materials [6,24,27].

5.2.6 Dufailly and Lemaitre Model

Dufailly and Lemaitre [28] developed a damage model for extreme cyclic conditions in the range of 1~20 cycles. This model was originally formulated under the framework of continuous damage mechanics, with full coupling between damage and plasticity. The damage kinematic equation is defined by:

$$\dot{D} = \left(\frac{G}{S}\right)^s \dot{\varepsilon}^P \tag{5.17}$$

$$\dot{\varepsilon}^P = \sqrt{2/3 \cdot \dot{\varepsilon}^P : \dot{\varepsilon}^P} \tag{5.18}$$

where S and s are material parameters; G represents the strain energy release rate, defined as:

$$G = \frac{q^2 R_v}{2E(1-D)^2} \tag{5.19}$$

$$R_v = \frac{2}{3}(1+v) + 3(1-v)\left(\frac{p}{q}\right)^2 \tag{5.20}$$

where p is the pressure, q is the von Mises equivalent stress, and v the Poisson ratio. The damage threshold is defined in the function of the accumulated plastic strain, i.e.:

$$\varepsilon_{\text{accumulated}}^P < \overline{\varepsilon}_{\text{accumulated}}^P \rightarrow \dot{D} = 0 \tag{5.21}$$

Assuming that under ULCF extreme conditions, the damage threshold is reached in the first cycle, $\overline{\varepsilon}_{\text{accumulated}}^P = 0$ and the critical damage, corresponding to the crack initiation is $D_c = 1$. According to the previous simplifications, one can derive the kinematic damage law as:

$$\dot{D} = \left(\frac{q^2 R_v}{2ES(1-D)^2} \right)^s \dot{\varepsilon}^P \tag{5.22}$$

Assuming a perfect plastic material, the yield function can be written as:

$$\frac{q}{1-D} - \sigma_s = 0 \tag{5.23}$$

where σ_s is the yield stress of the material, which can be assumed a constant for perfectly plastic material. In this case, Eq. (5.22) will be rewritten for the damage increment per cycle as:

$$\frac{\Delta D}{\Delta N} = \left(\frac{\sigma_s^2 R_v}{2ES} \right)^s \Delta\varepsilon^P \tag{5.24}$$

For the case of uniaxial tensile stress state ($R_v = 1$):

$$\frac{\Delta D}{\Delta N} = \left(\frac{\sigma_s^2}{2ES} \right)^s \Delta\varepsilon^P \tag{5.25}$$

Integrating the previous equation between $D=0$ ($N=0$) and $D=D_c=1$ ($N=N_f$) will result in failure cycles:

$$N_f = \left(\frac{2ES}{\sigma_c^2} \right)^s \frac{1}{\Delta\varepsilon^P} \tag{5.26}$$

5.2.7 Ohata-Toyoda Model

Ohata and Toyoda [29] developed a model for ULCF damage and in particular for seismic loading (variable increasing stress amplitudes), which combines a base monotonic ductile damage model with the equivalent plastic strain at failure, being related with the stress triaxiality (as proposed by Johnson-Cook [30]). An effective strain concept was proposed, supported by ductile damage curves for ductile cracking of structures under monotonic loading, for the prediction of failure by cyclic loading [29]. The key of the model is to perform an accurate elastoplastic cyclic analysis of the material. For that purpose, a nonlinear isotropic/kinematic hardening model is recommended with a J2 plasticity model:

$$f(\sigma - \alpha) - \bar{\sigma}(\bar{\varepsilon}_P) = 0 \tag{5.27}$$

where σ is the stress tensor, α is the back-stress tensor, geometrically representing the center of the yield surface, and the $\bar{\sigma}(\bar{\varepsilon}_P)$ represents the yield strength which is dependent on the equivalent plastic strain according to an isotropic hardening rule. With this plastic model, it is expected to represent the Bauschinger effects and the material hardening/softening effects [31]. The evolution of the J2 yield criterion back-stresses tensor, defining the yield function, will be determinant in the control of the material damage under cyclic loading. Following the Lemaitre and Chaboche [32] proposal, Ohata and Toyoda proposed a nonlinear evolution for the back-stress tensor, dependent on the plastic strain, and obtained by the superposition of several variables:

$$d\alpha^k = \frac{C^k}{\bar{\sigma}}(\sigma - \alpha)\, d\bar{\varepsilon}_P - \gamma^k \bar{\alpha}\, d\bar{\varepsilon}_P \qquad \alpha = \sum_{k=1}^{N} \alpha^k \tag{5.28}$$

It is proposed that the effective plastic strain will evolve during a cycle only if the back-stress exceeds the maximum back-stress under the preceding loading cycle, therefore contributing to the damage. The crack initiation will then happen when the accumulated effective plastic strain reaches the monotonic damage curve, which relates the plastic strain with the stress triaxiality from material small-scale specimens.

This model was demonstrated suitable for incremental loading (typical of seismic actions) when the cyclic plastic response is not stable by nature. If the cyclic response is stable, then the back-stress tensor becomes stationary, therefore, there will be no further effective plastic strain accumulation and the failure conditions will not be reached.

5.2.8 GTN-LPD Model

A coupled damage model was developed by Leblond, Perrin, and Devaux [15], known as the LPD model for ULCF damage. This model is framed under the background of the micromechanics-based models and is an extension of the GTN model, which was originally designed for isotropic plastic hardening materials with internal defects. The LPD model assumes the same base yield function of the GTN model, which is further modified to allow kinematic hardening, introducing the deviatoric stress tensor that is generated from the Cauchy and back-stress tensors [16], resulting in the following yield function:

$$F_{\text{LPD}} = \frac{3}{2}\frac{\boldsymbol{S}:\boldsymbol{S}}{\bar{\sigma}_1^2} + 2f_v^* \cosh\left(q_2 \frac{-3tr(\boldsymbol{S})}{2\bar{\sigma}_2^2}\right) - \left(1 + q_3 f_v^{*2}\right) = 0 \tag{5.29}$$

5.2.9 Event Independent Cumulative Damage

Jiang et al. [33] proposed a damage model for fatigue under general plastic cyclic loading, without the need for cycle counting and covering multiaxial loading conditions. Their modeling approach is based on a cyclic plasticity model and a damage kinematic equation defined as:

$$dD = \xi dW^P$$
$$D = \int dD \tag{5.30}$$

where D is the fatigue damage, W^P is the plastic strain energy density, and ξ is a calibration parameter related to the plastic deformation of the material. Fatigue failure is foreseen when the damage variable reaches unity. The ξ parameter accounts for the effects of the plastic deformation and mean stress as:

$$\xi = \alpha(h) + \beta(h)\frac{\langle \sigma_n \rangle}{E} \tag{5.31}$$

where $\alpha(h)$ and $\beta(h)$ are dependent on the plastic modulus, h, and σ_n is the hydrostatic stress or the normal stress to the maximum shear plane (multiaxial fatigue criterion). $\alpha(h)$ and $\beta(h)$ are calibrated with fully reversed and torsion test data of smooth specimens. This damage model is applied to evaluate the fatigue using an uncoupled model with stress-strain histories estimated with state-of-the-art cyclic plasticity models [33].

5.2.10 Barcelona Damage Model

As an outcome of the European project ULCF [13], a coupled damage model was developed for monotonic and ULCF damage [34]. This model, called Barcelona plastic damage model, considers an isotropic hardening law that combines both material strain hardening followed by softening due to damage (refer to Figure 5.2). Both hardening and softening isotropic hardenings of the plastic model are related to the fracture energy, which is then related to the finite element size, to minimize mesh size pathological influence. The model is supported by a classical J2 plasticity model, with kinematic hardening to represent cyclic plasticity and isotropic hardening. The model can then be calibrated from tensile test data for the isotropic hardening and cyclic tests of smooth specimens for the kinematic hardening. Afterward, the model can be applied to model ULCF problems with full-time integration.

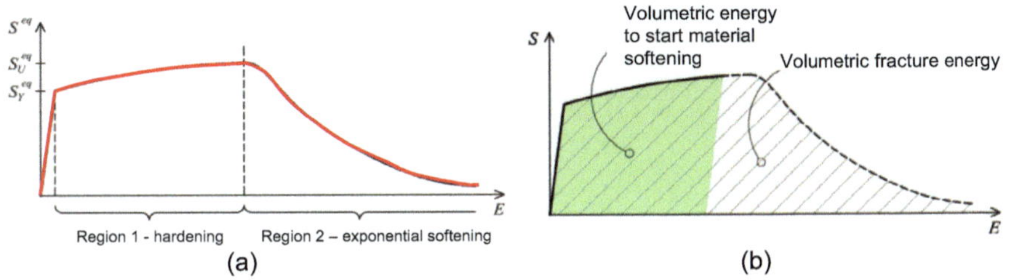

FIGURE 5.2 Isotropic hardening model defined in the Barcelona model. (Adapted from Ref. [34].)

5.3 A NEW DAMAGE MODEL FOR DUCTILE DAMAGE AND ULCF

ULCF is characterized by cyclic plastic deformations of high intensity, which may lead to a combination of ductile plastic damage and short fatigue crack propagation. These combined damage micro-mechanisms are responsible for the fatigue life overestimation using the Coffin-Manson approach for LCF, reported by some literature [19,20,22]. In effect, the Coffin-Manson approach does not account for the monotonic ductile damage that may be induced by the high-intensity cyclic loadings. Figure 5.3 illustrates typical fracture surfaces of monotonic ductile damage (Figure 5.3a1 and a2) and ULCF damage (Figure 5.3b1 and b2). While the pictures corresponding to the monotonic ductile damage show a dimple formation, the ULCF pictures show partial dimple formation regions with some smoother regions corresponding to cracked surfaces. In ULCF, each cycle may lead to some ductile failure followed by some crack propagation. Some authors, when formulating their models for ULCF damage, have also recognized the existence of different damage mechanisms involved in ULCF. This is the case of the Kuroda model [19], which recognized that ULCF damage involves the short fatigue crack propagation typical of LCF and the ductility exhaustion typical from a ductile damage. In addition, Tateishi included a ductile damage degradation term in the Coffin-Manson relation to allow for combined damage mechanisms [20]. Kermajani et al. [35] carried out a microstructural analysis of failure surfaces of specimens subjected to ULCF and concluded that a model to predict ULCF must account for both monotonic and fatigue fracture mechanisms, proposing a parameter to weigh the different contributions to the final failure.

Some models for ULCF have been just an extension of ductile damage models, which are integrated along the cyclic loading histories [23]. These models, originally developed for monotonic ductile damage, have been extended for cyclic loading with some adaptations (e.g., cyclic plasticity models).

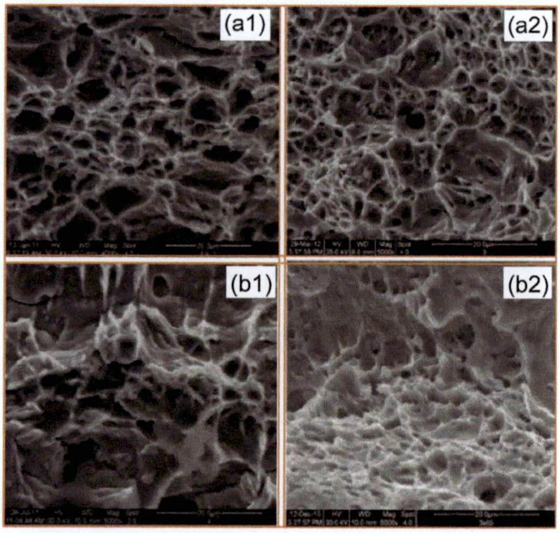

FIGURE 5.3 Fracture surfaces of monotonic ductile (a1) and (a2) and ULCF (b1) and (b2) damage. (Adapted from Ref. [8].)

A new model was proposed by Pereira et al. [8,12], which consisted of an update of Xue's model [21], which is reprinted here for cyclic loading:

$$N_i = \frac{1}{2}\frac{e^{\lambda}-1}{e^{\lambda\left(\frac{\Delta\varepsilon_{eq}^{P}}{\overline{\varepsilon_f}}\right)^{m}-1}} \tag{5.32}$$

where $\Delta\varepsilon_{eq}^{P}$ is the equivalent plastic strain range, which is defined according to the ASME B&PVC [25]:

$$\Delta\varepsilon_{eq}^{P} = \frac{\sqrt{2}}{3}\sqrt{\begin{aligned}&(\Delta p_{11}-\Delta p_{22})^2+(\Delta p_{22}-\Delta p_{33})^2+(\Delta p_{33}-\Delta p_{11})^2\\ &+\frac{3}{2}\left(\Delta p_{12}^2+\Delta p_{23}^2+\Delta p_{31}^2\right)\end{aligned}} \tag{5.33}$$

and Δp_{ij} denotes the variation of the plastic strain components, between two consecutive load reversal points; $\overline{\varepsilon_f}$ stands for the equivalent ductile fracture strain which in turn is computed using Bai and Wierzbicki [36] model, the later being sensitive to the triaxiality and Lode angle parameters:

$$\overline{\varepsilon_f}(\eta,\overline{\theta}) = \left[\frac{1}{2}\left(D_1e^{-D_2\eta}+D_5e^{-D_6\eta}\right)-D_3e^{-D_4\eta}\right]\overline{\theta}^2+\frac{1}{2}\left(D_1e^{-D_2\eta}-D_5e^{-D_6\eta}\right)\overline{\theta}+D_3e^{-D_4\eta} \tag{5.34}$$

Average values for the triaxiality and Lode angle parameters along the loading should be used:

$$\eta_{avg} = \frac{1}{\overline{\varepsilon_f}}\int_0^{\overline{\varepsilon_f}}\eta(\overline{\varepsilon})d\overline{\varepsilon} \tag{5.35}$$

$$\overline{\theta}_{avg} = \frac{1}{\overline{\varepsilon_f}}\int_0^{\overline{\varepsilon_f}}\overline{\theta}(\overline{\varepsilon})d\overline{\varepsilon} \tag{5.36}$$

The triaxiality η and Lode Angle $\overline{\theta}$ are defined as follows:

$$\eta = \frac{p}{q} = \frac{\frac{1}{3}(\sigma_1+\sigma_2+\sigma_3)}{\sqrt{\frac{1}{2}\left[(\sigma_1-\sigma_2)^2+(\sigma_2-\sigma_3)^2+(\sigma_3-\sigma_1)^2\right]}} \tag{5.37}$$

$$\overline{\theta} = 1-\frac{2}{\pi}\arccos\xi = 1-\frac{2}{\pi}\arccos\left[\left(\frac{r}{q}\right)^3\right] \tag{5.38}$$

$$r = \sqrt[3]{\frac{27}{2}S_1S_2S_3} \tag{5.39}$$

where σ_1, σ_2, and σ_3 are the principal stresses and S_1, S_2 and S_3 are the principal Chauchy and deviatoric stresses; p is the pressure (related to first stress tensor invariant); q is second invariant of the deviatoric stress tensor and r is related to the third invariant of the deviatoric stress tensor.

Equation (5.34) represents a failure surface, which must be identified for a given material using specific monotonic failure tests covering different stress triaxialities and Lode angle parameters. The triaxiality and Lode angle dependencies of the ULCF damage model is a novel aspect of the model that has been followed by recent authors [27].

Regarding Eq. (5.32), the two constants involved can be identified by means of cyclic tests in the ULCF regime. For each ULCF test, the ductile failure strain must be evaluated, which will be based on previous failure surface and the stress state at the maximum of the cyclic load applied to the specimen. In the following subsection, the model will be identified for a pipeline steel and the results presented.

The proposed modified Xue model allows the extension of the original model for multiaxial loading conditions. In addition, it redefines $\overline{\varepsilon_f}$ as a function of the stress state instead of being a material constant, as the fracture strain depends on the stress state conditions, as defined by Bai and Wierzbicki [36] model. In the proposed model, the equivalent plastic strain range, $\Delta\varepsilon_{eq}^{P}$, is normalized by the static fracture strain, $\overline{\varepsilon_f}$, which is a similar concept as explored by Nam et al. [37], but using energy considerations:

$$\omega = \sum \Delta\omega = \sum \frac{\Delta w_P}{w_f} \leq \omega_c \tag{5.40}$$

where Δw_P and w_f denote the equivalent plastic strain energy of the cycle, and the multiaxial fracture strain energy, respectively, the latter being dependent on stress triaxiality.

The proposed model could be applied in an uncoupled version, with the stress-strain history being defined as much accurately as possible, using a proper cyclic plasticity model.

5.4 APPLICATION OF MODIFIED XUE MODEL TO X60 PIPELINE STEEL

This section provides an overview of the modified Xue model identification procedure, demonstrating the results for the X60 pipeline steel. Only an overview of the workflow and some general results are presented. More details, including about the challenging ULCF testing, can be found elsewhere [7–9,12]. Figure 5.4 shows the workflow required to identify the model. Firstly, the material ductile failure properties must be evaluated resorting to a specific program of monotonic failure tests. These tests should cover different triaxialities and Lode angle parameters. Upper and lower bound limits of the surface loci are recommended to be evaluated, which corresponds to the Lode angle parameter values of 1 and 0, respectively. In this work, round smooth and notched specimens were considered to represent the upper bound and flat grooved plane strain specimens to represent the lower bound of the ductile

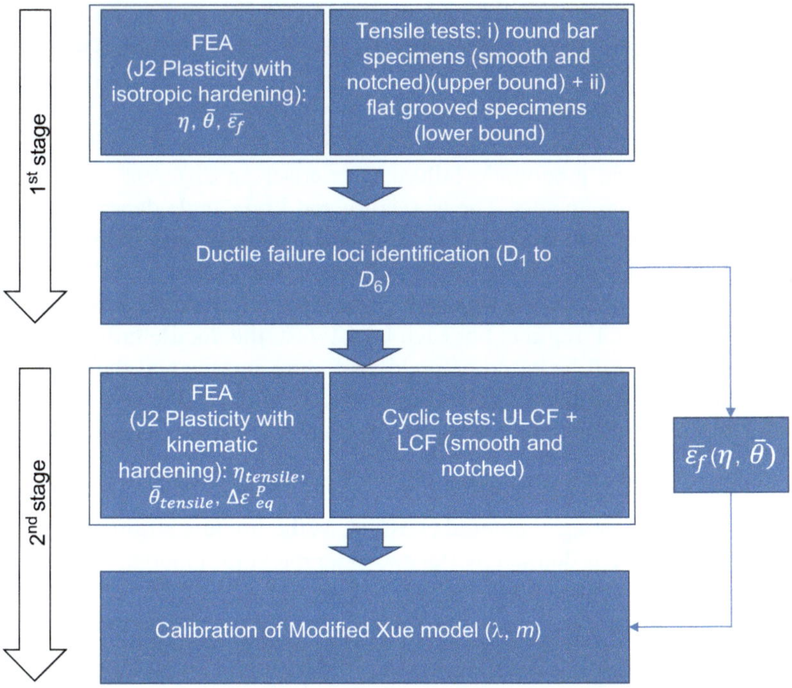

FIGURE 5.4 Workflow for the identification of the modified Xue model for ULCF.

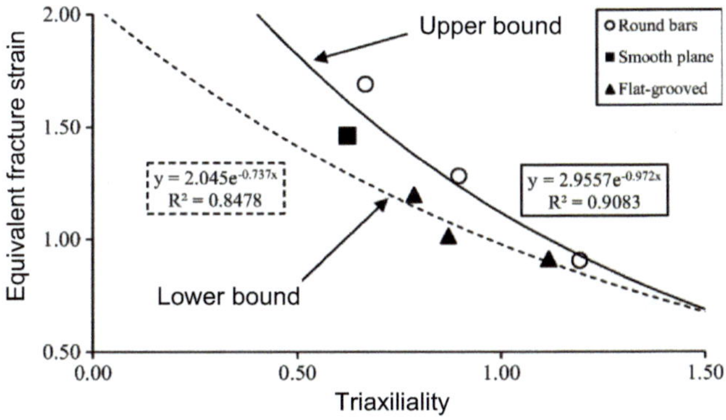

FIGURE 5.5 Upper and lower bound ductility limits of the X60 pipeline steel in the strain - stress triaxiality space. (Reprinted with permission from Ref. [8]. 2014, Elsevier.)

failure loci. The test results are modeled using a plasticity model with isotropic hardening to evaluate the stress triaxiality and Lode angles at failure.

Figure 5.5 illustrates the lower and upper bound limits of the X60 pipeline steel. Figure 5.6 represents the respective ductile failure surface.

The second stage of the model identification (refer to Figure 5.4) consists of the modified Xue model constants identification, which should be based on ULCF test data. The ULCF test data should also cover different specimen geometries

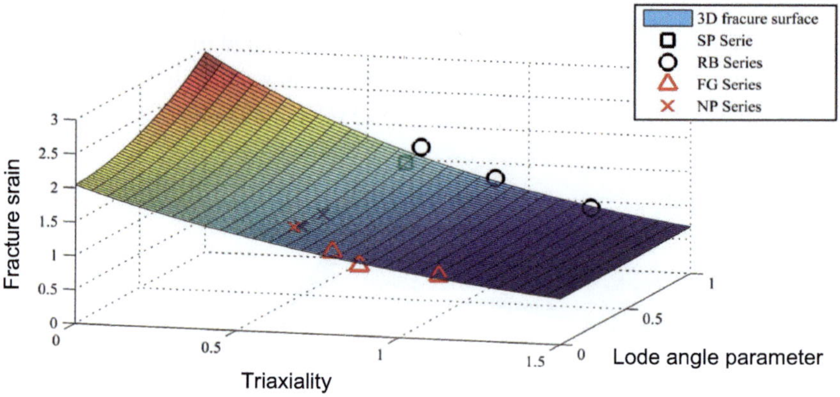

FIGURE 5.6 Ductile failure loci of the X60 piping steel (SP: smooth plane; RB: round bar; FG: flat grooved; NP: notched plane) (D_1=2.9557; D_2=0.972; D_3=2.045; D_4=0.737; D_5=2.9557; D_6=0.972). (Reprinted with permission from Ref. [8]. 2014, Elsevier.)

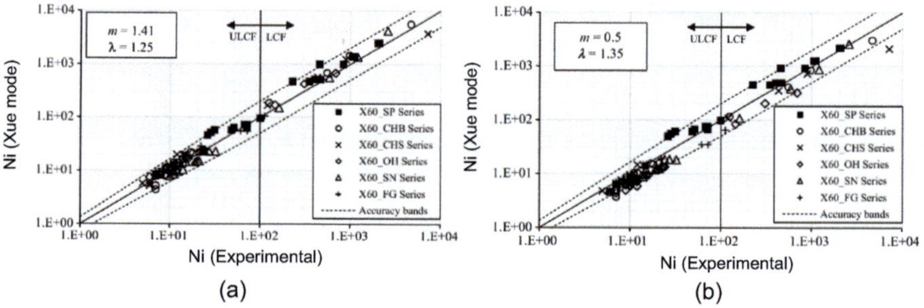

FIGURE 5.7 Comparison of experimental ULCF data and the modified Xue model predictions for the X60 piping steel: (a) using ductile failure strains computed from numerical simulation; (b) using failure strains from fracture locus. (SP: Smooth plane; CHB and CHS: central circular holes; OH: central oval hole; SN: side notch; FG: flat grooved). (Reprinted with permission from Ref. [8]. 2014, Elsevier.)

representing distinct stress triaxialities and Lode angle parameters, which could be different from those specimens from the first stage. Each test should be modeled aiming the stress fatigue parameters evaluation at tensile peaks of the loading as well as the equivalent plastic strain. Considering only the tensile peaks and neglecting the compressive parts of the loading is consistent with some literature [38]. All data is used to identify the two parameters of the modified Xue model.

Figure 5.7 illustrates the result of the model fitting to the experimental data of the X60 pipeline steels [8,9,12].

After the ULCF model constants identification, one can generate S-N surfaces representing the ULCF behavior of the steel under different stress parameter combinations. For example, Figure 5.8 represents the S-N surfaces for a fixed stress triaxiality ($\Delta\varepsilon_{eq}^{P} - N_i - \bar{\theta}$ field). In this representation, one can get different S-N behaviors dependent on the Lode angle parameter, in correspondence to the ductility failure loci; an S-N curve can be established for the lower and higher ductility, represented the lower and upper bounds of the S-N field (Figure 5.8b and d).

Figure 5.9a and c represent for the same X60 pipeline steel, the S-N surfaces for fixed Lode angle parameter and variable stress triaxialities ($\Delta\varepsilon_{eq}^{P} - N_i - \eta$ field).

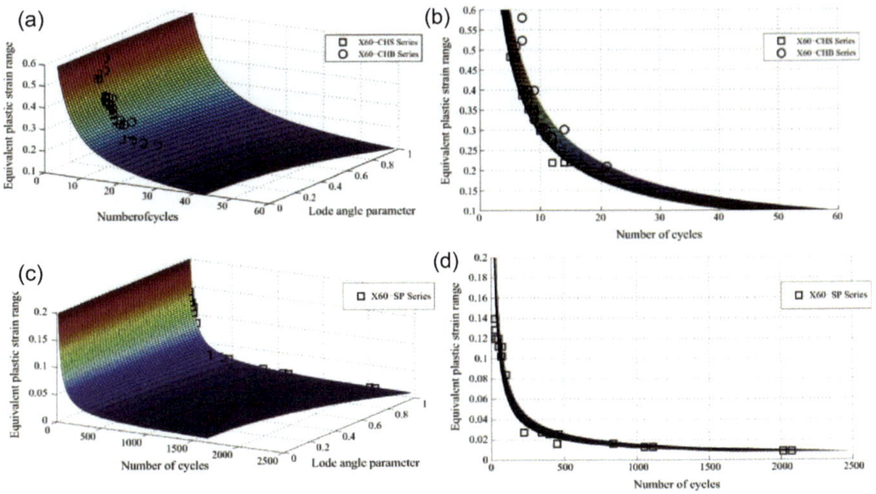

FIGURE 5.8 Strain-life surfaces of the X60 piping steel: (a) in the $\Delta\varepsilon_{eq}^{P} - N_i - \overline{\theta}$ space for $\eta = 0.575$; (b) on the plane $\Delta\varepsilon_{eq}^{P} - N_i$ plane for $\eta = 0.575$; (c) in the $\Delta\varepsilon_{eq}^{P} - N_i - \overline{\theta}$ space for $\eta = 0.333$ (smooth cylindrical); (d) on the $\Delta\varepsilon_{eq}^{P} - N_i$ plane for $\eta = 0.333$ (smooth cylindrical). (Reprinted with permission from Ref. [8]. 2014, Elsevier.)

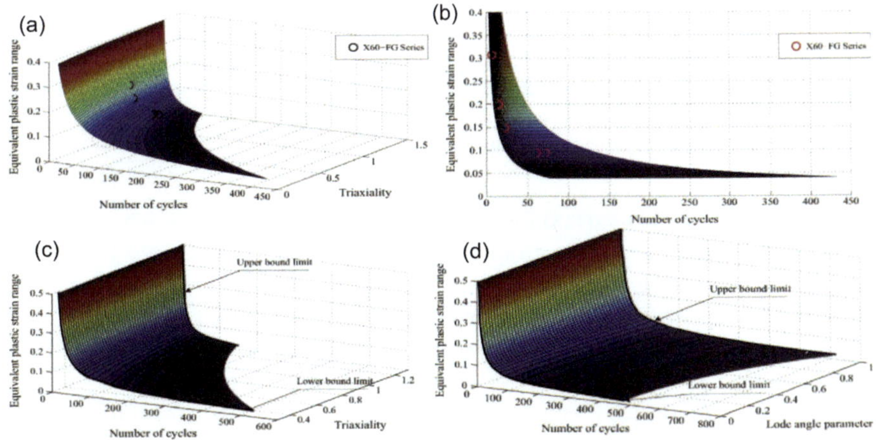

FIGURE 5.9 Strain-life surfaces of the X60 piping steel: (a) in the $\Delta\varepsilon_{eq}^{P} - N_i - \eta$ space for $\overline{\theta} = 0$ (plane strain); (b) on the $\Delta\varepsilon_{eq}^{P} - N_i$ plane for $\overline{\theta} = 0$ (plane strain); (c) in the $\Delta\varepsilon_{eq}^{P} - N_i - \eta$ space for $\overline{\theta} = 1$; (d) in the $\Delta\varepsilon_{eq}^{P} - N_i - \overline{\theta}$ space for $\eta = 0.333$. (Reprinted with permission from Ref. [8]. 2014, Elsevier.)

The modified Xue model allows an innovative representation of the ULCF data and consequently performing damage predictions. It makes the transition of the ductile damage failure to the LCF damage. The $\Delta\varepsilon_{eq}^{P}$ parameter will be more or less critical to fatigue damage, depending on the ductility limit of the material under the given stress conditions of the applied cyclic loading. This idea is similarly explored by Nam et al. [37].

5.5 ULCF MODELING OF A FULL-SCALE STRAIGHT PIPE

The ULCF model presented in the previous section will be applied to model the ULCF behaviors of a full-scale pipe which was tested under extreme cyclic bending conditions with the ULCF project [10,13]. An X60 straight pipe with 16″ outside diameter/9.5 mm wall thickness was tested under pure cyclic bending with internal pressure, on the laboratory floor, as illustrated by Figure 5.10a. In order to localize the failure location, the mid-section of the pipe was ground to reproduce a flat with dimensions illustrated in Figure 5.10b. The pipe was subjected to cyclic bending with an increasing end displacements history, induced by hydraulic cylinders, and keeping an internal hydraulic pressure, as illustrated in Figure 5.11. While the internal pressure is kept at a constant value, the cylinder displacements are increased until a constant plateau, which was held until pipe leak.

The tested pipe suffer a leak after 39 cycles, which was anticipated by previous buckling verified after 33 cycles. The plastic buckling was responsible for plastic strain localization and then for fatigue damage evolution, ended with a crack initiation responsible for pipe leaking.

In order to simulate the damage of the pipe under cyclic loading, a finite element model of the pipe was built using ABAQUS software. Besides the pipe, the

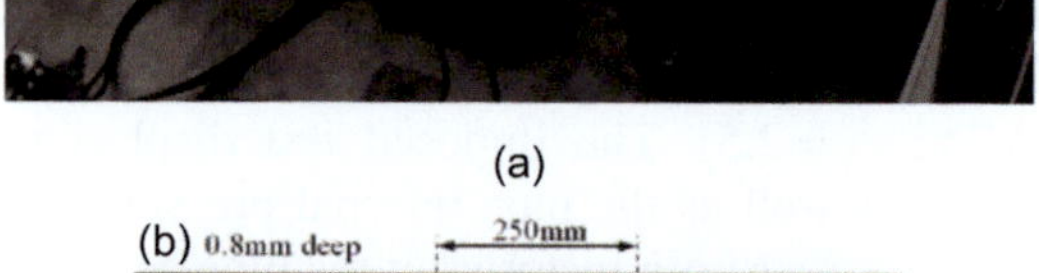

(a)

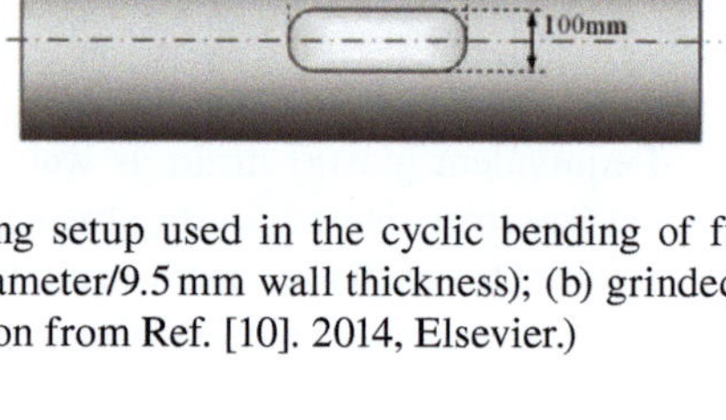

FIGURE 5.10 (a) Testing setup used in the cyclic bending of full-size straight pipes of X60 steel (16″ outside diameter/9.5 mm wall thickness); (b) grinded flat face on pipe center. (Reprinted with permission from Ref. [10]. 2014, Elsevier.)

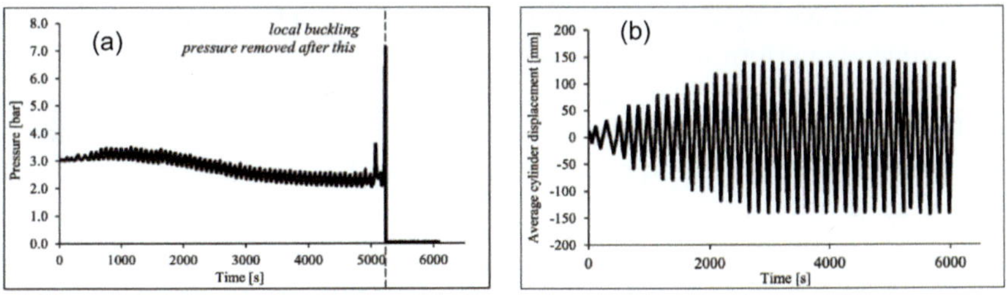

FIGURE 5.11 Testing loading: (a) internal pressure; (b) cylinders displacement. (Reprinted with permission from Ref. [10]. 2014, Elsevier.)

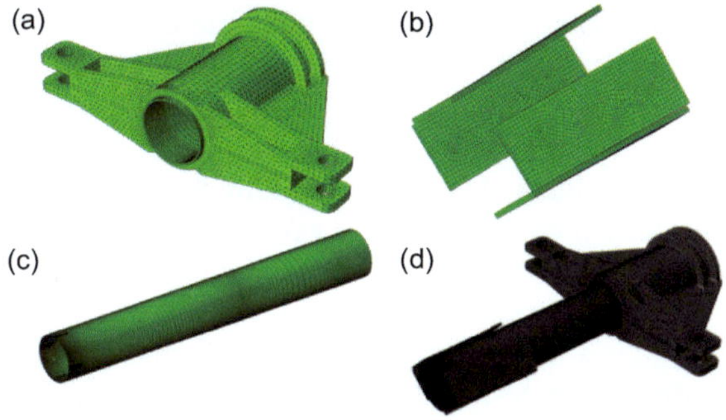

FIGURE 5.12 FEM model of the full-scale test of the pipe: (a) tubeholder; (b) central guiding plates; (c) pipe and (d) full assembled model. (Reprinted with permission from Ref. [10]. 2014, Elsevier.)

end tubeholder and central bend plates were modeled with contact elements. Half of the test was simulated taking advantage of symmetry. The following elements were used in the model: C3D8R for the pipe, C3D4 for the tubeholder and R3D4 for the central bend pipes. Figure 5.12 illustrates the FEM model with the meshes, which were very refined at the regions of interest. The success of the damage modeling depends on the accurate simulation of the cyclic plastic behavior of the pipe. For this respect, a J2 plasticity constitutive model was selected to model the pipe using a kinematic hardening model with Chaboche multiple nonlinear kinematic hardening (σ_{yield} = 410 MPa; C_1 = 52 GPa; γ_1 = 450; C_2 = 2 GPa; γ_2 = 80; C_3 = 450 MPa; γ_3 = 2.5). The imposed test displacements were applied in the numerical model as well as the pipe internal pressure (Figure 5.11) and the reaction moments and stresses/strain histories at the pipe center were simulated and analyzed. Figure 5.13a represents the bending moment–displacement history where it is visible an almost perfect match between the experimental and numerical results. This is a good indication of the calibration of the numerical model. Figure 5.13b illustrates the accumulated equivalent plastic strain as well as the failure location. Each point of the FEM model was investigated for the stress histories (stress triaxiality and Lode angle were computed for each cycle) and equivalent plastic strain range. Then, resorting to the linear damage accumulation, the fatigue life was computed using the modified Xue model.

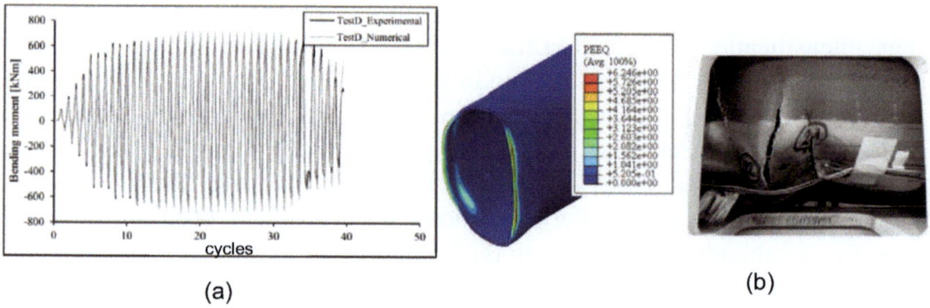

FIGURE 5.13 Elastoplastic numerical simulation results: (a) reactions history; (b) equivalent plastic strain field and failure location and appearance. (Reprinted with permission from Ref. [10]. 2014, Elsevier.)

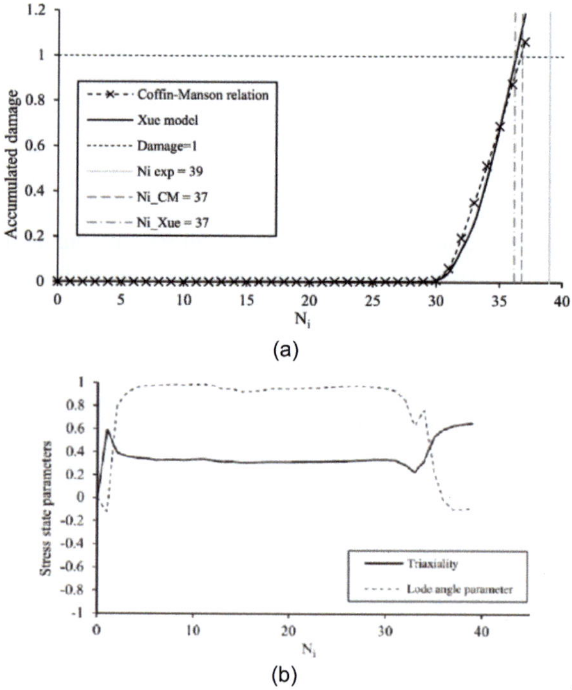

FIGURE 5.14 Simulation results: (a) ULCF damage evolution; (b) triaxiality and Lode angle parameter histories. (Reprinted with permission from Ref. [10]. 2014, Elsevier.)

Resorting to the linear damage accumulation, the fatigue damage was computed cycle-by-cycle using the modified Xue model. The damage evolution results are presented in Figure 5.14a, where it is possible to find the number of cycles to failure of 37, two cycles below the observed experimental pipe leaking. The sudden damage increase after the 30 cycles is due to the plastic buckling. This is corroborated by the triaxialitity and Lode angle histories (refer to Figure 5.14b), which after the 30th cycle also experience significant changes, with the increase of stress triaxiality and decrease in the Lode angle parameter, both variations are aligned with the ULCF damage intensification. Figure 5.14a also shows the Coffin-Manson prediction, which resulted in very similar results as the modified Xue model. This could seem negative for the modified Xue model, which requires a very significant effort to be implemented. Nevertheless, Pereira et al. [10] also presented two additional full-scale test results for straight pipes of X65 pipeline steel, and in this case, the modified Xue model provided significantly better predictions, with Coffin-Manson overestimating fatigue lives as already documented in the literature. The experimental fatigue lives were 11 and 17 cycles, significantly lower than observed for the X60 pipe. This corroborates that the benefits of the modified Xue model are higher when the fatigue lives are lower, where a mix of damage mechanisms are clearer.

5.6 SUMMARY

This chapter presented an overview of the ongoing research in the last two decades on the ELCF regime. The investigation on this fatigue regime has raised with the need to develop more resilient structures, namely to resist extreme events like earthquakes, without sudden failures. After the pioneer works in Japan and California, the ULCF European project brought attention to the topic. Recently, a first literature review was published with a total of 88 references, which is an indication of the increasing interest of the topic.

The ULCF has been considered a transition damage mechanism, and therefore its investigation is affected with this perspective. Some models proposed in the literature are extensions of ductile damage models and others are extension of LCF models. Also, regarding the models, some are based on CDM coupled models, which can be still applied to the short duration of the cyclic loading and others are empirical phenomenological relations that combine the damage parameters in an appropriate functional form to correlate the experimental data.

This chapter provided an overview of the main models proposed in the literature for the ULCF and selected one simple, but innovative model developed in the ULCF project and illustrated the steps of its identification with test data and also demonstrated its application to a case study of a full-scale testing of a pipeline.

This chapter should serve as the motivation for future work in the field as there is not a consensual model for ULCF. Also, fatigue testing in this fatigue regime, not covered in this book chapter, is very challenging requiring the development of best practices that could later result in standardized practices.

ABBREVIATIONS

ASME	The American Society of Mechanical Engineers
CDM	Continuous Damage Mechanics
ELCF	Extremely Low-Cycle Fatigue
GCF	Giga-Cycle Fatigue
GTN	Gurson-Tvergaard-Needleman
HCF	High-Cycle Fatigue
LCF	Low-Cycle Fatigue

LPD	Leblond-Perrin-Devaux
RFCS	Reseach Fund for Coal and Steel
UHCF	Ultra High-Cycle Fatigue
ULCF	Ultra Low-Cycle Fatigue
VHCF	Very High-Cycle Fatigue
$\bar{\varepsilon}_P$	Equivalent plastic strain
$\bar{\varepsilon}_f$	Critical fracture strain
$\bar{\theta}_{avg}$	Average Lode Angle
C^k	Chaboche hardening model constant
C_m	Tateishi model coefficient
D_1 **to** D_6	Bai and Wierzbicki model constants
D_{crack}	Damage associated to cracking
D_{cyclic}	Damage associated to cyclic exhaustion
$D_{tensile}$	Ductile damage
F_{LPD}	Leblond-Perrin-Devaux yield function
N_f	Cycles to failure
N_i	Number of cycles to crack initiation
R_v	Triaxiality function
S_1, S_2, S_3	Deviatoric principal stresses
$\mathbf{VGI}_{cyclic}$	Void Growth Index under cyclic loading
$\mathbf{VGI}_{cyclic}^{critical}$	Critical Void Growth Index under cyclic loading
$\mathbf{VGI}_{monotonic}$	Void Growth Index under monotonic loading
$\mathbf{VGI}_{monotonic}^{critical}$	Critical Void Growth Index under monotonic loading
W^P	Plastic strain energy density
f_v^*	Volume fraction of voids
q_2, q_3	GTN model constants
w_f	Fracture strain energy
γ^k	Chaboche hardening model coefficient
$\varepsilon_1, \varepsilon_2, \varepsilon_3$	Principal strains
$\varepsilon_P^{accumulated}$	Accumulated plastic strain
$\varepsilon_{P\max}$	Applied maximum plastic strain
ε_d	Equivalent plastic distortion
ε_f'	Fatigue ductility coefficient
$\varepsilon_{\max}$	Maximum strain
ε_{pD}	Accumulated plastic strain damage threshold
$\varepsilon_p^{critical}$	Critical plastic strain
$\underline{\eta}_{avg}$	Average triaxiality
$\bar{\theta}$	Lode angle
$\bar{\sigma}$	Yield strength
σ_n	Hydrostatic pressure
σ_s	Yield stress
σ_{yield}	Yield strength
ω_c	Critical normalized fracture energy
ε^P	Plastic stain tensor
$\Delta\varepsilon^P$	Plastic strain range

c	Fatigue ductility exponent
$C_1 = C_2 = C$	Kanvinde and Deierlein model constants
D_c	Critical damage
G	Strain energy release rate
h	Plastic modulus
k	Coffin-Manson constant
p	Pressure
q	von Mises equivalent stress
S	Material constant (Lemaite model)
s	Material constant (Lemaite model)
Δp_{ij}	Range of plastic strain components
ψ	Damage potential function
C	Coffin-Manson constant
C'	Kuroda model constant
D	Damage
E	Young's modulus
R	Void radius
VGI	Void Growth Index
a'	Kuroda model constant
m	Xue model constant
r	Third invariant related function
S	Deviatoric stress tensor
α'	Kuroda model constant
η	Stress triaxiality
λ	Xue model constant
λ	Kanvinde and Deierlein model constant
ν	Poisson ratio
ξ	Event independent cumulative damage constant
α	Back-stress tensor
σ	Stress tensor

REFERENCES

1. Bathias, C., Paris, P.C. (2004). *Gigacycle Fatigue in Mechanical Practice*; 1st edition. CRC Press, Boca Raton, FL.
2. Coffin Jr, L.F. (1971). A note on low cycle fatigue laws. *Journal of Materials JMLSA*, 6(2):388–402.
3. Manson, S.S. (1954). *Behavior of Materials under Conditions of Thermal Stress*. NACA-TR-1170. National Advisory Committee for Aeronautics, Washington, D.C.
4. de Castro e Sousa, A., Nussbaumer, A. (2019). Multiaxial ultra low cycle fatigue in welded high strength steel structural components. *Journal of Constructional Steel Research*, 153:473–482.
5. Fernández Canteli, A., Castillo, E., Blasón, S., Correia, J.A.F.O., De Jesus, A.M.P. (2022). Generalization of the Weibull probabilistic compatible model to assess fatigue data into three domains: LCF, HCF and VHCF. *International Journal of Fatigue*, 159:106771, 18 pages.
6. Pereira, J.C., De Jesus, A.M.P., Xavier, J., Fernandes, A.A. (2014). Ultra low-cycle fatigue behaviour of a structural steel. *Engineering Structures*, 60:214–222.

7. Pereira, J.C.R., De Jesus, A.M.P., Fernandes, A.A., Varelis, G. (2016). Monotonic, low-cycle fatigue and ultra-low cycle fatigue behaviors of the X52, X60 and X65 piping steel grades. *Journal of Pressure Vessel Technology*, 138(3):031403, 9 pages.
8. Pereira, J.C.R., de Jesus, A.M.P., Fernandes, A.A. (2016). A new ultra-low cycle fatigue model applied to the X60 piping steel. *International Journal of Fatigue*, 93(1):201–213.
9. Pereira, J., De Jesus, A.M.P., Xavier, J., Fernandes, A.A. (2017). ULCF assessment of X52 piping steel by means of cyclic bending tests. *Journal of Constructional Steel Research*, 138:663–674.
10. Pereira, J., Van Wittenberghe, J., De Jesus, A., Thibaux, P., Correia, J.A.F.O., Fernandes, A.A. (2018). Damage behaviour of full-scale straight pipes under extreme cyclic bending conditions. *Journal of Constructional Steel Research*, 143:97–109.
11. Pereira, J.C.R., de Jesus, A.M.P., Xavier, J., Correia, J.A.F.O., Susmel, L., Fernandes, A.A. (2019). Low and ultra-low-cycle fatigue behavior of X52 piping steel based on theory of critical distances. *International Journal of Fatigue*, 134:105482, 9 pages.
12. Pereira, J.C.R. (2016). Characterization of X52, X60 and X65 steel grades under monotonic and ultra-low-cycle fatigue loading. PhD Thesis, University of Porto.
13. Fernandes, A.A., de Jesus, A.M.P., Natal Jorge, R. (2018). *Monotonic and Ultra-Low-Cycle Fatigue Behaviour of Pipeline Steels - Experimental and Numerical Approaches.* Springer International Publishing AG, Cham, 525 pages. ISBN: 978-3-319-78095-5 (Print) 978-3-319-78096-2 (Online). https://doi.org/10.1007/978-3-319-78096-2
14. Lemaitre, J. (1985). A continuous damage mechanics model for ductile fracture. *Journal of Engineering Materials and Technology – Transactions ASME*, 107:83–89.
15. Leblond, J.B., Perrin, G., Devaux, J. (1995). An improved Gurson-type model for hardenable ductile metals. *European Journal of Mechanics - A/Solids*, 14:499–527.
16. Steglich, D., Pirondi, A., Bonora, N., Brocks, W. (2005). Micromechanical modelling of cyclic plasticity incorporating damage. *International Journal of Solids and Structures*, 42:337–351.
17. Komotori, J., Shimizu, M. (1998). Fracture mechanism of ferritic ductile cast iron in extremely low cycle fatigue. In: Rie, K.T., Portella, P.D., editors. *Low Cycle Fatigue and Elasto–Plastic Behavior of Materials.* Elsevier Science Ltd, Amsterdam, pp. 39–44.
18. Pereira, J.C.R., Xavier, J.M.C., de Jesus, A.M.P. (2014). Ultra low-cycle fatigue behavior of a structural steel. *Engineering Structures*, 20:214–222.
19. Kuroda, M. (2001). Extremely low cycle fatigue life prediction based on a new cumulative fatigue damage model. *International Journal of Fatigue*, 24:699–703.
20. Tateishi, K., Hanji, T. (2004). Low cycle fatigue strength of butt-welded steel joint by means of new testing system with image technique. *International Journal of Fatigue*, 26:1349–1356.
21. Xue, L. (2007). A unified expression for low cycle fatigue and extremely low-cycle fatigue and its implication for monotonic loading. *International Journal of Fatigue*, 30:1691–1698.
22. Hatanaka, K., Fujimitsu, T. (1984). Some considerations on cyclic stress–strain relation and low cycle fatigue life. *Transactions of the Japan Society of Mechanical Engineering - Part A*, 50(451):291–300.
23. Kanvinde, A.M., Deierlein, G.G. (2007). Cyclic void growth model to assess ductile fracture initiation in structural steels due to ultra low cycle fatigue. *Journal of Engineering Mechanics*, 133:701–702.
24. Xu, Y., Li, X., Zhang, Y., Yang, J. (2023). Ultra-low cycle fatigue life prediction model—a review. *Metals*, 13:1142.
25. ASME Boiler & Pressure Vessel Code, Section VIII. (2013). *Rules for Construction of Pressure Vessels, Division 2 – Alternative Rules.* The American Society of Mechanical Engineers (ASME), New York, NY.
26. Rice, J.R., Tracey, D.M. (1969). On the ductile enlargement of voids in triaxial stress fields. *Journal of the Mechanics and Physics of Solids*, 35:201–217.

27. Yin, Y., Li, S., Han, Q., Lei, P. (2019). Calibration and verification of cyclic void growth model for G20Mn5QT cast steel. *Engineering Fracture Mechanics*, 206:310–329.
28. Dufailly, J., Lemaitre, J. (1995). Modeling very low cycle fatigue. *International Journal of Damage Mechanics*, 4(2):153–170.
29. Ohata, M., Toyoda, M. (2004). Damage concept for evaluating ductile cracking of steel structure subjected to large-scale cyclic straining. *Science and Technology of Advanced Materials*, 5:241–249.
30. Johnson, G.R., Cook, W.H. (1985). Fracture characteristics of three metals subjected to various strains, strains rates, temperatures and pressures. *Engineering Fractures Mechanics*, 21:33–48.
31. Socie, D.F., Marquis, G.M. (2000). *Multiaxial Fatigue.* Society of Automotive Engineers, Inc., Warrendale, PA.
32. Lemaitre, J., Chaboche, J.L. (1990). *Mechanics of Solid Materials.* Cambridge University Press, Cambridge, MA.
33. Jiang, Y., Ott, W., Baum, C., Vormwald, M., Nowack, H. (2009). Fatigue life predictions by integrating EVICD fatigue damage model and an advanced cyclic plasticity theory. *International Journal of Plasticity*, 25(5):780–801.
34. Martinez, X., Oller, S., Barbu, L.G., Barbat, A.H., De Jesus, A.M.P. (2015). Analysis of ultra low cycle fatigue problems with the Barcelona plastic damage model and a new isotropic hardening law. *International Journal of Fatigue*, 73:132–142.
35. Kermajani, M., Malek Ghaini, F., Miresmaeili, R., Baghi-Abadi, M.K., Mousavi-Nasab, M. (2020). Damage mechanisms in the ultra-low cycle fatigue loading. *Engineering Fracture Mechanics*, 223:106772.
36. Bai, Y., Wierzbicki, T. (2008). A new model of metal plasticity and fracture with pressure and Lode dependence. *International Journal of Plasticity,* 24:1071–1096.
37. Nam, H.-S., Lee, J.-M., Kim, Y.-J., Kim, J.-W. (2018). Numerical ductile fracture prediction of circumferential through-wall cracked pipes under very low cycle fatigue loading condition. *Engineering Fracture Mechanics*, 194:175–189.
38. Tian, Q., Xie, X., Li, S. (2021). A model for ultra low cycle fatigue damage prediction of structural steel. *Journal of Constructional Steel Research*, 187:106956.

6 Very High Cycle Fatigue

Qingyuan Wang and Chao He

6.1 VERY HIGH CYCLE FATIGUE DAMAGE REGIMES-FUNDAMENTALS

With the advancement of science and technology, achieving ultra-long life and high reliability has become a critical goal and urgent need for national major technical equipment. Traditional fatigue theory posits that if a material does not fail after 10 million cycles, it has an infinite fatigue life, implying the existence of a "fatigue limit." However, extensive research in recent years has revealed that metal materials can still experience fatigue failure beyond 10 million cycles. This discovery not only challenges the "infinite life design criterion" of traditional fatigue theory but also opens up a new research area: very high cycle fatigue (VHCF). Studies have shown that VHCF is not merely a quantitative extension of traditional high-cycle fatigue (HCF), but more importantly, it involves a qualitative change in the failure mechanism. Given the ultra-long service life demands of modern mechanical equipment, researching the VHCF behavior and failure mechanisms of structural materials holds significant engineering importance.

Fatigue life is one of the most important parameters for evaluating the fatigue performance of metallic materials, and it is closely related to the specific service life of the material. Based on fatigue life, the fatigue issues of metallic materials can be divided into three categories: when the fatigue life is less than 10^5 cycles, it is low cycle fatigue (LCF); when the fatigue life is between 10^5 and 10^7 cycles, it is HCF; and when the fatigue life is greater than 10^7 cycles, it is VHCF.

With the advancement of science and technology, the service life of some modern mechanical equipment has exceeded 10 million cycles, entering the VHCF range. Currently, in fields such as aerospace, rail transit, and passenger vehicles, VHCF has become one of the primary design standards. The modern challenge of VHCF in mechanical equipment can be exemplified by aircraft engines. The development of aircraft engines pushes materials toward higher temperatures, longer life, and higher performance, with longevity and reliability being key to their safe operation. As a critical component of aircraft engines, compressor rotor blades play a significant role in enhancing overall engine performance. Aerodynamic excitations can easily cause high-frequency resonance or forced vibration in rotor blades at low stress amplitudes, subjecting them to cyclic loads during operation. Additionally, the service environment temperature for engine disks and blades ranges from 300°C to 600°C, further degrading the mechanical strength of materials under high-temperature conditions. In actual service, materials face significant risks of high-temperature fatigue failure. Fatigue failure is the primary failure mode for rotor blades and other rotating components in aircraft, with blade fractures being one of the main causes of major

DOI: 10.1201/9781003405122-6

engine incidents, posing a great threat to the reliable operation of aircraft engines. For example, on February 20, 2021, a United Airlines Boeing 777 experienced a right engine fire and explosion, causing external components to detach and resulting in an emergency return to the airport, fortunately with no casualties. Subsequently, United Airlines grounded 24 Boeing 777 aircraft with PW4000 series engines. Japan also required planes with these engines to be grounded. Preliminary investigations showed that two engine blades had fractured, with other blades damaged at the tips and leading edges, and one of the fractured blades was found to have failed due to metal fatigue. In recent years, incidents caused by metal fatigue failures in aircraft engine blades have been frequent, resulting in irreparable personal injuries and economic losses. Blades are the most delicate parts of aircraft engines; if a fatigue fracture occurs during service, the damage to the engine and the entire aircraft can be catastrophic, potentially leading to immeasurable losses.

Currently, modern standards have clearly set requirements for aircraft engine components to have an ultra-long service life of over 10^9 cycles. In 1999, the United States issued the "Engine Structural Integrity Program," specifying that titanium alloy components should have a fatigue life of 10^9 cycles. The revised guidelines in 2001 further mandated that all engine components must have a fatigue life of at least 10^9 cycles. In China, the Guidelines for the Structural Integrity of Aircraft Engines (issued in 1997) stated that aviation materials should meet a fatigue life of over 10^7 cycles. In 2010, China released the General Specification for Aero Turbofan and Turbojet Engines, which explicitly required titanium alloy components to achieve a HCF life of at least 10^9 cycles. VHCF loading stress is only 30%–50% of the yield strength, lacks a traditional fatigue limit, and materials do not show significant surface plastic deformation before failure. However, under prolonged dynamic loading, fatigue damage often initiates cracks from within the material, with no apparent warning signs before structural failure; thus, VHCF is considered the "cancer" of structural materials. In the context of "carbon peaking" and "carbon neutrality," reducing energy consumption and extending service life have become urgent needs and developmental goals for major engineering projects as well as key technological issues and research directions in significant fields such as aerospace.

Under low-stress conditions, the VHCF failure behavior of materials is completely different from conventional LCF and HCF. Countries like the United States, Japan, Germany, and China have conducted nearly three decades of research in this field and have currently developed a preliminary understanding of the S-N curve characteristics and the modes of internal fatigue crack initiation in VHCF. Most metallic materials (such as high-strength steel, titanium alloys, aluminum alloys, magnesium alloys, etc.) will experience fatigue fracture in the VHCF regime (ranging from 10^7 to 10^{10} cycles), and fatigue strength will decrease again after enduring a sufficient number of cyclic loads. Key components requiring VHCF resistance, such as those in aircraft engines, high-speed trains, and nuclear power plants, face severe risks of ultra-long life fatigue failure. Therefore, understanding the fatigue characteristics in this regime is crucial for the fatigue design of components demanding a long service life.

VHCF failure of metallic materials is typically characterized by internal crack initiation. In high-strength steel, such cracks are primarily caused by internal inclusions, forming a special "fish-eye" fracture appearance. In the central region of the fatigue crack initiation site, a dark ring-shaped appearance can be observed under an optical microscope, while a rough fine granular morphology is visible under an electron microscope, and this is called the Fine Granular Area (FGA). The formation mechanism of this area is critical to revealing the mechanisms of VHCF failure. The mechanisms of fatigue crack initiation using advanced micro-nano characterization techniques were explored, and several mechanism models have been proposed. However, these models' explanations of the initiating area formation are based on post-fracture analysis and have yet to clarify the connection among early damage nucleation, crack initiation, and fine granule formation. Presently, scholars worldwide remain divided on the formation mechanism of the FGA characteristic area. The characteristic area of ultra-long life fatigue crack initiation typically measures within tens of micrometers, approaching or reaching the average grain diameter of the material. The formation of the initiation area can no longer be viewed as a mechanical problem within a homogeneous material. Research must fully integrate detailed material characterization and corresponding micro-mechanical analysis methods. Many critical issues regarding the mechanisms of ultra-long life fatigue crack initiation and propagation remain unresolved. Besides, ultrasonic vibration fatigue testing technology is the most direct and effective method for conducting ultra-long life fatigue research. While systematic and in-depth studies on the fatigue performance and failure of metallic materials under room-temperature and symmetric stress loading conditions exist, research on fatigue strength and damage mechanisms under complex service conditions is still lacking. This inadequacy leads to insufficient grounds for evaluating the long-life service capabilities of critical components in major equipment.

6.2 ULTRASONIC VIBRATION FATIGUE TESTING METHOD

Based on extensive experimental data, the concept of "fatigue limit" seems to be less applicable. A more appropriate term would be "fatigue strength at a given number of cycles." Therefore, it is necessary to establish a reliable database on fatigue strength that extends fatigue data for different materials into the VHCF regime. Currently, existing fatigue testing standards for fatigue testing specify the equipment and methods for these tests. However, up to now, stress-controlled fatigue tests still commonly use 10^7 cycles as the test termination condition. Consequently, relevant standards have yet to address the VHCF domain. This indicates that, for fatigue testing in the VHCF regime, there are currently no corresponding standards, including those for testing equipment and methods. Ultrasonic fatigue testing systems are particularly well-suited for VHCF research. Here, a brief introduction and comparison of ultrasonic fatigue testing systems and conventional fatigue testing systems are provided.

6.2.1 Principles of the Ultrasonic Fatigue Test System

The VHCF testing system is usually implemented using accelerated testing methods. Figure 6.1 is a VHCF testing system that uses piezoelectric ceramics to drive the vibration of specimens and measure the fatigue durability performance of metals and their alloy materials. Ultrasonic fatigue is a displacement-controlled resonant

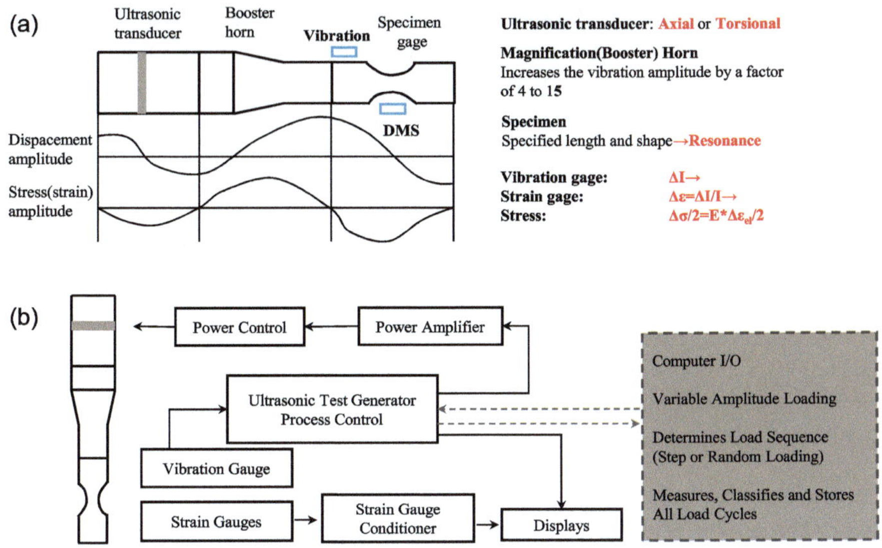

FIGURE 6.1 Ultrasonic fatigue test system and the stress displacement distribution [1,2]. (Reprinted with permission from Ref. [2], Elsevier.)

experimental method, which essentially utilizes axial mechanical resonance waves to load the specimen. The main equipment and functions of the ultrasonic fatigue testing system are:

- **Ultrasonic generator:** providing power and high-frequency vibration electrical signals, converting 50 or 60 Hz sinusoidal AC signals into 20 ± 0.5 kHz ultrasonic signals.
- **Converter:** Convert the electrical signals generated by the signal generator into axial mechanical vibration waves of the same frequency.
- **Ultrasonic horn:** Usually in the shape of a horn. The maximum amplitude of axial mechanical vibration obtained by piezoelectric transducer is usually less than 10 μm, so the amplitude of mechanical vibration is increased by an amplitude amplifier so that the resonant sample can achieve the required loading strain.
- **Displacement control system:** Mainly composed of computer systems, data acquisition devices, as well as digital-to-analog and analog-to-digital converters, the entire displacement control system is used to control the precise output of vibration displacement amplitude and real-time acquisition of fatigue cycle frequency and vibration frequency.

These four parts are the most basic devices for fatigue testing. Figure 6.1 provides a schematic diagram of the structure of the ultrasonic fatigue testing machine, and the entire system is controlled by a computer. A complete ultrasonic fatigue experimental system also includes a compressed air-cooling system and an infrared non-contact temperature measurement system to control the temperature of the specimen during the testing process.

The normal operating frequency of the ultrasonic fatigue testing system is around 20 kHz, but during the loading process, after the temperature rise caused by the loading of the amplifier and the specimen tends to stabilize, the appearance of internal fatigue cracks in the specimen will reduce the stiffness of the vibration system, leading to a decrease in the operating frequency. Therefore, frequency analysis and control can achieve automatic stopping of the testing system when fatigue cracks propagate to a certain length. Usually, when the operating frequency of the test system is below 19.5 kHz, it is considered that the specimen has undergone fatigue failure.

The essence of the ultrasonic fatigue vibration experiment is to establish a mechanical resonance wave on the tested sample. Its frequency is about 20 kHz, which is much higher than that of conventional fatigue testing systems, and can be used to test the VHCF performance of materials at 10^7 cycles or more. VHCF testing can effectively shorten the fatigue testing cycle, improve efficiency, and save time and testing costs.

6.2.2 Ultrasonic Vibration Fatigue Test under Complex Loads

6.2.2.1 Asymmetric Ultrasonic Fatigue Experimental Method ($R < -1$)

Traditional ultrasonic vibration fatigue testing machines can only perform symmetric tensile and compressive vibration fatigue (stress ratio $R = -1$). By integrating with static tensile testing machines and increasing static forces, asymmetric ultrasonic vibration fatigue loading can be achieved, as shown in Figure 6.2.

The asymmetric ultrasonic fatigue testing machine is more complex than the conventional fatigue testing machine, as shown in Figure 6.3. Added static average stress and other control devices, an optical displacement measuring instrument was installed and aimed at the bottom of the amplitude modulation rod for measuring amplitude.

The amplitude of the specimen is determined by the amplitude of the transducer and the amplification factor of the amplification rod. The asymmetric ultrasonic fatigue testing machine requires static tension to be applied under stress ratio. The piezoelectric ceramic in the transducer that converts electrical energy into mechanical energy can't bear large static tension and pressure. To avoid directly applying

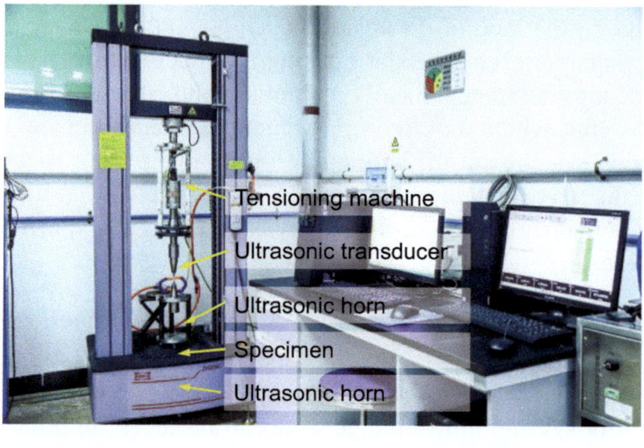

FIGURE 6.2 Asymmetric ultrasonic fatigue experimental system.

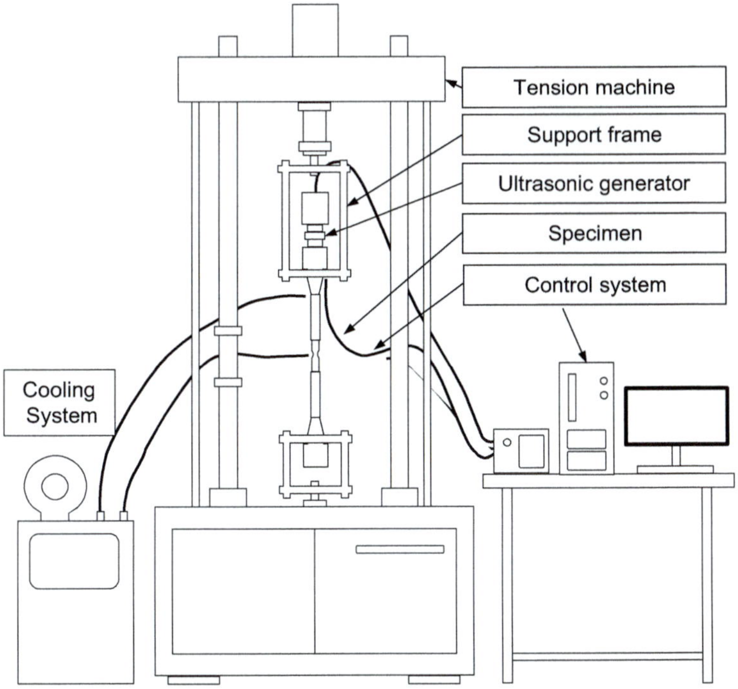

FIGURE 6.3 Schematic diagram of $R > -1$ ultrasonic fatigue testing machine.

static tension or pressure to the piezoelectric ceramic and causing failure, an amplitude modulation rod and two brackets are used to bear external forces. The amplification rod must be designed with an appropriate amplification factor to meet the experimental requirements. The amplitude of the contact position between the bracket and the amplitude modulation rod should be designed to be close to 0 and vibration isolation measures should be added to avoid resonance caused by the vibration excitation of the amplitude modulation rod on the bracket, and a gasket should be added for isolation.

However, due to the lack of disclosure of equipment interfaces by experimental equipment manufacturers, the static stretching machine and ultrasonic fatigue testing machine are only mechanically assembled and controlled by two independent control systems, which require improvement in integration and compatibility. There are no major technical difficulties in highly integrating such experimental systems. With the development of the market, it is expected that highly integrated commercial asymmetric-loaded ultrasonic vibration fatigue experimental systems will soon emerge.

6.2.2.2 Ultrasonic Torsional Alternating Stress Loading Method

The composition of the ultrasonic torsion fatigue test system is similar to that of the axial load ultrasonic fatigue test system, as shown in Figure 6.4. Both systems comprise a control system, ultrasonic generator, transducer, amplification rod, and

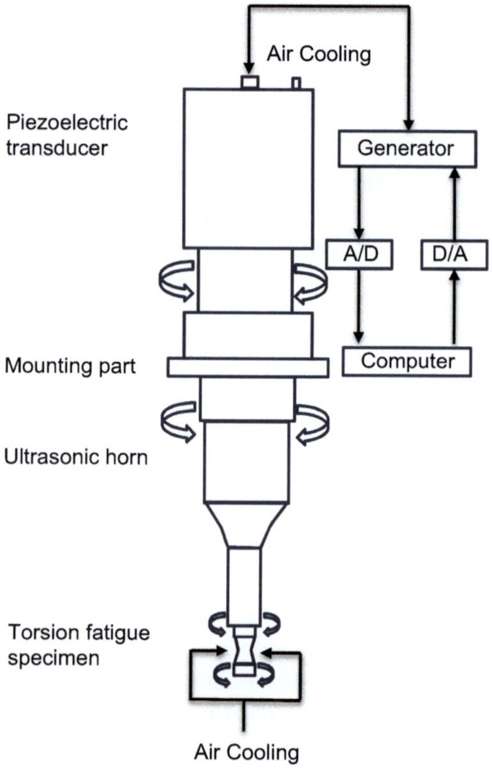

FIGURE 6.4 Ultrasonic torsional fatigue testing system.

fatigue specimen. During operation, the control system converts digital signals into analog signals via a digital-to-analog converter, driving the ultrasonic generator to produce a specific power output ultrasonic signal. This signal is transformed into a mechanical wave by the transducer and transmitted to the ultrasonic resonance module. The ultrasonic resonance module consists of two parts: an amplification rod connected to the transducer and a fatigue specimen. The first-order torsional resonance frequency of each component is approximately 20 kHz. The amplification rod transmits the mechanical waves from the transducer to the fatigue specimen, thereby applying a 20 kHz torsional load to the fatigue specimen.

The finite element analysis results indicate that the input and output amplitudes of the amplification rod are in opposite directions. Consequently, a zero-displacement point occurs on the surface of the amplification rod during the experiment. This finding enables the implementation of asymmetric ultrasonic torsional loads. Based on this observation, displacement constraints were applied to the zero-amplitude position of the upper amplification rod during the experiment, as shown in Figure 6.5. Additionally, static torque was applied to the zero-amplitude position of the lower amplification rod. This system can conduct fatigue failure tests on various materials, including metals and composites, under different stress ratios.

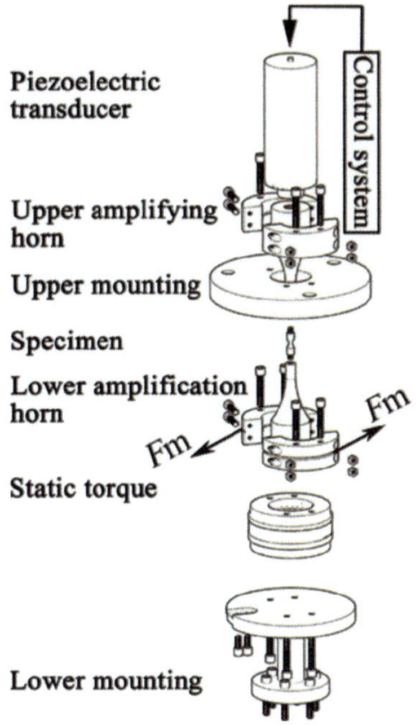

FIGURE 6.5 The main components of the ultrasonic torsion fatigue testing machine. (Reprinted with permission from Ref. [47], Elsevier.)

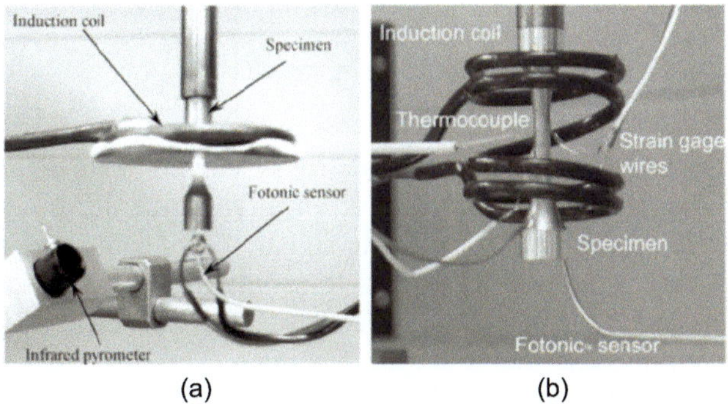

FIGURE 6.6 High-temperature ultrasonic fatigue experimental system: (a) infrared non-contact temperature measurement method [3]; (b) thermocouple contact temperature measurement method [4]. (Reprinted with permission from Refs. [3,4], Elsevier.)

6.2.3 Ultrasonic Fatigue Experimental System in Complex Environments

6.2.3.1 High-Temperature Ultrasonic Fatigue Experimental System

By using traditional resistance-type high-temperature furnaces or high-frequency induction heaters, high-temperature environmental conditions can be achieved. Due to the high efficiency and lightweight of induction heating equipment, it is easy to integrate with existing ultrasonic fatigue testing machines, as shown in Figure 6.6.

For temperature monitoring, contact-type thermocouples and non-contact infrared thermometers can be used. With the development of electronic technology, the performance of non-contact infrared thermometers has been greatly improved, which can quickly and accurately monitor the temperature of specimens. Due to the challenges of accurate temperature monitoring and control, the technical difficulty in developing a high-temperature ultrasonic fatigue experimental system is very high, and only a few laboratories have successfully developed such systems.

For high-temperature VHCF experiments in the air, as the elastic modulus is a fundamental material property, it is a temperature-dependent function, meaning that the elastic modulus decreases almost linearly with increasing experimental temperature [5]. In this case, it is inevitable that the elastic modulus of the material will be unevenly distributed on the surface of the specimen. At high temperatures, the elastic modulus of materials changes, so the size of fatigue specimens at room temperature cannot be simply used for ultrasonic fatigue experiments in high-temperature environments. However, this point is often overlooked in most studies on high-temperature ultrasonic fatigue. Although in some cases, the specimen size at room temperature can still undergo fatigue resonance loading at 20 kHz in high-temperature environments, the elastic modulus of the specimen decreases with increasing temperature. The density, elastic modulus, and size of the material jointly determine the stress displacement C_s of the specimen [1]. When the elastic modulus changes, it will affect the C_s of the sample. That is, as the temperature increases and the elastic modulus decreases, the C_s will also decrease, thereby affecting the results of high-temperature fatigue experiments. This causes the stress amplitude applied to both ends of the sample to be larger than the true stress amplitude, especially under high-temperature conditions, where this deviation is more pronounced. Considering the variation of elastic modulus, it is necessary to redesign the size of fatigue specimens at high temperatures to obtain C_s values at high temperatures in order to obtain the true stress amplitude more accurately. While the temperature of the straight section in the middle of the sample reaches the target temperature required for the experiment, the temperature at both ends of the sample is slightly lower than the target temperature, resulting in an uneven distribution of the elastic modulus of the sample. Therefore, it is difficult to obtain the size of ultrasonic fatigue specimens using traditional formula calculation methods to meet the resonance requirements at ultrasonic frequencies. Using modal analysis in the finite element software ABAQUS to calculate the size of the specimen at high temperatures can effectively solve this problem. Using an infrared camera, the temperature field on the entire surface of the sample after induction heating is obtained, which is then converted into the corresponding elastic modulus and assigned to the sample model established by ABAQUS. Modal analysis is then performed to obtain the final sample size. It is worth mentioning that the fatigue specimen was designed in a dog bone shape instead of the commonly used funnel shape, mainly considering that the heating coil can be placed in an equal straight section. For funnel-shaped arc segments, it is difficult to place the heating coil in the middle of the sample, which may cause the temperature at the center of the sample arc segment to be lower than the target temperature, thereby affecting the experimental accuracy and precision.

6.2.3.2 Vacuum Ultrasonic Fatigue Experimental System

The equipment mainly provides vacuum and temperature environments for VHCF test of materials. Composed of container system, vacuum system, temperature system, and measurement and control system, as shown in Figure 6.7.

The main body of the equipment adopts a square box structure vacuum container; The crude extraction adopts a vortex pump, with a molecular pump as the main pumping pump to establish vacuum experimental conditions; using a refrigeration machine and an electric heating coil to provide a high- and low-temperature environment for the experiment; adopting automatic/manual control modes.

The container system, vacuum system, temperature system, and measurement and control system are assembled in one rack. Except for the container installed on the side of the rack, all other systems are integrated inside the rack box. The vacuum container is a horizontal box structure, with an integral front door and a molecular pump, refrigeration machine, and vacuum gauge interface arranged on the rear sealing plate; The upper part provides a corrugated pipe sealing interface for external loading purposes; on both sides are test flange interfaces; set up observation windows at the entrance and upper part of the container, and install the entire container on the box bracket through sliding rails, as shown in Figure 6.8.

Vacuum measurement is carried out using a 2-in-1 composite vacuum gauge with a resistance and cold cathode ionization gauge. The container is equipped with two vacuum measurement ports, each equipped with one ZJ-52T resistance gauge and one ZJ-14 ionization gauge, for detecting the vacuum degree of the container.

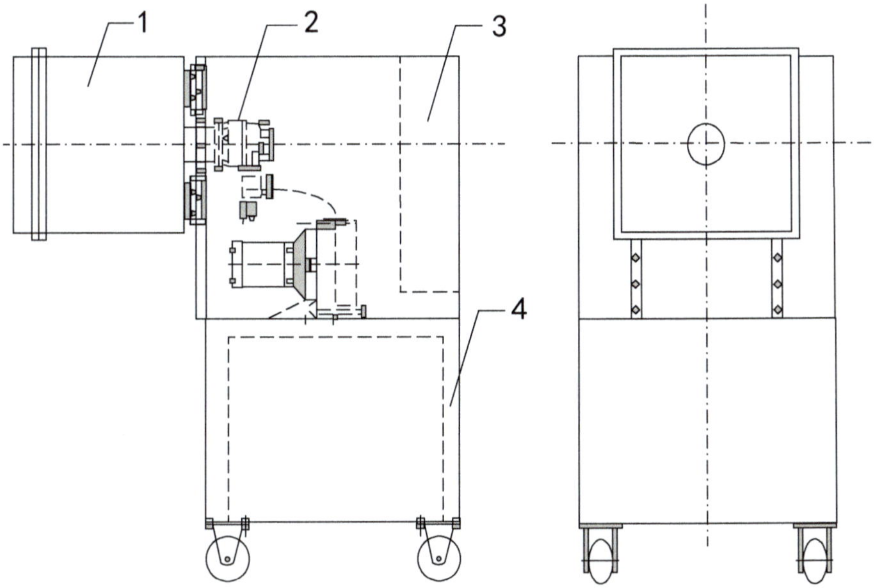

FIGURE 6.7 Equipment composition diagram (1. Vacuum container 2. Vacuum system 3. Measurement and control system 4. Temperature system).

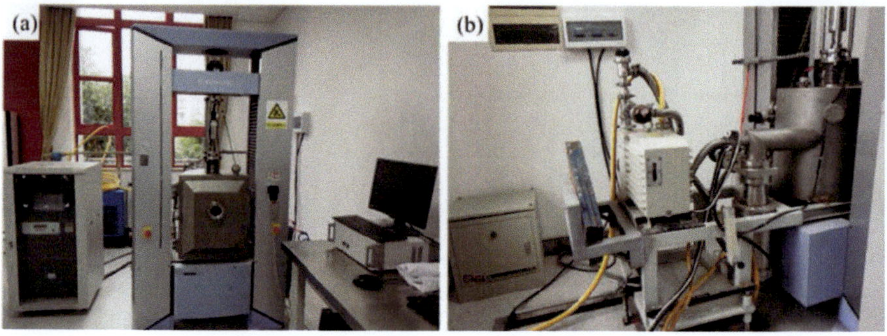

FIGURE 6.8 Vacuum-ultrasonic fatigue system: (a) vacuum system assembly and integration; (b) vacuum pump and tank connection.

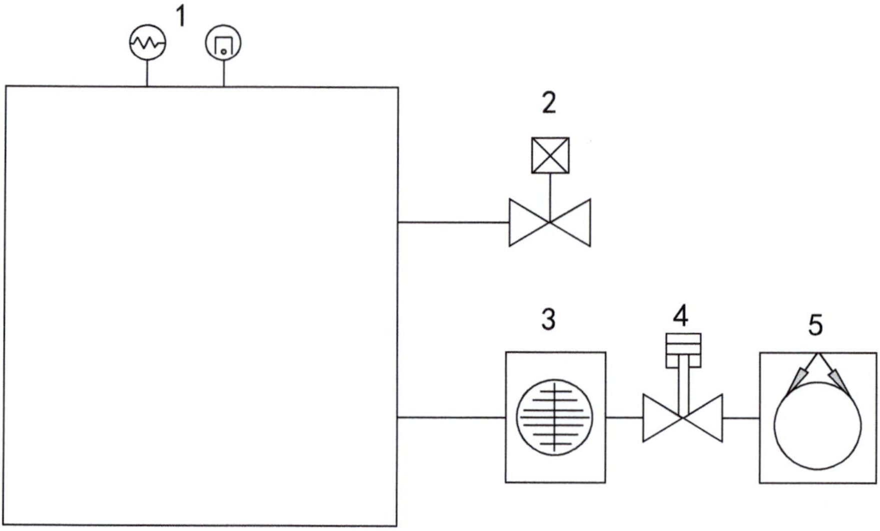

FIGURE 6.9 Vacuum system flowchart (1. Resistance gauge and ionization gauge; 2. Electromagnetic baffle valve/inflation valve; 3. Molecular pump; 4. Pneumatic baffle valve; 5. Vortex pump).

The measurement range of the ZDF-5214 composite vacuum gauge is 1.0×10^{5} to 1.0×10^{-5} (Pa), which can meet the experimental requirements. The vacuum system configuration is shown in Figure 6.9.

6.2.3.3 Corrosion Ultrasonic Fatigue Experimental System

Due to the transmission of ultrasonic wave at the solid-liquid gas interface and the requirement for free vibration conditions of the specimen, the specimen cannot be directly immersed in corrosive solution. The method of continuous spraying of the solution is adopted to allow the surface of the specimen to come into good

FIGURE 6.10 Corrosion ultrasonic fatigue experimental system at 20 kHz. (Reprinted with permission from Ref. [6], Elsevier.)

contact with the solution, and a micro water pump is used to achieve the flow and stirring of the liquid body, ensuring uniform solution composition, as shown in Figure 6.10.

The VHCF specimens were tested under three different conditions: (1) without any corrosion (virgin state), (2) after pre-corrosion, and (3) under real-time artificial sea water flow. The pre-corrosion of the specimens (before any mechanical loading) was carried out according to the ASTM G85 standard. The specimens stayed 600 hours in a salt fog corrosion chamber under temperature and humidity control with the following conditions: 35°C with 95% of humidity. The salt solution contains 5% of NaCl, its pH is 6.6 and it is applied in the chamber with a rate flow of 1.52 mL/h. After the pre-corrosion process, the specimens were removed from the corrosion chamber, first chemically cleaned and then cleaned with emery paper to remove the oxide layer.

6.2.4 Design of Ultrasonic Fatigue Specimen Design

In ultrasonic fatigue testing, the geometric dimensions of the specimen must adhere to strict requirements: (1) Resonance condition: The specimen must meet the resonance condition of 20 kHz; (2) Stress and Displacement Ratio: The specimen must establish an appropriate coefficient of stress to displacement to achieve the required stress levels for the test. To ensure that specimens of different materials can achieve the desired stress values during testing, ultrasonic fatigue specimens are often designed with varying cross-sectional shapes. Typically, the cross-sectional area of the middle part of the specimen is reduced. This design strategy not only reduces the energy input required to achieve resonance but also improves the ratio of stress to displacement. Additionally, it helps to lower the heat generated within the specimen, thereby reducing the cooling requirements.

6.2.4.1 Hourglass-Shaped Specimen

Symmetric tension compression ultrasonic vibration fatigue experiments often use hourglass-shaped specimens, which are composed of cylindrical segments at both ends and variable cross-section segments in the middle, as shown in Figure 6.11.

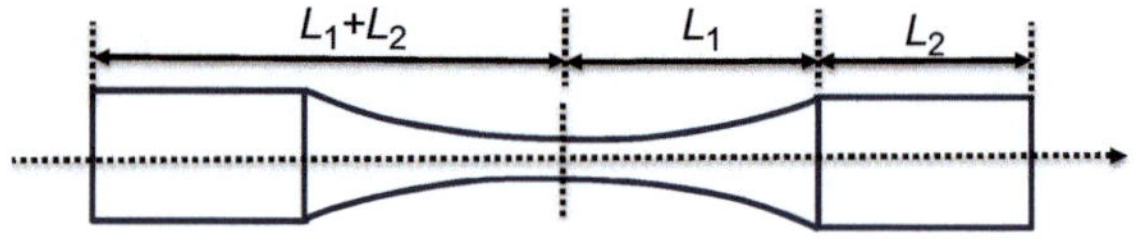

FIGURE 6.11 Hourglass-shaped specimen.

According to the theory of solid vibration, it is assumed that: (1) the material is continuous, uniform, and isotropic; (2) deformation follows Hooke's law; (3) excluding volumetric forces; (4) ignoring plastic deformation, the axial vibration can be simplified into a one-dimensional vibration form, and the dynamic equation is:

$$\rho S(x)\frac{\partial^2 u(x,t)}{\partial t^2} = \frac{\partial F(x,t)}{\partial x} \tag{6.1}$$

where ρ is the material density, $S(x)$ is the cross-sectional area at the x coordinate and $F(x,t)$ is the force acting on that cross-section. $F(x,t)$ can be represented by the following equation:

$$F(x,t) = -E_d S(x)\frac{\partial u(x,t)}{\partial x} \tag{6.2}$$

where E_d is the dynamic elastic modulus of the material.

Substituting Eq. (6.2) into Eq. (6.1) yields the one-dimensional longitudinal wave equation:

$$\rho S(x)\frac{\partial^2 u(x,t)}{\partial t^2} + E_d\left[S'(x)\frac{\partial u(x,t)}{\partial x} + S(x,t)\frac{\partial^2 u(x,t)}{\partial x^2}\right] = 0 \tag{6.3}$$

When the vibration satisfies the resonance condition, Eq. (6.3) can be solved using the decomposition variable method. Let the displacement function $u(x,t)$ be:

$$u(x,t) = U(x)e^{i\omega t} \tag{6.4}$$

Then:

$$\begin{cases} \dfrac{\partial^2 u(x,t)}{\partial t^2} = -\omega^2 U(x)e^{i\omega t} \\ \dfrac{\partial^2 u(x,t)}{\partial x^2} = U''(x)e^{i\omega t} \\ \dfrac{\partial u(x,t)}{\partial x} = U'(x)e^{i\omega t} \end{cases} \tag{6.5}$$

Substituting Eq. (6.5) into Eq. (6.3) yields the differential equation regarding displacement:

$$U''(x) + p(x)U'(x) + k^2 U(x) = 0 \tag{6.6}$$

Among them, $p(x)=\frac{S'(x)}{S(x)}$, $k=\frac{\omega}{c}$. c is the propagation speed of resonant waves in materials, and $c=\sqrt{\frac{E_d}{\rho}}$, ω is the angular frequency.

If the contour line in the middle of the sample shown in Figure 6.11 is a catenary line, R_1 and R_2 are the cross-sectional radii at $x = 0$ and $x = L$, respectively, are the length of the catenary line-shaped variable cross-section. When the resonance condition is met, L_2 is the resonance length. Therefore, the axial contour line equation is:

$$\begin{cases} y(x) = R_2, & L_1 < |x| < L \\ y(x) = R_1 \cosh(\alpha x), & |x| < L_1 \end{cases} \tag{6.7}$$

In the formula:

$$\alpha = \frac{1}{L_1} a\cosh\left(\frac{R_2}{R_1}\right) \tag{6.8}$$

The cross-sectional areas are represented as follows:

$$\begin{cases} S(x) = \pi y^2 = \pi R_2^2, & L_1 < |x| < L \\ S(x) = \pi R_1^2 \cosh^2(\alpha x), & |x| < L_1 \end{cases} \tag{6.9}$$

The $S(x)$ cross-sectional area of the cylindrical part is the same, therefore:

$$p(x) = \frac{S'(x)}{S(x)} = 0 \tag{6.10}$$

Substitute Eq. (6.10) into Eq. (6.6), where

$$U''(x) + k^2 U(x) = 0 \tag{6.11}$$

Equation (6.11) is a quadratic ordinary differential equation, and the general solution is:

$$U(x) = C_1 \cos(kx) + C_2 \sin(kx), \quad L_1 < |x| < L \tag{6.12}$$

For variable cross-section sections:

$$p(x) = \frac{S'(x)}{S(x)} = 2\alpha \tanh(\alpha x) \tag{6.13}$$

By substituting Eq. (6.13) into Eq. (6.6), it can be obtained that:

$$U''(x) + 2\alpha \tanh(\alpha x) U'(x) + k^2 U(x) = 0 \tag{6.14}$$

Introducing functions:

$$W(x) = \cosh(\alpha x) U(x) \tag{6.15}$$

There are:

$$W''(x) = \left[U''(x) + 2\alpha \tanh(\alpha x)U'(x) + \alpha^2 U(x)\right]\cosh(\alpha x) \tag{6.16}$$

From Eqs. (6.14) and (6.16), it can be concluded that:

$$W''(x) - \left(\alpha^2 - k^2\right)W(x) = 0 \tag{6.17}$$

The solution of Eq. (6.17) is:

$$W(x) = C_3 \exp(\beta x) + C_4 \exp(-\beta x) \tag{6.18}$$

In equation, $\beta = \sqrt{\alpha^2 - k^2}$.

Substituting Eq. (6.18) into (6.15) yields:

$$U(x) = \frac{C_3 \exp(\beta x) + C_4 \exp(-\beta x)}{\cosh(\alpha x)}, \quad |x| < L_1 \tag{6.19}$$

According to the boundary conditions of the sample vibration, the middle part is a symmetrical boundary and the end strain is zero, resulting in:

$$U\big|_{x=0} = 0, \quad U'\big|_{x=L} = 0 \tag{6.20}$$

The longitudinal vibration displacement amplitude of the constrained end of the sample is the same as the output longitudinal vibration displacement amplitude U_0 of the experimental machine, with

$$U\big|_{x=l} = U_0 \tag{6.21}$$

Continuity conditions:

$$U\big|_{x\to L_1-} = U\big|_{x\to L_1+}, \quad U'\big|_{x\to L_1-} = U'\big|_{x\to L_1+} \tag{6.22}$$

By substituting Eqs. (6.20)–(6.22) into Eqs. (6.12) and (6.19), the constants C_1, C_2, C_3, C_4 and harmonic length L_2 can be obtained. Therefore, the longitudinal vibration displacement amplitude function of the specimen cross-section is obtained as follows:

$$U(x) = \begin{cases} U_0 \cos\left[k(L-x)\right], & L_1 < |x| < L \\ U_0 \dfrac{\cos(kL)\cosh(\alpha L_1)\sinh(\beta x)}{\sinh(\beta L_1)\cosh(\alpha x)}, & |x| < L_1 \end{cases} \tag{6.23}$$

The resonance length L_2 satisfies:

$$L_2 = \frac{1}{k}\arctan\left\{\frac{1}{k}\left[\frac{\beta}{\tanh(\beta L_1)} - \alpha \tanh(\alpha L_1)\right]\right\} \tag{6.24}$$

Taking the derivative of the displacement function can obtain the distribution of strain along the axial direction:

$$\begin{cases} \varepsilon(x) = U_0\varphi(L_1, L_2)\dfrac{\left[\beta\cosh(\beta x)\cosh(\alpha x) - \alpha\sinh(\beta x)\sinh(\alpha x)\right]}{\cosh^2(\alpha x)}, & |x| < L_1 \\ \varepsilon(x) = U_0\sin\left[k(L - x)\right], & L_1 < |x| < L \end{cases} \tag{6.25}$$

and $\varphi(L_1, L_2) = \dfrac{\cos(KL_2)\cosh(\alpha L_1)}{\sinh(\beta L_1)}$.

According to Hooke's law, the distribution of stress along the axial direction can be obtained from Eq. (6.25).

The distribution of axial vibration displacement amplitude and stress amplitude of the hourglass-shaped specimen is shown in Figure 6.12. From the figure, the maximum stress in the sample occurs at the mid-section ($x = 0$).

When $x = 0$, from Eq. (6.25) and Hooke's law, it can be obtained that

$$\sigma_{\max} = E_d\varepsilon_{\max} = \beta E_d\varphi(L_1, L_2)U_0 \tag{6.26}$$

The C_s of the specimen can be obtained from Eq. (6.26):

$$C_s = \frac{\sigma_{\max}}{U_0} = \beta E_d\varphi(L_1, L_2) \tag{6.27}$$

In the actual design of ultrasonic vibration fatigue specimens, considering the difficulties of mechanical processing, arc lines are commonly used instead of catenary

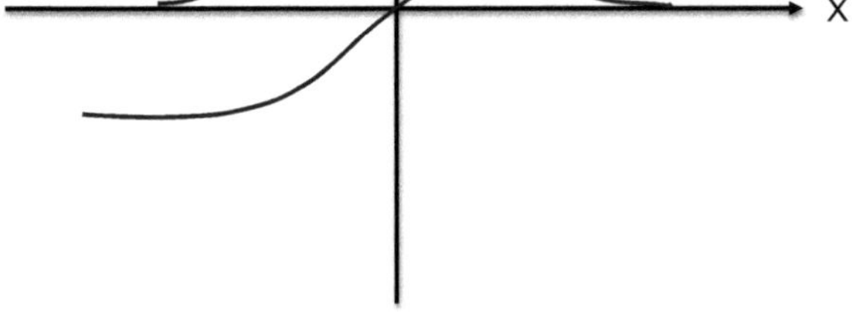

FIGURE 6.12 Schematic diagram of displacement amplitude and stress amplitude distribution in hourglass-shaped specimens.

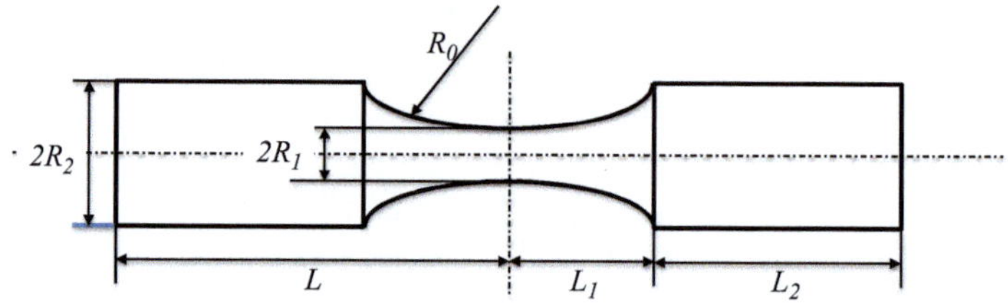

FIGURE 6.13 Hourglass shaped ultrasonic vibration fatigue specimens.

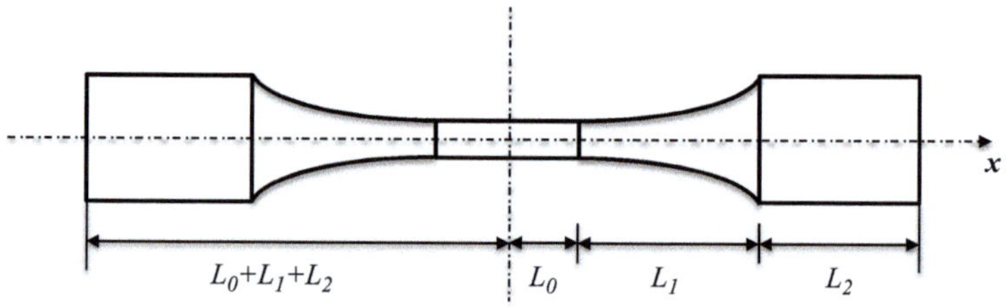

FIGURE 6.14 Dog bone-shaped specimen.

lines. Through finite element simulation analysis, it can be concluded that replacing the catenary with a circular arc has little effect on the frequency and stress displacement coefficient of the specimen, with an error of less than 1%. Therefore, the ultrasonic vibration fatigue specimen can be designed in the following shape, as shown in Figure 6.13.

6.2.4.2 Dog Bone-Shaped Specimen

Through the observation of the hourglass-shaped specimens after the experiment, it was found that many of the fractures did not occur in the cross-section where the maximum stress was designed. Therefore, to further ensure that fatigue fracture occurs at the maximum stress section, the shape of the specimen was improved based on the hourglass shape by adding an equal straight section in the middle of the specimen, as shown in Figure 6.14.

Due to symmetry, only the right half of the sample is considered for analysis. The contour line of the sample along the axis can be expressed as:

$$\begin{cases} y(x) = R_1, & \text{when } x \le L_0 \\ y(x) = R_1 \cosh\left[\alpha(x - L_0)\right], & \text{when } L_0 \le x \le L_0 + L_1 \\ y(x) = R_2, & \text{when } L_0 + L_1 \le x \le L_0 + L_1 + L_2 \end{cases} \tag{6.28}$$

and $\alpha = \dfrac{1}{L_1} \operatorname{arccosh}\left(\dfrac{R_2}{R_1}\right)$.

Wave equation and its solution:

Assuming the axis is the X-axis of the coordinate, $u(x,t)$ represents the axial displacement at the X section at time t, $A(x)$ is the corresponding cross-sectional area. So, the wave equation can be expressed as:

$$\frac{\partial^2 u}{\partial t^2} = \frac{E_d}{\rho}\left(\frac{\partial^2 u}{\partial x^2} + \frac{A'(x)}{A(x)}\frac{\partial u}{\partial x}\right) \tag{6.29}$$

where ρ is density, E_d is Dynamic modulus of elasticity.

If the resonance condition is met, using the separation of variables method, $u(x,t)$ can be expressed as $u(x,t) = U(x)e^{i\omega t}$. Therefore, the equation can be simplified as:

$$\frac{\partial^2 U(x)}{\partial x^2} + B(x)\frac{\partial U(x)}{\partial x} + k^2 U(x) = 0 \tag{6.30}$$

where $B(x) = \frac{A'(x)}{A(x)}$; $c = \sqrt{\frac{E_d}{\rho}}$ is wave velocity; $k = \frac{\omega}{c}$; $\omega = 2\pi f$; f is frequency.

The wave equations of each part of the sample are therefore expressed as:

$$\begin{cases} \frac{\partial U^2(x)}{\partial x^2} + k^2 U(x) = 0\,, & \text{when } x \le L_0 \\ \frac{\partial U^2(x)}{\partial x^2} + 2\alpha \tanh\left[\alpha(x - L_0)\right]\frac{\partial U(x)}{\partial x} + k^2 U(x) = 0, & \text{when } L_0 \le x \le L_0 + L_1 \\ \frac{\partial U^2(x)}{\partial x^2} + k^2 U(x) = 0, & \text{when } L_0 + L_1 \le x \le L_0 + L_1 + L_2 \end{cases} \tag{6.31}$$

Equation (6.31) can be solved mathematically to obtain the displacement distribution.

$$U(x) = \begin{cases} A_1 \cos(kx) + A_2 \sin(kx), & \text{when } x \le L_0 \\ \frac{B_1 \cosh(\beta x) + B_2 \sinh(\beta x)}{\cosh\left[\alpha(x - L_0)\right]}, & \text{when } L_0 \le x \le L_0 + L_1 \\ C_1 \cos(kx) + C_2 \sin(kx), & \text{when } L_0 + L_1 \le x \le L_0 + L_1 + L_2 \end{cases} \tag{6.32}$$

Based on boundary conditions and the continuity of displacement and stress, there are

$$\begin{cases} \left.\frac{\partial^2 U(x)}{\partial x^2}\right|_{x=0} = 0 \\ U(0) = 0 \end{cases}, \quad \begin{cases} U(L_0 + L_1 + L_2) = U_0 \\ \left.\frac{\partial U(x)}{\partial x}\right|_{x=L_0+L_1+L_2} = 0 \end{cases} \tag{6.33}$$

$$\begin{cases} U(L_0^-) = U(L_0^+), & U\left[(L_0 + L_1)^-\right] = U\left[(L_0 + L_1)^+\right] \\ U'(L_0^-) = U'(L_0^+), & U'\left[(L_0 + L_1)^-\right] = U'\left[(L_0 + L_1)^+\right] \end{cases}$$

By substituting the above conditions into the equation, we can obtain

$$\begin{cases} A_1 = 0, \quad A_2 = U_0 \dfrac{\beta \cos(kL_2)\cosh(\alpha L_1)}{\beta \sin(kL_0)\cosh(\beta L_1) + k\cos(kL_0)\sinh(\beta L_1)} \\ B_1 = A_2\left[\sin(kL_0)\cosh(\beta L_0) - \dfrac{k}{\beta}\cos(kL_0)\sinh(\beta L_0)\right] \\ B_2 = A_2\left[-\sin(kL_0)\sinh(\beta L_0) + \dfrac{k}{\beta}\cos(kL_0)\cosh(\beta L_0)\right] \\ C_1 = U_0\cos\left[k(L_0 + L_1 + L_2)\right], \quad C_2 = U_0\sin\left[k(L_0 + L_1 + L_2)\right] \end{cases} \tag{6.34}$$

So, the displacement distribution function can be expressed as:

$$U(x) = \begin{cases} \dfrac{U_0\beta\cos(kL_2)\cosh(\alpha L_1)}{\beta\sin(kL_0)\cosh(\beta L_1) + k\cos(kL_0)\sinh(\beta L_1)}\sin(kx), & x \le L_0 \\ \dfrac{U_0\cos(kL_2)\cosh(\alpha L_1)\begin{Bmatrix}\beta\cosh\left[\beta(x - L_0)\right] \\ +k\sinh\left[\beta(x - L_0)\right]\end{Bmatrix}}{\beta\sin(kL_2)\cosh(\beta L_1) + k\cos(kL_0)\sinh(\beta L_1)}, & x \le L_0 + L_1 \\ \left\{\cosh[\alpha(x - L_0)]\right\} & \\ U_0\cos\left\{k(L_0 + L_1 + L_2 - x)\right\}, & L_0 + L_1 \le x \le L_0 + L_1 + L_2 \end{cases} \tag{6.35}$$

The resonant length L_2 of the sample can be calculated using the following formula:

$$L_2 = \frac{1}{k}\arctan\left\{\frac{1}{k}\left[\frac{\beta k + \beta^2\tan(kL_0)\tanh(\beta L_1)}{\beta\tan(kL_0) + k\tanh(\beta L_1)} - \alpha\tanh(\alpha L_1)\right]\right\} \tag{6.36}$$

Based on the relationship between displacement and stress $\sigma = E_d\varepsilon = E_d\dfrac{\partial U(x)}{\partial x}$, the stress distribution function can be easily obtained, as shown in Figure 6.15. From the figure, the stress is highest in the middle straight section of the dog bone-shaped specimen and uniformly distributed.

The maximum stress in the sample is:

$$\sigma_{\max} = E_d\left.\frac{\partial U(x)}{\partial x}\right|_{x=0} = \frac{U_0E_dk\beta\cos(kL_2)\cosh(\alpha L_1)}{\beta\sin(kL_0)\cosh(\beta L_1) + k\cos(kL_0)\sinh(\beta L_1)} \tag{6.37}$$

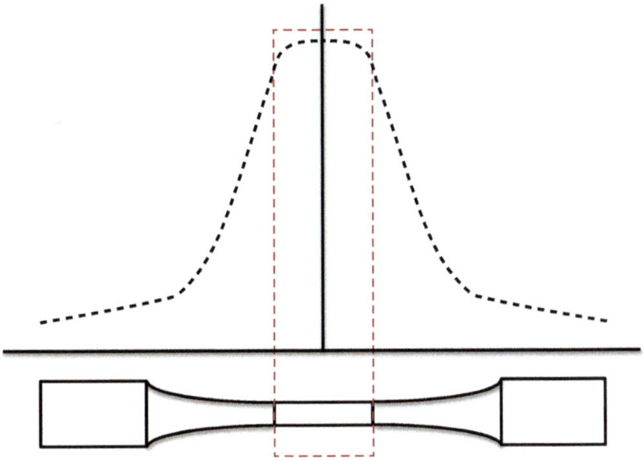

FIGURE 6.15 Axial stress distribution diagram of dog bone-shaped specimen.

C_s can be represented as

$$C_S = \frac{\sigma_{\max}}{U_0} = \frac{E_d k\beta \cos(kL_2)\cosh(\alpha L_1)}{\beta \sin(kL_0)\cosh(\beta L_1) + k\cos(kL_0)\sinh(\beta L_1)} \tag{6.38}$$

6.3 FAILURE MECHANISMS OF VHCF

6.3.1 VHCF S-N Curve

The S-N curve, also known as the Wöhler curve, represents the relationship between cyclic stress and fatigue life. This curve can reveal the competitive behavior between different fatigue failure mechanisms in metallic materials. With the development of the VHCF regime, the characteristic "step-wise" or "duplex" S-N curves have been discovered.

The VHCF life of three Cr-Si high-strength spring steels under ultrasonic frequency (20 kHz) loading with a stress ratio of −1 is shown in Figure 6.16a. The test results show that when the fatigue life exceeds 10^7 cycles, the specimens still experience fatigue failure, indicating that there is no traditional fatigue limit. The S-N curves of the three high-strength steels all exhibit a "step-wise" characteristic. In the first stage, as fatigue life increases, the cyclic stress gradually decreases, and fatigue cracks initiate from the surface of the specimen in this region. In the second stage, as the fatigue life further increases, the cyclic stress no longer decreases but forms a horizontal plateau, where surface crack initiation and internal crack initiation coexist. In the third stage, as the fatigue life continues to increase in the VHCF range, the cyclic stress decreases a second time, with fatigue cracks mainly initiating from the interior of the specimen. The "step-wise" S-N curve is thus named. It can be seen that the competition between surface and internal crack initiation failure modes leads to the appearance of this stress plateau, and their competition mainly occurs

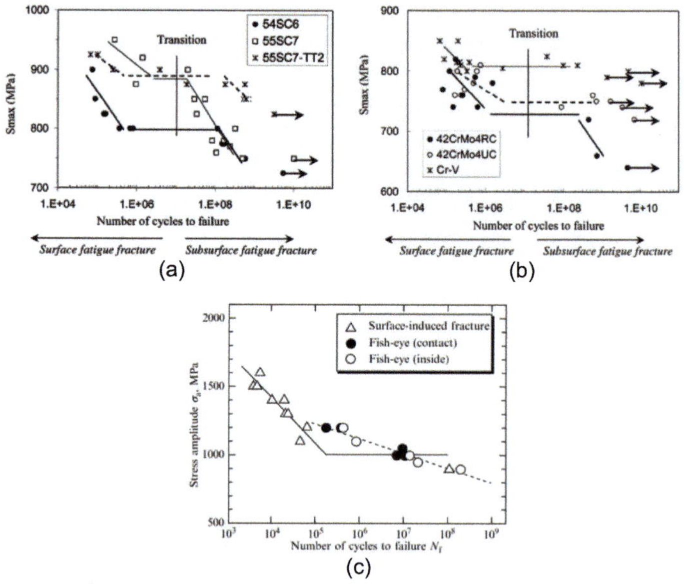

FIGURE 6.16 "Step-wise" or "duplex" S-N curves in VHCF regime ($R=-1$). (a) High-strength steel, 20 kHz [7], (b) titanium alloy, 20 kHz [8], (c) high-strength steel, 50 Hz [9]. (Reprinted with permission from Refs [8,9], Elsevier.)

around a fatigue life of 10^7 cycles. Since the surface crack initiation failure mode and the internal crack initiation failure mode each have their own S-N curves, it is also referred to as a "duplex" curve.

There are significant differences in the material composition of high-strength steel and titanium alloy, which are two typical types of alloy materials. The differences in material composition also lead to different fatigue crack initiation characteristics. Because high-strength steel generally contains non-metallic inclusions, its fatigue cracks often initiate from the inclusions. However, titanium alloys do not have inclusions or defects, and their fatigue crack initiation points are often located in areas with heterogeneous microstructures. The VHCF life of a titanium alloy under ultrasonic frequency (20 kHz) loading with a stress ratio of −1 is shown in Figure 6.16b. Compared to high-strength steel, although they have different fatigue crack initiation mechanisms, the competition between surface and internal crack initiation failure modes similarly leads to the appearance of a "duplex" S-N curve. Additionally, the development trend of the S-N curve after a fatigue life exceeding 10^9 cycles showed that after the third stage, the S-N curve exhibited another stress plateau, which is defined as the fourth stage of the S-N curve. The stress plateau in the second stage of the S-N curve corresponds to the stress limit for surface-induced fatigue failure, while the stress plateau in the fourth stage corresponds to the stress limit for internally induced fatigue failure.

The VHCF study on a high-strength bearing steel with a stress ratio of −1 using a conventional loading frequency (50 Hz) and the S-N curve is shown in Figure 6.16c.

The experimental results also revealed a "duplex" S-N curve composed of surface crack initiation and internal crack initiation failure modes. In summary, under different materials and loading frequencies, the competition between surface and internal crack initiation failure modes can lead to the appearance of "step-wise" or "duplex" S-N curves. Additionally, the competition between these two failure modes can also lead to the emergence of other types of S-N curves. Regarding the influence of these two failure modes on the shape of the S-N curve, it was explained from the perspective of the relative position of their respective occurrence probabilities, as shown in Figure 6.17.

As shown in Figure 6.17a, with the variation of cyclic stress, the relative positions of the probability distributions of surface and internal failure modes will change. When internal failure occurs at infinite life, the S-N curve will exhibit a classical horizontal asymptotic line shape, with the fatigue limit corresponding to the surface crack failure limit, as in class A in Figure 6.17b; When the probabilities of the two failure modes are separated, the classic "step-wise" S-N curve of the VHCF domain will appear, with a clear plateau transition between surface crack initiation and internal crack initiation, as in class B in the figure; When the probabilities of the two failure modes partially overlap, the S-N curve will exhibit a "duplex" shape, with an unclear transition between the two failure modes, where surface and internal failures occur simultaneously within a specific range of stress amplitudes, as in class C in the

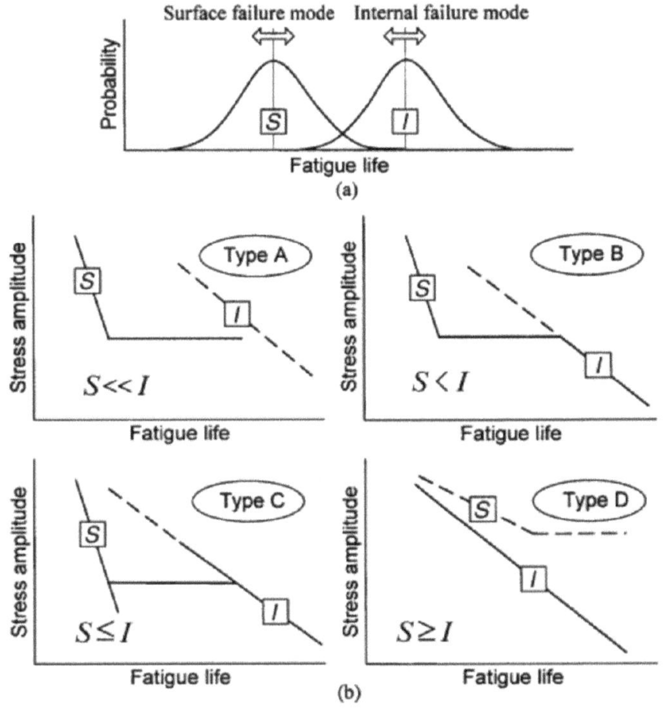

FIGURE 6.17 The effects of surface and interior crack initiation on the shapes of S-N curves in VHCF regime [10].

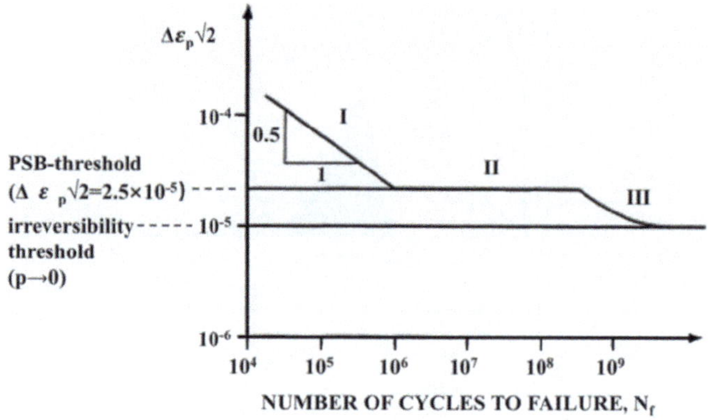

FIGURE 6.18 The S-N curve in VHCF regime of single-phase ductile material without inclusion [11,12]. (Reprinted with permission from Ref. [11], Elsevier.)

figure; When internal failure occurs completely earlier than surface failure, the S-N curve will exhibit a continuous descent, as in class D in the figure. This phenomenon generally occurs in fatigue tests after surface strengthening, with the first three classes being the most common shapes in conventional tests.

The competition between surface and internal failure modes is just one of the most common competitive mechanisms leading to the appearance of "step-wise" or "duplex" S-N curves in very HCF. The essence of the appearance of the horizontal stress plateau in "duplex" S-N curves is that the lifetimes required for fatigue failure of the two competing failure mechanisms at the same cyclic stress level differ significantly. It was pointed out that for monophasic ductile metal materials without internal inclusions, their S-N curves for VHCF also exhibit multi-stage descent characteristics, as shown in Figure 6.18. At this time, fatigue cracks all initiate from the surface, so the reason for the appearance of the "duplex" S-N curve is not the competition between surface and internal fatigue failure modes.

As shown in Figure 6.18, the S-N curve for monophasic ductile metal materials in VHCF also exhibits four stages. In the first stage, the S-N curve shows a continuous descent characteristic, corresponding to the LCF stage following the Coffin–Manson criterion. With the accumulation of cyclic loading, persistent slip bands (PSBs) induce intrusion-extrusion bands on the surface of the specimen, from which fatigue microcracks initiate and ultimately lead to fatigue failure; In the second stage, the S-N curve exhibits a horizontal plateau. This plateau corresponds to the threshold for directly forming PSBs. When the cyclic load is below this threshold, PSBs are no longer directly formed on the surface of the specimen. This threshold corresponds to the plastic strain fatigue limit of the HCF stage; In the third stage, the S-N curve exhibits a secondary descent. Although the cyclic load is below the threshold for PSB formation, with increasing cycles, the surface of the specimen accumulates irreversible random slip. After long-term cyclic loading, a large amount of irreversible slip accumulates on the surface of the specimen, increasing its surface roughness, thereby re-inducing PSBs, ultimately leading to the initiation and propagation of fatigue cracks from the formed PSBs; In the fourth stage, the S-N curve exhibits a

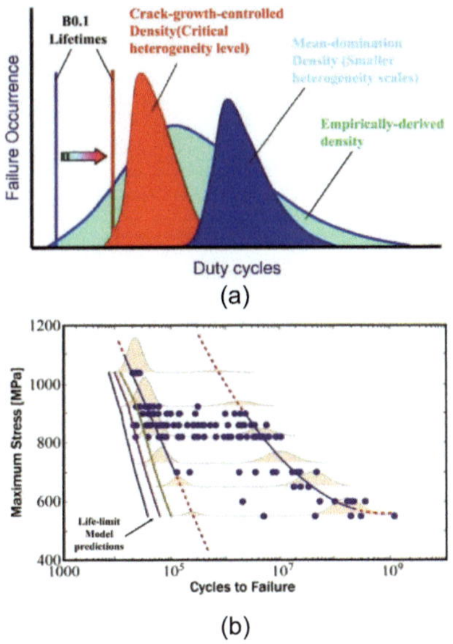

FIGURE 6.19 The "duplex" S-N curve of Ti-6246 titanium alloy (R=0.05) [13]. (a) Probability density distribution of two failure models, (b) S-N curves at 20 Hz and 20 kHz. (Reprinted with permission from Ref. [13], Elsevier.)

second horizontal plateau, corresponding to the threshold for irreversible slip formation. When the fatigue load is below this threshold, permanent damage no longer occurs on the surface of the specimen, and fatigue failure does not occur again. Therefore, this stage corresponds to the limit of VHCF. In summary, the competition of the surface PSB formation mechanism in monophasic ductile metal materials without inclusions leads to the appearance of their "duplex" S-N curves.

The fatigue life of Ti-6246 titanium alloy at 9 stress levels using a large number of fatigue specimens is shown in Figure 6.19. After performing probability statistical analysis on the fatigue life at a certain stress level, it was found that the fatigue life varied with two completely different distributions, namely, the "life-limiting behavior" and "mean lifetime behavior", as shown in Figure 6.19a. The microstructure at the crack initiation point of the life-limiting behavior often exhibits extremely high deformation non-uniformity, and after a small number of cycles of cyclic loading, the size of fatigue microcracks reaches the threshold for crack propagation, so its fatigue life is mainly composed of crack propagation life. Therefore, under the same fatigue load, the life-limiting behavior mainly occurs in a lower life interval, as shown by the curve on the left side of Figure 6.19b. However, the microstructure at the crack initiation point of the mean lifetime behavior often exhibits smaller deformation non-uniformity, and the process of fatigue crack initiation corresponding to the threshold size of fatigue crack propagation requires a large number of cycles to complete, that is to say, the fatigue life is composed of both crack initiation life and crack propagation life. The deformation non-uniformity of the microstructure is not only affected by the microstructure itself but also closely related to the loading stress. As the fatigue loading stress decreases, the probability of occurrence of highly deformed microstructures decreases, while the probability of occurrence of less deformed microstructures increases. This leads to a decrease in the probability of occurrence of limit life behavior with decreasing fatigue loading stress, while the probability of occurrence of average life behavior increases with decreasing fatigue loading stress. The competition between these two failure mechanisms results in significant dispersion in fatigue life data in the high-cycle and very high-cycle fatigue regimes, with the S-N curve exhibiting a characteristic "duplex" shape overall, as depicted in Figure 6.19b.

6.3.2 Investigation into Fatigue Failure Characteristics

Fatigue failure progresses through distinct stages: microcrack nucleation and initiation, microcrack propagation, steady-state propagation of macro-cracks, and the unstable propagation of macro-cracks. By employing fatigue crack propagation tests and quantitative fracture surface analysis, in conjunction with fracture mechanics theory, it is possible to estimate the fatigue life at each stage. Extensive research indicates that for VHCF failure, the crack initiation phase accounts for the majority of the total fatigue life, often exceeding 90%. Understanding the mechanisms behind crack initiation in VHCF failure is therefore of paramount importance. Engineering metals can generally be categorized into two types: those without micro-defects and those with micro-defects. Each category exhibits distinct mechanisms for crack initiation under VHCF conditions.

6.3.2.1 Micro-Defects Induced VHCF Crack Initiation

In metals containing micro-defects, such as carbon steel, VHCF cracks predominantly initiate at these micro-defects. These defects include inclusions and manufacturing-induced micro-defects (e.g., Al_2O_3, SiO_2, MgO, CaO, and sulfides found at initiation sites in high-strength steels), as well as process-related defects like pores introduced during casting and additive manufacturing. Research indicates that the size, distribution, quantity, and chemical composition of these inclusions and defects significantly influence the material's VHCF behavior. Moreover, mismatches in the mechanical properties (e.g., hardness) between inclusions and the matrix, as well as geometric discontinuities, lead to non-uniform plastic deformation at the interfaces, which would therefore induce fatigue crack initiation at these sites.

Taking carbon steel as an example, Figure 6.20a and b depict typical fracture surfaces after VHCF failure of carbon steel materials and the local fracture morphology near inclusions that induce fatigue crack initiation. As shown in Figure 6.20c, the fatigue life of the material can be divided into three stages: crack nucleation stage (a_0), early short crack propagation stage (a_i), and long crack propagation stage (a_c). The corresponding fatigue crack propagation rates for each stage are shown in Figure 6.20d. During crack nucleation and early short crack propagation stages, the influence of the material's microstructure on the fatigue crack propagation rate is more pronounced, resulting in relatively low and fluctuating crack propagation rates. With the accumulation of fatigue loading cycles, the fatigue crack length slowly increases, and the stress intensity factor at the crack tip gradually increases. When

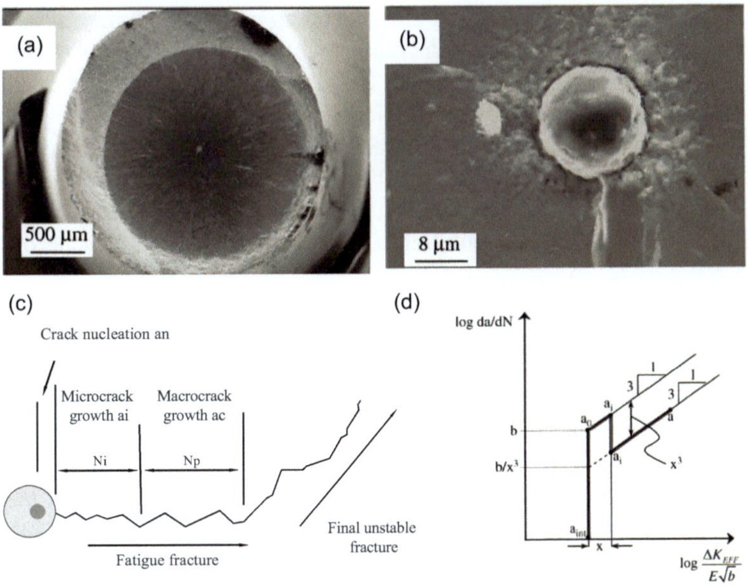

FIGURE 6.20 (a) Fracture surface displaying the characteristic fish-eye feature associated with fatigue failure, (b) high-resolution image detailing the fish-eye feature at the fatigue crack initiation site [15], (c) schematic illustrating the initiation of a fatigue crack at inclusion and its subsequent propagation characteristics [16], and (d) schematic representation of the crack growth rate across different stages as depicted in (c). (Reprinted with permission from Ref. [17], Elsevier.)

the driving force at the fatigue crack tip rises sufficiently to overcome the obstacles of the material's microstructure, the crack propagation transitions to the long crack propagation stage, during which the fatigue crack exhibits steady-state propagation until final instantaneous fracture. During crack nucleation and early propagation stages, a relatively rough fine grain area (FGA) forms around the inclusions, referred to as the optical dark area (ODA) by Murakami et al. [14]. From the fracture morphology perspective, crack initiation and initial propagation exhibit fish-eye (FiE) characteristics, including the FGA region. The outer region of the FiE zone beyond the FGA is relatively smooth, while the area outside the FiE zone represents the steady-state propagation region of the crack, exhibiting a greater surface roughness. Figure 6.21 schematically illustrates the entire process of fatigue crack initiation to fracture of the surface of VHCF of high-strength steel.

The mechanism underlying the formation of FGA during the fatigue crack initiation stage in carbon steel materials remains a subject of diverse interpretations (Figure 6.21). Murakami et al. [14] proposed a correlation between FGA formation and the enrichment of hydrogen near inclusions during the steel smelting process. Under cyclic loading, stress concentrations near internal inclusions induce brittle fractures owing to the elevated hydrogen content. Limited by crack length, the initial stage witnesses relatively weak driving forces and sluggish crack propagation rates. During slow crack propagation, hydrogen assists in overcoming resistance, leading

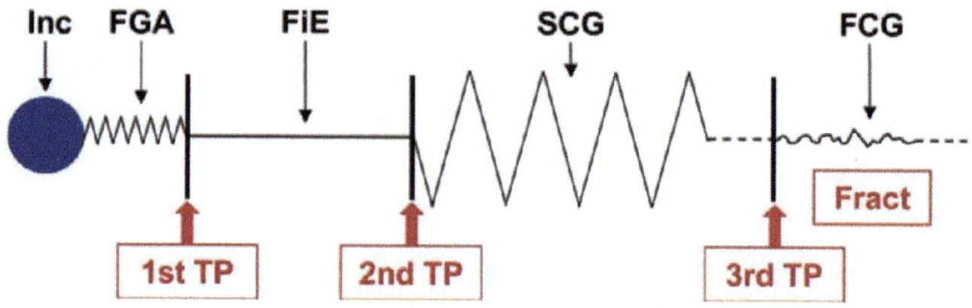

FIGURE 6.21 A schematic representation of the entire process from the internal initiation to fracture of fatigue cracks under ultra-high-cycle conditions in high-strength steel. (Reprinted with permission from Ref. [18], Elsevier.)

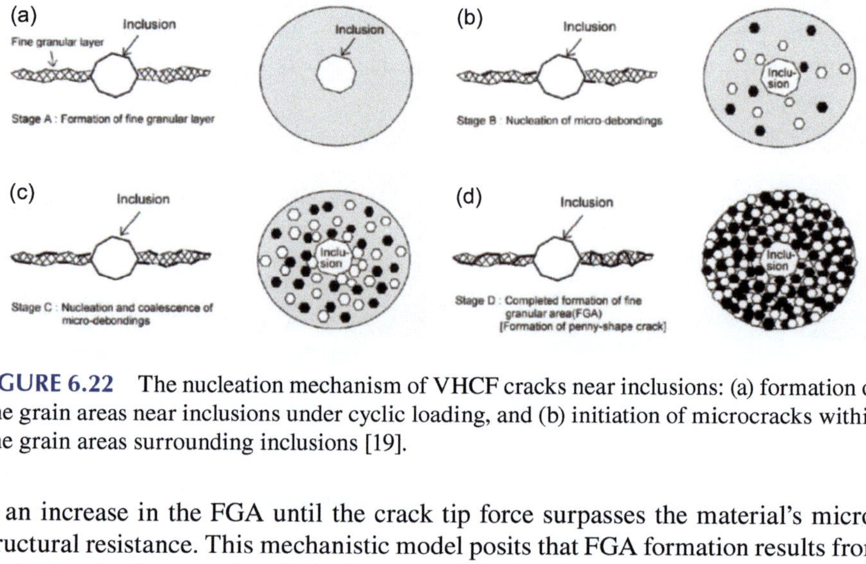

FIGURE 6.22 The nucleation mechanism of VHCF cracks near inclusions: (a) formation of fine grain areas near inclusions under cyclic loading, and (b) initiation of microcracks within fine grain areas surrounding inclusions [19].

to an increase in the FGA until the crack tip force surpasses the material's microstructural resistance. This mechanistic model posits that FGA formation results from the interaction between locally hydrogen-enriched regions near inclusions and cyclic strain during steel smelting. Sakai [19] observed a gradual decrease in grain refinement within FGA with increasing distance from the fracture surface with the help of transmission electron microscopy. Based on this, a nucleation mechanism for VHCF cracks near inclusions is proposed (Figure 6.22). Long-period low-amplitude cyclic loading induces crystal structure damage near inclusions, resulting in grain refinement and FGA formation (Stage A). Subsequent cyclic loading leads to microcrack formation via delamination between adjacent crystal structures within FGAs. As loading cycles increase, delamination sites multiply (Stage B), and microcracks propagate until they intersect, forming macroscopic cracks (Stage D). Therefore, according to this model, fatigue failure occurring within the VHCF regime in carbon steel materials due to inclusions is primarily attributed to stress concentration at the inclusion sites under cyclic loading, leading to refinement of the initial microstructure and the formation of FGA near the inclusions. This localized increase in anisotropy weakens the material's ability to self-consolidate under cyclic loading, resulting in the nucleation of microcracks in this region, followed by microcrack propagation and the formation of macroscopic cracks. Grad et al. [20] further proposed that during the FGA formation stage, grain refinement is caused by the entanglement of dislocations within the microstructure near the inclusions under long-period low-amplitude cyclic loading, forming dislocation cells and subsequent sub-grain structures, thus completing the grain refinement process. The threshold characteristic value for crack propagation is related to the size of the material's microstructure. Grain refinement reduces the threshold value for crack propagation, leading to crack initiation from a static crack tip under certain driving force conditions. The slow propagation of the crack then triggers the refinement process within the microstructure near the crack tip until entering the stage of long crack propagation, ultimately resulting in material fatigue failure. Therefore, the vast majority of the material's fatigue life is consumed by the process of grain refinement. Hong et al. [21] suggested that FGAs form during the stage of short crack propagation in the fatigue failure process, driven by numerous cyclic pressing (NCP) between the two sides of the crack surface. As shown in Figure 6.23, under cyclic loading, stress concentration near the inclusions leads

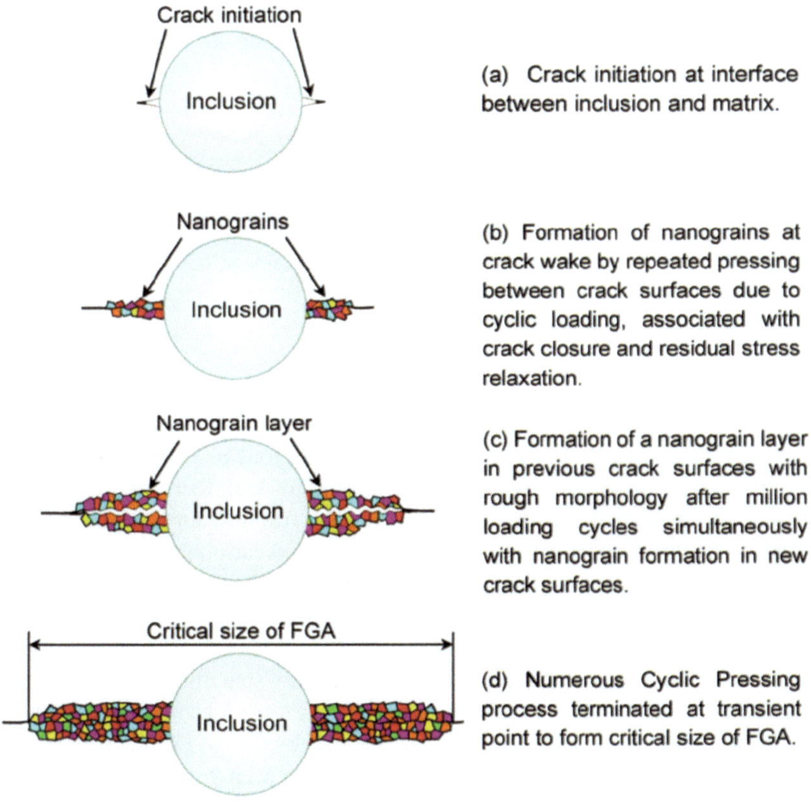

FIGURE 6.23 The progression of fatigue crack nucleation and propagation near inclusions under cyclic loading: (a) nucleation of microcracks near inclusions under cyclic loading, (b) expansion of microcracks and mutual compression, resulting in nanocrystals formation near the crack surface, (c) increase in nanocrystal density leading to the formation of a nanocrystal layer, and (d) formation of fine grain areas (FGA) and progression of cracks into the long crack propagation stage. (Reprinted with permission from Ref. [21], Elsevier.)

to crack nucleation near the inclusions, where plastic accumulation occurs in the adjacent matrix material, initiating the initial crack. Under symmetric loading conditions, high-frequency long-period crack closure results in severe plastic deformation in the local surface layer of the crack surface, causing refinement of the material's microstructure and the formation of fine-grained layers near the crack surface until the crack enters the stage of long crack propagation.

In practical applications, structural metal materials, such as titanium alloys, are often processed or joined employing welding/additive manufacturing methods. During the welding process of metal materials, the coupling of high melting temperatures and low thermal conductivity leads to steep temperature gradients in the adjacent regions of the fusion zone (FZ), resulting in differential microstructures in the heat-affected zone (HAZ). Consequently, the entire weld joint exhibits heterogeneous microstructures from the FZ to the base material. Welded joints are inevitably prone to the formation of welding defects, such as pores, slag inclusions, and microcracks, primarily dominated by welding pores, due to factors including welding process, plate condition, and cleanliness. The geometric discontinuities caused by welding pores result in localized stress concentration in critical regions of the weld joint, critically influencing its fatigue behavior. Welding pores are prone to inducing crack initiation, rendering them weak spots in the weld joint. As shown in Figure 6.24, microstructural changes resembling the initiation of ultra-high-cycle fatigue cracks near inclusions in carbon steel materials are observed around welding pores, forming FGA. Fracture morphology resulting from mutual compression and friction between the upper and lower surfaces is observed within the FGA, highlighting the significance of mutual compression between sections during FGA formation,

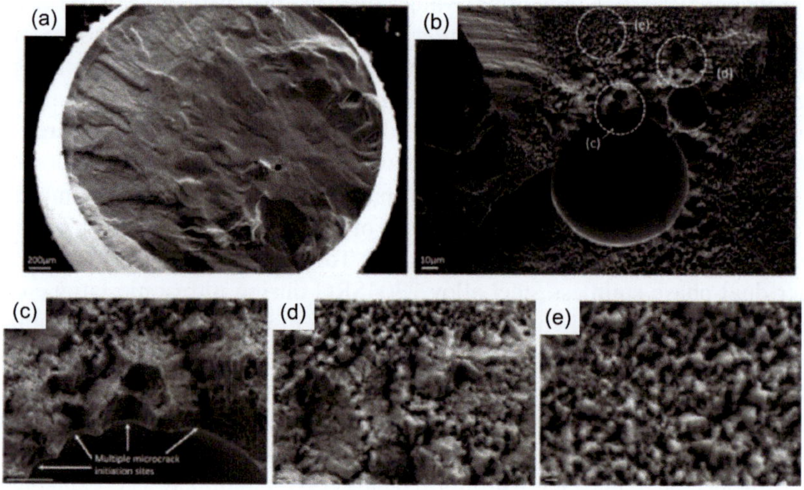

FIGURE 6.24 The morphology of fatigue failure caused by welding porosity: (a) macroscopic appearance of the fracture surface, (b) FGA in the crack initiation zone, (c) morphology of the porosity edge in the crack initiation zone, (d) friction marks on the inner surface of FGA, and (e) fine grain at the edge of FGA. (Reprinted with permission from Ref. [22], Elsevier.)

consistent with the mechanisms proposed by Hong et al. [21] and Pan et al. [23] for the formation of FGA in carbon steel materials and dual-phase titanium alloys under symmetric loading conditions. On the other hand, when the stress ratio is positive, welding pores still exert a direct influence on the fatigue failure of electron beam welded joints. In such cases, fatigue failure of the weld joint still predominantly originates from welding pores, but FGA is not formed within the crack initiation zone. Liu et al. [24] conducted statistical analysis on the sizes of welding pores causing fatigue failure in titanium alloy electron beam welded joints and found that the sizes of pores serving as crack initiation sites were mostly larger than 50 μm but smaller than 100 μm. When the fatigue crack initiation occurs near the surface of the specimen, the size of welding pores inducing fatigue crack initiation may decrease to below 50 μm.

6.3.2.2 Microstructural Heterogeneity Induced VHCF Crack Initiation

As for the second category of material, where no micro-defects exist internally, the intrinsic structure of first-phase materials plays a crucial role in VHCF failure processes, and its characteristics differ from those induced by micro-defects.

For dual-phase steels, such as martensite-ferrite low-alloy steels and austenite-ferrite stainless steels, the initiation and propagation of VHCF cracks are significantly influenced by the fundamental mechanical properties of the two phases (such as hardness, elastic modulus, etc.) and local microstructures (such as grain boundaries, phase boundaries, grain orientations, etc.). Specifically, when there is a significant difference in hardness between the two phases, fatigue cracks tend to initiate from the softer phase and then propagate. For instance, Tofique et al. [25] observed during ultrasonic fatigue experiments on dual-phase stainless steel LDX2101 that the significant difference in hardness values between the austenite and ferrite phases results in irreversible plastic deformation induced by cyclic fatigue loading (such as persistent slip bands, PSBs) predominantly occurring in the softer α-ferrite phase, as shown in Figure 6.25. Consequently, fatigue cracks are more likely to initiate and propagate within the PSBs that have already formed. However, when there is little difference in hardness values between the two phases, the cyclic deformation behavior of the material and the initiation and propagation of fatigue cracks are more likely to be influenced by the local microstructure of the material, such as grain boundaries and grain orientations of the two phases. Tofique et al. [25], in their study on another dual-phase stainless steel alloy 2304SRG during ultrasonic fatigue experiments and microstructural characterization, found that when there is no significant difference in hardness value between the α-ferrite and γ-austenite phases, no distinct strong and weak phases exist, and irreversible plastic deformation and fatigue cracks tend to occur along the grain boundaries of the two phases. Moreover, Sahu et al. [26], in their fatigue loading experiments on dual-phase stainless steel, found that fatigue cracks tend to nucleate within slip bands corresponding to {111} slip planes in the γ-austenite phase, and the phase boundary between the α-phase and the γ-phase significantly impedes the propagation of the nucleated fatigue cracks. Additionally, as shown in Figure 6.26, when fatigue cracks propagate within the α-ferrite phase, their propagation direction also follows a specific crystal plane of the α-ferrite phase. Krupp et al. [27], in their study on plastic deformation and fatigue crack propagation

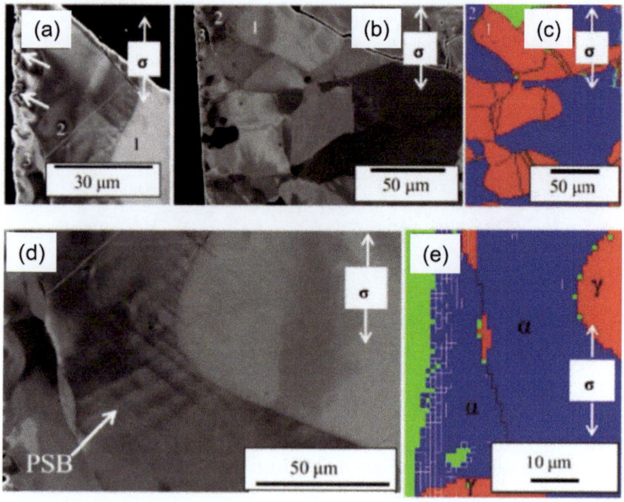

FIGURE 6.25 The cyclic plastic deformation behavior and fatigue crack initiation in dual-phase stainless steel LDX2101, indicate that persistent slip bands and fatigue cracks tend to initiate within the softer α-ferrite phase. (Reprinted with permission from Ref. [25], Elsevier.)

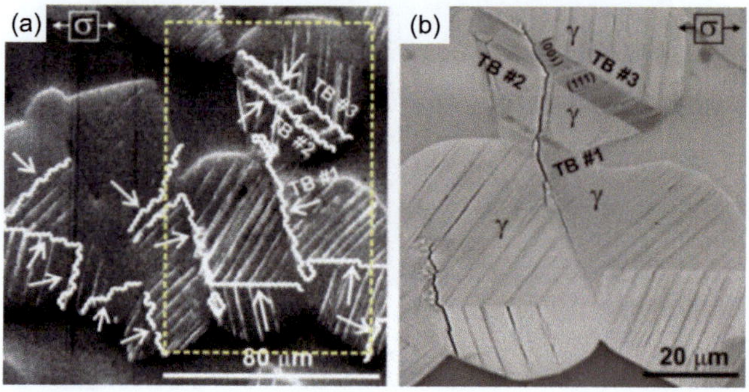

FIGURE 6.26 The SEM image of secondary cracks on the surface of dual-phase stainless steel. (Reprinted with permission from Ref. [26], Elsevier.)

behavior of dual-phase stainless steel, found that the transfer of slip deformation within the alloy is significantly influenced by twist misorientations between different grain pairs and between identical grain pairs. When there is no twist misorientation between adjacent grains and only an angular deviation exists, dislocation deformation can be effectively transferred between adjacent grains.

For microstructurally flawless dual-phase titanium alloys, the microstructure depends on the volume fraction and morphology of the α-phase (hexagonal close-packed structure, HCP) and β-phase (body-centered cubic structure, BCC). Figure 6.27 presents a typical scanning electron microscope image of the microstructure of dual-phase $\alpha+\beta$ titanium alloy, where the blocky black regions

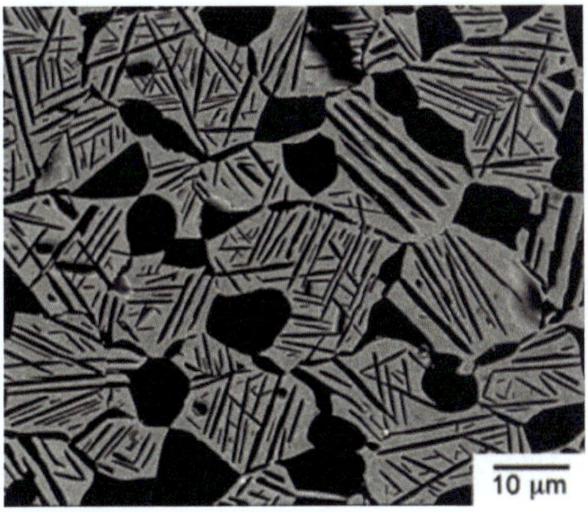

FIGURE 6.27 The microstructure of $\alpha+\beta$ titanium alloy. (Reprinted with permission from Ref. [29], Elsevier.)

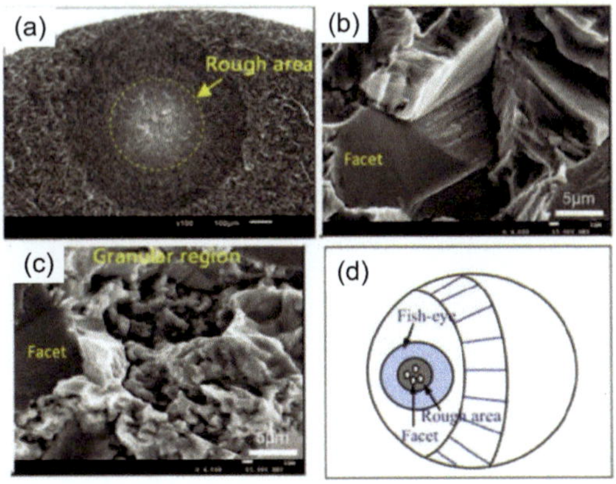

FIGURE 6.28 Typical fracture morphology of internal crack initiation in titanium alloys: (a) "fish-eye" morphology; (b) flat surface within the rough area; (c) grain area within the rough area; (d) schematic diagram of the entire fracture morphology [28,34–36]. (Reprinted with permission from Ref. [28], Elsevier.)

represent equiaxed primary α-phase (α_P), the lamellar black regions represent secondary α-phase (α_L), and the light gray regions represent the matrix β-phase. In the VHCF regime, due to the relatively soft α-phase in the bimodal microstructure of titanium alloys, microplastic deformation accumulates in the softer phase, making fatigue damage more prone to occur in the softer α-phase. As the number of cycles increases, dislocation slip gradually accumulates, leading to crack initiation from the α-phase, resulting in the formation of facets. Moreover, due to the mismatch in hardness between the α and β-phases in dual-phase titanium alloys, significant plastic deformation is also prone to occur at phase boundaries during cyclic loading, causing a high density of dislocation pileups at phase boundaries, thereby leading to crack initiation from phase boundaries. Figure 6.28 demonstrates the typical

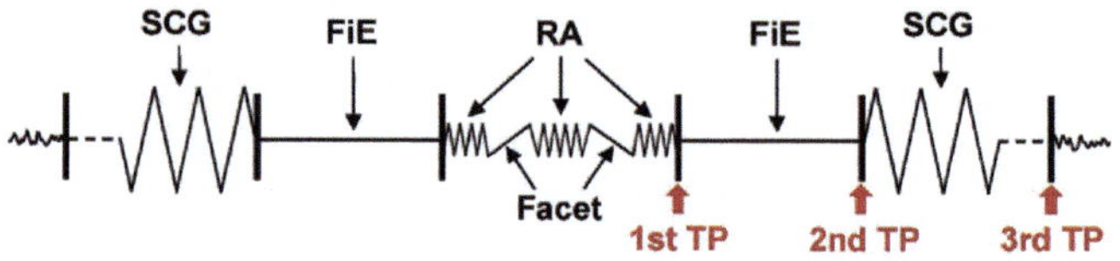

FIGURE 6.29 Schematic diagram of the entire process from crack initiation to final fracture in titanium alloys during VHCF; RA: rough area, FiE: fish eye, SCG: stable crack growth, TP: transition point, (In fact, FiE includes RA and facets, here FiE only represents the portion outside RA). (Reprinted with permission from Ref. [18], Elsevier.)

fracture surface morphology characteristics of internal crack initiation in titanium alloys. It is observed that the fracture surface morphology characteristics of VHCF are consistent with the internal crack initiation characteristics of high-strength steels; the crack initiation zones are relatively rough. However, no inclusions or defects are found in the crack initiation zones of titanium alloys, but numerous facet clusters and fine granular regions are observed. The unique area in the crack initiation zone of titanium alloy VHCF is often referred to as the rough area (RA). Liu et al. [28] conducted atomic force microscopy measurements on Ti-6Al-4V specimens subjected to ultrasonic axial loading (20 kHz) with a fracture cycle of $N_f = 2.02 \times 10^7$, showing that the facet surface has minimal roughness, with a maximum value of 14 nm, while the localized roughness connecting the facets is significant, with a maximum value of 2217 nm, and the maximum roughness of the FiE area is 209 nm, indicating that the roughness of the RA area is an order of magnitude larger than that of the FiE area. Figure 6.29 illustrates the schematic of crack propagation in titanium alloys during the entire VHCF process: cracks initiate from the RA area and form facet morphology in the RA, then the cracks stably propagate to the FiE in the RA area and subsequently enter the rapid crack propagation zone until instantaneous fracture occurs. The process of crack propagation on the entire fracture surface of titanium alloys is similar to that of high-strength steels; FiE and stable crack propagation zones are formed internally in the specimens. The RA area in titanium alloys is similar to the FGA in high-strength steels, consuming the vast majority of the total fatigue life. Therefore, research on the VHCF behavior of titanium alloys mainly focuses on the crack initiation zone as well.

For the formation mechanism of fine grains in the RA, Oguma and Nakamura [30] proposed the "cold-welding-induced grain refinement" model, suggesting that in the VHCF process under vacuum conditions, microscopic structures may undergo reciprocal contact, resulting in "cold welding" on the contact surfaces. During this process, microstructural refinement near the crack surface may occur, leading to the formation of fine grains in the RA. The model proposed by Oguma and Nakamura [30] for the formation of RA in titanium alloys is similar to the model proposed by Hong et al. [18,21] for the formation of FGA in high-strength steel, both based on mutual squeezing between the upper and lower crack surfaces, causing microstructural refinement and resulting in the formation of fine grains. Heinz and Eifler [31] observed microstructural grain refinement and found that microcrack formation occurred within primary α-phase grains in their study of titanium alloys under VHCF, which then aggregated into crack clusters during subsequent cycles.

The phenomenon of grain refinement in the RA of titanium alloys, similar to the formation of FGA in high-strength steel, suggests that both materials may undergo the same physical process during grain refinement, which may originate from external loads. As for the characteristic facets on the fracture surface of titanium alloys, most researchers believe that they result from cleavage fracture of the softer α-phase, which is either primary α-phase or secondary α-phase embedded in the β-phase. The crystallographic planes of the α_p facets are basal or close to basal, and these facets are believed to form on a hard grain with the c-axis close to or parallel to the direction of the local maximum principal stress, adjacent to a soft grain favorable for sliding. There are also related studies reporting that the formation of flat surfaces can also occur along prism slip. Furthermore, the initiation of fatigue cracks is also influenced by the angle between the loading direction and the basal planes of the α_p grains. Ward-Close and Beevers [32] found that when the angle between the c-axis of the α_p grain and the dynamic loading axis is between 40° and 60°, the α_p grains are prone to slip fracture to form facets. Regarding the formation process of α_p facets, Jha et al. [33] summarized three mechanisms, among which pure slip is considered the main factor. The mechanisms for the formation of α_p facets include pure slip, basal slip on α_p planes combined with normal stress, and quasi-cleavage. In addition, the formation mechanism of facets not only involves the slip fracture of α_p grains under cyclic loading but also includes the fracture of α_p grains caused by the propagation of short cracks.

For materials of the first category (i.e., those without micro-defects), the mechanism of crack initiation is more complex compared to materials of the second category (those with micro-defects), and there are fewer relevant theoretical models. From the current research results, it appears that these materials tend to initiate cracks from locations with uneven phase distribution or microstructures in the VHCF regime.

6.4 VHCF LIFE PREDICTIONS: CASE STUDY

With the development and refinement of experimental testing and analysis methods, computers, and finite element stress analysis, fatigue science has made rapid advancements. The understanding of fatigue phenomena has become increasingly comprehensive, leading to the emergence of numerous theories and predictive models concerning fatigue life. However, these models are unable to accurately predict the VHCF life of metallic materials. From the perspective of the failure process, VHCF is a nonlinear and uncertain process. To date, accurate prediction of VHCF life remains a global challenge that requires further in-depth research.

The total fatigue life comprises crack initiation life and crack propagation life. Fracture mechanics can effectively analyze and predict LCF life. In the process of VHCF failure, crack propagation life constitutes only a small part of the total fatigue life [37,38]. Therefore, for VHCF, the primary focus is on the crack initiation life. The ability to accurately estimate the fatigue life of crack initiation stage becomes the key to predicting fatigue life. The fatigue cumulative damage theory can be used to predict fatigue crack initiation life by defining damage variables and studying the damage evolution law, which allows for fatigue life prediction. There are many methods for predicting fatigue crack initiation life, which can be broadly categorized based on the basic assumptions and control parameters of fatigue crack initiation life prediction: the nominal stress method, the local stress-strain method, the energy method, and the field intensity method. The following sections will introduce examples of recent fatigue life prediction models.

6.4.1 Prediction of VHCF Life for High-Strength Steel

Wang et al. [38] conducted experimental research and analysis on six types of high-strength low-alloy steel. They established a relationship between inclusion size and the fatigue life of high-strength steel, resulting in the development of a quantitative model capable of predicting the fatigue life of high-strength steel.

$$N_f = \frac{9GW_S}{\left(\Delta\sigma - \Delta\sigma_D^R\right)^2 a_0} + \frac{a_0^{(1-n/2)}}{C\Delta\sigma^n \beta_1^n \pi^{n/2}\left(\frac{n}{2}-1\right)} \tag{6.39}$$

$$W_s = \frac{\Delta K_{\text{th}}^2}{2E} \tag{6.40}$$

where G is the shear modulus, W_s is the fracture energy, $\Delta\sigma$ is the stress amplitude, $\Delta\sigma_D^R$ is the fatigue limit of the material at a stress ratio R, a_0 is the initial inclusion size, β_1 is a geometric constant, and C and n are material properties. The validation results show that the model predictions are highly consistent with the experimental results for several high-strength steels, as shown in Table 6.1.

TABLE 6.1
Fatigue Life and Fracture Surface Data [39]

Material and Specimens	Stress (MPa)	Inclusion Size a_0, (μm)	Experiment N_f	Prediction N_f	Prediction N_i/N_f (%)
Cr–Si-55SC7-1	800	19	3.3×10^8	2.8×10^7	99.53
Cr–Si-55SC7-7	780	12	1.7×10^8	1.2×10^8	99.88
Cr–Si-55SC7-13	760	23	1.1×10^8	5.8×10^8	99.97
Cr–Si-55SC7-TT2-1	850	8	5.5×10^8	3.6×10^8	99.96
Cr–Si-55SC7-TT2-2	875	22	1.8×10^8	4.3×10^7	99.76
Cr–Si-54SC6-1	775	12	1.2×10^8	1.6×10^8	99.90
Cr–Si-54SC6-2	800	11	1.1×10^8	5.9×10^7	99.76
Cr–Si-54SC6-5	775	28	6.7×10^7	5.2×10^7	99.75
Cr–V-1	825	12	3.8×10^7	8.8×10^7	99.76
Cr–V-3	810	13	2.3×10^8	1.1×10^8	99.79
Cr–V-4	810	22	9.8×10^7	3.7×10^7	99.40
Cr–Mo-UC-1	760	18	5.8×10^8	1.3×10^8	99.95
Cr–Mo-UC-2	750	13	7.1×10^8	3.4×10^8	99.98
Cr–Mo-UC-4	740	16	8.8×10^8	5.3×10^8	99.99

6.4.2 Prediction of VHCF Life for Titanium Alloys

In the VHCF regime, the total fatigue life of titanium alloys is predominantly consumed during the crack initiation stage [34], which is correlated with crack size. When the crack size extends into the rough region, the consumed life almost equals the total life of the specimen. To emphasize that VHCF life depends on the formation of the crack initiation zone, a fatigue indicator parameter (FIP) based on the critical plane method is used to predict fatigue life independent of environmental conditions. This parameter model was initially proposed by Fatemi and Shamsaei [40] to express the strain-life relationship, with the critical plane defined as the plane of maximum shear strain range. The equation is as follows:

$$\mathrm{FIP} = \Delta\gamma_{\max}\left(1 + k\frac{\sigma_{n,\max}}{\sigma_y}\right) \tag{6.41}$$

where $\Delta\gamma_{\max}$ is the maximum shear strain range, k is taken equal to 1 to obtain similar mechanical and plasticity contributions, and $\sigma_{n,\max}$ is the maximum normal stress in the plane of the maximum shear range. To maintain the dimensionless feature of strain, the maximum normal stress $\sigma_{n,\max}$ is normalized by yield strength σ_y.

Based on this, Steuer et al. [41] redefined the FIP parameter to consider the influence of crack size on crack nucleation under uniaxial loading. They argued that, according to Schmid's law, the value of $\Delta\gamma_{\max}$ in Eq. (6.41) can be calculated using the following formula:

$$\Delta\gamma_{\max} = \frac{\mu\Delta\sigma}{E} \tag{6.42}$$

where $\Delta\sigma$ is the applied tensile stress amplitude, and E is the elastic modulus of the titanium alloy.

In the study by Liu et al. [42], it was noted that although the elastic modulus distribution throughout the specimen is uneven, internal cracks initiate at a particular point within the specimen, where the elastic modulus is 99 GPa at 400°C. The value μ represents the Schmid factor, obtainable from EBSD analysis. Due to the predominant slip modes being prism and basal slip under loading at high-temperature stress ratios, its average value is approximately 0.47. To further analyze the dependency of initial crack size under axial loading, the stress intensity factor ΔK_{facet} of the facet and the threshold value $\Delta K_{\text{th-long}}$ for long crack propagation are respectively used to replace $\sigma_{n,\max}$ and σ_y.

$$\mathrm{FIP} = \frac{\mu\Delta\sigma}{E}\left(1 + k\frac{\Delta K_{\text{facet}}}{\Delta K_{\text{th-long}}}\right) \tag{6.43}$$

$$\Delta K_{\text{facet}} = \frac{2}{\pi}\Delta\sigma\sqrt{\pi r_{\text{facet}}} \tag{6.44}$$

Substituting the formula for ΔK_{facet} (Eq. 6.44) and the value of the threshold $\Delta K_{th\text{-}long}$ (4.4 MPa·m$^{1/2}$) into Eq. (6.43), the FIP for internal crack initiation within the specimen at 400°C under a stress ratio ($R = 0.5$) is as follows:

$$\text{FIP} = \frac{\mu\Delta\sigma}{E}\left(1 + k\frac{0.5\Delta\sigma\sqrt{\pi r_{facet}}}{4.4}\right) \tag{6.45}$$

For specimens under axial loading, combined with the Manson–Coffin equation, the total strain-life relationship is expressed as:

$$\frac{\Delta\varepsilon}{2} = \frac{\sigma'_f}{E}(2N_i)^b + \varepsilon'_f(2N_i)^c \tag{6.46}$$

where $\Delta\varepsilon/2$ is the total strain amplitude, σ'_f is the fatigue strength coefficient, b is the fatigue strength exponent, ε'_f is the fatigue ductility coefficient, and c is the fatigue ductility exponent. For VHCF, deformation is mainly controlled by elastic strain, and the plastic stage can be neglected. Combining Eqs. (6.45) and (6.46), the relationship between FIP and crack initiation life is:

$$N_i = \frac{1}{2}\left[\frac{\mu\Delta\sigma}{2\sigma'_f}\left(1 + k\frac{0.5\Delta\sigma\sqrt{\pi r_{facet}}}{4.4}\right)\right]^{\frac{1}{b}} \tag{6.47}$$

Therefore, by combining mechanical properties and the S-N curve, fitting parameters σ'_f and b can be obtained to establish the crack initiation life curve. Figure 6.30 illustrates the relationship between internal crack initiation life and FIP, where it is evident that the higher the fatigue life, the lower the corresponding FIP value. This result is also consistent with the trend of FIP versus fatigue life found in other literature on titanium alloys. Comparing the predicted fatigue life with the actual fatigue life results using Eq. (6.47) is shown in Figure 6.31. Within three times the curve, the

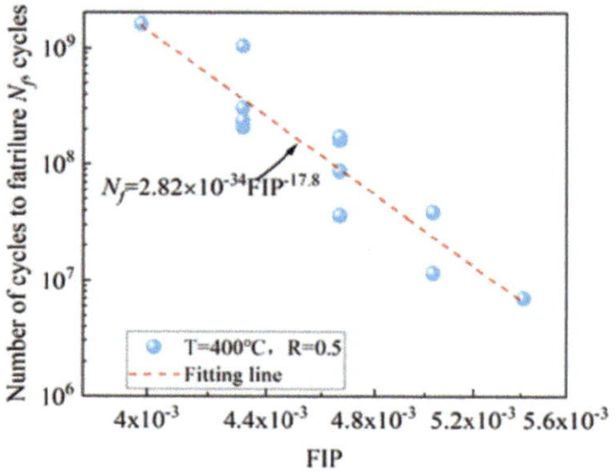

FIGURE 6.30 Relationship between fatigue life N_f and FIP for titanium alloy [42].

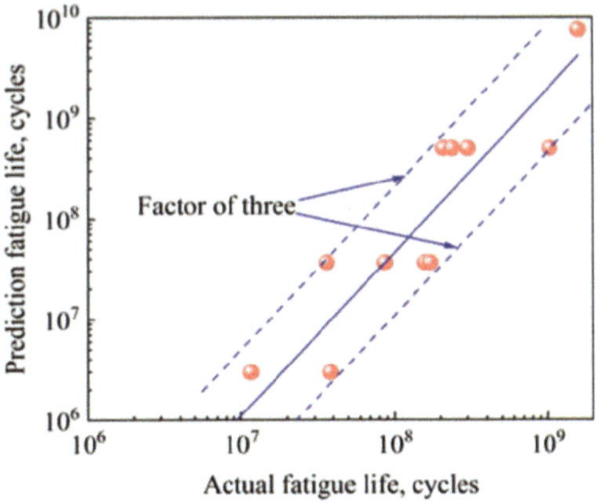

FIGURE 6.31 Comparison of the predicted and experimental results for titanium alloy [42].

consistency is quite satisfactory. In terms of crack initiation process, the proposed fatigue life assessment method is reasonably justified.

6.4.3 Prediction of VHCF Life Based on Micro Scratch

Ding et al. [43] considered integrating the area of surface scratches with existing theoretical foundations to establish a VHCF fatigue life model that accounts for the influence of surface micro scratches. It is well known that fatigue crack propagation on fracture surfaces follows the following power-law behavior:

$$\frac{da}{dN} = C(\Delta K)^m \tag{6.48}$$

where $\frac{da}{dN}$ represents the crack propagation rate, ΔK denotes the amplitude of the effective stress intensity factor, m and C are material-dependent constants. When fatigue cracks initiate from surface defects to fatigue failure, the relationship between defect area and fatigue life can be expressed as Shiozawa et al. [44].

$$(K_{\mathrm{I\,max}})^m \frac{N_f}{\sqrt{\mathrm{area}}} = \frac{2}{(m-2)C} \tag{6.49}$$

If $\sqrt{\mathrm{area}}$ represents the crack area, then the maximum stress intensity factor K_{Imax} along its leading edge is approximately [45]:

$$\Delta K_{\mathrm{Imax}} = 0.65\sigma_a\sqrt{\pi\sqrt{\mathrm{area}}} \tag{6.50}$$

Combining the results of the Murakami theory with the Paris formula, the relationship between surface microgrooving and fatigue life of TC17 titanium alloy is revealed:

$$N_f = \frac{0.229\times10^{12}\sqrt{\mathrm{area}}}{(0.65\sigma_a\sqrt{\pi\sqrt{\mathrm{area}}})^{3.9}} \tag{6.51}$$

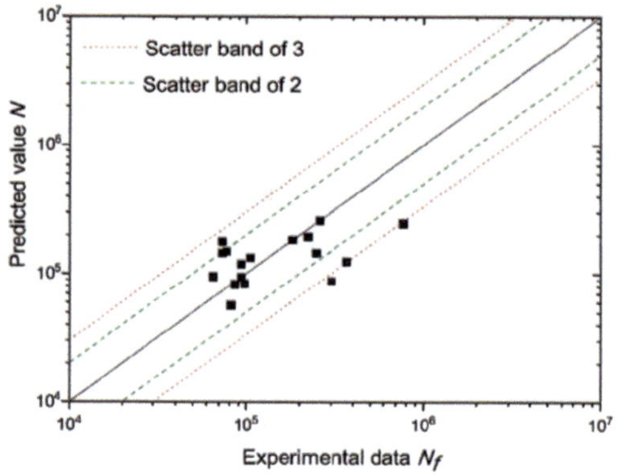

FIGURE 6.32 Result of fatigue life prediction. (Reprinted with permission from Ref. [43], Elsevier.)

From Figure 6.32, it can be observed that the predicted life data mainly falls within a scatter band of factor 2. It is noted that the error is within an acceptable range.

6.4.4 Prediction of VHCF Life Based on FGA

In Sun et al. [46], the Paris law is not exploited, but the stress-life relation is obtained starting from considerations on the crack propagation within the FGA. For this reason, this model has been included in this section. In particular, an equivalent crack length after a number of cycles N is computed by assuming that "the equivalent crack growth rate in the FGA region is related to the maximum size of the plastic zone at the crack tip" and that the number of cycles consumed within the FGA can approximate the total life with negligible differences. The stress-life relation for high-strength steels failing in the VHCF region becomes

$$N_f = \frac{1}{\alpha}\left(\frac{S_a}{\sigma_Y}\right)^{-l} \ln\left(\frac{a_{\text{FGA}}}{a_0}\right) \tag{6.52}$$

where σ_Y the yield stress, α and l are two parameters that have to be estimated from the experimental data, and an FGA is the size of the FGA. According to Eq. (6.52), the fatigue life increases with the ratio between the FGA size and the initial inclusion size, in agreement with the experimental evidence, as shown in Figure 6.33.

6.5 CURRENT PROBLEMS AND FUTURE CHALLENGES

Fatigue experiment method based on ultrasonic vibration: The ultrasonic fatigue method has been widely used in the testing of the VHCF strength of metallic materials and has established corresponding industry or national testing standards in countries such as China and Japan. However, applying ultrasonic fatigue testing methods to non-metallic materials, such as polymers, biomaterials, and composites, still presents numerous challenges. The broad promotion and application of ultrasonic vibration fatigue testing technology in the future still has a long way to go.

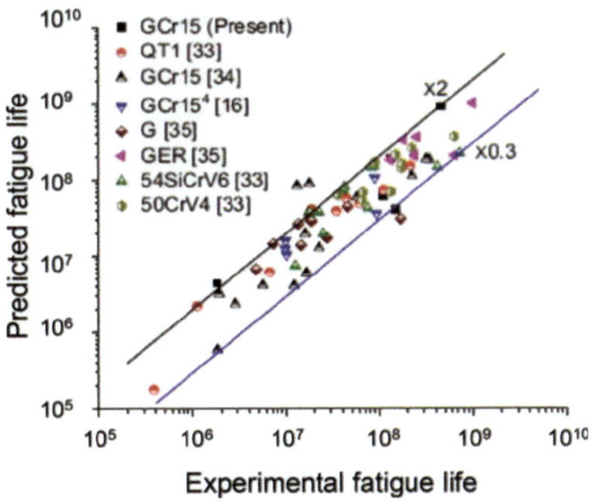

FIGURE 6.33 Predicted fatigue life with respect to the experimental fatigue life [46].

Evolution process of the VHCF crack initiation and initial propagation: By utilizing continuously developing advanced experimental equipment and technologies, and designing unique testing methods, the initiation and initial propagation process of VHCF cracks can be detected. This is particularly important for the evolution and formation process of FGA crack initiation characteristic zones within the range of several tens of micrometers. Once the formation process of the FGA crack initiation characteristic zone is directly revealed, it will signify a new breakthrough in VHCF research.

Life prediction model driven by physical mechanisms and data: Starting from physical mechanisms, including the microscopic mechanisms of microstructure stress evolution and the microscopic mechanisms of defect formation and growth, mechanical models are established to quantitatively describe the damage behavior of VHCF and predict fatigue performance. Additionally, machine learning methods are used to establish the influence relationships among additive manufacturing service environments/loads, defects and microstructures, failure characteristic parameters, and macroscopic service performance/life. Thus, a life prediction method driven by physical mechanisms and data is developed.

6.6 SUMMARY

The VHCF is a critical consideration in modern engineering due to the extended service life of equipment, often exceeding 10^9 cycles. Traditional "fatigue limit" are challenged by findings that metals can fail beyond this point. Ultrasonic fatigue testing systems have been successfully developed for various load forms such as bending, asymmetric stress ratios, and torsion, as well as under different environmental conditions including high and low temperatures, corrosion, and vacuum. The advancement of fatigue testing systems has propelled deeper research into VHCF. For most high-strength steels and some non-ferrous metal materials, VHCF S-N curves typically exhibit a stepped pattern. Within the VHCF range, fatigue failure of materials is primarily dominated by the initiation of internal cracks, forming an

FGA near the crack initiation site. The formation of the FGA characteristic area occupies the majority of the total fatigue life and its mechanism is closely related to local microstructure.

ABBREVIATIONS

FGA Fine granular area
FiE Fish eye
FIP Fatigue indicator parameter
HCF High cycle fatigue
LCF Low cycle fatigue
ODA Optical dark area
PSB Persistent slip band
VHCF Very high cycle fatigue

REFERENCES

1. Bathias C, Paris PC. *Gigacycle Fatigue in Mechanical Practice.* Boca Raton, FL: CRC Press; 2004. https://doi.org/10.1201/9780203020609
2. Stanzl-Tschegg S. Very high cycle fatigue measuring techniques. *Int J Fatigue* 2014;60:2–17. https://doi.org/10.1016/j.ijfatigue.2012.11.016
3. Yi JZ, Torbet CJ, Feng Q, et al. Ultrasonic fatigue of a single crystal Ni-base superalloy at 1000°C. *Mater Sci Eng A* 2007;443:142–149.
4. Zhu X, Shyam A, Jones J, Mayer H, Lasecki J, Allison J. Effects of microstructure and temperature on fatigue behavior of E319-T7 cast aluminum alloy in very long life cycles. *Int J Fatigue* 2006;28:1566–71. https://doi.org/10.1016/j.ijfatigue.2005.04.016
5. Welsch G, Boyer R, Collings EW. *Materials Properties Handbook: Titanium Alloys.* Materials Park: ASM International; 1993.
6. Palin-Luc T, Pérez-Mora R, Bathias C, Domínguez G, Paris PC, Luis Arana J. Fatigue crack initiation and growth on a steel in the very high cycle regime with sea water corrosion. *Eng Fract Mech* 2010;74:156.
7. Wang Q, Bathias C, Kawagoishi N, Chen Q. Effect of inclusion on subsurface crack initiation and gigacycle fatigue strength. *Int J Fatigue* 2002;24:1269–74. https://doi.org/10.1016/S0142-1123(02)00037-3
8. Heinz S, Balle F, Wagner G, Eifler D. Analysis of fatigue properties and failure mechanisms of Ti6Al4V in the very high cycle fatigue regime using ultrasonic technology and 3D laser scanning vibrometry. *Ultrasonics*, 2013;53:1433–40. https://doi.org/10.1016/j.ultras.2013.03.002
9. Sakai T, Sato Y, Oguma N. Characteristic S–N properties of high-carbon–chromium-bearing steel under axial loading in long-life fatigue. *Fatigue Fract Eng Mat Struct* 2002;25:765–73. https://doi.org/10.1046/j.1460-2695.2002.00574.x
10. Shiozawa K, Lu L, Ishihara S. S–N curve characteristics and subsurface crack initiation behaviour in ultra-long life fatigue of a high carbon-chromium bearing steel. *Fatigue Fract Eng M* 2001;24:781–90. https://doi.org/10.1046/j.1460-2695.2001.00459.x
11. Mughrabi H. Specific features and mechanisms of fatigue in the ultrahigh-cycle regime. *Int J Fatigue* 2006;28:1501–8. https://doi.org/10.1016/j.ijfatigue.2005.05.018
12. Mughrabi. On the life-controlling microstructural fatigue mechanisms in ductile metals and alloys in the gigacycle regime. *Fatigue Fract Eng M* 1999;22:633–41. https://doi.org/10.1046/j.1460-2695.1999.00186.x

13. Larsen JM. Reducing uncertainty in fatigue life limits of turbine engine alloys. *Int J Fatigue* 2013;57:103. https://doi.org/10.1016/j.ijfatigue.2013.01.012
14. Murakami Y, Nomoto T, Ueda T. Factors influencing the mechanism of superlong fatigue failure in steels. *Fatigue Fract Eng Mater Struct* 1999;22:581.
15. Wang QY, Berard JY, Dubarre 1 A., Baudry 2 G., Rathery S, Bathias C. Gigacycle fatigue of ferrous alloys. *Fatigue Fract Eng Mater Struct* 1999;22:667–72.
16. Wang QY, Berard JY, Rathery S, Bathias C. High-cycle fatigue crack initiation and propagation behaviour of high-strength spring steel wires. *Fatigue Fract Eng Mater Struct* 1999;22:673.
17. Marines-Garcia I, Paris PC, Tada H, Bathias C. Fatigue crack growth from small to long cracks in VHCF with surface initiations. *Int J Fatigue* 2007;29:2072–8.
18. Hong Y, Sun C. The nature and the mechanism of crack initiation and early growth for very-high-cycle fatigue of metallic materials–an overview. *Theor Appl Fracture Mech* 2017;92:331–50.
19. Sakai T. Review and prospects for current studies on very high cycle fatigue of metallic materials for machine structural use. *J Solid Mech Mater Eng* 2009;3:425–39.
20. Grad P, Reuscher B, Brodyanski A, Kopnarski M, Kerscher E. Mechanism of fatigue crack initiation and propagation in the very high cycle fatigue regime of high-strength steels. *Scripta Mater* 2012;67:838–41.
21. Hong Y, Liu X, Lei Z, Sun C. The formation mechanism of characteristic region at crack initiation for very-high-cycle fatigue of high-strength steels. *Int J Fatigue* 2016;89:108–18.
22. Zheng Y, Zhao Z, Zhang Z, Zong W, Dong C. Internal crack initiation characteristics and early growth behaviors for very-high-cycle fatigue of a titanium alloy electron beam welded joints. *Mater Sci Eng A* 2017;706:311–8.
23. Pan X, Su H, Sun C, Hong Y. The behavior of crack initiation and early growth in high-cycle and very-high-cycle fatigue regimes for a titanium alloy. *Int J Fatigue* 2018;115:67–78.
24. Liu F, Zhang H, Liu H, Chen Y, Muhammad Kashif K, Wang Q, Liu Y. Influence of welded pores on very long-life fatigue failure of the electron beam welding joint of TC17 titanium alloy. *Materials* 2019;12:1825.
25. Tofique MW, Bergström J, Svensson K, Johansson S, Peng RL. ECCI/EBSD and TEM analysis of plastic fatigue damage accumulation responsible for fatigue crack initiation and propagation in VHCF of duplex stainless steels. *Int J Fatigue* 2017;100:251–62.
26. Sahu JK, Krupp U, Christ H-J. Fatigue crack initiation behavior in embrittled austenitic–ferritic stainless steel. *Int J Fatigue* 2012;45:8–14.
27. Krupp U, Knobbe H, Christ H-J, Köster P, Fritzen C-P. The significance of microstructural barriers during fatigue of a duplex steel in the high-and very-high-cycle-fatigue (HCF/VHCF) regime. *Int J Fatigue* 2010;32:914–20.
28. Liu X, Sun C, Zhou Y, Hong Y. Effects of microstructure and stress ratio on high-cycle and very-high-cycle fatigue behavior of Ti-6Al-4V alloy. *Acta Metall Sin* 2016;52:923–30.
29. Szczepanski CJ, Jha SK, Shade PA, Wheeler R, Larsen JM. Demonstration of an in situ microscale fatigue testing technique on a titanium alloy. *Int J Fatigue* 2013;57:131–9.
30. Oguma H, Nakamura T. Fatigue crack propagation properties of Ti-6Al-4V in vacuum environments. *Int J Fatigue* 2013;50:89–93.
31. Heinz S, Eifler D. Crack initiation mechanisms of Ti6Al4V in the very high cycle fatigue regime. *Int J Fatigue* 2016;93:301–8.
32. Ward-Close CM, Beevers CJ. The influence of grain orientation on the mode and rate of fatigue crack growth in α-titanium. *Metall Trans A* 1980;11:1007–17.

33. Jha SK, Szczepanski CJ, Golden PJ, Porter III WJ, John R. Characterization of fatigue crack-initiation facets in relation to lifetime variability in Ti-6Al-4V. *Int J Fatigue* 2012;42:248–57.
34. Yang K, He C, Huang Q, Huang ZY, Wang C, Wang Q, Liu YJ, Zhong B. Very high cycle fatigue behaviors of a turbine engine blade alloy at various stress ratios. *Int J Fatigue* 2017;99:35–43.
35. Huang ZY, Liu HQ, Wang HM, Wagner D, Khan MK, Wang QY. Effect of stress ratio on VHCF behavior for a compressor blade titanium alloy. *Int J Fatigue* 2016;93:232–7.
36. Liu X, Sun C, Hong Y. Faceted crack initiation characteristics for high-cycle and very-high-cycle fatigue of a titanium alloy under different stress ratios. *Int J Fatigue* 2016;92:434–41.
37. Wang QY, Berard JY, Dubarre A., Baudry G., Rathery S, Bathias C. Gigacycle fatigue of ferrous alloys. *Fatigue Fract Eng Mater Struct* 1999;22:667–72.
38. Wang QY, Bathias C, Kawagoishi N, Chen Q. Effect of inclusion on subsurface crack initiation and gigacycle fatigue strength. *Int J Fatigue* 2002;24:1269–74.
39. Wang QY, Bathias C, Kawagoishi N, Chen Q. Effect of inclusion on subsurface crack initiation and gigacycle fatigue strength. *Int J Fatigue* 2002;24:1269–74. https://doi.org/10.1016/S0142-1123(02)00037-3.
40. Fatemi A, Shamsaei N. Multiaxial fatigue: an overview and some approximation models for life estimation. *Int J Fatigue* 2011;33:948–58.
41. Steuer S, Villechaise P, Pollock TM, Cormier J. Benefits of high gradient solidification for creep and low cycle fatigue of AM1 single crystal superalloy. *Mater Sci Eng A* 2015;645:109–15.
42. Liu F. Investigations on very long life fatigue behaviors of TC17 titanium alloy and its electron beam welded joints. Ph.D. Thesis, 2023, Sichuan University, China.
43. Ding MC, Zhang YL, Lu HT. Fatigue life prediction of TC17 titanium alloy based on micro scratch. *Int J Fatigue* 2020;139:105793.
44. Shiozawa K, Murai M, Shimatani Y, Yoshimoto T. Transition of fatigue failure mode of Ni–Cr–Mo low-alloy steel in very high cycle regime. *Int J Fatigue* 2010;32:541–50. https://doi.org/10.1016/j.ijfatigue.2009.06.011
45. Murakami K, Shimoda K, Enkhbaatar P, Traber L, Salzman A, Traber D. Inhibition of poly(ADP-ribose) synthase (pars) attenuates ovine acute lung injury following smoke inhalation and pneumonia. *Shock* 2002;43:18.
46. Sun C, Liu X, Hong Y. A two-parameter model to predict fatigue life of high-strength steels in a very high cycle fatigue regime. *Acta Mech Sinica* 2015;31:383–91.
47. Liu H, Wang H, Huang Z, Wang Q, Chen Q. Comparative study of very high cycle tensile and torsional fatigue in TC17 titanium alloy. *Int J Fatigue* 2020;139:105720.

7 Fatigue Reliability

Xiaopeng Niu, Shun-Peng Zhu, and Qingyuan Wang

7.1 INTRODUCTION

Reliability usually refers to the ability of a structure to finish its aimed functions without failure during operation, which is influenced by materials, loadings, structures, and service conditions. It is usually a probabilistic definition in practice such as civil, aerospace, and offshore engineering, which can ensure structures to meet safety and performance requirements under complicated service conditions. Many different approaches, such as probabilistic methods, safety factor, and Monte Carlo simulation (MCs), have been employed to evaluate this ability.

In the procedure of reliability analysis, the strength and stress play crucial roles. Strength means the ability of a material or structure to sustain loads without failure, including fatigue strength, yield strength, ultimate strength, etc. It is a fundamental property that varies based on the material nature properties, design, and manufacturing processes. Stress is the internal force within structures, typically expressed as force per unit area. It is a critical parameter to reflect the response of structures to withstand external loads.

Strength analysis is through comparing the calculated stresses with the allowable stresses to ensure safety operation of structures. Allowable stress is the maximum stress that a structure is allowed to experience under normal operating conditions. It is a conservative value that lower than the material's strength to provide a safety margin. The design process compares the calculated stresses in a structure with the allowable stress. If the calculated stress is below the allowable stress, it is considered safe; if it exceeds the allowable stress, the structure may need to be redesigned. Recently, the allowable stresses are still determined on basis of the data of structures during the service period or the results of full-scale testing. In summary, understanding the relationship between the strength and stress is critical to ensure the safety of structures. It requires to consider the uncertainties and variations from loads, material properties, and environments through probabilistic methods to contribute to a comprehensive reliability assessment.

The well-known bath-tub curve illustrates the relation between sources of variation and reliability, see Figure 7.1. Failures in the first stage of the utilization period, infant mortality period, are usually due to potential manufacturing defects or design issues. Within the normal lifespan of the product, its reliability is relatively high and remains stable. This phase is depicted by a relatively flat, horizontal segment of the curve, indicating stable performance with few failures. At the end of the stage, wear-out period, the reliability of the equipment gradually decreases as components undergo wear, aging, or deterioration. In this stage, fatigue may become a primary

DOI: 10.1201/9781003405122-7

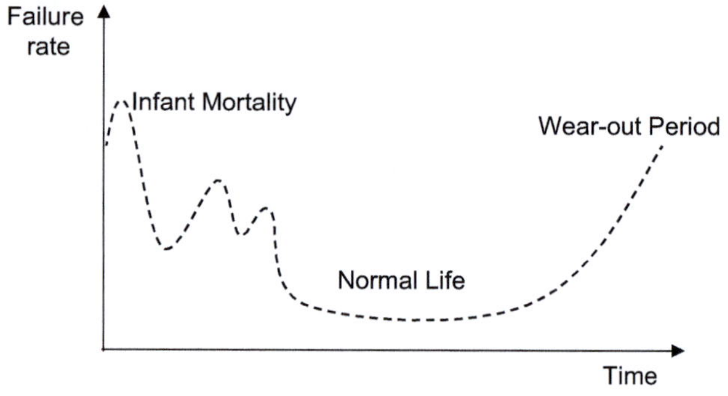

FIGURE 7.1 Bath-tub curve.

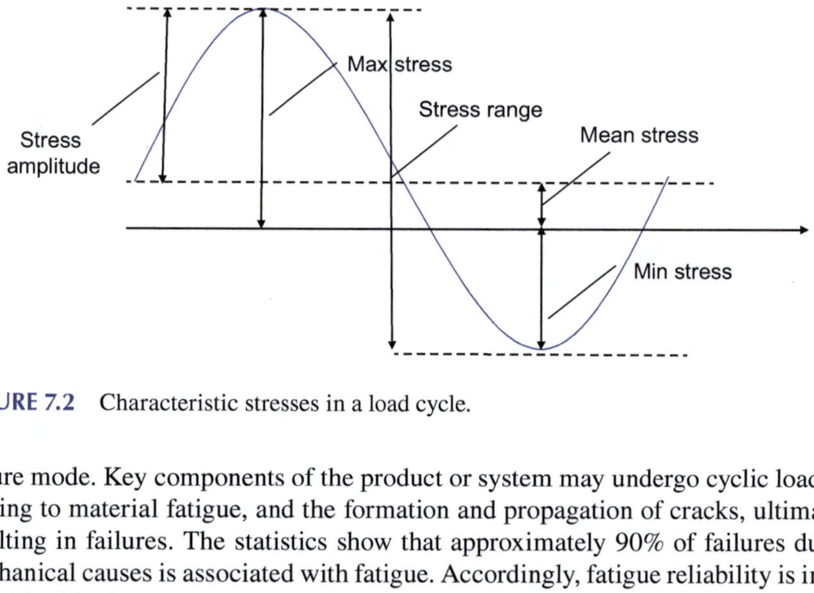

FIGURE 7.2 Characteristic stresses in a load cycle.

failure mode. Key components of the product or system may undergo cyclic loading, leading to material fatigue, and the formation and propagation of cracks, ultimately resulting in failures. The statistics show that approximately 90% of failures due to mechanical causes is associated with fatigue. Accordingly, fatigue reliability is introduced in this chapter.

One of the earliest studies on fatigue issues can be traced back to 1842 when the failure of railway carriage wheels happened. In the subsequent decades, engineers began to take note of fatigue problems arising from cyclic loadings. It was systematically investigated by Woehler in the mid-19th century. The famous Woehler curve, namely stress–life (S–N) curve, depicts the number of cycles a material can endure before fatigue failure at different stress levels. The concepts of stress including maximum and minimum stresses, mean stress, stress amplitude, and stress range in a load cycle are shown in Figure 7.2.

Traditional fatigue design approaches, e.g. infinite-life design, safe-life design, and damage tolerance design, are applied in engineering practice. Among them, infinite-life design takes into account a loading lower than the fatigue limit, then the components won't fail under this stress level. Safe-life design assumes the structure to be flawless and the lifetime is described through traditional S–N curve based on fatigue testing. These curves are commonly used during fatigue design in order to determine for the structure a lifetime that will not result in its failure below the static strength. The material parameters fitted by testing data rely on specimens and material types, but the curve fitting procedure lacks discussion on fatigue mechanism and belongs to the experienced approach. In this chapter, we introduce fatigue reliability from the point of view of variations, uncertainties, and S–N-based safe-life design. Finally, there are two cases, fatigue assessment of turbine-bladed disk and relationship quantification between fatigue reliability and safety factor, which are evaluated by using probabilistic approaches.

7.2 ENGINEERING STATISTICS AND PROBABILISTIC METHODS – FUNDAMENTALS

Due to the complex service environment, large amounts of structures are subjected to variable loads. In practice, these alternating service loads are always considered as continuously varying, with random amplitude and period of variation. Also, the natural properties of materials have shown the dispersion characteristic. Therefore, statistical analysis is often applied to experimental data to fit S–N curves, considering variability in material behavior. Such as mean and standard deviation, which are used to characterize the distribution of fatigue data. This statistical information is crucial for making predictions and estimating the variability in fatigue life. Therefore, the discussions with respect to statistics and probability theory are necessary.

If one experiment can be replicated many times, with each repetition termed a "trial". By observing the frequency of event A, the probability of event A occurring can be assessed. This frequency exhibits a "statistical regularity" as the number of trials increases, showing a gradual reduction of fluctuations and indicating that the probability of event occurrence approaches a specific value with an increasing number of trials. If the number of trials is large enough, the probability tends to be:

$$\lim_{n \to \infty} \left(\frac{m}{n} \right) = P\{A\} \tag{7.1}$$

where m is the recorded number of event A occurs, and n is the total number of trials, and the equation represents the "probability of the event A". The probability is nonnegative, and an event that must occur is a deterministic event. The probability of such a deterministic event occurring is equal to one. Conversely, an event that never occurs in any number of trials is an impossible event, and its probability is equal to zero.

Random variables usually include discrete random variables and continuous random variables. For continuous random variables (see Figure 7.3), its probability density function is expressed as:

$$P\{x < \delta < x + dx\} = P(x)dx \quad \Delta x \to dx \quad P\{\} \geq 0 \tag{7.2}$$

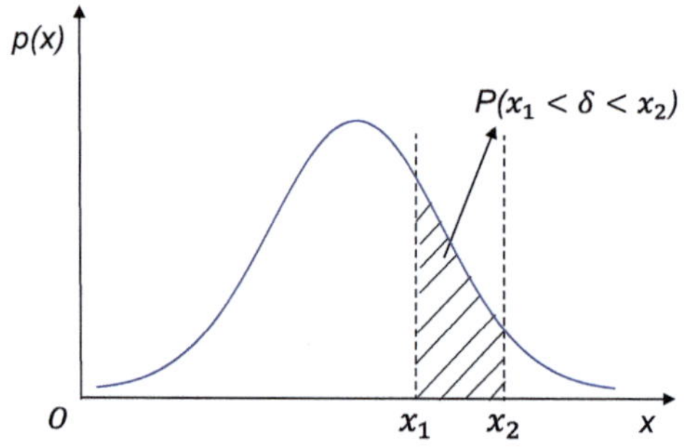

FIGURE 7.3 The PDF and integral from x_1 to x_2 of random variable x.

Available literature studies often report that the fatigue life/strength always follows normal, lognormal, and Weibull distribution (probability density function, PDF):
Normal:

$$P(x;\sigma,\mu)=\frac{1}{\sigma\sqrt{2\pi}}e^{-\frac{(x-\mu)^2}{2\sigma^2}} \tag{7.3}$$

Lognormal:

$$P(x;\sigma,\mu)=\frac{1}{x\sigma\sqrt{2\pi}}e^{-\frac{(\ln(x)-\mu)^2}{2\sigma^2}} \tag{7.4}$$

Weibull ($\lambda>0,\ k>0$):

$$P(x;\lambda,k)=\frac{k}{\lambda}\left(\frac{x}{\lambda}\right)^{k-1}e^{-(x/\lambda)^k} \tag{7.5}$$

where σ, μ are the mean value and standard deviation of normal or lognormal distribution, λ, k are the scale and shape parameters of Weibull distribution.

Integrating $p(x)$ within the total range of values x results in $\int_{-\infty}^{+\infty}p(x)=1$. Cumulative Distribution Function (CDF) of these distributions can be written as:

$$C(x)=\int_{-\infty}^{x}P(t)dt \tag{7.6}$$

As mentioned in the introduction, the reliability of structure is closely related to stress and strength. The stress-strength interference model is widely used for structural reliability analysis and remains a popular choice in various industries. It is assumed that the strength and stress are statistically independent, then reliability can be computed easily and analytical solutions are accurate for many actual situations. The "stress" should be interpreted broadly to include the applied load or load-induced response (e.g., stress, force, strain, deformation), which has the potential to induce failure. Similarly, the "strength" can be explained by a broader concept as the capacity to withstand the applied loads. It includes yield stress, ultimate stress, fatigue limit, and elongation that depend on the material's

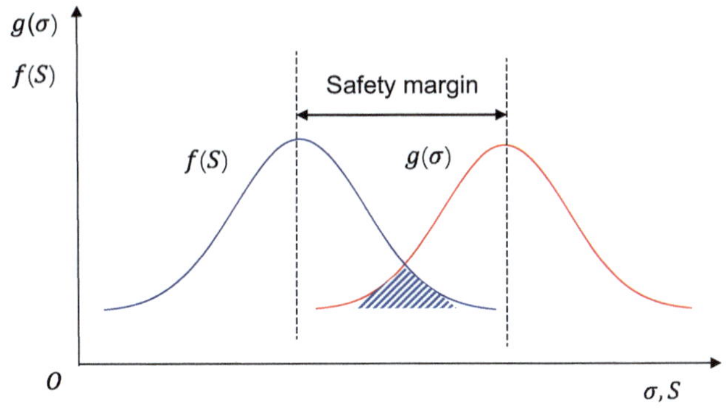

FIGURE 7.4 Stress-strength interference model.

individual properties. The selection of stress and strength depends on the type of load or load-induced response and the failure criteria (yield failure, collapse, fatigue, excessive deformation, etc.).

For the stress–strength interference method, it is thought that failure occurs when the stress is over the strength. It can be applied to various failure modes including yielding and fatigue. Take fatigue as an example, the stress can be applied cyclic loading and the strength is fatigue strength. It should be noted that the interference region does not represent the real failure probability in Figure 7.4, and the unified equation of failure probability is introduced in Section 7.4.3.

7.3 PROBABILISTIC PROPERTY OF THE UNCERTAINTIES

The intended life of a structure during the design stage used to be assumed as a deterministic value. Safety factor is an applicable way to account for uncertainties in foreseeing in-service loads. For example, ASME Section III [1] recommends a safety factor of 20 on mean experimental fatigue life and 2 on stress/strain. However, it should be considered from a probabilistic point of view when evaluating the service life. Deterministic methodologies, supported by experimental data, cannot guarantee that fatigue failure of structures can be completely avoided during the whole service period. The uncertainties can arise from any stages, such as design and assessment procedures, due to a lack of knowledge, errors of modeling approximations, or differences between laboratory tests and as-operated components. Table 7.1 offers an overview of the different variables affecting fatigue life and safety factor design.

These uncertainties and variabilities can be comprehended and quantified through the application of probabilistic approaches. Probabilistic structural integrity assessments enable the consideration and reflection of uncertainties inherent in fatigue design. In contrast to more conservative deterministic assessments that use single-point input parameters, probabilistic assessments assign multiple possible magnitudes to parameters. This is achieved through the use of parameter distributions or when the ample sufficient data can be obtained when the data itself can directly show uncertainty. Consequently, the combinations of these random input parameters,

TABLE 7.1
Variables Affecting the Fatigue Behavior in Air and Light Water Reactor Environments [2]

Material	Composition. Metallurgy: grain size, inclusions, orientation within a forging or plate. Processing: cold work, heat treatment. Size and geometry. Surface finish: fabrication surface condition. Surface preparation: surface work hardening
Loading	Strain rate: rise time. Sequence: load history; Mean stress; Biaxial effects: constraints
Environment	Water chemistry: lithium hydroxide, boric acid concentrations. Temperature flow rate

sampled from these distributions or data, are utilized to generate a range of quantities against reliability probability. Such assessments allow a shift from empirical method towards a more accuracy probability-based approach and decision-making.

Scatter generated by physical uncertainty is associated with the inherent random or statistical nature of structures, such as the variation of strength and external load. In fatigue, the substantial scatter observed can be attributed to its localized nature. The initiation of the crack leading to final failure is influenced by various factors, including microstructural inhomogeneities, and defects (inclusions, scratches, and pores). These features together with loads are critical in three dimensions. While microstructures may impact mechanical properties and compete with defect-induced cracks at low stress levels, defects are believed to have a more significant influence on the fatigue properties of components. Microstructure, on the other hand, tends to exert greater influences on static properties. The stochastic distribution of defects, encompassing various features, contributes to the apparent scatter in fatigue, significantly complicating the determination of the lifetime of actual components. In addition to scatter resulting from defects and microstructure variability in materials, errors in testing machines and specimens can also contribute to fatigue scatter. The stress/strain-life curve derived from basic tests exhibits apparent scatter in fatigue life and fatigue limit, as illustrated in Figure 7.5.

In certain applications, the loading history consists of a series of loads with randomly varied time intervals and amplitudes (refer to Figure 7.6). Service load histories are typically of variable amplitude in nature when considering the loading history. Load sequence and interactions play a crucial role in such loadings, where cumulative fatigue damage evolution is a significant consideration, especially in the presence of multiaxial stresses. In engineering practice, random fatigue tests can be conducted by applying a suitable stress-time spectrum collected from actual random signals. This approach allows for a more realistic simulation of the variable amplitude loading conditions encountered in real applications. In general, one structure involves numerous geometric sizes (see Figure 7.7), and the allowable tolerances between actual and aimed sizes are introduced during the fabricated and machining process. Even minor changes in dimensional characteristics could have a critical impact on the mechanical response of the actual structure, leading to variations in fatigue performance. To ensure the service safety of engineering components, geometric tolerances must be controlled within an acceptable range, commonly referred to as dimensional tolerance.

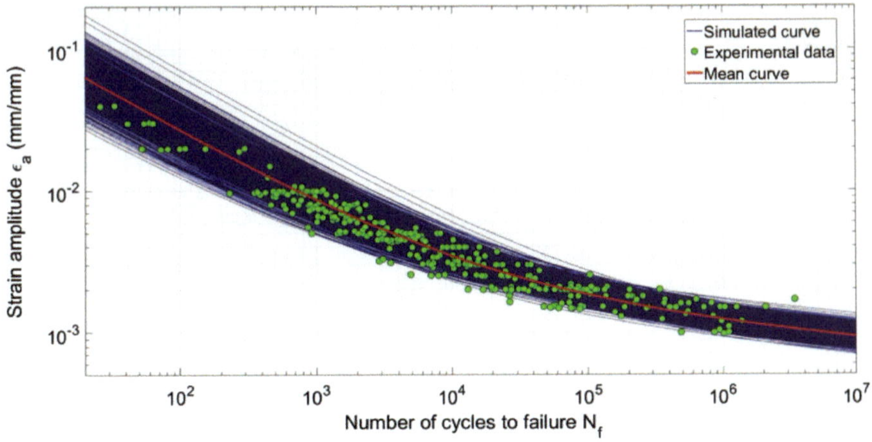

FIGURE 7.5 Fatigue scatter of Strain-life for 950X steel. (Reprinted with permission from Ref. [3], Elsevier.)

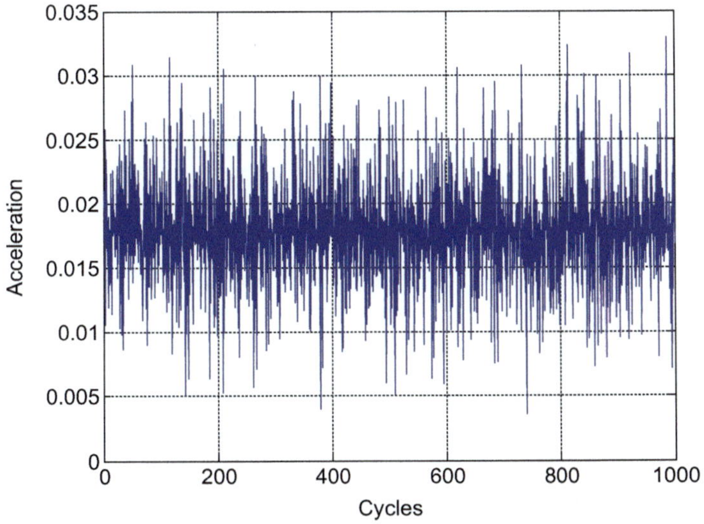

FIGURE 7.6 Actual random load (acceleration) spectrum.

Probability distribution is employed to quantify and describe these uncertainties. The probability of an event occurring or the likelihood of a random variable falling within a specific range can be confirmed if their distributions are known. For instance, if the dimensions of a structure follow a specific distribution, the probability of the product being outside the tolerance limits can be calculated.

The various features of defect are largely responsible for the fatigue scatter, especially in emerged additive manufacturing. Statistical analysis of defects mainly includes the defect shape and size. It is shown that several distributions properly

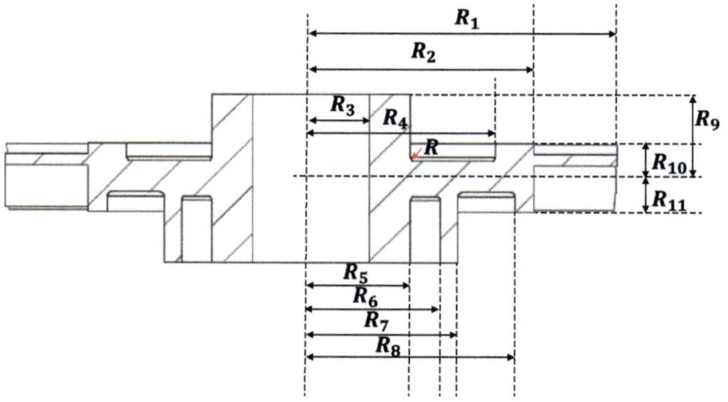

FIGURE 7.7 The main geometric size of turbine-bladed disk.

describe the defects features: a normal distribution for the sphericity; distributions in the domain of extreme value statistics, such as a lognormal, exponential or Weibull distribution, for defects size. Generally, the upper tail of the initial defect distribution is related to the distribution of fatigue life.

The mechanical properties of material, load fluctuation, and geometrical tolerances usually follow normal distribution. Of course, sometimes the loadings are not random, but regular and periodic in nature because of the application environments and human subjective factors. For example, the record of loading for a car that running on the same road should be similar. The above-mentioned uncertainties are transmitted to fatigue performance through a complex mapping relationship.

In general, the probabilistic distribution of cycles to failure for a given stress level can be regarded as lognormal distribution, and the distribution parameters can be estimated through a statistical analysis based on laboratory tests. The quantification of the distribution parameters includes: the maximum likelihood estimation method, least squares method, the method of moments and the Bayesian estimation method. After that, checking the validity of a given distribution model requires a goodness-of-fit analysis before probabilistic fatigue modeling. Goodness-of-fit analysis can be performed by: Chi-Square test, T-test, Kolmogorov-Smirnov test, Anderson-Darling test, Residual analysis, Coefficient of Determination, etc.

It is worth pointing out that the different methods fit different problems; for example, Anderson-Darling test usually checks normal distribution and focuses on the distribution tails, which makes it more sensitive to extreme values. For defects statistics, the maximum defect always leads to the final failure, which is appropriate for extreme value statistics and Anderson-Darling test. Besides, the number of samples is important with respect to the confidence level. Chi-Square test is suitable for large samples; however, the samples are not always abundant due to the costs and T-test is suitable for small samples. The quantification of uncertainties, as the inputs influencing fatigue scatter, is critical in fatigue reliability assessment. Considering the aforementioned random uncertainties, probabilistic assessments can effectively reflect their most crucial statistical properties, providing several benefits. By representing design parameters as random variables, probabilistic design approaches offer a more physically accurate representation than the deterministic method. This not only decreases economic costs by enabling the safety factor design but also provides optimization capabilities by identifying the largest contributors to overall risk.

7.4 FATIGUE ASSESSMENT AND RELIABILITY ASSESSMENT

7.4.1 S–N Curve, Fatigue Life Models and Fatigue Cumulative Damage

In classical fatigue theory, continuum damage mechanics is raised to deal with the fatigue crack initiation life. Fatigue life is estimated by cycle-by-cycle computing of an empirical cumulative damage parameter, which usually utilizes empirical models such as the Basquin equation, Coffin–Manson equation, and Palmgren–Miner's rule.

The S–N curve illustrates the relationship between applied stress and fatigue life. It describes the fatigue lives, spanning from quasi-static failure to high cycle fatigue failure. It exhibits a strong curvature and a horizontal plateau in the high cycle fatigue region, known as the knee point, as shown in Figure 7.8. The knee point represents the asymptotic value of stress corresponding to the traditional fatigue limit. In essence, the fatigue limit implies that if the specimen is exposed to a stress level lower than the fatigue limit, it can sustain an infinite life or, more precisely, would not fail within a specific life (for instance, 10^7 cycles). The initial interpretation of the fatigue limit was the "limit of crack initiation under cyclic stress". This interpretation is still true for some metals. However, as highlighted by Murakami [4], it is asserted that the fatigue limit is not a critical stress for crack initiation but the stress over which an initiated micro-crack cannot continue to propagate. It is found that a crack started from the defect edge with extremely high stress intensity factor in the initial stage and subsequently stopped propagating when subjected stress is lower than fatigue limit.

Basquin [5] gave an exponential equation to describe the S–N curve:

$$S = A\left(N_f\right)^B \tag{7.7}$$

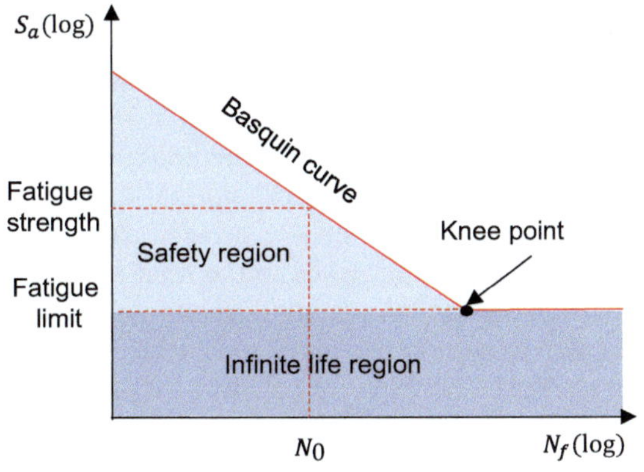

FIGURE 7.8 The typical S–N curves in fatigue: curves fitting of Basquin equation.

where S represents the stress amplitude or range, N_f represents the number of cycles to failure, A and B are material constants.

Consider fatigue performance under strain-controlled conditions, especially when material deformation is significant. Coffin and Manson [6,7] raised an empirical equation that quantifying the cycles of crack initiation versus the plastic strain. Integrate with the Basquin equation, the Coffin–Manson–Basquin equation is developed by Morrow:

$$\varepsilon = \frac{\sigma_f'}{E}\left(2N_f\right)^b + \varepsilon_f'\left(2N_f\right)^c \tag{7.8}$$

where ε is the total strain amplitude, and separated into elastic and plastic phases. Elastic phase includes the fatigue strength coefficient σ_f', fatigue strength exponent b and Young's modulus E; plastic phase includes fatigue ductility coefficient ε_f' and fatigue ductility exponent c;

On the basis of the aforementioned stress or strain-life models, many fatigue life models are developed, such as Smith–Watson–Topper (SWT) and Fatemi–Socie (FS) models that address the multiaxial fatigue loads. Among them, SWT damage criterion assumes that the critical damage plane includes the maximum principal strain amplitude, and the damage parameter is expressed by the product of $\varepsilon_{n,a}$ and the maximum normal stress $\sigma_{n,\max}$ on this plane:

$$\varepsilon_{n,a}\sigma_{n,\max} = \frac{\sigma_f'^2}{E}\left(2N_f\right)^{2b} + \varepsilon_f'\sigma_f'\left(2N_f\right)^{b+c} \tag{7.9}$$

The plane with the maximum equivalent shear strain amplitude $\gamma_{a,\mathrm{eq}}$, in the Fatemi–Socie (FS) model, is taken into account as the most critical plane where fatigue failure occurs, and the value of maximum equivalent shear strain amplitude can be estimated by:

$$\gamma_{a,\mathrm{eq}} = \frac{\Delta\gamma_{\max}}{2}\left(1 + k\frac{\sigma_{n,\max}}{\sigma_y}\right) = \frac{\tau_f'}{G}\left(2N_f\right)^{b_\gamma} + \gamma_f'\left(2N_f\right)^{c_\gamma} \tag{7.10}$$

where $\Delta\gamma_{\max}$ and $\sigma_{n,\max}$ are the maximum shear strain range and the maximum normal stress on the critical plane, respectively; τ_f' and b_γ are the shear fatigue strength coefficient and exponent, respectively; γ_f' and c_γ are the shear fatigue ductility coefficient and exponent, respectively; G is the shear modulus; k is the normal stress sensitivity coefficient, which characterizes the influence of normal stress on fatigue lifetime and can be identified based on axial and torsional tests.

For the issues of variable amplitude load, S–N curve-based linear, double-linear and nonlinear models have been developed. Among them, the linear Palmgren–Miner's rule [8] has been widely adopted to deal with variable amplitude load. If there are several stages of loading history for a component, its total accumulated damage can be expressed as:

$$D = \sum_{i=1}^{k}\frac{n_i}{N_{f,i}} \tag{7.11}$$

where $N_{f,i}$ is fatigue lifetime in i_{th} loading stage; n_i is the actual service number of cycles in i_{th} loading stage; the ratio $n_i/N_{f,i}$ is the fatigue damage accumulated in i_{th} loading stage; k denotes the number of loading levels; D denotes the cumulative fatigue damage, and it is usually thought the fatigue failure occurs when its value exceeds 1.

However, sometimes Palmgren–Miner's rule is not a universal solution for all cases, which neglects the load frequence and sequence. Other types of models, such as energy-based and fracture mechanisms (stress intensity factor, SIF)-based fatigue life models, have also been developed. Among them, SIF-based crack growth models focus on the stage of fatigue crack propagation, while the above stress or strain-life curve-based models concentrate on the initial stage of crack formation.

When dealing with random load and stress derived from full-scale data records, determining the cyclic loading parameters such as cycle, amplitude, and range was not straightforward. With this regard, the rainflow method gives accurate results together with the Palmgren–Miner's rule, which was first introduced originally by Matsuishi and Endo [9]. The steps of rainflow counting for calculating amplitude and frequency are as follows: Collect actual loading history data, obtained from experiments, practical measurements, or simulations; identify peaks and valleys in the loading history by detecting maximum and minimum values, determining the starting and ending points of load cycles; conduct rainflow counting on the identified peaks and valleys, partitioning the loading history into different cycles, and assigning equivalent cycle counts; categorize cycles from rainflow counting based on amplitudes, shapes, etc.; calculate equivalent cycles for each category, representing theoretical cycles with the same fatigue damage effect; perform fatigue analysis using equivalent cycles, calculating fatigue life or damage.

In summary, the rainflow counting method is effective in handling random loading histories and provides a powerful tool for fatigue life analysis of engineering structures. However, the rainflow filtering neglects the stress interaction effects, for example, the cyclic curve sensitive to the "low-high" and "high-low" transitions still remains a problem.

7.4.2 Factors Affecting Fatigue Reliability of Structures

In the current methodology for fatigue assessment of structures, mechanical properties are examined through testing of small specimens under cyclic loading, but many influencing factors must be considered for the actual conditions of structures during the service period. This chapter explores the impact of key factors on the fatigue reliability of structures.

From the early investigations for the fatigue of components, it was found that fatigue cracks typically initiate at structural discontinuities, where local stress rises in comparison to nominal stress due to variations in the stress field around these structural irregularities. Notch features, encompassing holes, grooves, fillets, and shoulders, represent crucial geometries with extensive applications in engineering practice.

The local stress near notch can be easily described by a stress concentration factor, which is defined as the ratio of the maximum local stress to the nominal stress:

$$K_t = \frac{S_{\max}}{S_{\text{nom}}} \tag{7.12}$$

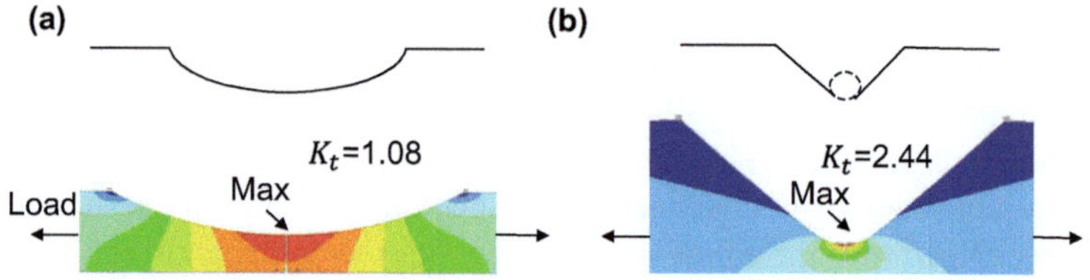

FIGURE 7.9 Notch structures with different stress concentration factors.

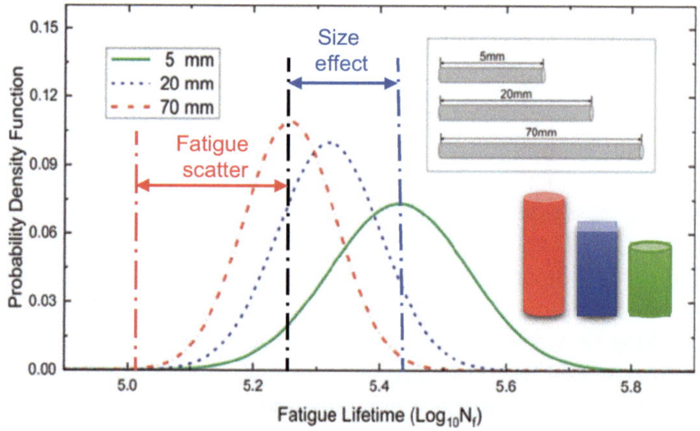

FIGURE 7.10 The fatigue scatter and size effect. (Reprinted with permission from Ref. [10], Elsevier.)

For structure with complex geometry, effective quantification of stress concentration and fatigue behavior induced by notch structures can be conducted through numerical simulation methods. Finite element method (FEA) can compute the stress concentration and figure out the stress/strain response of interested parts with complex geometry, as shown in Figure 7.9.

In traditional metal fatigue analysis, fundamental fatigue properties are typically collected from small-scale tests and subsequently applied to the fatigue assessment of components. However, it has been observed that the fatigue behaviors of small-scale specimens and full-scale components is different, even when subjected to the same nominal load. For instance, there is a decrease in fatigue strength with increasing volume, known as the statistical size effect, attributed to a higher probability of larger defects.

Currently, the statistical size effect is being progressively expanded to describe fatigue strength under different stress levels, not confined to the fatigue limit under infinite life. As illustrated in Figure 7.10, fatigue life distributions of three groups of specimens with the same radius but different lengths are presented, highlighting the evident presence of statistical size effects.

According to various sources, other size effects include geometrical, production, and surface effects. Geometrical size effect stems from geometrical discontinuities, depending on the stress gradient, and often works in conjunction with the statistical size effect. Actually, the concept of geometrical size effect appears to overlap with the notch effect, which is considered as the coupling effect between inhomogeneous stress and statistical size effect. Production processes, post-processing treatments, and surface machining change the defect population, as production and surface size effects, consequently influence the fatigue performance. These size effects can be summarized into the influence of defect density in critical region on fatigue performance.

In a narrow sense, the size effect is a comparative concept, whereas fatigue scatter arises from multiple factors, such as material variability, load fluctuation, and human errors, presenting a coefficient of variation in comparison to the average property. As illustrated in Figure 7.10, scatter and size effect govern the "scale" and "location or average" of fatigue property distribution, respectively. In a broader sense, the size effect can also be considered as one component of fatigue scatter.

Besides, the strength, hardness, toughness, and other mechanical properties of a material directly impact its ability to resist fatigue cracking. Materials with high strength and toughness often exhibit better fatigue performance. The magnitude, frequency, and duration of cyclic loading affect the material's fatigue life. Environmental conditions, such as temperature, humidity, corrosion, etc., significantly affect the fatigue performance of materials. Some environmental conditions may lead to stress corrosion cracking or hydrogen embrittlement, exacerbating fatigue failure. Surface treatment methods, such as shot peening, coatings, etc., can improve the surface quality of materials, slowing surface cracking and corrosion and enhancing fatigue performance. Material performance changes at high or low temperatures, impacting fatigue life. High temperatures may induce creep and thermal fatigue, while low temperatures can lead to brittle fracture. The microstructure of a material, including grain structure, internal defects, and inclusions, also influences fatigue performance. These microstructural features can serve as initiation points for cracks during loading cycles.

Considering these factors, appropriate fatigue assessment and design approaches should be chosen by analyzing the main influence factors, since it is impossible to find a perfect model that covering all the factors. By comprehensively considering these critical factors, effective fatigue assessment and design strategies can be used to ensure that structures have sufficient fatigue life and performance in practice.

7.4.3 Reliability Assessment and Sensitivity Analysis

Fatigue reliability is typically defined as the probability that the component will survive a specified number of loading cycles. A widely used strategy is to integrate FEA with MCs to estimate the failure probability under specified loading cycles. MCs is to perform groups of deterministic simulation with different combinations of random distributed parameters. This approach approximates the probability distribution of an output parameter by conducting repeated computations. In the process of fatigue reliability assessment, MCs can be utilized to simulate different uncertainty factors through repeated random sampling from the distribution of random parameters.

S–N curves, simulated by experimental data, are often employed to predict the fatigue life of materials. FEA tools can be employed to simulate structural stress

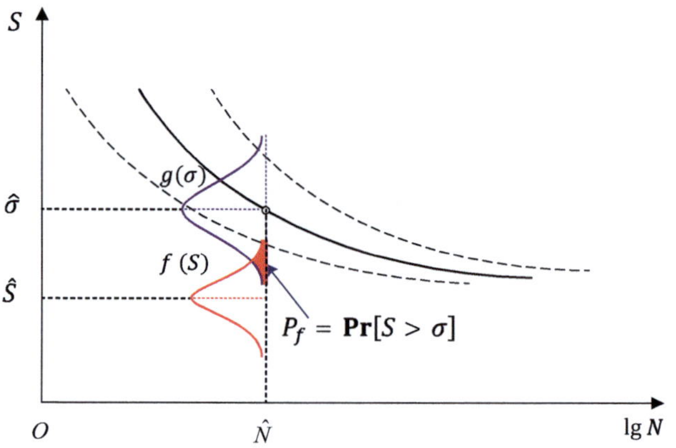

FIGURE 7.11 Probabilistic formats for fatigue design: stress–strength interference model. (Reprinted with permission from Ref. [11], Elsevier.)

under actual operating conditions. By analyzing stress distribution, fatigue life of complex structures can be assessed by using fatigue life models or fitted S–N curves. The difference between estimated fatigue life and design/intended fatigue life can account for uncertainties and reliability based on three interference models.

Material strength (stress under estimated fatigue life) and load (stress under designed/intended fatigue life) can be regarded as distributed independently. As shown in Figure 7.11, $g(\sigma)$ and $f(S)$ is strength and stress distributions, respectively. The failure probability P_f is estimated by using the stress–strength interference theory. Its limit state function and failure probability can be expressed as:

$$G_1 = \sigma - S \tag{7.13}$$

$$P_f = P_r[G_1 \le 0] = \int_{-\infty}^{+\infty}\left[f(S)\cdot\left[\int_S^{+\infty} g(\sigma)d\sigma\right]dS\right] \tag{7.14}$$

When $g(\sigma)$ and $f(S)$ follows the normal distribution, the failure probability P_f can be simplified as:

$$P_f = P_r[G_1 \le 0] = \phi\left(-\frac{\mu_{\lg\sigma} - \mu_{\lg S}}{\sqrt{\sigma_{\lg\sigma}^2 + \sigma_{\lg S}^2}}\right) \tag{7.15}$$

where $[\mu_{\lg\sigma}, \mu_{\lg S}, \sigma_{\lg\sigma}, \sigma_{\lg S}]$ are the mean and standard deviation fitted by the strength and stress distributions.

To calculate the fatigue reliability, another limit state function, according to the load–life interference (Figure 7.12), is expressed as:

$$G_2 = \lg N - \lg N_0 \tag{7.16}$$

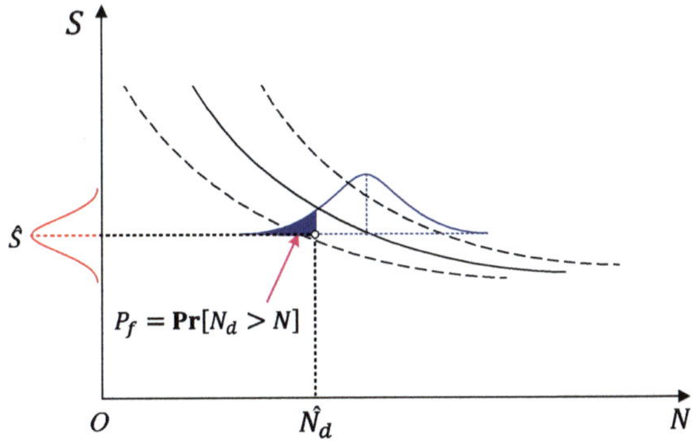

FIGURE 7.12 Probabilistic formats for fatigue design: load–life interference model. (Reprinted with permission from Ref. [11], Elsevier.)

The structure is thought to be unsafe if the actual work time exceeds the designed life. The failure probability P_f can be calculated by:

$$P_f = P(G_2 \le 0) = \phi\left(\frac{\lg N - \mu_{\lg N_0}}{\sigma_{\lg N_0}}\right) \tag{7.17}$$

where N represents the number of cycles to failure and is obtained from tests, predicted life or actual records of components, N_0 denotes the designed life.

From the point of view of fatigue damage accumulation, the fatigue reliability, according to the strength–damage interference (Figure 7.13) is expressed as:

$$G_3 = \lg D_c - \lg D \tag{7.18}$$

$$P_f = P(G_3 \le 0) = \phi\left(\frac{\mu_{\lg D_c} - \mu_{\lg D}}{\sqrt{\sigma^2_{\lg D_c} + \sigma^2_{\lg D}}}\right) \tag{7.19}$$

where D denotes the random cumulative damage under several stages of fatigue loadings, and D_C is the critical damage and can be quantified by:

$$\begin{cases} D = \dfrac{N_d}{N} \\ D_c = \dfrac{N_{ft}}{\overline{N_f}} \end{cases} \tag{7.20}$$

where N_{ft} represents the tested fatigue life, and $\overline{N_f}$ denotes the median fatigue life, which can be obtained from the fitted S–N curve. It is worth pointing out that the D_C is also considered as random in this chapter.

In order to quantify the effects of random variables on the fatigue reliability of engineering components, sensitivity analysis is required. Local sensitivity analysis

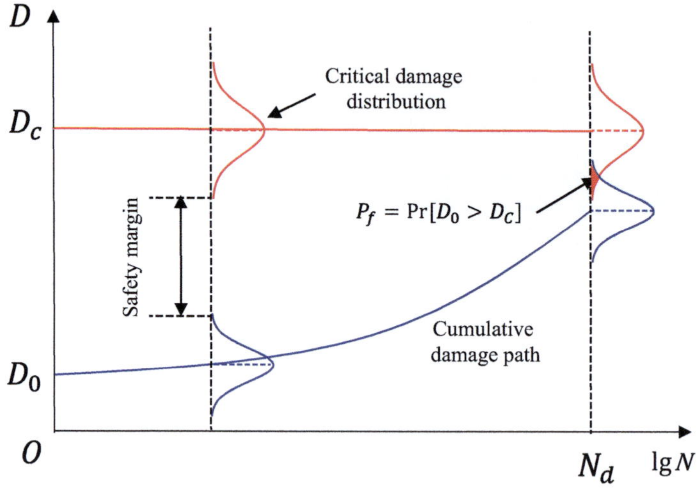

FIGURE 7.13 Probabilistic formats for fatigue design: strength–damage interference model. (Reprinted with permission from Ref. [11], Elsevier.)

uses mathematical tools (such as partial derivatives or gradients) to calculate the local sensitivity of the model output to changes in input parameters. It is based on the concept of differentiation and provides sensitivity information at a specific point; global sensitivity analysis aims to evaluate the sensitivity of the model across the entire input parameter space. The MCs, belongs to global sensitivity analysis, estimates the impact of parameters on the output by generating a large number of random samples. It involves random sampling of input parameters, running the model multiple times, and statistically analyzing the output variations. One example of MCs-based sensitivity analysis is presented in Section 7.5.1.4.

7.5 APPLICATIONS OF FATIGUE RELIABILITY ANALYSIS TO DESIGN

7.5.1 Fatigue Assessment of Turbine-Bladed Disk

Fatigue failure may lead to damage in engine components, resulting in accidents that pose a threat to flight safety. The "Comet 1" aircraft (1954) experienced a sudden mid-air explosion shortly after takeoff, plummeting from an altitude of 9000 m into the Mediterranean Sea. In United Airlines Flights 585 and 427 accidents (1994 and 1996), both accidents involved fatal crashes of aircraft due to fatigue failure in the tail control systems. These cases highlight the severe consequences that are caused by aircraft engine fatigue failure.

Bladed disks, as the key structures of turbine systems, are subjected to rapid acceleration during startup and operate under multiaxial loadings with various uncertainties. Probabilistic fatigue analysis for such critical structures has become essential. It is necessary to quantify these uncertainties for the aim that ensuring the structural

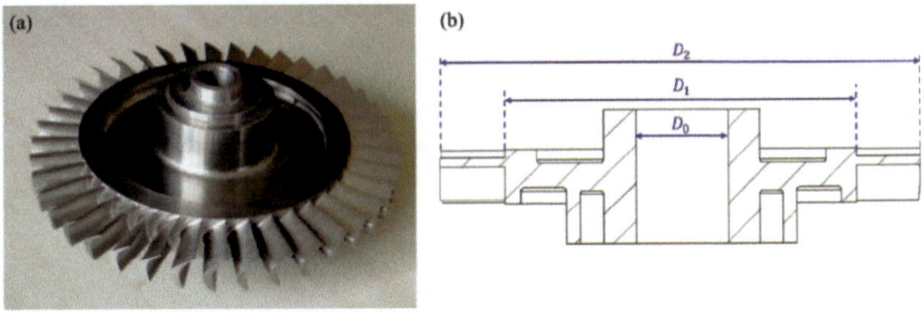

FIGURE 7.14 The sketch map of the turbine-bladed disk. (a) Structure, (b) cross-section. (Reprinted with permission from Ref. [12], Elsevier.)

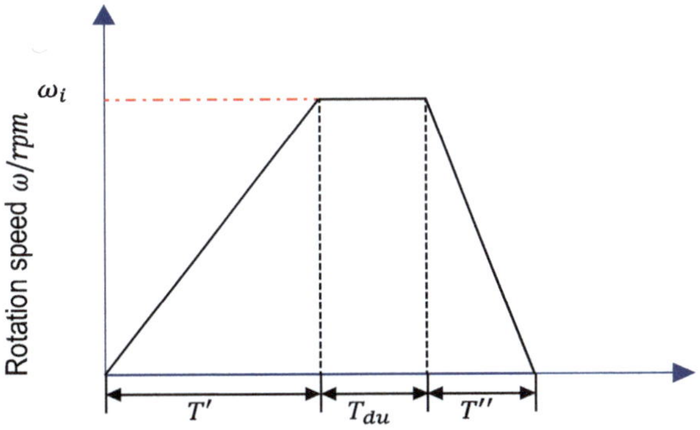

FIGURE 7.15 The load pattern for the bladed disk. (Reprinted with permission from Ref. [12], Elsevier.)

integrity and operational safety of components. In this section, a comprehensive probabilistic framework is presented. This framework covers material variability, load variation, and geometrical uncertainty, where FEA and MCs are connected to simulate the statistical properties and fatigue reliability.

7.5.1.1 Experiments and Finite Element Analysis

A full-scale bladed disk, with its main dimensions shown in Figure 7.14, is made of titanium alloy TC4. Its main mechanical properties (room temperatures) consist of Young's modulus, $E = 108$ GPa, Poisson's ratio, $\nu = 0.3$, density, $\rho = 4500$ kg/m^3, specific heat capacity, $C = 544$ J/kg$\cdot$K, thermal expansion coefficient, $\alpha = 8.2\times10^{-6}$/°C and yield strength, $\sigma_y = 800$ MPa.

The rotating speed of the bladed disk has a significant impact on its working performance and lifetime, which are usually limited by their allowable stresses. If this limit is exceeded, it works in an overspeed situation. The overspeed test of full-scale bladed disk ZUST1 is performed through a rotor high-speed spin tester designed by Zhejiang University, which belongs to one type of accelerated life testings. The overspeed test was performed with a monotonically increasing step load spectrum with five speeds ω_i $(i = 1,2,\ldots,5)$, sequentially 1969, 2514, 3142, 4216, and 5238 rad/s. As shown in Figure 7.15, T', T_{du} and T'' represent the hold time at the maximum rotation speed ω_i, increasing and decreasing time respectively, and their values are $T' = 120$ seconds,

TABLE 7.2
The Dimensional Changes of the Turbine-Bladed Disk

Dimensions	Before Experiment (mm)	After Experiment (mm)	After FEA (mm)
Disk diameter D_1	228	230	229.36
Blade tip diameter D_2	312.8	315	313.82

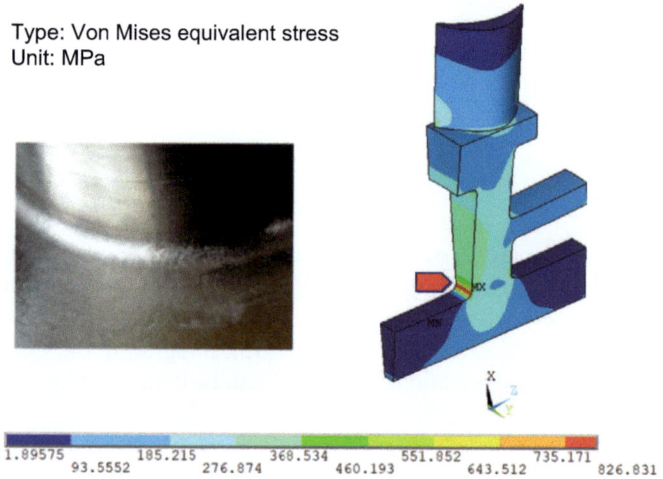

FIGURE 7.16 Von Mises equivalent stress nephogram of 1/40 turbine-bladed disk under $\omega_3 = 3143$ rad/s.

$T_{du} = 60$ seconds and $T'' = 60$ seconds. The structural dimensions were measured after applying a load spectrum, then the next higher load spectrum was applied until deformations appear. Specially, it is assumed that the bladed disk will be out of date if the crack length grows to approximately 0.8 mm, more testing details can be found in Zhu et al. [12]. The overspeed spin test is different from fatigue test, and the former concerns for the static strength. However, it can still provide the evidence of failure location and prove the credibility of mechanical response obtained from FEA. The dimensional information with and without applying all load spectrums are listed in Table 7.2, which indicates that FEA deformation results are almost identical to the measured experimental results after testing. The failure locations of the experimental and FEA results both locate at the disk root region (see Figure 7.16).

7.5.1.2 Material Variability, Load Fluctuation and Geometrical Uncertainty

In general, the fatigue lifetime and reliability index of engineering components are closely associated with the mechanical response, such as stress and strain. As previously mentioned, variations in applied loads, mechanical properties, and geometrical sizes significantly influence the cyclic response of the structures, especially their scatter is obvious owing to multi-source uncertainties. Therefore, it is necessary to quantify the fatigue dispersion induced by multi-source uncertainties (as stochastic inputs) for comprehensive fatigue reliability assessment. In this section, the FS damage criterion, as one of the commonly used models in fatigue modeling, is employed to compute the fatigue lifetime. It is assumed that the critical plane where fatigue failure occurs is located at the plane with the maximum equivalent shear strain amplitude $\gamma_{a,\mathrm{eq}}$ in Eq. (7.10), wherein the normal stress sensitivity coefficient can be approximated by Eq. (7.21).

$$k = \frac{\sigma_y}{\sigma'_f} \tag{7.21}$$

where the cyclic yield stress σ_y is estimated according to 0.05% offset rule and expressed as:

$$\sigma_y = K'\left(\varepsilon_{\mathrm{pa}}\right)^{n'} \tag{7.22}$$

where $\varepsilon_{\mathrm{pa}}$ denotes the plastic strain amplitude of 0.05%; K' and n' is the cyclic strength coefficient and cyclic strain hardening exponent, respectively. The stochastic mechanical response can be well described by the four Coffin–Manson parameters $\{\sigma'_f, \varepsilon'_f, b, c\}$. According to Dowling's research [13], integrating the Coffin–Manson with the Ramberg–Osgood equations, K' and n' can be computed by:

$$K' = \frac{\sigma'_f}{\left(\varepsilon'_f\right)^{n'}} \tag{7.23}$$

$$n' = \frac{b}{c} \tag{7.24}$$

when torsional fatigue data is lacking, the shear fatigue properties $\left\{\tau'_f,\ \gamma'_f,\ b_\gamma,\ c_\gamma, G\right\}$ in the FS model can be approximated based on the axial fatigue properties $\left\{\sigma'_f, \varepsilon'_f, b, c, E\right\}$:

$$\begin{cases} G = E/2(1+v) \\ \tau'_f \approx \dfrac{\sigma'_f}{\sqrt{3}} \\ \gamma'_f \approx \sqrt{3}\varepsilon'_f \\ b_\gamma \approx b \\ c_\gamma \approx c \end{cases} \tag{7.25}$$

There are many related material parameters; therefore, the analysis of correlations is carried out by using the R-software (see Figure 7.17, the more centralized samples indicate a stronger correlation of the two material parameters). Accordingly, it is found that parameters $\{E, \sigma'_f, \varepsilon'_f\}$ can be appropriate enough to describe the fatigue dispersion of TC4 alloy. The $\{k, \sigma_y, K'\}$ are treated as intermediate variables and $\{n',\ b, c\}$ are constant.

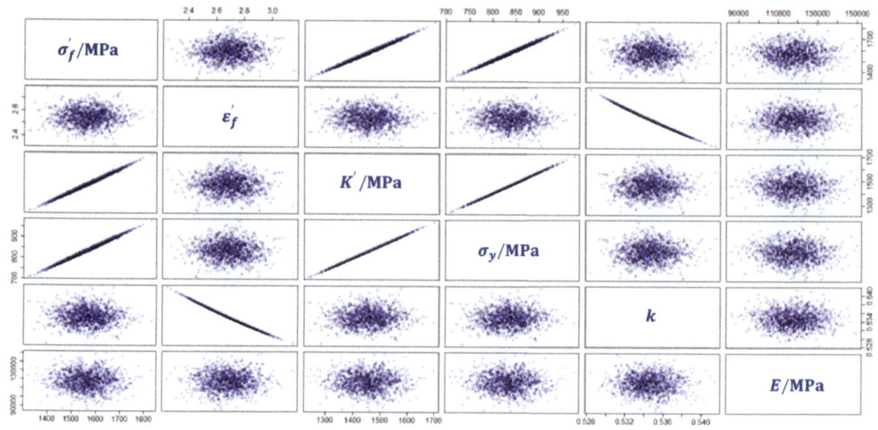

FIGURE 7.17 Correlation analysis among material parameters of TC4. (Reprinted with permission from Ref. [12], Elsevier.)

To characterize the elasto-plastic performance of the TC4 alloy, the three-term Chaboche nonlinear kinematic hardening rule is employed as a constitutive model when carrying out FEA. This model has abundant variations to capture the nonlinear behavior and can effectively describe ratcheting phenomena. Each of the back-stress terms within the Chaboche model adheres to an Armstrong–Frederick evolution law, where the stress range $\Delta\sigma$ can be formulated using the plastic strain range $\Delta\varepsilon_p$ as follows:

$$\frac{\Delta\sigma}{2} = \sigma_y + \sum_{i=1}^{3} \frac{C_i}{\gamma_i} \tanh\left(\gamma_i \frac{\Delta\varepsilon_p}{2}\right) \tag{7.26}$$

where σ_y is fatigue yield strength determined from uniaxial tests, $\{C_i, \gamma_i\}$ $(i = 1,2,3)$ are experimentally fitted material constants, wherein it is assumed that C_i are random and γ_i are constant. The fitted procedure is shown as below:

Step 1: Approximate the cyclic stress–strain response of materials by using the Ramberg–Osgood:

$$\varepsilon = \frac{\sigma}{E} + \left(\frac{\sigma}{K'}\right)^{\frac{1}{n'}} \tag{7.27}$$

Step 2: On the basis of material properties $\{E, \sigma_f', \varepsilon_f'\}$ listed in Table 7.3, the intermediate variable K' in Eq. (7.27) can be dispersed and the plastic phase of probabilistic stress–strain curves (red one is the mean stress–strain curve) are shown in Figure 7.18a. $\{C_1, C_2, C_3\}$ of the Chaboche model is fitted by using the family of cyclic stress–strain curve and their probability distribution is shown in Figure 7.18b–d. In this work, the uncertainties generated by material properties are characterized through the stress–strain response being simulated by the material constitutive parameters $\{C_1, C_2, C_3\}$.

TABLE 7.3
Input Random Variables

Parameters	Distributions	Mean	CV
Young's modulus E	Normal	1.08×10^5 MPa	0.05
Material density ρ	Normal	4.5×10^3 kg/m^3	0.01
		1969 rad/s	
Rotating speed ω_i		2514 rad/s	
	Normal	3143 rad/s	0.01
$(i = 1, 2, \ldots, 5)$		4216 rad/s	
		5238 rad/s	
σ'_f	Normal	1564 MPa	0.05
ε'_f	Normal	2.69	0.05
R_2	Normal	114.5 mm	0.00033
R_5	Normal	29.5 mm	0.00033
R_{10}	Normal	18 mm	0.00033

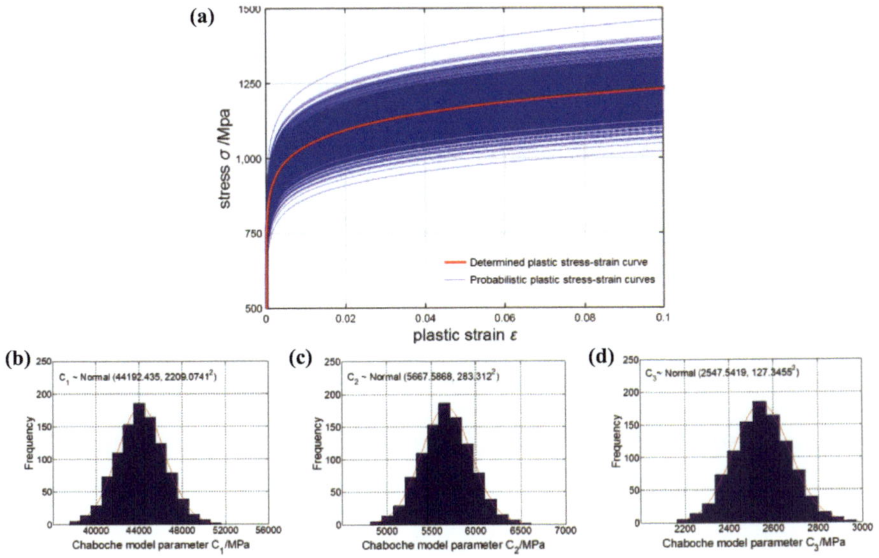

FIGURE 7.18 Stochastic stress–strain response of TC4 alloy: (a) plastic stress–strain curves obtained using Ramberg–Osgood model, (b–d) probability distribution of sampling parameters C_i $(i = 1, 2, 3)$. (Reprinted with permission from Ref. [14], Elsevier.)

For the turbine-bladed disk, its dominant subjected load is the centrifugal force generated from bladed rotating, which mainly depends on the rotating speed. During operation, the unstable rotation speed yields the load variation, and the coefficients of speed variation are listed in Table 7.3 to model the stress and strain response.

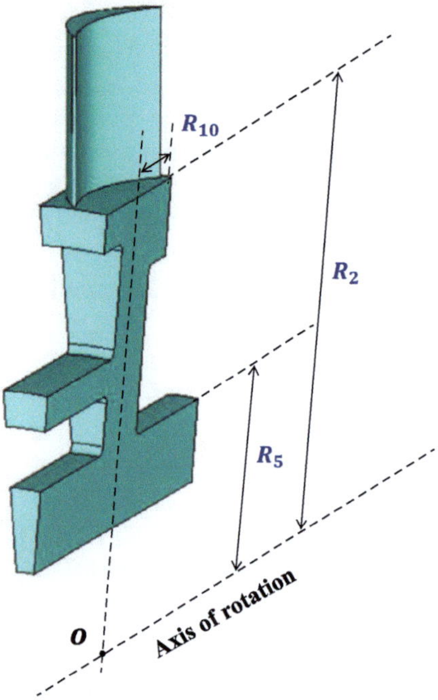

FIGURE 7.19 Three main dimensions of the turbine-bladed disk. (Reprinted with permission from Ref. [14], Elsevier.)

Mechanical response of structures is also influenced by its geometrical uncertainty. After determining the tolerance range of each geometrical size, sampling was performed based on the distributions of geometrical sizes. The sampled geometrical sizes are then utilized for three-dimensional modeling, and geometry updating is achieved through integrating FE modeling with MATLAB® coding. There are so many geometrical sizes in turbine-bladed disk, but it is time-wasting and unnecessary if all the size changes are taken into account during FE analysis. With this regard, the sensitivity analysis for mechanical response/life/reliability of turbine-bladed disks is a good way for the selection of main geometric sizes. For example, the application of stress altering factor is a convenient and practical approach [15]. As depicted in Figure 7.19, three critical sizes of the turbine-bladed disk are chosen as the inputs of geometrical uncertainty, which is assumed to follow normal distribution. The ranges of geometrical tolerances are approximated as ±0.1% variations of the standard designed size. The range of $\mu \pm 3\sigma$ of a normal distribution covers a proportion of 99.73%; therefore, the coefficients of variation of three geometrical sizes were taken as 0.033%.

The aforementioned random variables are separated into three categories, including rotating speeds $\{\omega_1, \omega_2, \omega_3, \omega_4, \omega_5\}$, geometrical dimensions $\{R_2, R_5, R_{10}\}$, and material properties $\{\sigma'_f, \varepsilon'_f, \rho, E\}$. These inputted parameters, as stochastic variables with variation coefficient listed in Table 7.3, are sampled through the Latin hypercube sampling (LHS) method. As a stratified sampling technique, LHS is efficient to sample random variables from their distributions. By incorporating LHS into the MCs, the sampling process is optimized and more efficient, thereby computational time cost is reduced.

7.5.1.3 Fatigue Life Assessment

The procedure of fatigue lifetime and reliability assessment follows the below steps (Figure 7.20):

1. LHS is applied to sample from the distributions of $\{E, \rho, \omega, R, \sigma'_f, \varepsilon'_f\}$.
2. FE three-dimensional model, with variable geometrical dimensions $\{R_2, R_5, R_{10}\}$, is programmatically updated by coupling ANSYS® with MATLAB®.
3. Chaboche model parameters $\{C_1, C_2, C_3\}$ are simulated based on stochastic material parameters $\{E, \sigma'_f, \varepsilon'_f\}$.

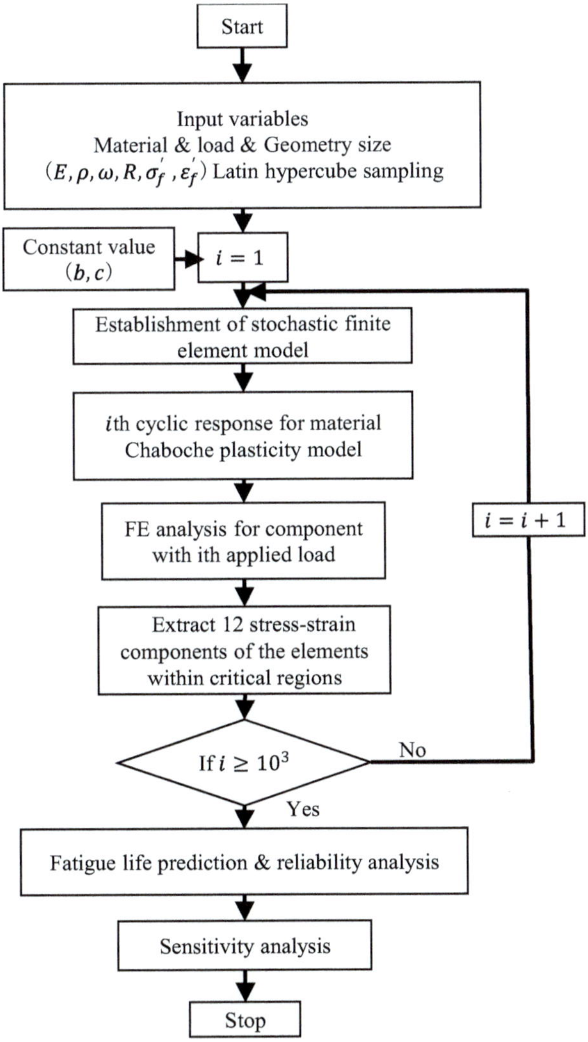

FIGURE 7.20 The probabilistic framework of fatigue life assessment and reliability analysis. (Reprinted with permission from Ref. [14], Elsevier.)

4. FE analysis of the turbine-bladed disk is carried out under given rotational speed ω.
5. The stress–strain components of the critical regions are extracted and applied to the FS criterion for fatigue lifetime prediction.
6. Calculate fatigue failure probability and the sensitivity to fatigue reliability.

Note from the results of FEA, the root of turbine-bladed disk is the critical region with largest local stress, as shown in Figure 7.16. The stress–strain responses at the disk root, under five rotational speeds, were substituted to FS model. The calculated results of fatigue lifetime within material variation, load fluctuation, and geometrical uncertainties are shown in Figures 7.21 and 7.22. The parameters, mean value, standard deviation, skewness, and kurtosis of fatigue life distribution are listed in Table 7.4.

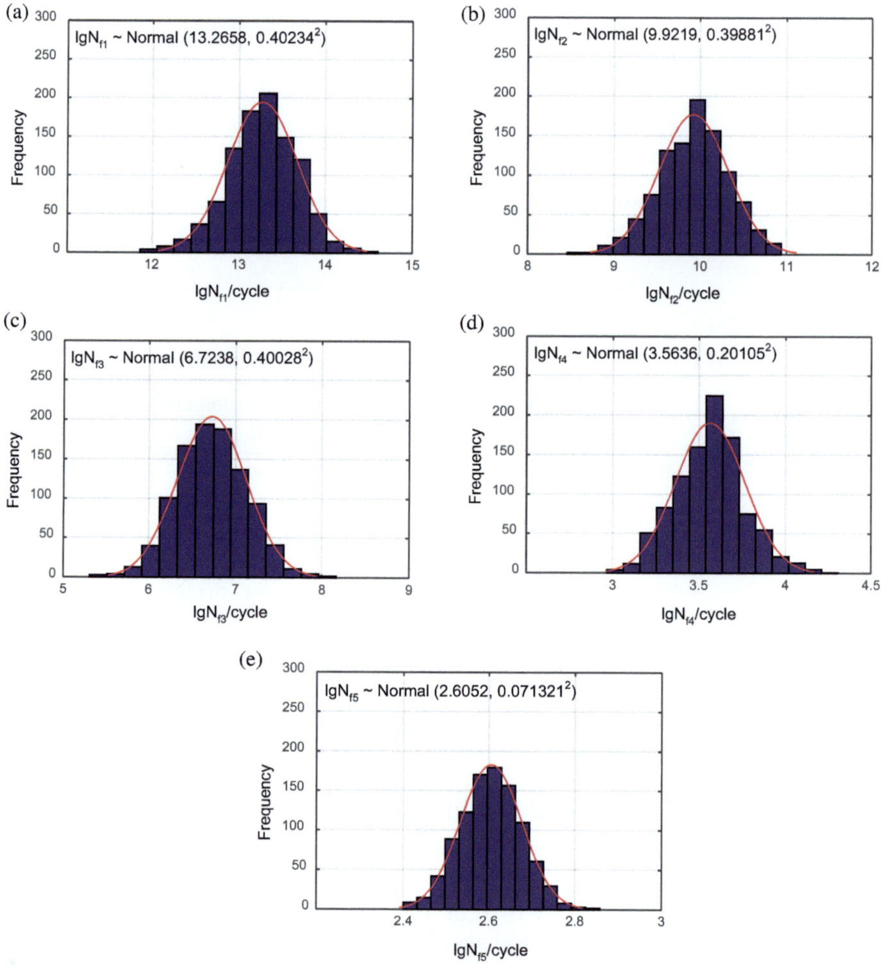

FIGURE 7.21 Predicted fatigue life distributions with material variation, load fluctuation and geometrical uncertainties. (Reprinted with permission from Ref. [14], Elsevier.)

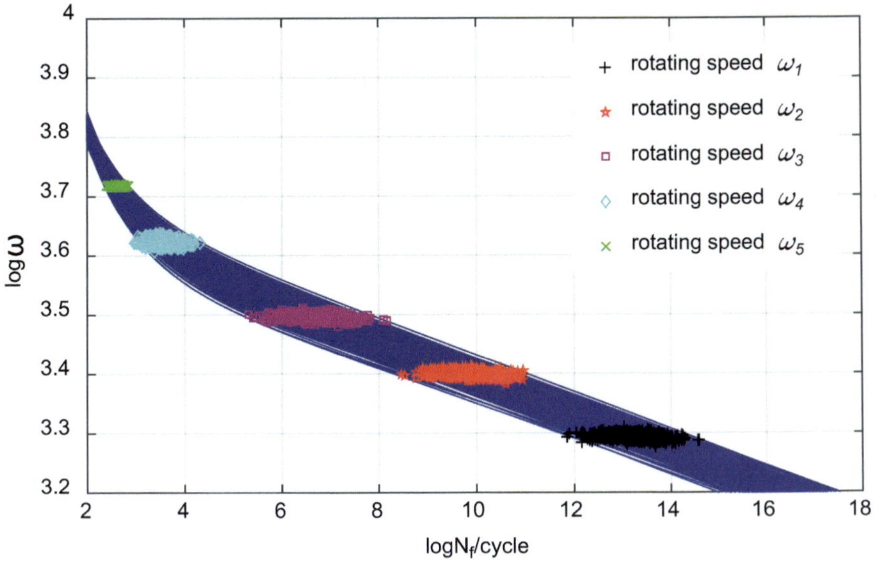

FIGURE 7.22 Probabilistic $\lg\omega - \lg N_f$ curves with material variation, load fluctuation and geometrical uncertainties. (Reprinted with permission from Ref. [14], Elsevier.)

TABLE 7.4
Statistical Parameters of Predicted Fatigue Life Distributions

Rotating Speed (rad/s)	With Material Variation, Load Fluctuation and Geometrical Uncertainties			
	Mean ($\lg N_f$)	Standard Deviation	Skewness	Kurtosis
1969	13.2658	0.402	−0.2305	3.3005
2514	9.9219	0.399	−0.1786	3.0469
3143	6.7238	0.400	0.0413	3.1423
4216	3.5636	0.201	0.1792	3.2637
5238	2.6052	0.071	0.0336	2.9657

The results indicate that the mean fatigue lifetime decreases with rotational speed, and the scatter increases firstly before 10^6 cycles and then tends to be steady.

7.5.1.4 Fatigue Reliability and Sensitivity Analysis

In order to estimate the fatigue reliability of turbine-bladed disk under specific designed life, the load–life and strength–damage interferences are introduced. As shown in Figure 7.23, there are two curves of failure probability, wherein the solid one is associated with strength–damage model and dashed line is simulated by the load–life model. To conclude, the two models exhibit the similar trend of predicted failure probability. However, in comparison with the load–life model, strength–damage interference-based method is more conservative.

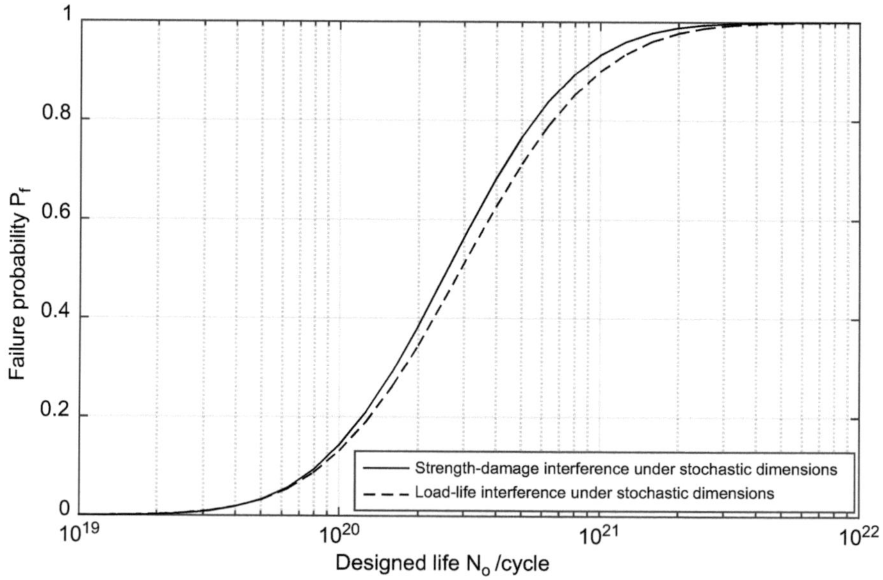

FIGURE 7.23 Failure probability of the disk versus the designed life. (Reprinted with permission from Ref. [14], Elsevier.)

To quantify the influences of these variables on reliability, the sensitivity analysis is carried out, and the sensitivity of each variable to reliability can be expressed as:

$$\frac{\partial P_f}{\partial \theta_{x_i}^{(j)}} = \int \cdots \int_{\Omega} \frac{\partial h_X(\boldsymbol{x})}{\partial \theta_{x_i}^{(j)}} dx \tag{7.28}$$

$$\frac{\partial P_f}{\partial \theta_{x_i}^{(j)}} = E\left[\frac{I[G_2 < 0]\partial h_{X_i}(x_i)}{h_{X_i}(x_i)\partial \theta_{x_i}^{(j)}}\right] \tag{7.29}$$

where $\theta_{x_i}^{(j)}$ is the distribution parameters of the basic random variable x_i, j denotes the number of the distribution parameters.

When the basic random variable x_i follows normal distribution, Eq. (7.30) can be written as:

$$\begin{cases} \dfrac{\partial h_{X_i}(x_i)}{h_{X_i}(x_i)\partial \mu_{x_i}} = \dfrac{x_i - \mu_{x_i}}{\sigma_{x_i}^2} \\ \dfrac{\partial h_{X_i}(x_i)}{h_{X_i}(x_i)\partial \sigma_{x_i}} = \dfrac{1}{\sigma_{x_i}}\left[\left(\dfrac{x_i - \mu_{x_i}}{\sigma_{x_i}}\right)^2 - 1\right] \end{cases} \tag{7.30}$$

The units of material parameters, loads, and geometrical sizes are different; therefore, their importance ranking cannot be directly compared. Therefore, a dimensionless process is carried out using Eqs. (7.31) and (7.32).

$$\begin{cases} \alpha_i = \dfrac{\partial P_f}{\partial \mu_i} \cdot \dfrac{\mu_i}{P_f} \\ \eta_i = \dfrac{\partial P_f}{\partial \sigma_i} \cdot \dfrac{\sigma_i}{P_f} \\ s_i = \sqrt{(\alpha_i)^2 + (\eta_i)^2} \end{cases} \tag{7.31}$$

$$\begin{cases} \lambda_{\alpha,i} = \dfrac{|\alpha_i|}{\sum_{k=1}^{n} |\alpha_k|} \times 100\% \\ \lambda_{\eta,i} = \dfrac{|\eta_i|}{\sum_{k=1}^{n} |\eta_k|} \times 100\% \\ \lambda_{s,i} = \dfrac{s_i}{\sum_{k=1}^{n} s_k} \times 100\% \end{cases} \tag{7.32}$$

where α_i and η_i respectively represent the sensitivities of mean/standard deviation of ith random variables to failure probability; s_i represents the sensitivity gradient of random variables, namely the combined sensitivity of mean and standard deviation; $\lambda_{\alpha,i}$, $\lambda_{\eta,i}$, and $\lambda_{s,i}$ denote the sensitivity factors, namely the proportion of importance. The sensitivity factors of random variables to failure probability are calculated based on Eqs. (7.31) and (7.32), as shown in Figure 7.24.

According to the sensitivity rank of all random variables in Figure 7.24a, it is found that the importance of R_1, R_2, R_3, ω and material properties are in a decreasing order. It indicates that the geometrical tolerance also shows a great influence on the fatigue performance of the turbine-bladed disk. Thus, geometrical sizes, as one of the critical factors impacting scatter and reliability, should also be considered during the design and manufacturing stage.

The quantification of stochastic uncertainties is a crucial aspect of fatigue reliability assessment, and probabilistic methods, particularly MCs, play a significant

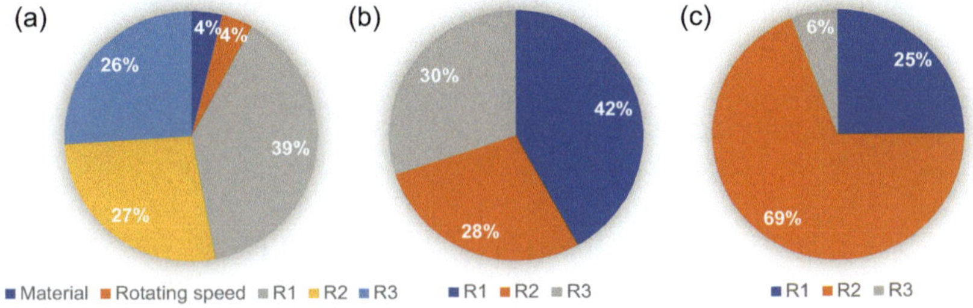

FIGURE 7.24 The sensitivity factors of random variables to failure probability at $N_d = 10^{19.6}$: (a) combined sensitivity factors, (b) sensitivity factors of mean value, (c) sensitivity factors of standard deviation. (Reprinted with permission from Ref. [14], Elsevier.)

role in this evaluating procedure. Future developments should involve more efficient MCs-based methods to decrease the costs and time of computational process. This further development of probability distribution models for material properties and structural parameters, including stochastic modeling of fatigue loads, material strengths, and geometric parameters, is necessary to comprehensively capture the stochastic nature of complex structures. Moreover, the development of more comprehensive reliability indices, such as failure probabilities and reliability indices, is essential to synthetically consider the impact of multiple stochastic uncertainties on fatigue performance. Advanced approaches of sensitivity analysis to reliability will identify parameters with the most significant impact on structure reliability and then to guide engineering decisions. These advancements are expected to enhance the application of probabilistic methods in fatigue reliability assessment, providing a more accurate understanding of the fatigue performance of structures under complex environmental conditions and multiple sources of uncertainties.

7.5.2 Fatigue Reliability and Safety Factor

During the design stage of structures, safety factor is usually used to explain uncertainties that are difficult to quantify. Its application acts as a margin, covering the effects of uncertainties, to keep the service safety of components. For example, empirical or semi-probabilistic approaches are suggested in many standards/codes, which usually give a partial safety factor accompanying with mean performance of components to characterize their limit states and achieve the target reliability. In ASME Section III [1], safety factor and mean curve fitted by experimental data are utilized to keep the enough safety margin of reactor pressure vessel and determine its allowable cycles for crack initiation (see Figure 7.25), where the safety factors are set as 20 on fatigue cycles and 2 on strain/stress. The FKM guideline [16] attracted much attention due to its convenient procedure being directly applicable in industrial design. It gives S–N diagrams with failure probability $P_f = 2.5\%$ and safety factor $\gamma_F = 1.5$ for the steel and aluminum components, where the statistical methods, e.g. Weibull distribution, are utilized to quantify the influences of surface material defects on fatigue crack initiation.

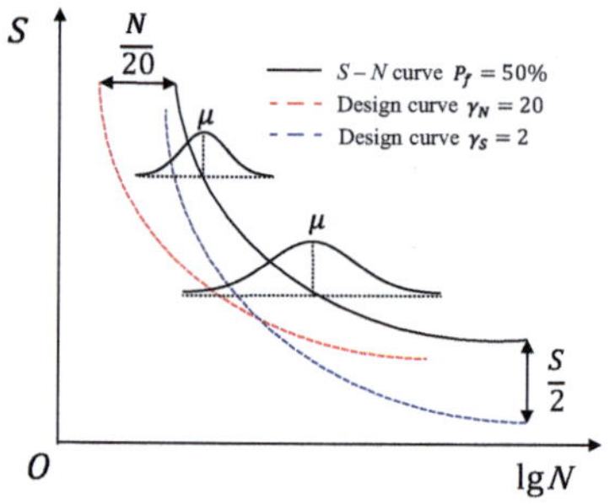

FIGURE 7.25 Design curves with safety factors of 2 on stress and 20 on life. (Reprinted with permission from Ref. [11], Elsevier.)

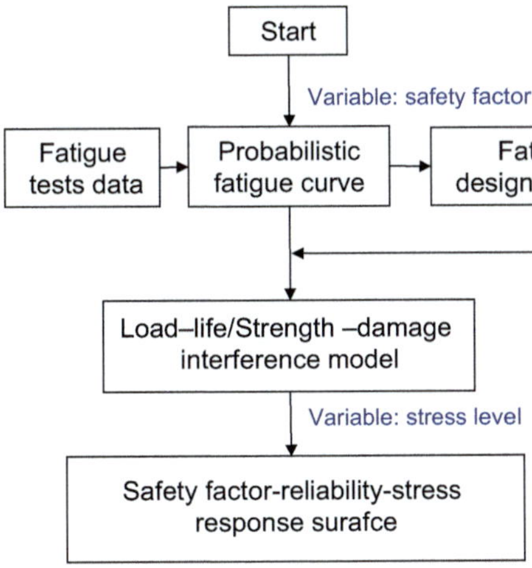

FIGURE 7.26 The flowchart of safety factor design with specific reliability.

Similarly, Eurocode 3 [17] raised a S–N diagram with a 5% failure probability for welded structures. It recommends to add a safety factor $\gamma_F = 1.35$ onto the S–N curve of the welded detail of critical components for the safety assessment. In the case of aero engine components, the range (i.e., $\mu - 3\sigma$) of S–N curve obtained from experimental data is considered as a reference for life assessment. The allowable stress and safety region are defined by three standard deviations below the medium S–N curve.

In essence, safety factors refer to a comprehensive presentation of many uncertainties. However, the direct application of safety factor always results in a conservative design, potentially leading to an overbuilt strength and excessive costs. Therefore, safety factor not only affects the reliability of the components but also the cost. Besides, the specifical value of reliability of a structure still keeps unknown when safety factor is given, potentially leading to an unsafe or inefficient design. In this section, two conversion models, relating safety factor and fatigue reliability, are raised based on load–life and strength–damage interference theories. The fatigue scatter is quantified and then applied to explain the reliability changing against safety factor. Furthermore, the safety factors with the aimed reliability level can be conveniently designed. The calculation framework is shown in Figure 7.26.

1. Fit the P–S–N curves based on the experimental data.
2. Simulate the designed fatigue curve by using mean S–N and safety factor.
3. Compute the failure probability through the interference of designed fatigue curve and P–S–N curves (one-by-one computing under different stress levels) through the load–life or strength–damage interference model.
4. Repeat the aforementioned computing procedure with different safety factors and construct the reliability response surface against safety factor and stress.

7.5.2.1 The Application of Load–Life Interference Theory

It is widely accepted that the fatigue lives of many materials follow lognormal distribution. Take low-alloy steels and austenitic stainless steels as examples, which are divided into two categories by taking into account the influences of temperature on material fatigue performance. The testing data and fitted P–S–N curves, under 25°C–200°C and 200°C–350°C, are shown in Figure 7.27.

In the computing procedure, designed life curve is demanded for reliability analysis. The designed life curve can be easily simulated by the below equation:

$$\begin{cases} \lg N_d = \mu_{\lg N} - \lg \gamma_N \\ \lg N_d = \mu_{\lg N} + \dfrac{\lg \gamma_S}{k} \end{cases} \tag{7.33}$$

where $\mu_{\lg N}$ is the mean fatigue life according to the mean S–N curve fitted by fatigue testing data; γ_N and γ_S are the safety factors on life and stress, respectively; k is the slope of mean S–N curve.

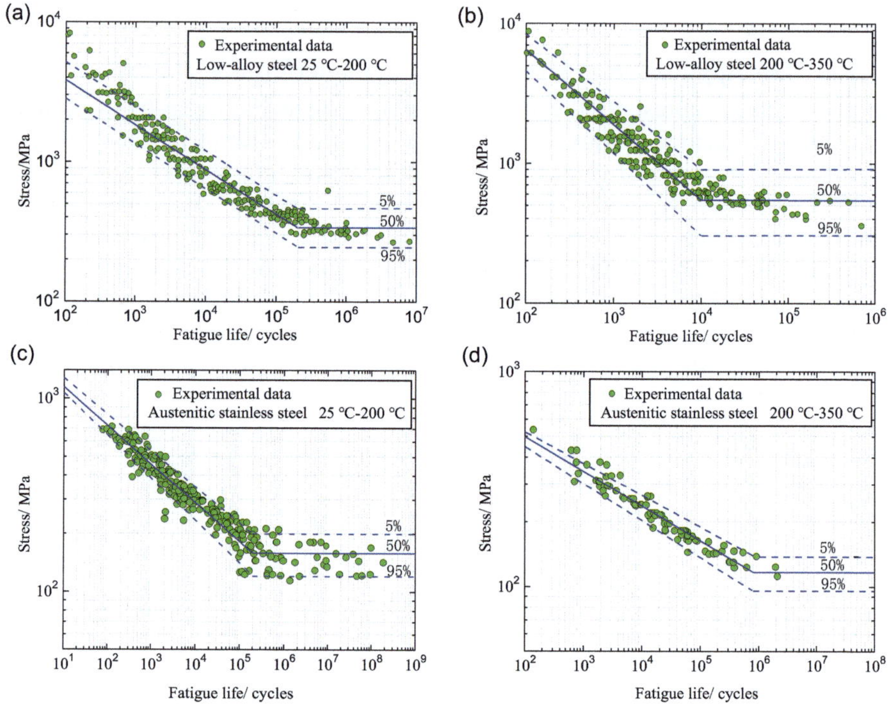

FIGURE 7.27 Fatigue data and fitted P–S–N curves (a) low-alloy steel under 25°C–200°C, (b) low-alloy steel under 200°C–350°C, (c) austenitic stainless steel under 25°C–200°C, (d) austenitic stainless steel under 200°C–350°C. (Reprinted with permission from Ref. [11], Elsevier.)

According to load–life interference in Eq. (7.17), except for designed life, the standard deviation $\sigma_{\lg N}$ representing the fatigue dispersion is required. The value of $\sigma_{\lg N}$ can be quantified by Eq. (7.34).

$$\sigma_{\lg N} = -\left[\frac{\lg N_{95\%} - \lg N_{5\%}}{\phi^{-}(0.95) - \phi^{-}(0.05)}\right] \tag{7.34}$$

where $\lg N_{95\%}$ and $\lg N_{5\%}$ are the logarithmic life with survival rate 95% and 5% at specific stress level, respectively; ϕ^{-} is the inverse function of standard normal distribution function ϕ, for example, $\phi^{-}(0.95)$ denotes the value when the integral of the standard normal PDF from negative infinity to this value is 0.95.

According to the load–life model, the failure probability P_f under specific designed life N_d and stress $\hat{S}$ can be calculated, which is analyzed result of single point. Moreover, the failure probability P_f under different stress levels and safety factors can be computed by repeating the process of load–life interference. Finally, the reliability response surface, which consists of failure probability changes with stress and safety factor, can be drawn. Because of the large fatigue dispersion near the fatigue limit, only the data with fatigue life less than 10^6 is considered.

7.5.2.2 The Application of Strength–Damage Interference Theory

Fatigue damage represents the state of material subjected cycle-by-cycle loading, which is a critical parameter to predict the fatigue life of mechanical components and structures. In practice, D_c and D in strength–damage interference model are random variables, which are assumed to follow lognormal distribution. Generally, the structure is thought of as unsafe if the cumulative damage D exceeds the critical damage D_c. According to the Eq. (7.34) and mean S–N curve in Figure 7.25, the distribution of logarithmic fatigue life can be expressed as $\lg N \sim \text{Normal}(\mu_{\lg N}, \sigma_{\lg N})$. Through analyzing experimental data based on Eqs. (7.20) and (7.33), the distribution of D_c and D utilized in this analysis can be easily expressed as:

$$\begin{cases} \lg D_c \sim \text{Normal}\left(0,\ \sigma_{\lg N}\right) & \\ \lg D \sim \text{Normal}\left(-\lg \gamma_N,\ \sigma_{\lg N}\right) & \text{For } \gamma \text{ on life} \\ \lg D \sim \text{Normal}\left(\dfrac{\lg \gamma_S}{k},\ \sigma_{\lg N}\right) & \text{For } \gamma \text{ on stress} \end{cases} \tag{7.35}$$

The distributions of critical damage D_c and cumulative damage D under different stress levels are easily simulated by using the safety factor and P–S–N curves. Moreover, the strength–damage interference theory can also be extended to the calculation of failure probability P_f versus safety factor and stress. In this way, the reliability response surface versus safety factors and stresses can be drawn.

7.5.2.3 Fatigue Reliability versus Safety Factor

The designed/safety curves are simulated by using a safety factor of 1.2–2.4 on the stress and 2–20 on the life. The response surface of reliability versus safety factor and stress are calculated based on the load-life interference theory, as shown in Figures 7.28 and 7.29. Moreover, the low-alloy steels and austenitic stainless steels, at temperatures of 25°C–200°C and 200°C–350°C, are employed to be analyzed and compared.

The reliability increases as expected with safety factor increasing. Because of the smaller fatigue dispersion at high stress levels, the reliability is relatively higher at low stress levels under the same safety factor according to Figures 7.28 and 7.29. Note from Figures 7.28a and 7.29a, the reliability does not change a lot with stress, because the change in material dispersion with stress is not as apparent as usual from Figure 7.27a. In Figures 7.28b–d and 7.29b–d, the reliability decreases with the decrease of stress and safety factor. It is reported that factor 20 on life consists of material dispersion, size effects, environments, etc., wherein the parts with respect to material is thought as 2 [18]. For the factor of 2 on stress, the value corresponding to material variables is 1.2–1.7 [2]. Only the fatigue scatter obtained from experimental data is taken into account in this section. Therefore, the safety factor of 2 on life and 1.7 on stress is concerned, instead of the safety factor of 20 on life and 2 on stress.

The reliability exceeds 85% under a safety factor of 2 on life, and the reliability is more than 99% under a safety factor of 1.7 on stress. It indicates that a safety factor

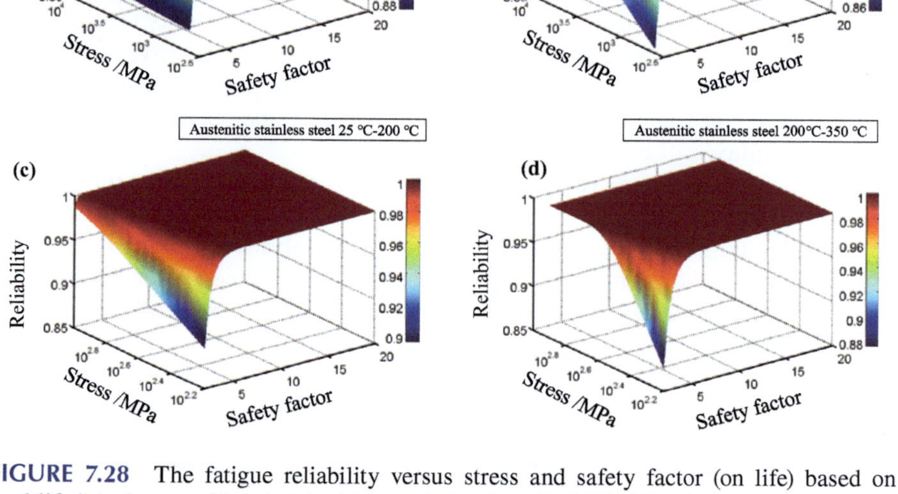

FIGURE 7.28 The fatigue reliability versus stress and safety factor (on life) based on load-life interference. (Reprinted with permission from Ref. [11], Elsevier.)

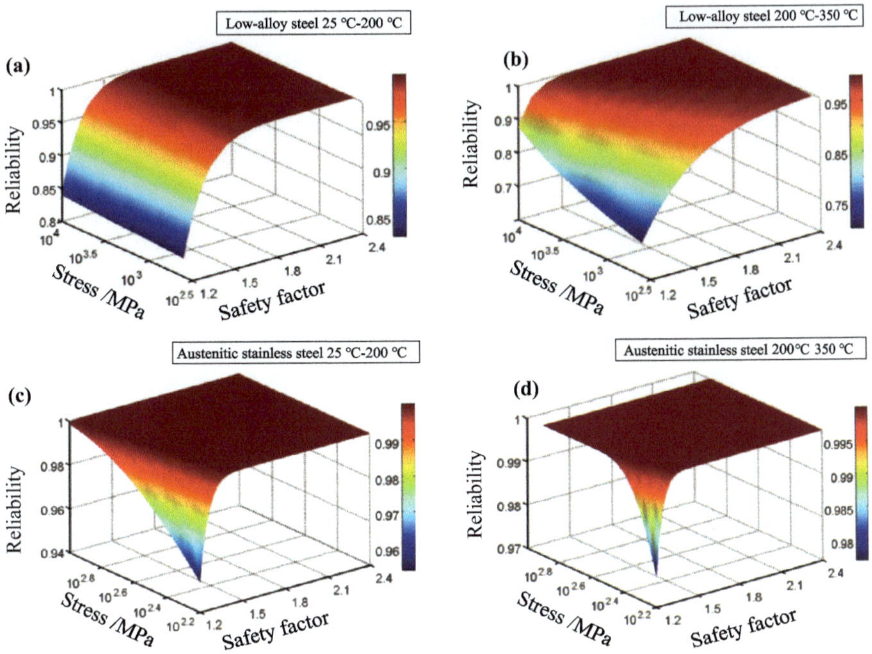

FIGURE 7.29 The fatigue reliability versus stress and safety factor (on stress) based on load-life interference. (Reprinted with permission from Ref. [11], Elsevier.)

of 1.7 on stress is more conservative. And the method setting safety factor on life is more robust, since the reliability keeps steady state even under different temperatures and materials. And the safety factor on stress is more fluctuating, which may explain why it changes from 1.2 to 1.7 instead of a constant. In addition, a rapid drop of reliability can be found within a certain range of safety factor. This range is significant for determining a cost-effective safety factor design. These reliability response surfaces, drawn based on two conversion models, illustrate a more intuitive relationship among reliability, stress, and safety factor. This procedure provides a practical way to safety design of components, especially, the safety can be obtained based on the desired reliability.

The reliability response versus safety factor and stress, using strength–damage interference theory, are shown in Figures 7.30 and 7.31. Because the same data, including N_d, $\mu_{\lg N}$, and $\sigma_{\lg N}$, are used for reliability analysis, the changing trend of response surfaces shown in Figures 7.30 and 7.31 is similar to the ones in Figures 7.28 and 7.29. And similar conclusions can be obtained: the reliability decreases with decrease of the stress and safety factor, wherein the stress level influences the reliability by dominating the life scatter while the safety factor is through changing the location of designed life curve. The reliability, with a safety factor of 2 on life and 1.7 on stress, is both over 80%. The reliability, with the safety factor on life, remains at a steady level even under the larger stress level for different material types. Nevertheless, the reliability for different material types varies a lot when using

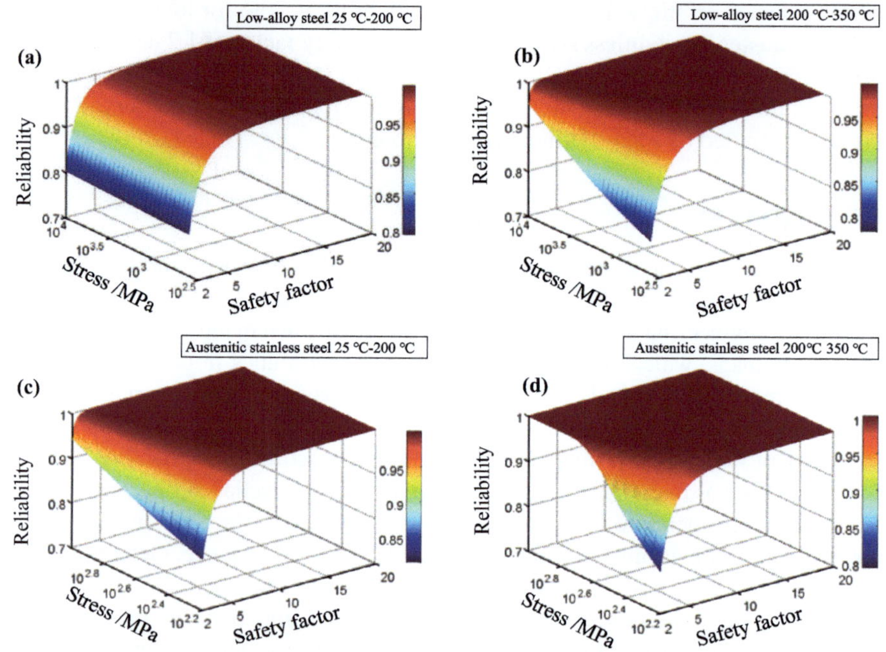

FIGURE 7.30 The fatigue reliability versus stress and safety factor (on life) based on strength–damage interference. (Reprinted with permission from Ref. [11], Elsevier)

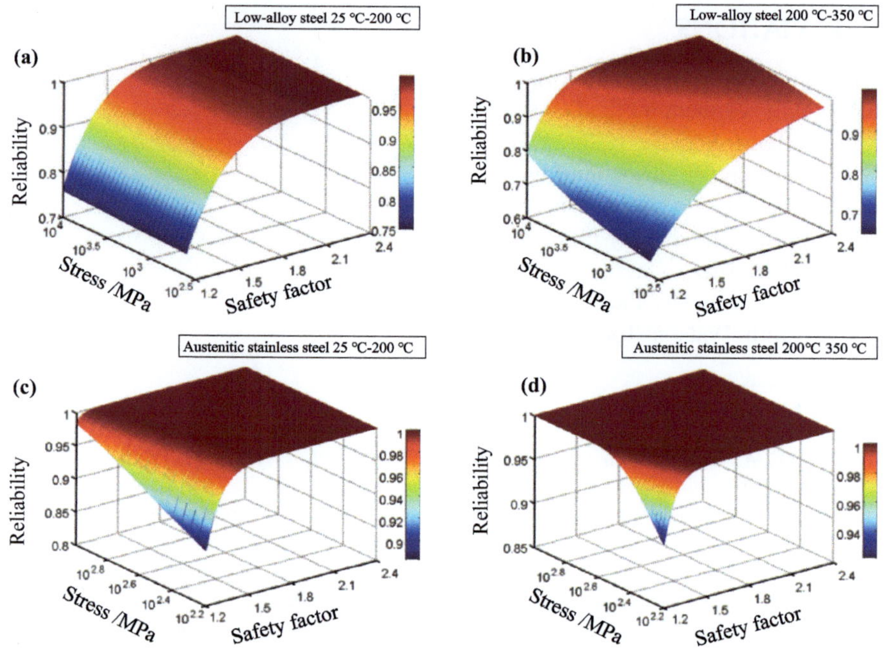

FIGURE 7.31 The fatigue reliability versus stress and safety factor (on stress) based on strength–damage. (Reprinted with permission from Ref. [11], Elsevier.)

the safety factor on stress. For instance, the reliability is 70% for low-alloy steel but is 90% for austenitic stainless steel with the same safety factor of 1.2 on stress by the comparison of Figures 7.31b and d. According to Figures 7.28–7.31, the reliability results obtained from strength–damage interference model is more conservative than the ones computed by load-life interference model.

7.6 SUMMARY

Material fatigue can cause mechanical failure even if the stress is lower than the static strength, which is a major safety concern for structures subjected to cyclic loads. Fatigue reliability in engineering denotes the ability of a structure achieving its aimed functions under cyclic loads, which could be influenced by uncertainties in design parameters, materials, and environmental conditions. The microstructure of a material, including grain structure and internal defects, influences fatigue performance, emphasizing the importance of statistical information for predictions and estimating variability in fatigue life. Factors such as magnitude, frequency, and duration of cyclic loading, environmental conditions (temperature, humidity, and corrosion), and surface treatment methods (shot peening, coatings) also significantly affect fatigue performance. Geometrical discontinuities, e.g. encompassing holes, grooves, fillets, and shoulders, in structures can lead to localized stress concentrations, accelerating the initiation and propagation of fatigue cracks. This chapter takes two cases as examples, including fatigue assessment of turbine-bladed disk and relationship quantification between fatigue reliability and safety factor, to analyze the fatigue reliability using probabilistic approaches.

ABBREVIATIONS

CDF cumulative distribution function
FEA finite element method
FS Fatemi–Socie
LHS Latin hypercube sampling
MCs Monte Carlo simulation
PDF probability density function
S–N stress–life
SIF stress intensity factor
SWT Smith–Watson–Topper

REFERENCES

1. ASME. *ASME BPVC Section III Division 1, Subsection NB*. ASME Boil Press Vessel Code; 2015.
2. Chopra OK, Shack WJ. Margins for ASME code fatigue design curve: effects of surface finish and material variability. *Press Vessel Pip Codes Stand* 2003; 453:71–85.
3. Zhu SP, Foletti S, Beretta S. Probabilistic framework for multiaxial LCF assessment under material variability. *Int J Fatigue* 2017;103:371–85.
4. Murakami Y. *Metal Fatigue: Effects of Small Defects and Nonmetallic Inclusions*. Academic Press; 2019.
5. Basquin OH. The exponential law of endurance tests. *Am Soc Test Mater Proc* 1910;10:625–30.

6. Manson SS. A complex subject-some simple approximations. *Exp Mech* 1965; 5:193–226.
7. Coffin LF. A study of the effects of cyclic thermal stresses on a ductile metal. *Trans Am Soc Mech Eng* 1954;76:931–50.
8. Miner MA. Cumulative damage in fatigue. *J Appl Mech Trans ASME* 1945; 12:A159–A164.
9. Matsuishi M, Endo T. Fatigue of metals subjected to varying stress. *Jpn Soc Mech Eng Fukuoka* Japan 1968;68:37–40.
10. Liao D, Zhu SP, Keshtegar B, Qian G, Wang Q. Probabilistic framework for fatigue life assessment of notched components under size effects. *Int J Mech Sci* 2020;181:105685.
11. Niu X, Zhu SP, He JC, Ai Y, Shi K, Zhang L. Fatigue reliability design and assessment of reactor pressure vessel structures: concepts and validation. *Int J Fatigue* 2021;153:106524.
12. Zhu SP, Liu Q, Peng W, Zhang XC. Computational-experimental approaches for fatigue reliability assessment of turbine bladed disks. *Int J Mech Sci* 2018;142–143:502–17.
13. Dowling NE. *Mechanical Behavior of Materials: Engineering Methods for Deformation, Fracture and Fatigue*, Third Edition. Pearson/Prentice Hall. Choice Rev Online 2007;46:46-6830–46-6830.
14. Niu X, Wang RZ, Liao D, Zhu SP, Zhang XC, Keshtegar B. Probabilistic modeling of uncertainties in fatigue reliability analysis of turbine bladed disks. *Int J Fatigue* 2021;142:105912.
15. Hu D, Su X, Liu X, Mao J, Shan X, Wang R. Bayesian-based probabilistic fatigue crack growth evaluation combined with machine-learning-assisted GPR. *Eng Fract Mech* 2020;229:106933.
16. Rennert R, Vormwald M, Esderts A. FKM Guideline. Analytical strength assessment of components in mechanical engineering. *Int J Fatigue* 2012;182:108165.
17. ISO EN 1993. Eurocode 3: design of steel structures. 2005.
18. Cooper WE. The initial scope and intent of the Section III fatigue design procedure. *Welding Research Council, Inc., Technical Information from Workshop on Cyclic Life and Environmental Effects in Nuclear Applications*, Clearwater, FL, January 20–21 1992.

8 Fatigue of Additively Manufactured Materials

Kun Yang and Qingyuan Wang

8.1 GENERAL CONCEPTS FOR FATIGUE OF AM METALS

8.1.1 AM Defect Type 1: Gas Pore

The additively manufacturing (AM) process of metal materials is a typical interdisciplinary field, which involves multiscale and multiphysical effects. This manufacturing process inevitably introduces two types of defects in the material [1]: the gas pore and the lack of fusion (LOF), as shown in Figure 8.1.

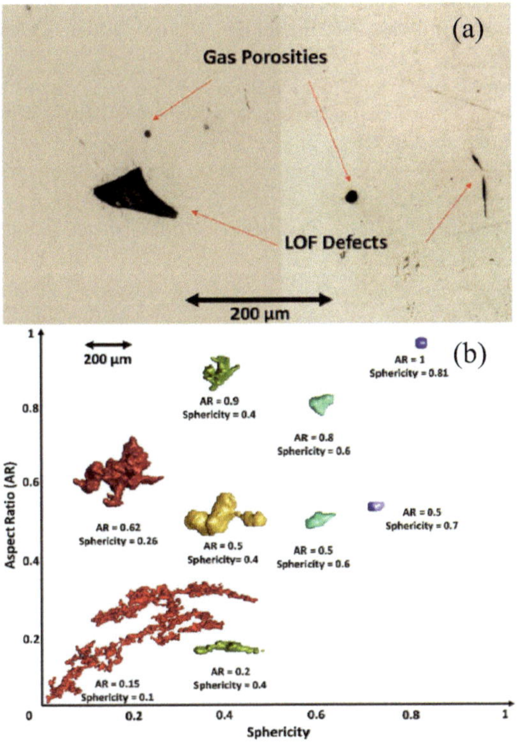

FIGURE 8.1 (a) 2D morphology and (b) 3D morphology of manufacturing defects in AM metals, mainly including the gas pore and the lack of fusion (LOF). (Reprinted with permission from Ref. [1], Elsevier.)

DOI: 10.1201/9781003405122-8

For the gas pore, it usually presents an ellipsoid or sphere shape with a higher aspect ratio (about 1), as shown in Figure 8.1. The formation of a gas pore with a nearly spherical shape can be due to the uniformly distributed pressure inside a pore [2]. It is speculated that the pressure is generated by the entrapped gas during the AM process. From the AM process point of view, the entrapped gas in a pore may result from two types of gas sources. On the one hand, due to their gas atomized fabrication process, some gas already exists in some initial powders. On the other hand, an inert atmosphere is employed during the AM process (such as laser powder bed fusion process), so that some gas (i.e., nitrogen or argon) fills the gap among the stacked powders. When a high-energy laser/electron beam strikes the top surface of a powder layer, the powder melts rapidly from top to bottom. Hence, the gas is trapped within a melt pool in the form of bubbles. Gas buoyancy and turbulence [3] push these bubbles due to the fluidity of molten metal. Then, most of the tiny bubbles bump into each other and subsequently merge into bigger bubbles. With cooling of a melt pool, molten metal fluidity rapidly decreases, and no escaped bubbles will be trapped in the form of gas pores. Excessive power of the beam often leads to a large number of gas pores.

It is known that the existence of gas pores can induce local stress concentration. Therefore, the gas pore is detrimental to the fatigue properties of AM materials. The results of finite element analysis [4] show that the degree of stress concentration induced by gas pores in the interior of the material is similar as presented in Figure 8.2. However, on the subsurface of the material, the stress concentration factor increases rapidly with the decrease of relative distance from free surface. Thus, gas pores on the surface or subsurface are more harmful to fatigue properties than those inside from a stress concentration point of view. The gas pore-nucleation fatigue mechanism will prevail if the gas pore is the main defect type of AM materials [2].

8.1.2 AM Defect Type 2: Lack of Fusion

For the AM defect type 2: Lack of fusion, it presents an irregularly shaped cavity as shown in Figure 8.1. Generally, an unstable processing parameter, such as the quality of the powder bed or stability of laser beam movements, results in local disturbances,

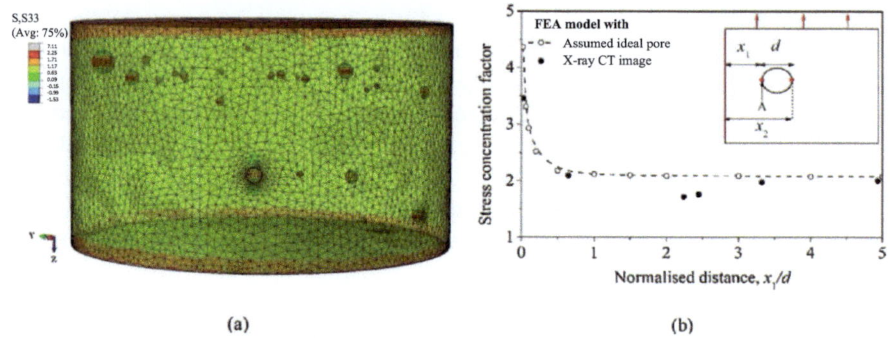

FIGURE 8.2 (a) Finite element model using X-ray computed tomography (CT) scanned geometry, (b) variation in stress concentration factor (Kt) at pore root point 'A' with distance "x1" from free surface, calculated using both idealized spherical pore geometry and X-ray CT scanned pore geometry. (Reprinted with permission from Ref. [4], Elsevier.)

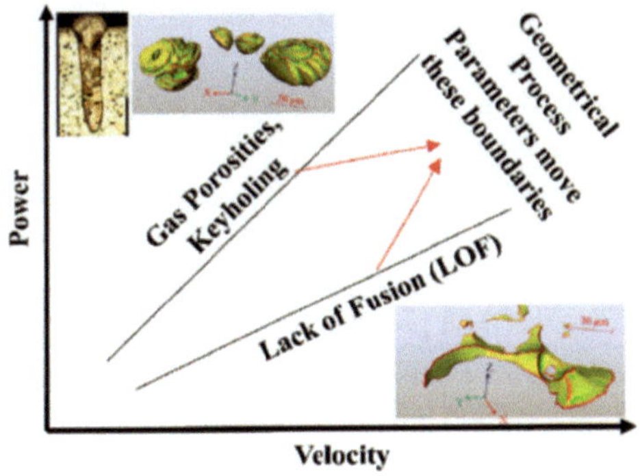

(a)

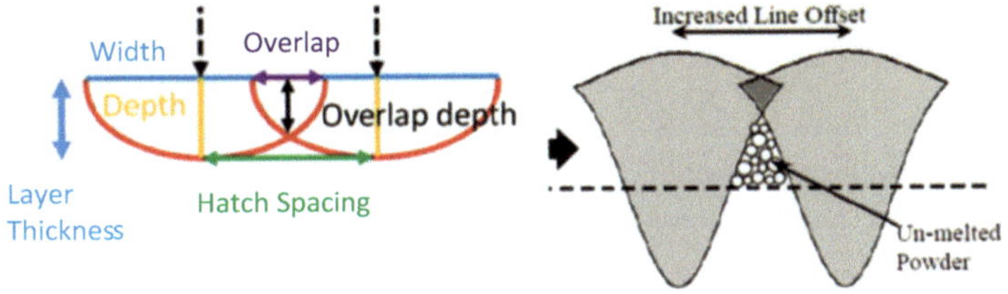

(b)

FIGURE 8.3 (a) The power and velocity of the beam can change the type of defects. (b) Improper layer thickness and hatch spacing value can also create the lack of fusion defect. (Reprinted with permission from Ref. [5], Elsevier.)

which may induce the LOF. Moreover, improper processing parameters can also cause a large amount of such defects. When the power of the beam is relatively low, an increase in the scan speed of the beam results in more LOF being generated [5], as exhibited in Figure 8.3a. Meanwhile, improper layer thickness and hatch spacing value can also create the LOF defect as shown in Figure 8.3b. Therefore, the appearance of this type of defect is rare and random, if an optimal processing parameter set has been used.

Problematically, this type of defect possesses sharp edges as shown in Figures 8.1 and 8.3a, which tend to induce severely local stress concentration at the tip of a sharp edge. Therefore, this type of defect is more critical for fatigue properties of the AM materials than the gas pore defect. In summary, manufacturing defects are unfavorable to the fatigue resistance of AM materials [2,6], so the defects are often suppressed by optimizing manufacturing parameters or applying additional post-treatment such as hot isostatic pressing, so as to achieve the purpose of improving the fatigue resistance of AM materials.

8.1.3 Matrix of AM Metals

The molten pool structure of the as-deposited AM alloy can be clearly observed in optical photographs with low magnification [2] as seen in Figure 8.4a and d. These vertical lines in Figure 88.4a result from the alternating parallel movement of a

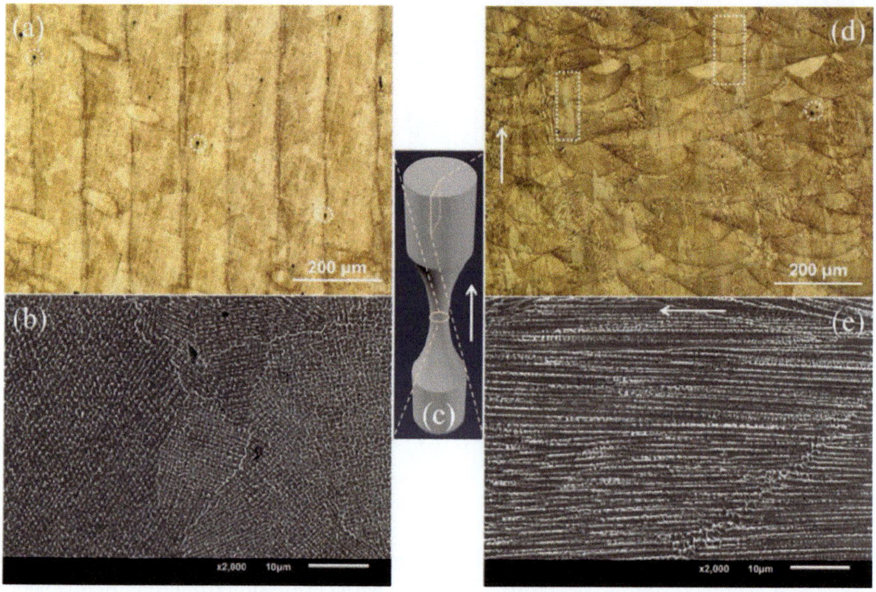

FIGURE 8.4 (a, d) Optical photographs with low magnification present the molten pool structure, (b, e) SEM photographs with high magnification reveal the dendritic crystal of an as-deposited AM alloy. (Reprinted with permission from Ref. [2], Elsevier.) Arrows indicate the building direction.

beam on the same layer. The distance between two adjacent lines is the same as the hatch spacing, which is one of the AM processing parameters. Two adjacent lines present the boundary of a single melt pool induced by the beam. Therefore, these lines are also called laser-melted tracks [7]. Many serried arcs were detected in the longitudinal section as shown in Figure 8.4d. The enveloped curve of each arc represents the boundary of a melt pool. Due to melt scans of every other layer, these melt pools present a stratified deposition feature. In general, the melt pool depth is more than the powder layer thickness, and the adjacent melt pools in the same layer partially overlap. These configurations help improve material quality and reduce manufacturing defects.

For the as-deposited nickel-based alloy, fine dendritic structures (a face-centered cubic phase), which appear gray under SEM, are detected as shown in Figure 8.4a and d. The size of the dendrite arm spacing of the AM material is significantly lower than the value of conventional cast materials due to the faster cooling rate during the AM process [2]. Most of the fine dendrites own a preferential growth direction, which is almost similar to the building direction. A large amount of granular precipitates, which appear white, are filled into the interdendritic regions. These precipitates mainly belong to Laves phases (a brittle, topologically close-packed phase) [3]. The extremely fast cooling rate of the AM process often generates a large number of intermediate phases and results in residual stress, hence, the post-heat treatment is often used to increase phase stability of AM materials and eliminate the residual stress.

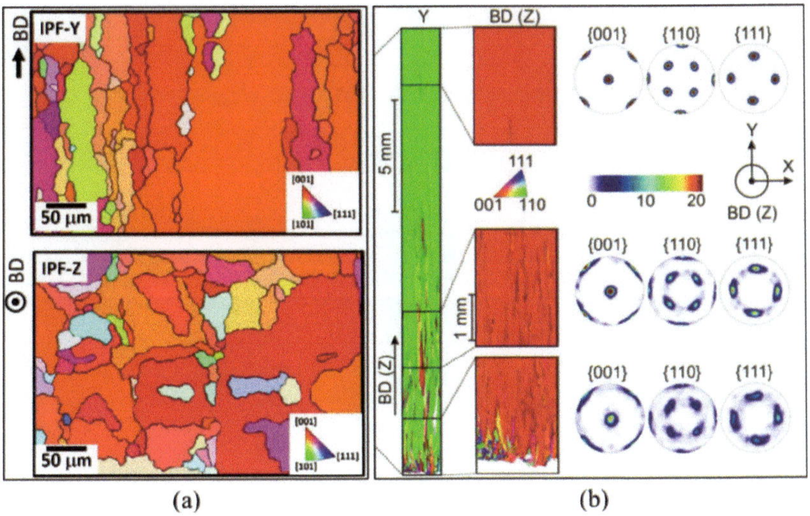

FIGURE 8.5 EBSD photographs reveal the inhomogeneous grain structure of AM alloys. (a) Typical AM polycrystalline structure. (Reprinted with permission from Ref. [8], Elsevier.); (b) AM single crystals emerge from an initial columnar grain structure. (With Permission from Springer.); BD: Building direction.

Columnar grain structures [8], as presented in Figure 8.5, prevail in the AM materials. The growth direction of the columnar grain is almost parallel to the building direction. The size of columnar grains depends on the AM parameters, and longer columnar grains generate under a higher power of beam [3]. The fatigue properties of the AM materials show obvious anisotropy due to columnar grain structures. High inhomogeneity of grain size and shape is commonly observed in the AM materials as shown in Figure 8.5a, and the inhomogeneity may result from specific localized solidification of the AM process. The inhomogeneity will reduce the deformation coordination of the AM materials. It makes tailoring the material microstructure possible due to the unique micro-zone melting layer-by-layer forming process of AM. The AM process can be used to produce polycrystalline structures and single crystals [9], as depicted in Figure 8.5. Therefore, the AM technology has great potential in the microstructural regulation of metals.

8.1.4 Multi-Source Competition Fatigue Failure

As summarized earlier, the AM metals mainly contain manufacturing defects (gas pores and LOF) and the matrix. Therefore, the limitations and optimization of manufacturing defects will directly affect the fatigue failure mechanism of AM metals.

If the AM manufacturing parameters are not optimal, manufacturing defects will prevail. Owing to the larger size and sharper shape of the LOF than the gas pore, the LOF will dominate fatigue failures of AM metals [6,10]. With the further optimization of manufacturing defects, the number and size of larger defects are controlled. Both manufacturing defects (e.g., gas pore and LOF) and matrix (crystallographic

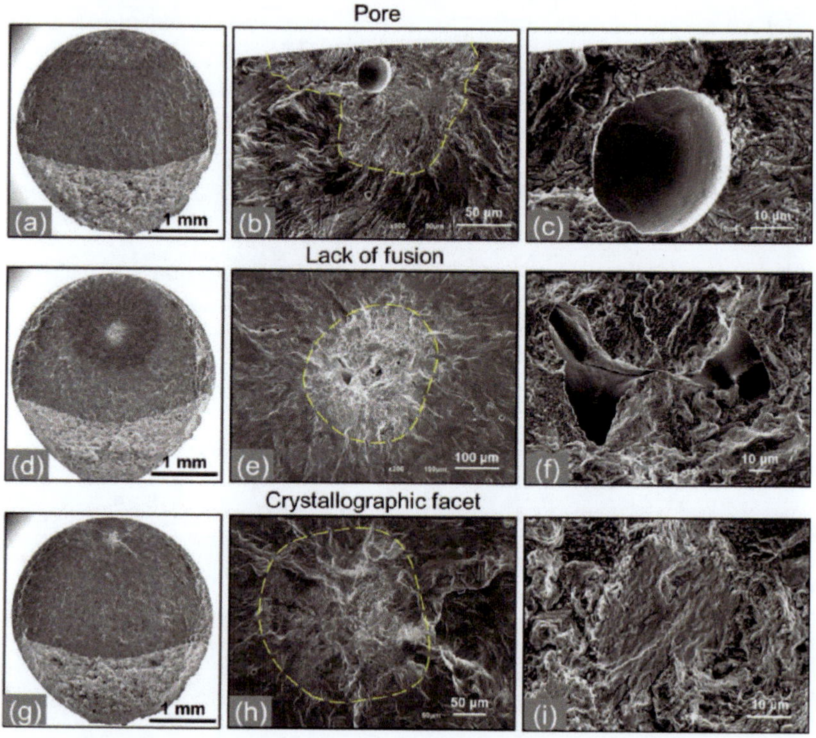

FIGURE 8.6 Three typical fatigue crack initiation modes in AM metals, namely (a–c) pore-assisted crack initiation, (d–f) crack initiation from lack of fusion, and (g–i) crack initiation induced by crystallographic facet (matrix). (Reprinted with permission from Ref. [2], Elsevier.)

facet) can act as the fatigue crack sources, as presented in Figure 8.6 [2]. At present, multi-source competition fatigue failure, namely (1) pore-assisted crack initiation, (2) crack initiation from LOF, and (3) crack initiation induced by crystallographic facet, will appear. In general, the preferred orientation of grains is responsible for fatigue crack initiation induced by crystallographic facet. With the application of some auxiliary control methods for the AM metals, the manufacturing defects will be greatly inhibited, and the fatigue cracks will mainly initiate from the gas pore or matrix. For some special metal materials, such as high-strength steels, aluminum alloys, and nickel-based alloys, they often have metallurgical defects (non-metallic inclusions). The metallurgical defect can also assist or directly induce the fatigue crack initiation of AM metals [11]. In summary, the introduction of manufacturing defects will increase the complexity of fatigue failure modes for AM metals.

8.1.5 Fatigue Performance Characteristics

The shape of the S-N curve presents single line characteristic [12] as depicted in Figure 8.7a, while a duplex S-N curve can also be detected in AM metals [2] as shown in Figure 8.7b. The competition between surface crack initiation and interior crack

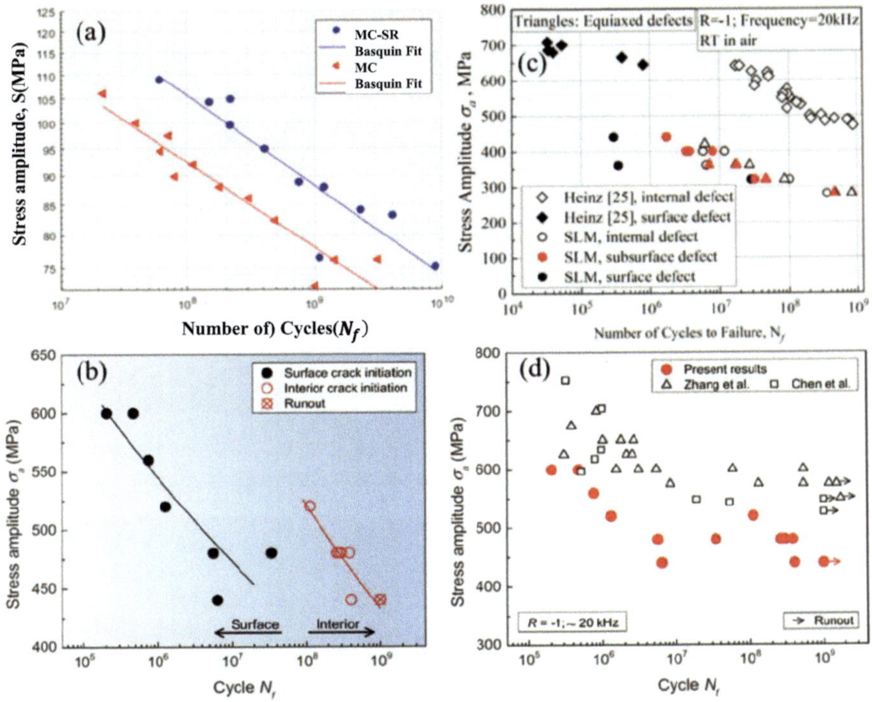

FIGURE 8.7 Fatigue performance characteristics of AM metals. (a) The S-N curve with single line shape for two kinds of AM metals, (b) the S-N curve with duplex shape. (Reprinted with permission from Ref. [2], Elsevier.), (c) the lower fatigue property of the AM metal than the conventionally processed titanium alloy. (Reprinted with permission from Ref. [6], Elsevier.), (d) the lower fatigue property of the AM metal than the conventionally processed nickel-based superalloy. (Reprinted with permission from Ref. [2], Elsevier.)

initiation directly results in the appearance of the duplex S-N curve. Moreover, the S-N curves present a continuous downward trend. Hence, the conventional fatigue limit does not exist in AM metals. In other words, the fatigue failure still occurs even if fatigue life exceeds 1E7 cycles for AM metals.

Through additional heat treatment after the AM process, the static strength of AM materials can be similar to that of traditional high-performance forging metals [13]. However, the fatigue performance of the material is more dependent on the local heterogeneity level of the microstructure. As discussed before, manufacturing defects are often inevitably introduced into AM materials, and conventional AM materials are mainly composed of non-uniform coarse columnar grains. These factors will cause the fatigue strength of AM materials to be obviously lower [2,6] as shown in Figure 8.7c and d, and the fatigue life scatter is greater than traditional high-performance metals. These weaknesses of AM materials have been the main bottlenecks in the application on key structural parts subjected to cyclic loading. In order to improve fatigue resistance and reduce fatigue life dispersion, both the microstructure and manufacturing defects must be optimized for AM metals.

At present, improving the fatigue property becomes an urgent need of AM metals, and the effective methods are powder alloying [14], optimization of AM parameters [15], post-heat treatment [12], hot isostatic pressing [16], etc.

8.2 AM PROCESS PARAMETERS

Although there are many technologies to realize additive manufacturing process, their processing principle is similar. AM processing is achieved by melting metal powders layer by layer. The AM process presents the multiscale and multiphysical features as shown in Figure 8.8 [17]. Hence, it is very difficult to correlate process parameters with fatigue behaviors of AM metals. The main factors affecting the performance of finally processed products include the heat source of molten metal powder, the characteristics of metal powder, and other process parameters. The fatigue performances of AM metals are often regulated by constantly adjusting these process parameters to change their microstructure and defects.

The effects of process parameters during additive manufacturing on the microstructure and manufacturing defect of materials are discussed to summarize the relationship between process parameters and fatigue behaviors in this section. The important parameters during the AM process are heat source type, heat source power (energy density), scanning speed, scanning spacing, powder layer thickness, scanning path, etc.

8.2.1 Heat Source Type

The use of heat source to melt metal powders is the key for additive manufacturing. For the AM process, the heat source is usually based on laser, electron beam, arc, etc. The most frequently heat sources used for additive manufacturing processes are the laser and electron beam. The working principles of electron beam and laser are different, so their effects on AM materials are also different.

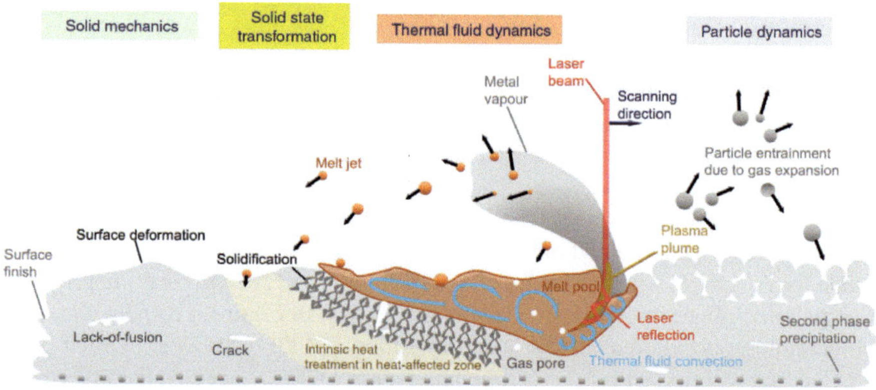

FIGURE 8.8 Schematic diagram depicting the multiscale and multiphysical processes in additive manufacturing. (With Permission from Springer.)

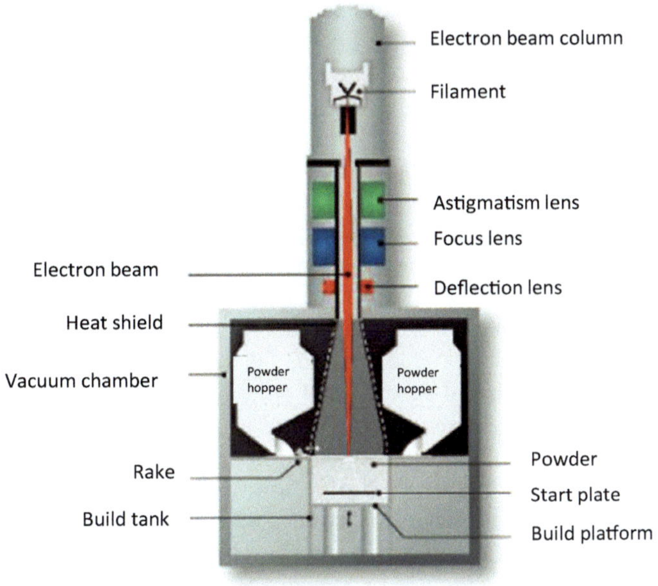

FIGURE 8.9 Electron beam melting processing schematic. (Reprinted with permission from Ref. [18], Elsevier.)

Electron beam melting (EBM) deposition technology is an additive manufacturing technology using electron beam as heat source. Electron beam is electron bunching, which has high energy density. Electron beam can convert the kinetic energy of electrons into thermal energy when it passes through metal materials. The generated thermal energy will be employed to melt the metal powders. The detailed schematic of the EBM process is shown in Figure 8.9 [18]. The electron beam heats the selected powders within the powder layer to melt it and subsequently moves according to the predetermined route. After one layer is completed, the next powder layer will be scanned and melted until the whole part is deposited.

Metal can be melted by the surface of metal material absorbing heat from the laser. Different materials have different absorption efficiency of laser thermal energy due to the difference in the surface state and size. There are mainly two AM technologies using laser as the heat source. The selective laser melting (SLM) forming technology is based on the powder bed and powder laying technology, which is similar to EBM technology. The selected metal powder is rapidly melted by the laser beam. It is suitable for AM parts with a complex shape owing to its very high forming precision. The schematic diagram of SLM technology is shown in Figure 8.10 [19]. The other AM technology is called as direct laser deposition (DLD) technology, which is like laser cladding technology for the component repair. DLD is based on the principle of surfacing. Raw powder materials are directly sent to the molten pool by powder jets instead of the powder bed. DLD possesses high forming efficiency and low forming

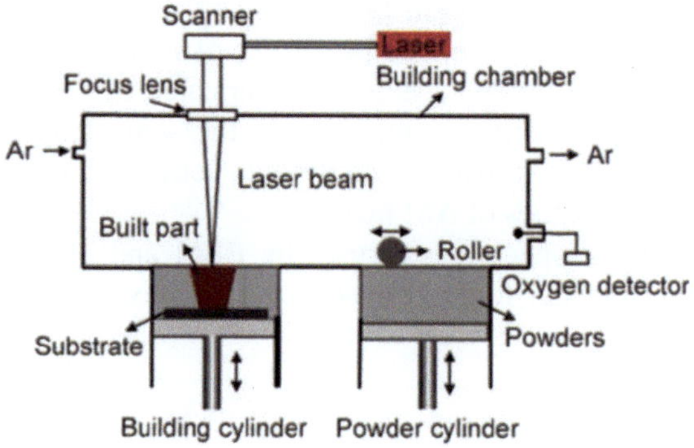

FIGURE 8.10 Schematic diagram of selective laser melting (SLM) forming technology. (Reprinted with permission from Ref. [19], Elsevier.)

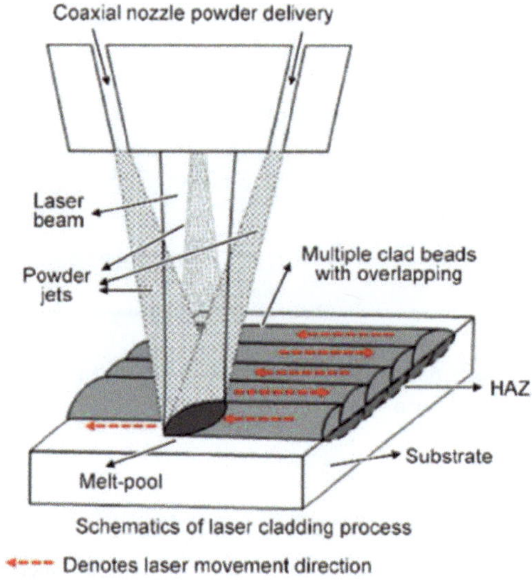

FIGURE 8.11 Schematic diagram of direct laser deposition (DLD) technology. (Reprinted with permission from Ref. [20], Elsevier.)

precision; hence, it is suitable for large structures, but not for complex parts. The schematic diagram of DLD is shown in Figure 8.11 [20].

EBM and SLM have similar forming process except for the heat source type (electron beam vs. laser) as mentioned before. To discuss the influence of heat source type on the fatigue properties of AM materials, EBM and SLM were selected to fabricate the fatigue specimens of Ti-6Al-4V alloy. Meanwhile, fatigue specimens in two directions (horizontal direction, namely *XY* axis specimen and vertical direction, namely Z axis specimen as shown in Figure 8.12 [21]) were prepared to study the effects of building direction on fatigue properties of AM materials. It can be seen from Figure 8.13 that the fatigue specimens, fabricated by EBM and SLM in the corresponding building direction, have roughly the same fatigue properties [21]. Therefore, these two different heat source types (electron beam vs. laser) have little effect on the fatigue properties of AM materials.

Moreover, the fatigue specimens formed by EBM and SLM in the horizontal direction have slightly better fatigue properties than those in the vertical direction

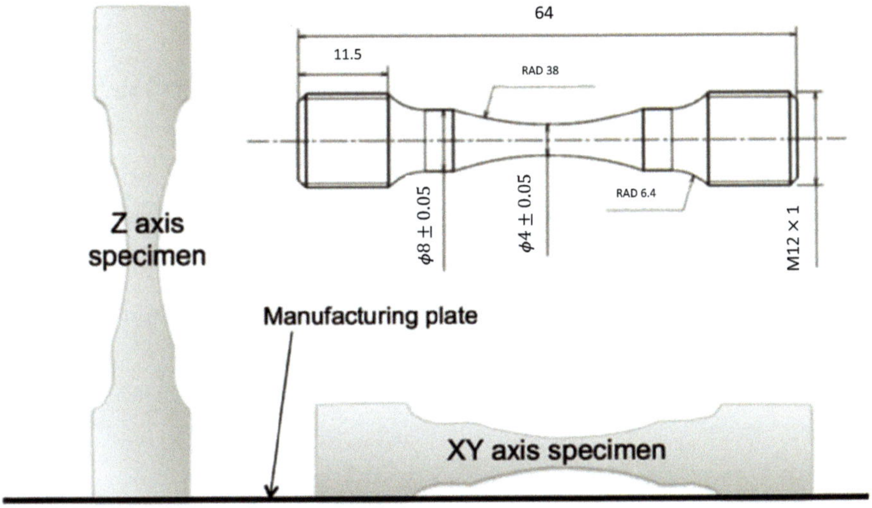

FIGURE 8.12 Fatigue specimens in two directions, the building direction is parallel to the Z axis. (Reprinted with permission from Ref. [21], Elsevier.)

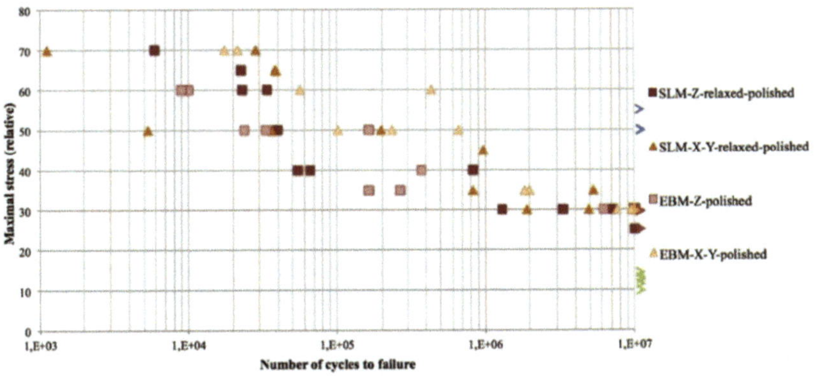

FIGURE 8.13 Comparison of fatigue properties between SLM and EBM processes for both vertical building direction (Z) and horizontal one (X-Y). (Reprinted with permission from Ref. [21], Elsevier.)

as depicted in Figure 8.13. This indicates that the fatigue performance of the AM material depends on the building direction, that is, the AM material has anisotropy. Microstructure inhomogeneity and 3D features of manufacturing defects of AM metals account for the fatigue performance anisotropy.

8.2.2 Powder

Raw materials are one of the key factors that affect the performance of AM parts. Metal powder is the main material used for additive manufacturing. Metal powder needs to meet the requirements of small particle diameter, low oxygen content, and good fluidity. The common metal powders used for the AM process are spherical or near spherical particles, and the size distribution of these powder particles ranges from about 15–53 μm, as shown in Figure 8.14 [19]. The diameter of the powder particles and the level of oxygen content have the important influence on the AM parts. The particle size of the powder affects the melting degree of the powder per unit time. When the particle size of the powder is small, the metal powder can melt more thoroughly at a lower heat source power, making the plastic fluidity of the molten pool more sufficient. In the additive manufacturing technology, the particle size of the powder required for SLM is about 20–45 μm. The powder diameter required for DLD can be larger, about 90–150 μm.

The powder with higher oxygen content is expected to form a material with more dispersed second phase, which may be beneficial to the improvement of material strength. However, in general, the oxygen content of metal powder should be reduced as much as possible during the preparation process because it is one of the main reasons that affects the porosity of the AM parts. Moreover, high oxygen content of metal powder can make the additive manufacturing process of active metals extremely dangerous. Therefore, according to the different properties of the material, inert gases such as nitrogen, argon, and helium are introduced into the process of preparing metal powder by the gas atomization method to reduce their oxygen content.

Ti-6Al-4V specimens were prepared using two powder materials with different particle sizes to study the effects of powder size on the fatigue properties of AM

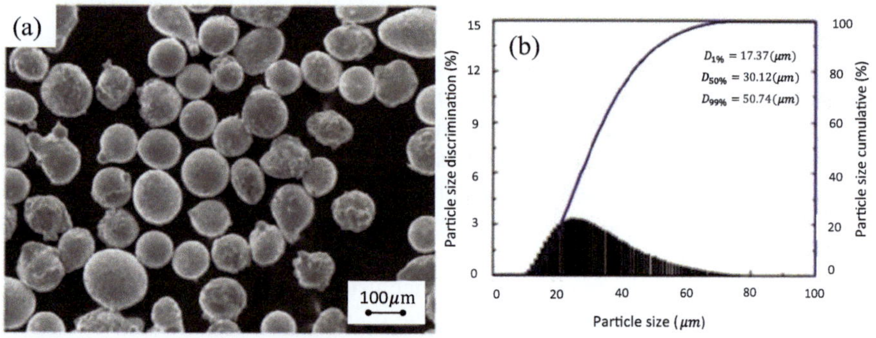

FIGURE 8.14 (a) Morphology of the metal powder particles; (b) particle size distribution. (Reprinted with permission from Ref. [19], Elsevier.)

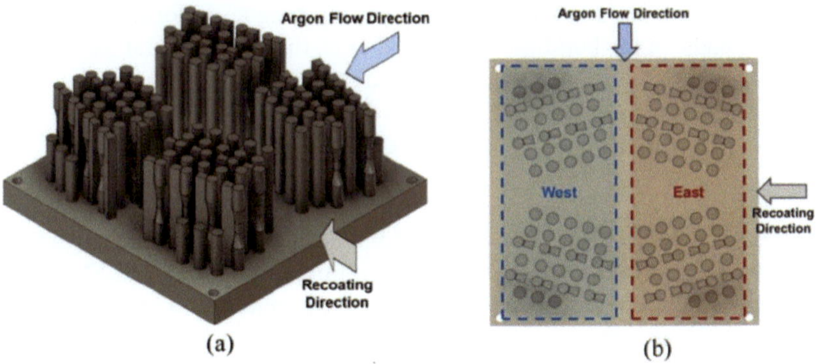

FIGURE 8.15 Fabrication layout of specimens, the East, namely closer to the feedstock bin, and the West, namely farther away from the feedstock bin. (Reprinted with permission from Ref. [22], Elsevier.)

materials [22]. The used particle sizes are 15–53 and 15–45 µm, and the two batches of powders are called as 'coarse' and 'fine', respectively. AM for these two batches of powders is performed using the same specimen layout as shown in Figure 8.15 and manufacturing parameters. To investigate the effects of powder location relative to feedstock bin on the fatigue properties, the fatigue specimens are specifically placed in two locations as shown in Figure 8.15.

The microstructures of AM materials [22] fabricated by course (Figure 8.16a) and fine (Figure 8.16b) powders under the non-heat treatment (NHT) condition, and course (Figure 8.16c) powders under the annealing condition indicate that the prior-β columnar grains are mainly elongated along the building direction depending on the path of heat dissipation. Moreover, the thickness of the lath martensite in the prior-β grains under NHT fine, NHT coarse, and annealing coarse conditions was measured according to the inverse pole figure. As shown in Figure 8.16d, the lath thickness of the fine powder batch is slightly thicker under NHT conditions. The average thickness of lath martensite increases slightly from NHT to annealing condition. Owing to the inferior packing state and lower heat absorptivity of coarser powders, more defects are detected in the coarse powder materials. However, the maximum size of fabrication defect is larger in the fine powder materials due to the powder spattering during the AM.

Under the surface machining conditions (eliminating the effect of surface roughness) as shown in Figure 8.17 [22], coarse powder materials have higher fatigue resistance than fine powder materials under medium cycle fatigue conditions, which is attributed to the higher ductility of coarse powder materials due to their smaller defects. For low-cycle fatigue and medium cycle fatigue with higher stress levels, crack propagation accounts for a more significant part of the overall fatigue life. Therefore, the higher ductility of coarse powder samples means higher fracture toughness and larger fracture crack size, which usually increases the crack propagation life. At a lower stress level, although the data overlap is more significant, the fatigue performance of coarse powder materials is still better than that of fine powder materials. This is because although more defects are detected in the coarse powder samples, the maximum defects in the fine powder samples are larger than those in

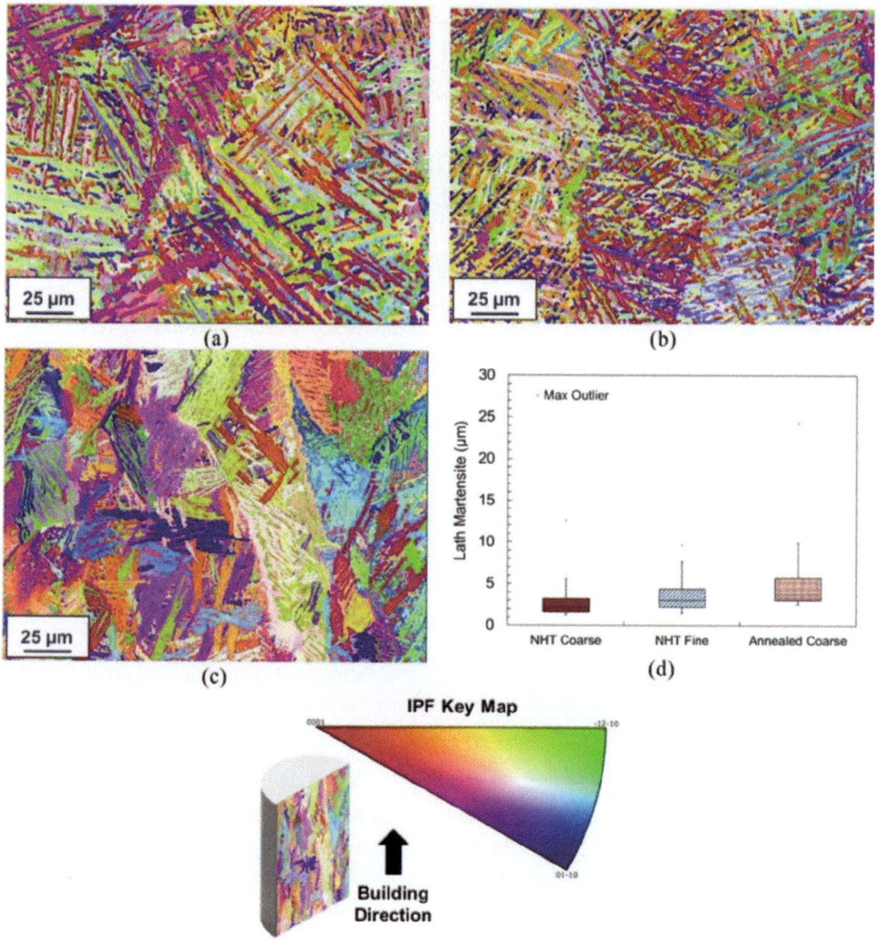

FIGURE 8.16 Microstructure features of the SLM Ti-6Al-4V metal for the (a) NHT coarse, (b) NHT fine, and (c) annealed coarse powder specimens, (d) thickness of lath martensite in the prior-β grains for all the conditions. (Reprinted with permission from Ref. [22], Elsevier.) NHT: non-heat treatment.

the coarse powder samples. Therefore, to obtain better fatigue performance by controlling the powder characteristics, the key is to control the maximum defects of AM materials [22]. According to Figure 8.17, no apparent trend is observed between the fatigue performance of East and West specimens. That is to say, there seems no location (relative to feedstock bin) dependency.

8.2.3 Substrate Heating

Substrate heating or substrate pre-heating is an important parameter in the AM process. Substrate heating refers to the heating of the manufacturing substrate to a certain temperature during the AM process to make the material easier to

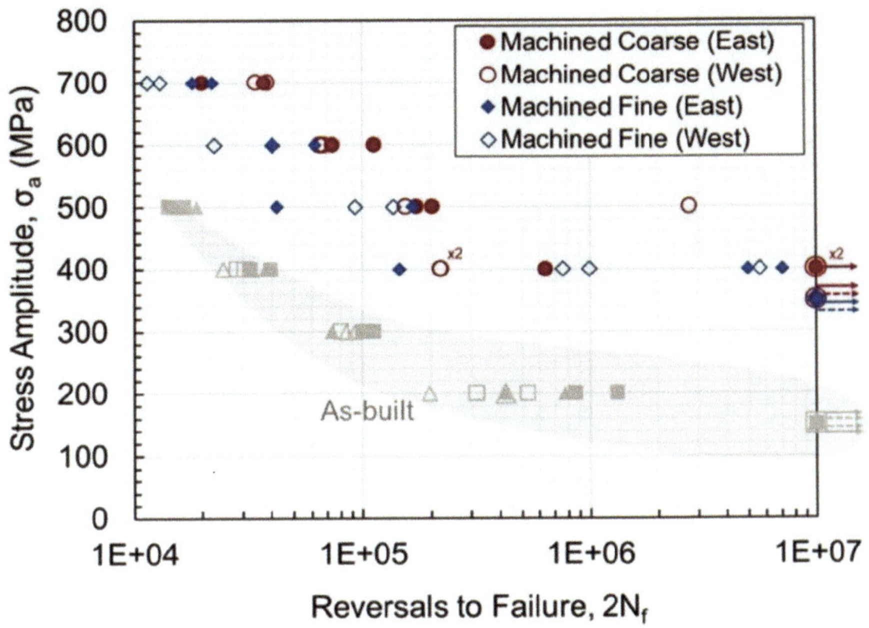

FIGURE 8.17 Fatigue property of SLM specimens fabricated from fine and coarse powder batches from the East and West locations in their machined surface and annealed conditions. (Reprinted with permission from Ref. [22], Elsevier.)

flow and closely integrate with the substrate to help the material adhere and fuse. Substrate heating is usually realized by heating elements (such as electric heating wire or heating plate) and temperature controllers. The temperature controller can monitor the temperature of the substrate surface and control the current or voltage of the heating element according to the preset parameters further to control the substrate temperature.

When the AM powders are melted, the heated substrate can improve the temperature and depth of the molten pool, thus promoting the powder melting of the AM material. On the other hand, the substrate heating can also change the temperature gradient and solidification rate of the molten pool, affect the grain size and grain boundary morphology, and optimize the defect density of the AM material. Overall, substrate heating can accelerate grain growth and grain boundary diffusion, improve grain size and crystallinity, control the crystal orientation of the AM material, reduce manufacturing defects, and finally affect the mechanical performance of the AM material.

In the process of metal AM, due to the thermal cycle, the material will undergo multiple thermal expansion and contraction, resulting in defects such as pores and thermal microcracks within the material [23]. Hence, reducing the thermal gradient should be helpful to control these defects. The samples were prepared by SLM under substrate heating at room temperature, 100°C, and 200°C using gas atomized Ti-15Mo powder [23]. Figure 8.18 shows the results of relative density

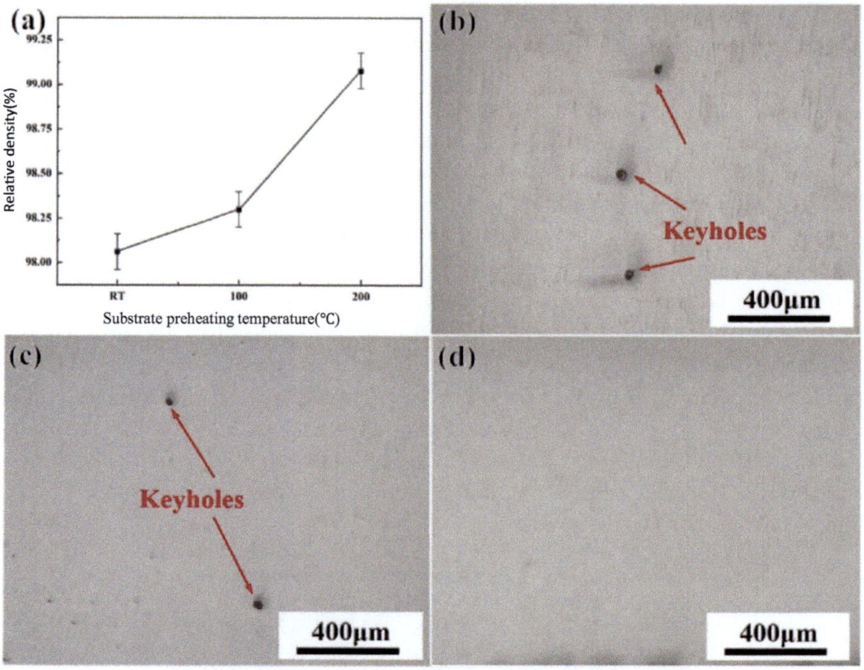

FIGURE 8.18 (a) There is clearly a positive relationship between substrate pre-heating temperature and relative density of AM metals, the corresponding defect characteristics of AM samples under different substrate pre-heating temperatures: (b) room temperature, (c) 100°C, (d) 200°C. (Reprinted with permission from Ref. [23], Elsevier.)

and defect characteristics of Ti-15Mo samples at different substrate temperatures [23]. As shown in Figure 8.18a, the relative density of the sample increases significantly with the increase of the substrate heating temperature. The main defect is gas pores as presented in Figure 8.18. These gas pores can also be defined as keyholes, which are easily formed in SLM-fabricated samples, owing to the recoil pressure toward the metal melt at the bottom. The temperature gradient is significantly reduced by substrate heating during the AM process, resulting in the reduction of the Marangoni effect. The reduced Marangoni effects lead to the decrease of interaction between the surface tension and recoil pressure, so that the pressure cannot push the liquid up, thereby failing to generate a keyhole defect at the molten pool bottom.

Moreover, substrate heating is very important for the control of residual stress. As the pre-heating temperature of the substrate increases from 100°C to 370°C, the residual stress decreases by 71% [24]. Increasing the pre-heat temperature can decrease the temperature gradient between the molten pool and the surrounding metal material, resulting in a more coordinated thermal deformation and finally leading to a reduction in residual stress. Whether the residual stress is harmful to the fatigue performance depends on (1) the orientation relationship between the residual stress and the loading direction of the cyclic stress, and (2) the state of residual stress, namely tensile or compressive stress. The residual compressive stress along the loading direction of the cyclic stress will be beneficial to improve the fatigue strength of the AM material.

Different batches of AlSi10Mg samples were prepared by SLM at 30°C (room temperature) and 300°C substrate pre-heating temperature along different building directions (0°, 45°, 90°) as shown in Figure 8.19 [25]. The stress-life data of fatigue specimens for each batch is shown in Figure 8.20 [25].

From the S-N curves along the 45° and 90° directions, the fatigue resistance at the substrate pre-heating temperature of 300°C is significantly higher than that at 30°C. This is because the laser beam can heat the material to a better temperature, the cooling rate and the possibility of distortion caused by the residual stress is reduced, so can reduce the appearance of defects (such as the LOF defect and thermal micro-crack) at the substrate pre-heating temperature of 300°C. A standard heat treatment (T6) was performed on these fatigue specimens, hence the effect of residual stress on fatigue resistance of the AM aluminum alloy will weaken or even disappear. Finally, less defects resulting from the substrate pre-heating should contribute to the higher fatigue resistance.

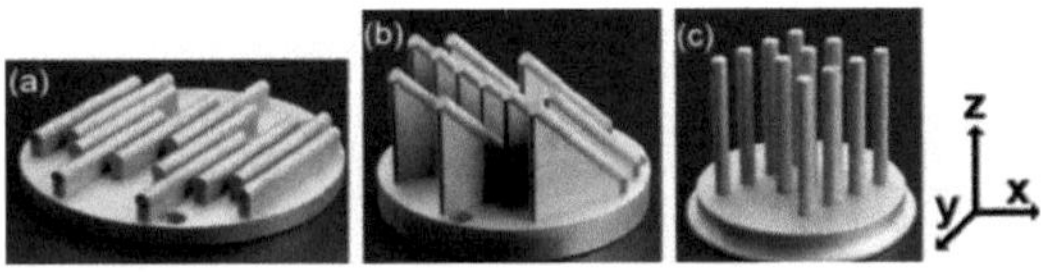

FIGURE 8.19 Samples built in different directions: (a) 0°, (b) 45°, and (c) 90°, the building directions is along the Z axis. (Reprinted with permission from Ref. [25], Elsevier.)

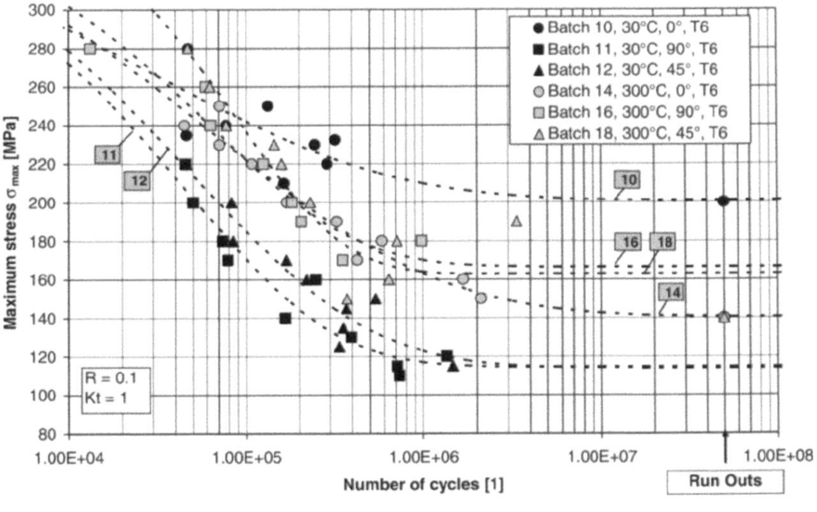

FIGURE 8.20 S-N curves of AM specimens with different substrate pre-heating temperatures and building directions, T6 refers to a standard heat treatment for aluminum alloys. (Reprinted with permission from Ref. [25], Elsevier.)

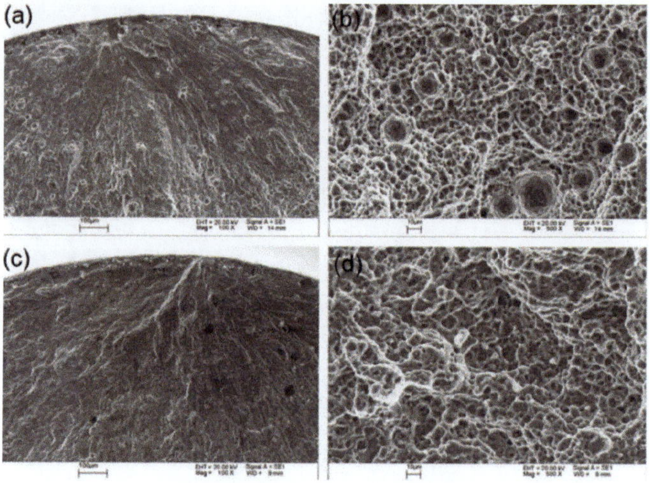

FIGURE 8.21 Fracture surfaces of two samples from batch 10 (30°C/0°/T6). (Reprinted with permission from Ref. [25], Elsevier.)

However, for the specimen constructed along 0° direction, an opposite result is obtained unexpectedly, the fatigue performance of the specimen at the substrate pre-heating temperature of 30°C is higher than that at 300°C. A positive effect of fewer defects becomes especially obvious in the 45° and 90° direction, while the effect is hard to be seen in the 0° direction. In general, defects are aligned parallel to the loading direction (smallest notch effect, Figure 8.21 [25]) in the 0° direction. AM parts generally show the highest fatigue strength in the 0° direction and present strong dependence on the building direction at 30°C substrate as shown in Figure 8.20 for this reason. The high substrate heating temperature may break the unique alignment of defects, further weaken or even disappear the smallest notch effect as shown in Figure 8.22 [25]. As a result, the fatigue property presents no dependence on the building direction at 300°C substrate as shown in Figure 8.20. But in general, the specimen pre-heated at high-temperature substrate has better fatigue performance than the specimen without pre-heating.

Substrate heating is one of the key factors affecting the performance of AM metals. Properly increasing the substrate temperature can effectively avoid defects such as pores and microcracks and reduce the residual stress [23]. Therefore, the heating for the bottom substrate can be employed to improve the fatigue properties of the AM material.

8.2.4 Scanning Strategy

Different scanning strategies during additive manufacturing result in different inter-layer temperature fields and heat dissipation directions of the molten pool, and then produce different microstructures. Common scanning strategies are shown in Figure 8.23, including (1) uni-directional scanning; (2) bi-directional/zigzag scanning; (3) island/chessboard scanning; (4) variation of scanning sequences based on

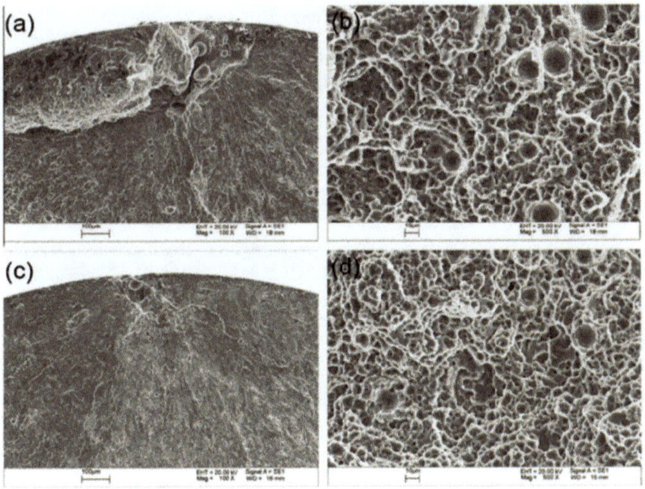

FIGURE 8.22 Fracture surfaces of two samples from batch 14 (300°C/0°/T6). (Reprinted with permission from Ref. [25], Elsevier.)

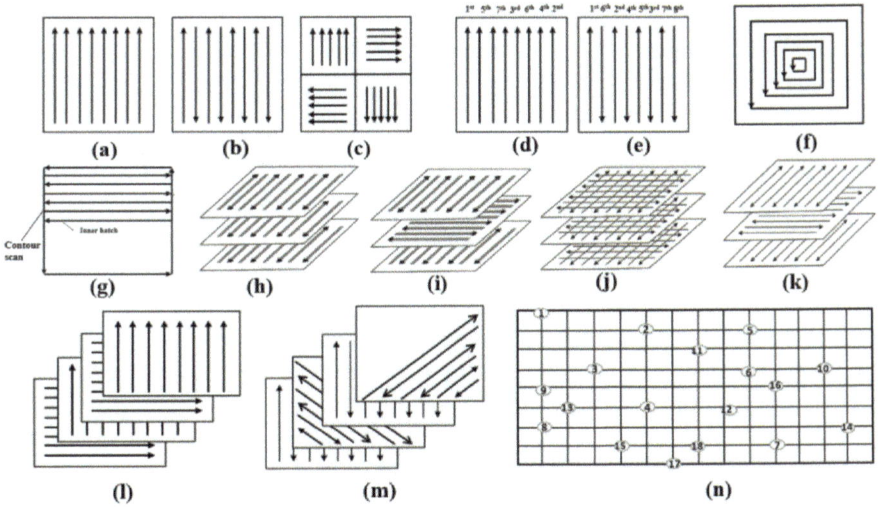

FIGURE 8.23 A summary of scanning strategies. (Reprinted with permission from Ref. [26], Elsevier.)

uni-directional scanning; (5) variation of scanning sequences based on bi-directional scanning; (6) helix scanning; (7) contour scanning; (8) bi-directional with double pass; (9) bi-directional with double pass, 90° rotation scanning between layers; (10) cross scanning; (11) bi-directional scanning, 90° rotation scanning between layers; (12) uni-directional scanning, 90° rotation between layers; (13) bi-directional scanning, 45° rotation between layers; (14) point scanning, etc. [26].

The chessboard scanning strategy is an alternative method to manufacture the large size parts due to its very short scan vector length [26]. It divides the scanning area into many small square areas to form a chessboard shape. The chessboard scanning strategy can make the received energy per unit area more uniform, and the more the number of chessboard unit, the more balanced the heat in the unit, to obtain a more uniform microstructure. The strategies of chessboard partition and rotating scanning between layers can also reduce the horizontal and vertical cyclic heat input to reduce the residual stress and plastic deformation. It can also effectively reduce the anisotropy between layers. The scanning strategy in additive manufacturing can effectively change the grain structure to form uniform and fine equiaxed grains, thereby optimizing the mechanical properties of the AM material.

Six types of samples were constructed by different scanning strategies as shown in Figure 8.24 [27]. As a result, the defect type, size, and distribution obviously depend on the scanning strategy, six kinds of intentionally designed defects were formed by these different scanning strategies. Fatigue experiment results [27] in Figure 8.25

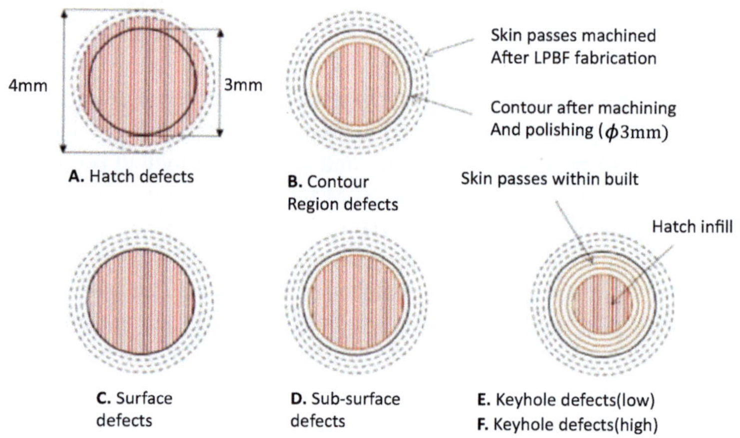

FIGURE 8.24 Six intentionally designed defects achieved by different scanning strategies.

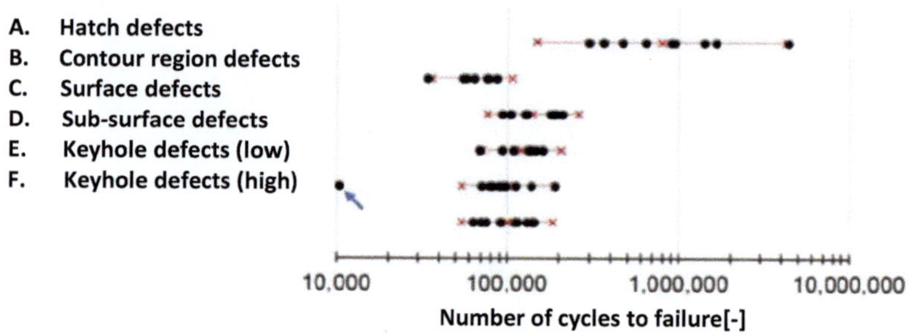

FIGURE 8.25 Fatigue life of each batch manufactured by six scanning strategies. (Reprinted with permission from Ref. [27], Elsevier.)

indicate that there are significant differences in the fatigue life, hatch defects present superior fatigue life, keyhole defects cause moderate life, and the LOF defect is detrimental.

There are many kinds of scanning strategies, which have great influence on the microstructure and defects of AM materials. The influence of the scanning strategy on the fatigue performance of the AM material should be considered from the perspective of both the defect and microstructure evolutions. Overall, it is an important path to optimize the fatigue properties of AM materials through scanning strategies.

8.2.5 Energy Density of Heat Source

The energy density of the heat source determines the amount of energy received by the metal powder per unit time. Therefore, the melting degree of metal powder is determined, which affects the forming quality of the AM material. When the heat source power is low, the metal powder absorbs less heat and melts less powders per unit time. Meanwhile, the temperature in the molten pool becomes lower, resulting in insufficient flow in the molten pool. Un-melted powder will appear near the molten pool to form spherical particles, which will lead to insufficient inter-layer microstructure connectivity and generate manufacturing defect. The relatively higher energy density will make the AM material form lower porosity as shown in Figure 8.26 [28]. The increase of energy density effectively increases the temperature and size of the molten pool, which improves its flowability and reduces un-melted powder.

The 316L alloys were prepared by the SLM technology under the different laser power [29]. Fifty percent (0.5P0), 70% (0.7P0), 100% (P0), 130% (1.3P0), and 150% (1.5P0) from the optimized laser power (P0) were selected. Figure 8.27 [29] is an optical microscopic image for the five batches of materials. The batch at 50% contains a large amount of irregular cavity, which is a void defect caused by LOF due to insufficient energy input. The void defects in other samples are mainly small and

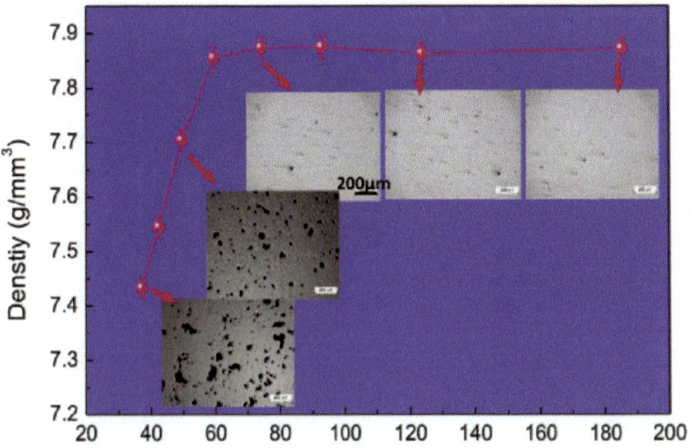

FIGURE 8.26 Effect of energy density on the density of the AM metal. (Reprinted with permission from Ref. [28], Elsevier.)

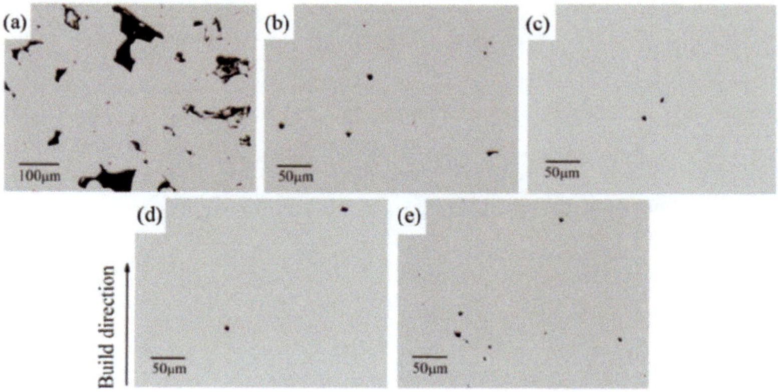

FIGURE 8.27 Optical micrographs (a–e) of AM metals with 50%, 70%, 100%, 130%, and 150% from optimized laser power, respectively. (Reprinted with permission from Ref. [29], Elsevier.)

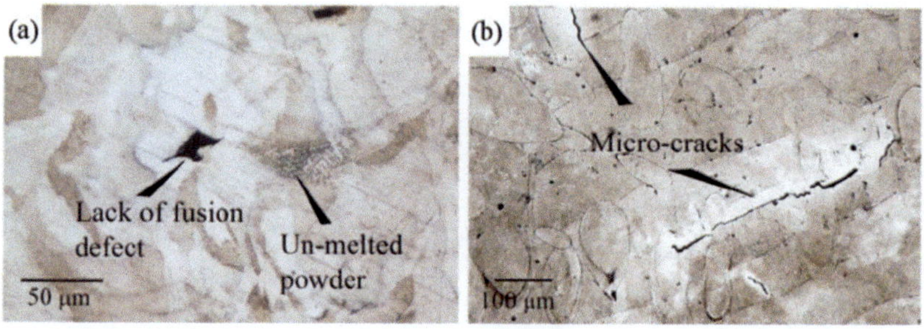

FIGURE 8.28 (a) Lack of fusion and un-melted powder in the batch at 70% and (b) cracks in the batch at 150%. (Reprinted with permission from Ref. [29], Elsevier.)

spherical pores. These two batches at 100% and 130% have the relatively better suppression for void defects than these two batches at 70% and 150%. Excessively high energy density will result in the increase of porosity again, which may be owing to the formation of Keyhole in the molten pool.

On closer inspection, the batch at 70% also shows the LOF defect and un-melted powder [29] (Figure 8.28a). The batch at 150% has randomly distributed long cracks (Figure 8.28b). These cracks are likely to be thermal cracks caused by excessively high laser power.

Because the fatigue performance of stainless steel 316L is not sensitive to residual stress, the influence of residual stress on fatigue performance can be eliminated [30]. Only the batch at 50% presents a significant reduction in fatigue strength as shown in Figure 8.29 [29]. This result is consistent with the observed results of voids in Figure 8.27. Figure 8.30 provides a detailed analysis of the experimental stress-life data at two selected cyclic stress levels [29]. It can be observed that the long fatigue life appears from the batch at 70% to the batch at 130%, and too high energy for the batch at 150% will reduce the fatigue performance due to many thermal cracks.

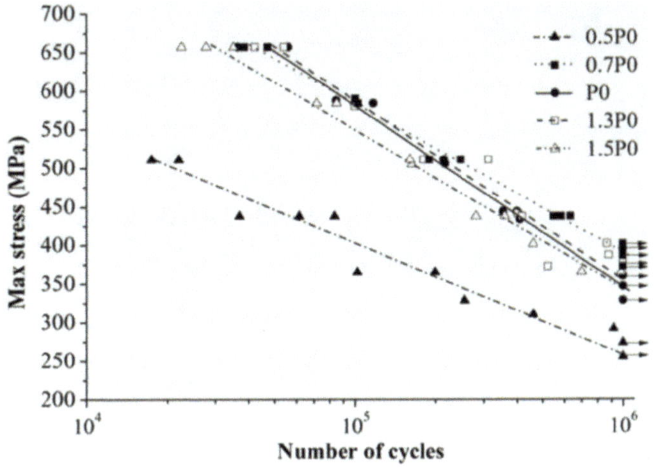

FIGURE 8.29 S-N curves of SLM 316L samples made with different laser powers. (Reprinted with permission from Ref. [29], Elsevier.)

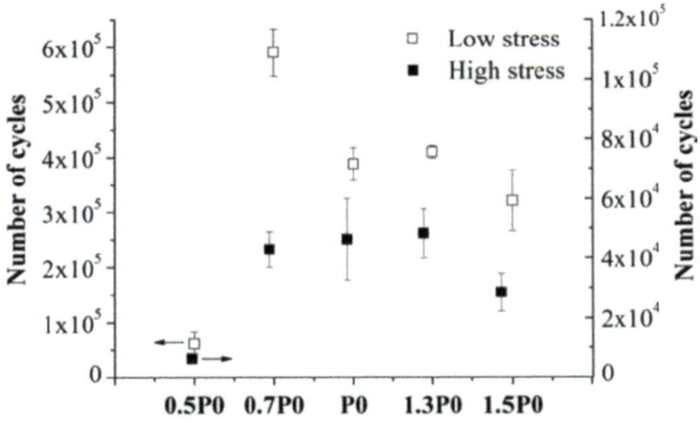

FIGURE 8.30 Fatigue life data of the samples with the maximum cyclic stress of 438 MPa (low stress) and 657 MPa (high stress). (Reprinted with permission from Ref. [29], Elsevier.)

In conclusion, with the increase of energy density, the porosity of the AM material will decrease at first and then increase. There will be a certain range of energy density, in which the porosity is relatively low and thermal cracks is well controlled. In a certain range, the fatigue performance of the AM metal will be optimal.

8.2.6 Powder Layer Thickness

AM with the powder bed is the process of stacking or curing materials layer by layer. Layer thickness refers to the thickness of the powder layer processed by each layer. For the SLM, the thickness can be selected according to the required part accuracy and surface quality, generally between 20 and 80 μm.

To understand the influence of layer thickness on fatigue strength, 316L alloy samples with 30 and 50 μm layer thickness were fabricated by the SLM technology [31]. As shown in Figure 8.31 [31], the thicker layer thickness means lower fatigue property both for the as-printed and post-machined materials. The main reason is that the increase of layer thickness will cause a rougher surface and the increase of porosity, which may result from the formation of more voids.

A thinner layer usually produces a more local microstructure. Because the melted liquid metal flow range of each layer is smaller, the formed solid metal structure is denser and finer, the grain size is smaller. More precipitated phases and smaller inclusions can be produced under thinner layer thickness. These microstructures have a positive effect on improving the mechanical properties and fatigue resistance of the AM material. On the contrary, the thicker layer thickness will produce a larger range of liquid metal flow during the forming process, making the formed metal structure

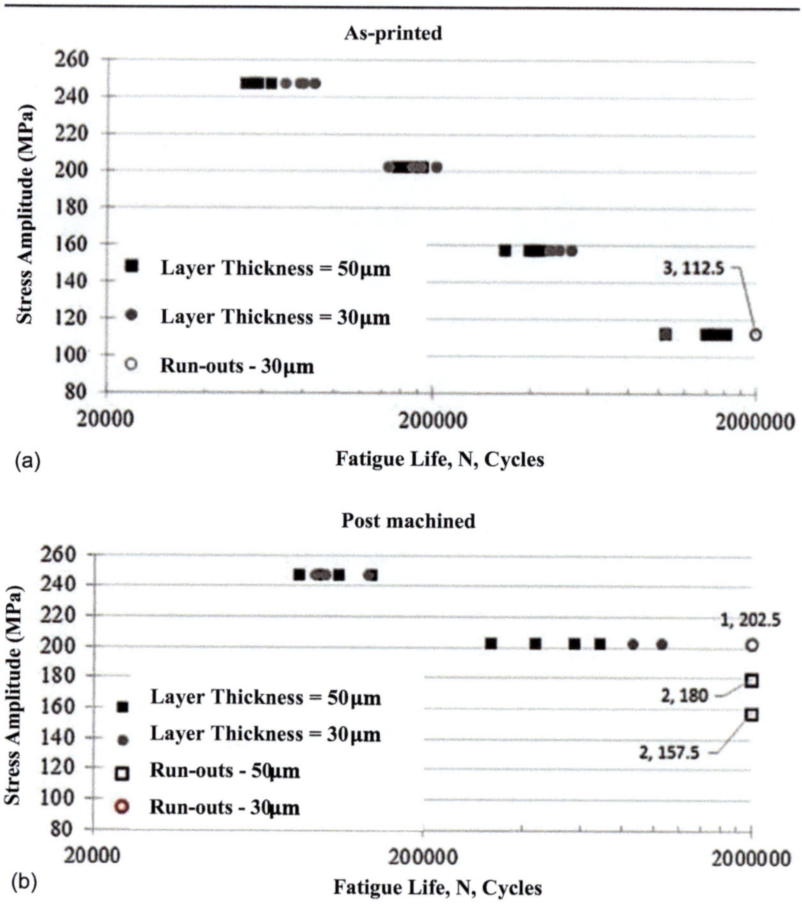

FIGURE 8.31 Fatigue data for AM 316L alloy with 30 and 50 μm layer thicknesses in (a) as-printed and (b) post-machined conditions. (With Permission from Springer.)

rougher and more voids. These weaknesses are easy to become the starting point of fatigue failure and reduce the fatigue resistance of the AM material.

Although a thinner layer thickness can produce a more optimal microstructure and better mechanical properties, a thinner layer thickness will increase the AM processing time and cost. Therefore, when selecting the layer thickness parameter during the AM process, it is necessary to balance various factors, such as manufacturing time, cost, surface quality, and material properties.

8.2.7 Scan Spacing

The scan spacing or hatch spacing refers to the distance between two adjacent scan lines of the laser beam or electron beam during the AM process, usually expressed in millimeters (mm) or microns (μm). The scanning spacing can affect the morphology of the AM materials. The microstructure of AM metal materials presents usually a fish scale shape formed by the mutual assembly of molten pools as shown in Figure 8.32 [32]. Each melting line section partially overlaps with the adjacent next melting line, which leads to the formation of this fish scale-like structure. The specific

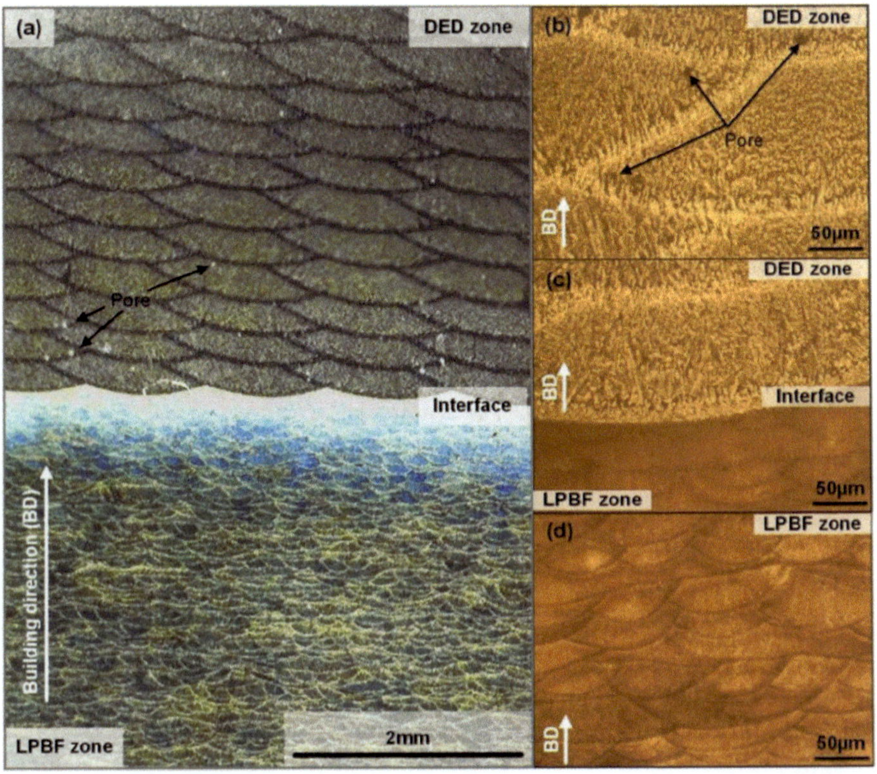

FIGURE 8.32 Optical micrographs of the AlSi10Mg samples built by SLM&DLD hybrid AM with different scan spacing. SLM corresponds to LPBF with small scan spacing, DLD corresponds to DED with large scan spacing. (Reprinted with permission from Ref. [32], Elsevier.)

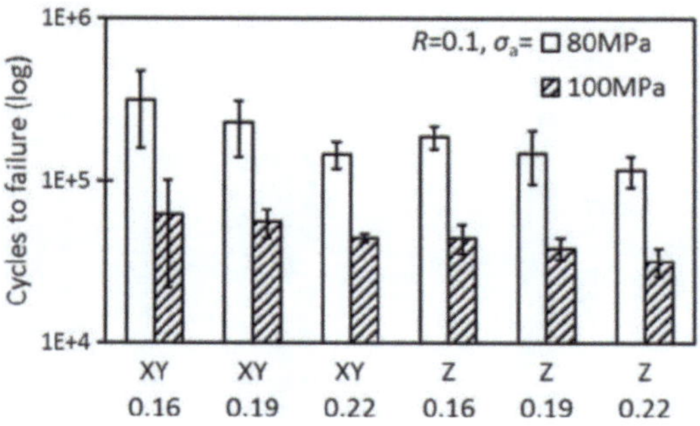

FIGURE 8.33 There is a negative correlation among hatch spacing and fatigue resistance regardless of building direction. Unit for the hatch spacing is mm. (Reprinted with permission from Ref. [33], Elsevier.)

formation process is that (1) the bottom of the current melting line is bowl-shaped currently, while the top surface is curved before the next melting line is formed. (2) When the next melting line is formed, part of the region of the previous melting line will be partially re-melted and involved in this melting line forming, causing the last molten pool trace to be partially erased.

The optical micrographs of the SLM&DLD hybrid AM AlSi10Mg samples with different scan spacing are illustrated in Figure 8.32 [32]. It can be observed from the figure that the molten pool stacks along the deposition direction. The size of the erasing part depends on the scanning spacing. The smaller the scanning spacing, the more parts of the mutual influence, that is, the more erasing parts. Therefore, the scanning spacing affects the coincidence degree of the molten pool stacking of the AM material. When the scanning spacing is small, the coincidence degree of each other is large, and the material is stacked more closely, which will decrease the manufacturing defects as shown in Figure 8.32 and further improve the mechanical properties of the AM material.

The size of the scanning spacing will also have an impact on the fatigue performance of the AM material. Figure 8.33 [33] gives the fatigue performances of AlSi10Mg specimens prepared by SLM with different scanning spacing (0.16, 0.19, 0.22 mm) and building direction (*XY*, *Z*). *XY* represents the horizontal direction, and *Z* represents the vertical direction along the building direction. It can be found that the fatigue life under the same cyclic stress decreases with the increase of scan spacing, this conclusion is not affected by the orientation of AM material. Smaller scanning spacing can promote the refinement of grains and reduce the internal defects and stress concentration of AM materials, thereby increasing the fatigue property of the material. However, if the scan spacing is too small, muti-melt and muti-solidification may cause the AM material produce thermal cracks, gas pores, and other defects, hence reducing the fatigue property of the material.

8.2.8 Scanning Speed

The scanning speed will determine the time that the heat source stays on the surface of the powder. Hence, it will affect the energy absorbed by the powder per unit volume, indirectly determine the powder thickness that can be melted, and finally

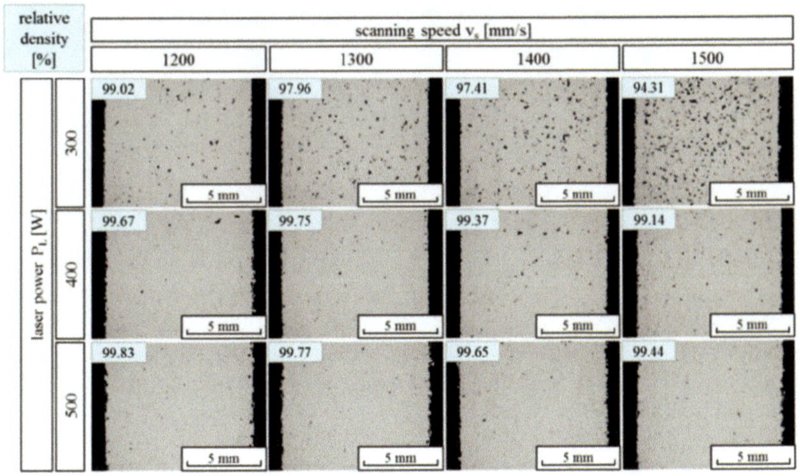

FIGURE 8.34 The lower the scanning speed, the higher the density of the AM sample. (Reprinted with permission from Ref. [34], Elsevier.)

affect the density of AM metals. As shown in Figure 8.34 [34], under the same laser power, the relative density of the AM material gradually decreases with the increase of the scanning speed. Moreover, the increase of the scanning speed at low laser power has a significant effect on the relative density of the AM material. In summary, the lower the scanning speed, the higher the density of the AM sample. When the scanning speed is too fast, the energy absorbed by the powder per unit time will be obviously reduced, and the thickness of the meltable powder layer will be decreased sharply. The metal powder does not melt sufficiently, so the density of the material will decrease with the increase of the scanning speed. The energy absorbed by the powder per unit time is relatively sufficient under high heat source power, hence the relative density of the AM material is not sensitive to the scanning speed at this time.

Hastelloy X alloys were prepared by SLM with high laser scanning speed (H-LSS) and low laser scanning speed (L-LSS), the other process parameters remain unchanged [35]. Figure 8.35 [35] is the EBSD image of the AM sample prepared by these two scanning speeds. By increasing the laser scanning speed, the grain size is decreased due to the smaller molten pool size, and a weaker texture appears. The smaller molten pool limits the grain growth, and the reduction of the energy inputted under the higher scanning speed makes it more likely that the partially melted particles act as crystal nucleuses during solidification, resulting in a finer grain structure.

The S-N curves of H-LSS and L-LSS specimens are shown in Figure 8.36 [35]. The H-LSS and L-LSS samples exhibit different behaviors at high and low stress levels. At high stress levels (low-cycle fatigue), the fatigue life of L-LSS samples is shorter than that of H-LSS samples, but at lower stress levels (high-cycle fatigue), the fatigue life of L-LSS samples is longer. This abnormal behavior results from the change in the dominant fatigue failure mechanism from the material microstructure to the surface profile of specimens. This tendency also indicates that the scanning speed needs to be adjusted to obtain better fatigue property for different service life requirements.

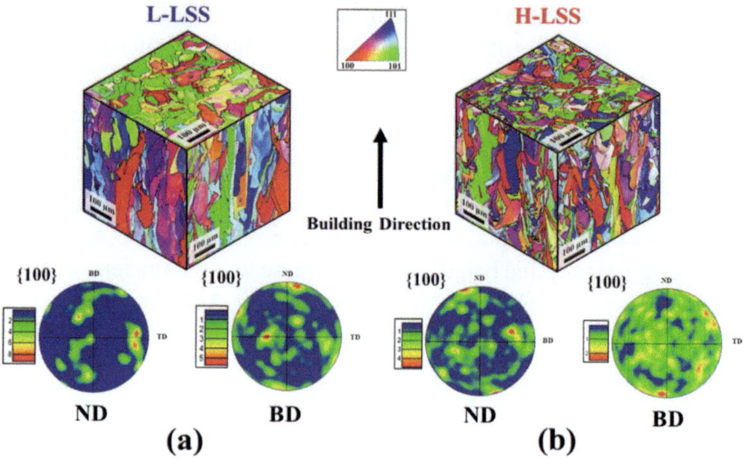

FIGURE 8.35 A finer grain structure and more randomized texture are detected at higher laser scanning speed, (a) L-LSS: low laser scanning speed; (b) H-LSS: high laser scanning speed. (Reprinted with permission from Ref. [35], Elsevier.)

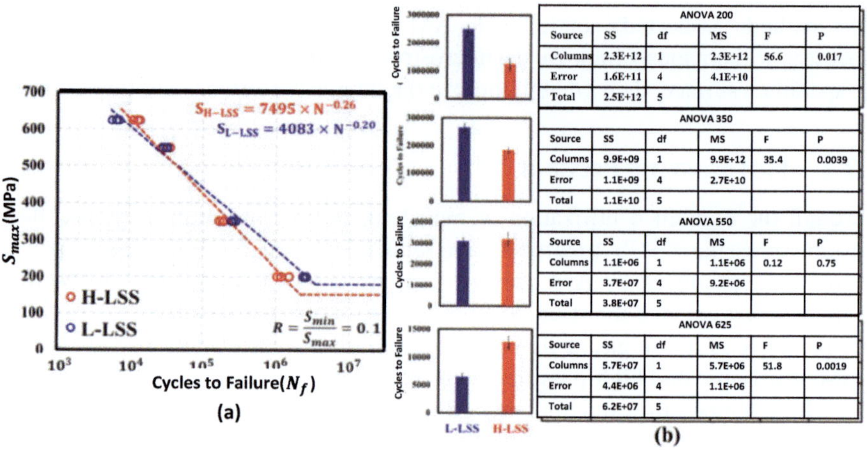

ANOVA 200					
Source	SS	df	MS	F	P
Columns	2.3E+12	1	2.3E+12	56.6	0.017
Error	1.6E+11	4	4.1E+10		
Total	2.5E+12	5			

ANOVA 350					
Source	SS	df	MS	F	P
Columns	9.9E+09	1	9.9E+12	35.4	0.0039
Error	1.1E+09	4	2.7E+10		
Total	1.1E+10	5			

ANOVA 550					
Source	SS	df	MS	F	P
Columns	1.1E+06	1	1.1E+06	0.12	0.75
Error	3.7E+07	4	9.2E+06		
Total	3.8E+07	5			

ANOVA 625					
Source	SS	df	MS	F	P
Columns	5.7E+07	1	5.7E+06	51.8	0.0019
Error	4.4E+06	4	1.1E+06		
Total	6.2E+07	5			

FIGURE 8.36 (a) S-N curves for different laser scanning speeds and (b) analysis for fatigue life data at different stress levels. (Reprinted with permission from Ref. [35], Elsevier.)

As we discussed before, the heat source power P, scanning speed v, powder layer thickness t, and hatch spacing s do affect the quality of AM materials to a certain extent. Sometimes, the energy density parameter E is used to comprehensively study the effects of heat source power P, scanning speed v, powder layer thickness t, and hatch spacing s on the microstructure and defects of the AM materials. Because there is a quantitative formula among these parameters, namely $\boldsymbol{E = P/(v \times t \times s)}$. In terms of AM defects, it's not that higher energy density is better. There is generally an optimal energy density interval in which the best defect control can be achieved. Even if an optimal energy density is adopted, it still makes sense to regulate P, v, t, and s. These regulations can achieve fine-tuning of porosity and control of microstructure, further realizes the optimization of fatigue performance of AM materials. When the energy density is lower than the certain threshold of AM material melting, the heat source power, scanning speed, powder layer thickness, and hatch spacing have the obvious influence on the quality of AM materials.

In summary, to obtain a material with very low porosity in the metal AM process, the optimal energy density for powder melting must be satisfied first. After reaching the threshold of energy density and balancing among these four parameters, a dense optimal AM microstructure will be obtained, then an anti-fatigue AM metal can be created.

8.3 POST-FABRICATION TREATMENTS

Due to the special forming process of additive manufacturing, it will inevitably bring manufacturing defects (such as gas pores, LOFs, thermal cracks, etc.) and residual stress into the metals. Therefore, it is necessary to reduce or eliminate the micro-defects and the residual stress of materials after additive manufacturing, then improve their comprehensive mechanical performances. At present, the commonly used post-treatment methods mainly include hot isostatic pressing (HIP), heat treatment, surface treatment, etc. In general, at least one post-treatment method or a combination of two post-treatment methods is selected to optimize the AM materials. The reasonable post-treatment processes will obviously improve the microstructure, increase the density, and enhance the specimen surface, then greatly improve the fatigue performance of the AM parts.

8.3.1 Hot Isostatic Pressing

Through the combined utilization of high temperature and high pressure, the hot isostatic pressing (HIP) treatment can reduce or even eliminate the voids and microcracks produced by AM and welding, effectively enhance the uniformity of microstructure, reduce the residual stress, finally improve the mechanical property. The principle is to use gas or liquid as the medium, the AM material is placed in a high-temperature and high-pressure environment, so that the material withstands the desire high temperature and high pressure. Owing to the necessary high-temperature environment during the HIP treatment, the AM metals may occur recrystallization, then resulting in the growth of grains, that is, increasing the grain size. In general, a larger grain size is harmful to the anti-fatigue property.

The SLM parts of CoCrFeMnNi alloy were treated by the HIP treatment [28]. It can be clearly observed by electron microscopy that after the HIP treatment, the micropores and microcracks are removed, hence the material is denser [28] as shown in Figure 8.37. Moreover, the molten pool boundary disappears, and the dendritic crystals in the microstructure are dissolved and homogenized, and the grain boundary is more obvious after the HIP treatment. Therefore, the proper HIP treatment can distinctly influence both the density and microstructure of AM metals, so it is used for improving the AM metals frequently.

The fatigue testing results of Ti-6Al-4V samples constructed in the vertical (V) and 45° directions using SLM are shown in Figure 8.38 [36]. The fatigue life data

FIGURE 8.37 Microstructure of high entropy alloy samples prepared by SLM before and after the HIP treatment. (Reprinted with permission from Ref. [28], Elsevier.)

of the specimens after the annealing treatment and HIP treatment were obtained. For the case of machined surface condition, the application of HIP treatment leads to a longer fatigue life regardless of the build orientation, and the fatigue life even is slightly longer than that of the forging material. Owing to the highly healing for the manufacturing defect, the fatigue crack initiation source will shift from defects to the matrix after the HIP treatment. However, the HIP treatment seems to have little positive effect on fatigue properties of AM materials for the case of as-built surface condition. This is because at this time, the surface profile dominates the fatigue failure of AM materials as shown in Figure 8.39 [36], while the HIP treatment could not change the surface profile. That is to say, a fine surface profile is necessary for the application of HIP treatment on the AM metals.

In summary, the HIP treatment is great for healing pores and cracks, which can also effectively enhance the microstructure uniformity and reduce the residual stress, thereby improving the fatigue property of AM metals. The HIP treatment can work well for increasing the fatigue performance of the AM titanium and nickel-based alloys, while it often does not achieve the ideal effect for aluminum alloys due to the significant change in the microstructure.

8.3.2 Heat Treatment

Heat treatment method generally contains the solution, aging, annealing, stress relief annealing treatment, etc. for the AM metals. Generally, the material bears additional post-heat treatment, which can effectively improve the strength or enhance the plasticity and toughness of the AM material.

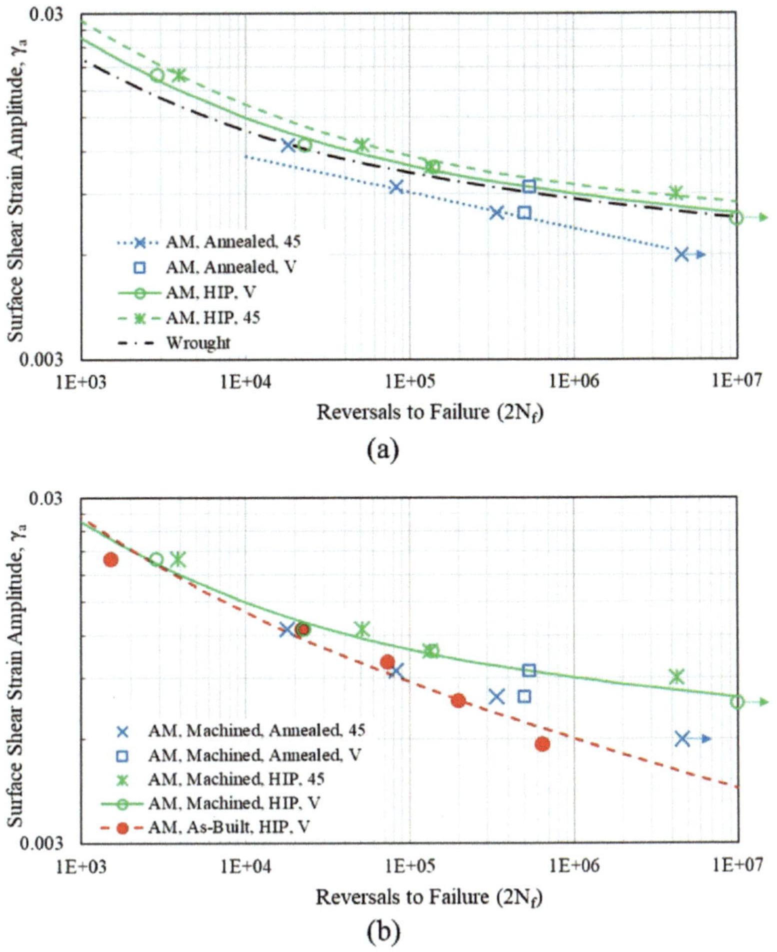

FIGURE 8.38 Fatigue life data of the SLM and wrought Ti-6Al-4V alloys under two build orientations with and without HIP treatments for the (a) machined and (b) as-built surface conditions. (Reprinted with permission from Ref. [36], Elsevier.)

The solution treatment is to heat the alloy to a certain temperature range, so that the solute atoms are further fully dissolved into the solid solution, and then cooling quickly to obtain the supersaturated solid solution. Due to its thermodynamic metastable state, dissolution or other transformations will occur under appropriate temperature or stress conditions. It is a preliminary heat treatment in most cases, which is used to prepare the best conditions for subsequent heat treatment. The dissolved solute atoms can increase the dislocation movement resistance within the solid solution, so that the alloy is strengthened. The solution treatment will cause the recrystallization of the microstructure, so that the dendritic crystals formed in the AM process will disappear and the equiaxed crystals may be formed. The solution treatment will further homogenize the microstructure, effectively eliminate element segregation in the microstructure, and enhance the performance of the AM alloy.

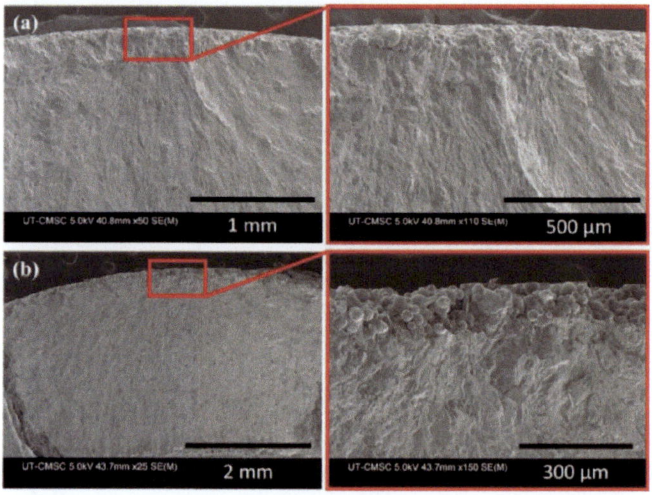

FIGURE 8.39 Fatigue cracks initiate from the rough outer surface under the as-built surface condition with the HIP treatment. (Reprinted with permission from Ref. [36], Elsevier.)

Aging treatment refers to metals placed at a relatively high temperature or room temperature to maintain its shape after the solution treatment, and its performance will change with time. The common treatment method is first solution treatment and then aging treatment. The effect of aging treatment alone on the performance is not obvious. Generally, the hardness and strength are increased, while the plastic, toughness, and residual stress are reduced after the aging treatment. The elements within the supersaturated solid solution are precipitated again into the matrix of the alloy during the aging treatment, thus forming a part of the smaller solute enrichment area, and then forming a lot of dispersive precipitate phases. The precipitate phase can effectively resist the movement of dislocations and then enhances the performance of the AM metals.

The SLM CoCrFeNiMn alloy was annealed at 990°C for 1 hour to tailor its microstructure [37]. The plasticity is significantly improved, and the tensile strength is slightly enhanced after the post-treatment as shown in Figure 8.40. However, the yield strength decreases owing to the recovery of the special cellular structures in AM metals after the heat treatment.

Heat treatments can mainly cause changes in the microstructure of the material, thus affecting the fatigue crack initiation and propagation behaviors, then affecting the fatigue property of the AM material. Generally, the use of heat treatment is beneficial to fatigue properties, while sometimes heat treatments do not improve the fatigue performance of AM materials. The SLM AlSi10Mg samples were prepared by different post-heat treatment methods, including stress relief annealing at 300°C for 2 hours, furnace cooling (SR) and solution treatment at 530°C for 1 hour, water quench (ST) [38]. Some of the solution treatment samples were followed by aging treatment at 170°C for 12 hours (T6, a standard heat treatment process). The fatigue properties of these specimens [38] are shown in Figure 8.41.

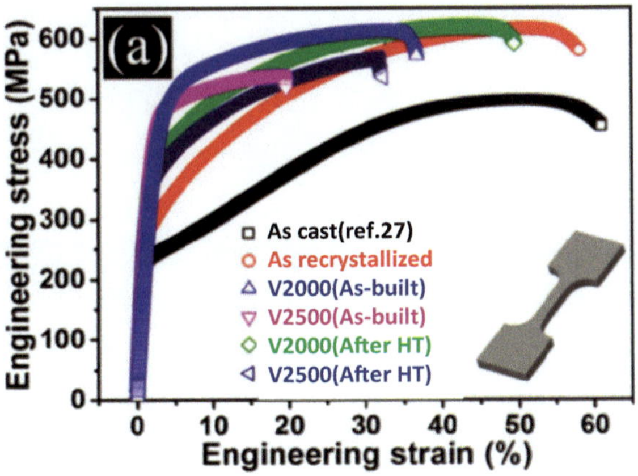

FIGURE 8.40 Comparison of strength and plasticity of AM high entropy alloy samples before and after the post-heat treatment. (Reprinted with permission from Ref. [37], Elsevier.)

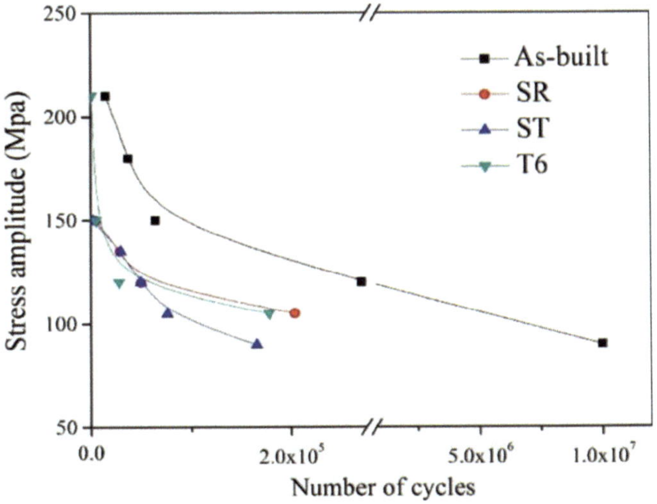

FIGURE 8.41 S-N curves of SLM AlSi10Mg samples with different post-heat treatments. (Reprinted with permission from Ref. [38], Elsevier.)

The results show that the fatigue performance of the as-built sample is the highest compared with the heat treatment samples, that is, the heat treatments reduce the fatigue performance of the AM metal. The fatigue life of T6 treated samples is slightly lower than that of SR treated samples, and the fatigue life of ST treated samples is the lowest. The fatigue crack initiation and crack propagation zones are shown in Figures 8.42 and 8.43, respectively [38]. These pores act as crack initiation sites for all batches. The fracture surface at crack propagation zone of the SR treatment sample presents microcleavage fracture. The specimens under ST and T6 treatments show micro-void coalescence fracture. That is to say, the total fatigue life

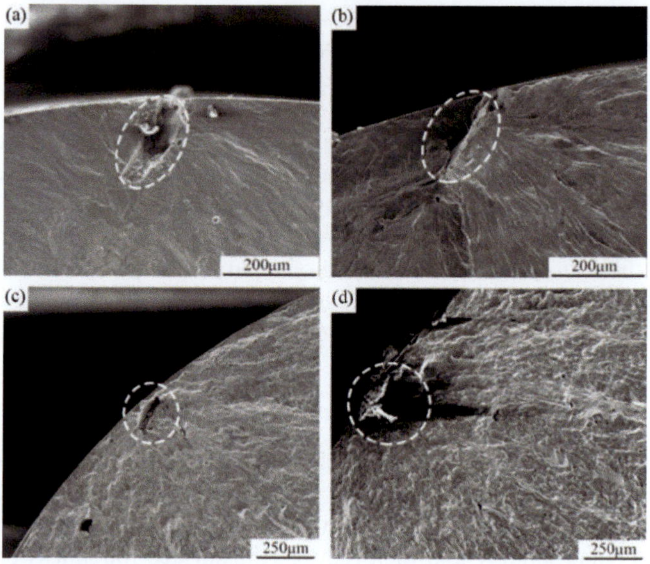

FIGURE 8.42 Morphology of fatigue crack initiation source of a AM aluminum alloy: (a) as-built; (b) SR; (c) ST; (d) T6. (Reprinted with permission from Ref. [38], Elsevier.)

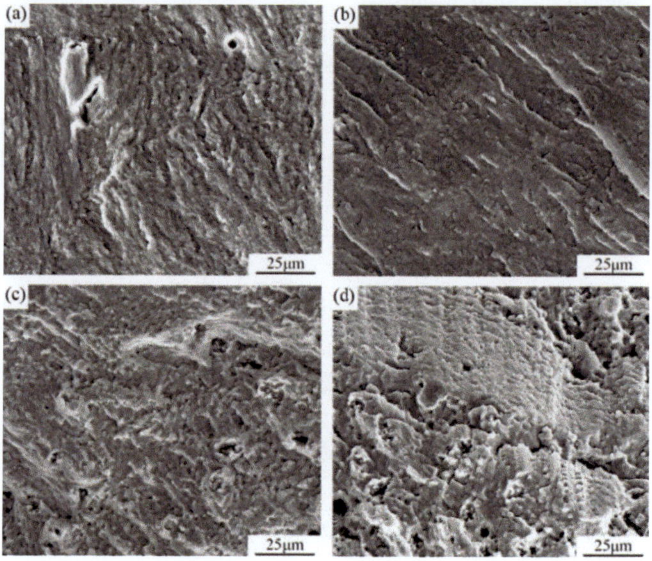

FIGURE 8.43 Morphology of fatigue crack propagation zone of a AM aluminum alloy: (a) as-built; (b) SR; (c) ST; (d) T6. (Reprinted with permission from Ref. [38], Elsevier.)

mainly consists of the fatigue crack propagation life, and different post-heat treatments lead to the change in microstructure, finally causing their difference in fatigue crack propagation behavior.

With the increase of heat treatment temperature, the morphology and size of dendrite Si in the AM microstructure change more greatly, which affects its static mechanical properties and fatigue properties. The continuous network dendrite Si is responsible for the highest fatigue property of the as-built samples. The dendrite Si breaks into separated particles after heat treatments, the decrease of fatigue property results from the formation and coarsening of separated Si particles. At present, the network/cellular microstructure formed during additive manufacturing is beneficial to mechanical properties, especially for fatigue properties.

8.3.3 Surface Mechanical Treatments

This kind of surface post-treatment is based on plastic deformation. Common treatment methods include shot peening, laser shock peening, cavitation peening, ultrasonic peening, etc. This type of treatment method produces more uniform and regular surface morphology and then reduces the surface roughness of AM parts. On the other hand, the treatment method based on plastic deformation can bring additional beneficial effects, such as grain refinement in the surface layer and the generation of compressive residual stress, both of which are helpful to improve the fatigue properties of metal materials.

Shot peening is a method of using compressed air to shoot small spherical particles onto the surface of the target parts as shown in Figure 8.44 [39]. The high-speed ball impact is employed to achieve cold plastic deformation and forging the parts to make the surface smooth. Its effect on surface morphology and plastic deformation is shown in Figure 8.45b. Shot peening has been applied to AM metal parts to control

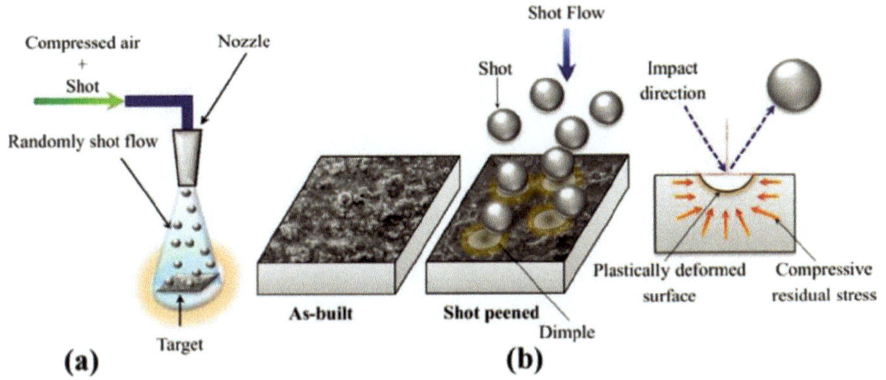

FIGURE 8.44 Schematic illustration of (a) shot peening process, (b) shot peening effect on the surface layer. (Reprinted with permission from Ref. [39], Elsevier.)

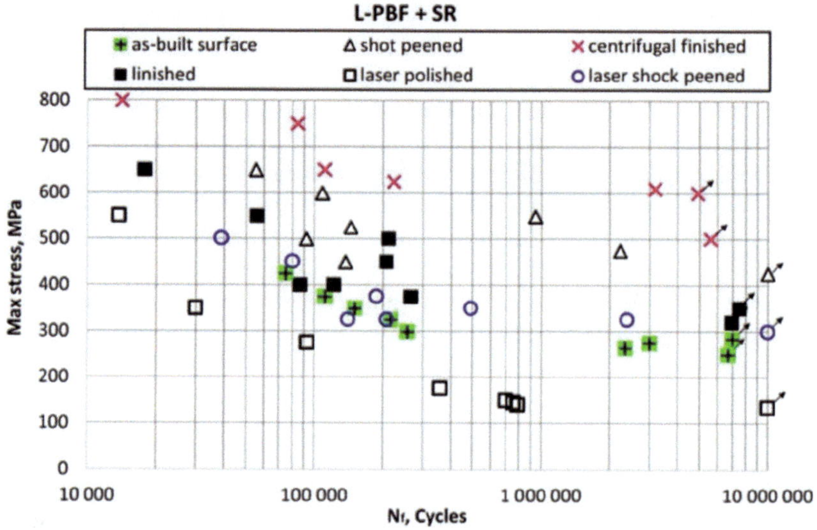

FIGURE 8.45 Fatigue life data for SLM Ti6Al4V with various post-surface processes. (Reprinted with permission from Ref. [41], Elsevier.)

surface roughness, improve mechanical properties, induce high compressive residual stress, and ultimately enhance their fatigue properties.

Laser shock peening is to use the shock wave formed by the explosion of high temperature and high-density plasma generated by pulsed laser to act on the surface of the target material. As a result, the surface of the material is plastically deformed, moreover a high and deep residual stress is generated below the material to inhibit or delay the generation and propagation of fatigue cracks.

Ti-6Al-4V samples were prepared by EBM, and three different surface treatments of shot peening, cavitation peening, and laser shock peening were applied to these samples [40]. The results show that the shot peening performs best in improving fatigue life, followed by laser shock peening and cavitation peening. Figure 8.45 is the fatigue life data of the specimens prepared by SLM after different post-surface treatments [41]. As seen, the fatigue performances of shot peening and laser shock peening are better than that of as-built surface. The shot peening also presents a better ability than the laser shock peening. A significant improvement in the fatigue property is detected for the centrifugal finished samples. This treatment method is mainly used to polish the AM surface without applying any surface plastic deformation. That is to say, a smooth surface is the key to enhance the fatigue property of AM metals.

8.3.4 Surface Finishing

Some common surface post-treatments include centrifugal finishing, linishing (abrasive finish), laser polishing, mechanical polishing, etc. These treatments can remove a layer of surface material, thereby reducing surface roughness and eliminating defects near the surface of AM metals. They induce almost no or low surface plastic deformation.

Centrifugal finishing refers to rotate a mixture of parts, abrasive media and carrying agents in a barrel. The surface of the component is abraded through the relative motion between the parts and the media. The process can smooth the surface to a polished state, hence finally reduces the surface roughness of components. Linishing is the high-speed abrasive brush used to remove peaks and smoothen valleys on the part surface to achieve a uniform surface condition. Laser polishing is used to reduce the surface roughness of AM materials by re-melting thin surface layer (depth: about 50–200 μm) of material. Any un-melted powders on the AM surface will be melted, and reflowing into the surface valley to achieve a finer surface. Mechanical polishing can achieve thick material removal and very smooth surface profile.

The surface of the AM material is much rougher than that of conventionally formed (such as casting, rolling, etc.) materials. Therefore, surface finishing is very necessary for the AM metals to achieve better structural integrity. Ti6Al4V materials prepared by SLM and EBM were applied with different surface post-treatments [41]. The fatigue properties after surface post-treatments are presented in Figures 8.45 and 8.46 [41]. The fatigue strengths of the material manufactured by SLM and EBM are significantly improved after centrifugal finishing or linishing. However, the fatigue strength of the AM material after laser polishing is lower than that of the as-built material. In summary, the key to improving the fatigue performance of AM materials by the surface finishing is surface material removal rather than re-melting surface layer.

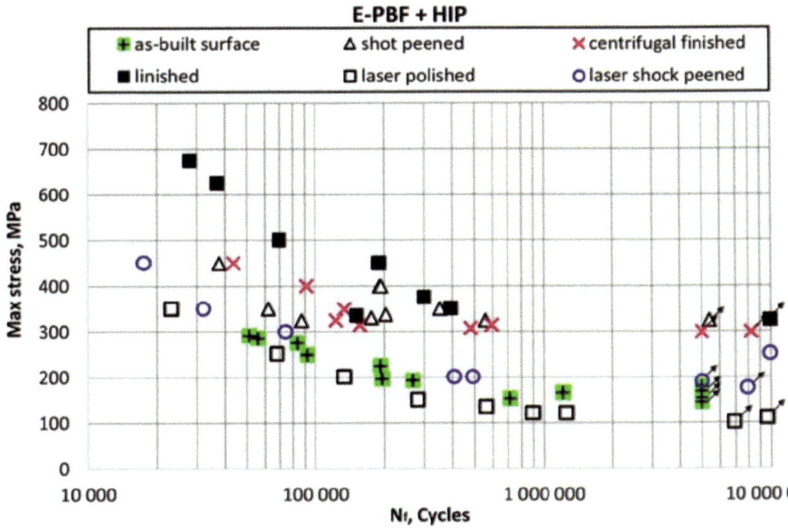

FIGURE 8.46 Fatigue life data for EBM Ti6Al4V with various post-surface processes. (Reprinted with permission from Ref. [41], Elsevier.)

8.4 LOADING EFFECT ON AM FATIGUE

AM metals have excellent potential for manufacturing load-bearing parts. Fatigue failure is the dominant failure mode of load-bearing components. Therefore, it is necessary to consider the fatigue failure behavior of AM materials under different cyclic loading modes. This section mainly pertains to the cyclic loading conditions that a component experiences during its service life. These conditions include loading frequency, variable amplitude loading, notch effect (stress gradient), stress/strain ratio, size effect, etc. The influence of each of these factors on the fatigue failure behavior of an AM material is different, and they also affect each other in various ways. Overall, it is essential to consider all these factors when evaluating the fatigue performance of AM materials and to take appropriate measures to improve their fatigue life under specific cyclic loading conditions that they will experience in their service life.

8.4.1 Loading Frequency

In general, materials subjected to higher frequency cyclic loading are expected to exhibit a higher fatigue property. However, high-frequency loading may also accelerate the fatigue crack evolution due to the temperature rise by self-heating, leading to fewer cycles for fatigue failure. Meanwhile, the frequency effect is almost unobservable for some specific metallic materials. What is known is that the frequency effect depends on the type of material, the type of phase, etc. Consequently, the effect of loading frequency on fatigue performance has not been conclusively determined. Further research is needed to reconcile these conflicting results from the perspective of the physical mechanism.

The loading frequency of fatigue tests can impact the deformation mechanism of AM materials and subsequently affect their fatigue performance. Slip-related events such as planner-slips and cross-slips, the main deformation mechanism of cyclic damage accumulation, are loading rate-dependent and thermally activated. The occurrences of these slips may be limited, leading to less damage accumulation under higher loading rates (for example ultrasonic fatigue loading, about 20 kHz). On the other hand, the change in the loading rate may result in the change in deformation mechanism, such as changing from shear bands to deformation twinning or even slip with the increase of loading rate. Loading frequency effect is particularly complex for AM materials since their diversity of composition and the heterogeneity of material microstructure than conventional manufacturing materials.

IN718 specimens were prepared by SLM in the vertical and diagonal directions relative to the substrate plate [42]. The difference in fatigue performance at 20 kHz and 5 Hz loading frequencies was studied. The fatigue life data of IN718 alloy under non-notched conditions were collected from the Metal Material Performance Development and Standardization (MMPDS) for comparison. It can be observed in Figure 8.47 [42] that the fatigue performance of the specimens tested at 20 kHz is higher than that of the specimens tested at 5 Hz for the wrought and AM IN718 alloy, and it is more obvious for the specimens constructed vertically. In conclusion, the SLM IN718 specimen has better fatigue performance in the ultrasonic fatigue test than the traditional fatigue test regardless of the building direction.

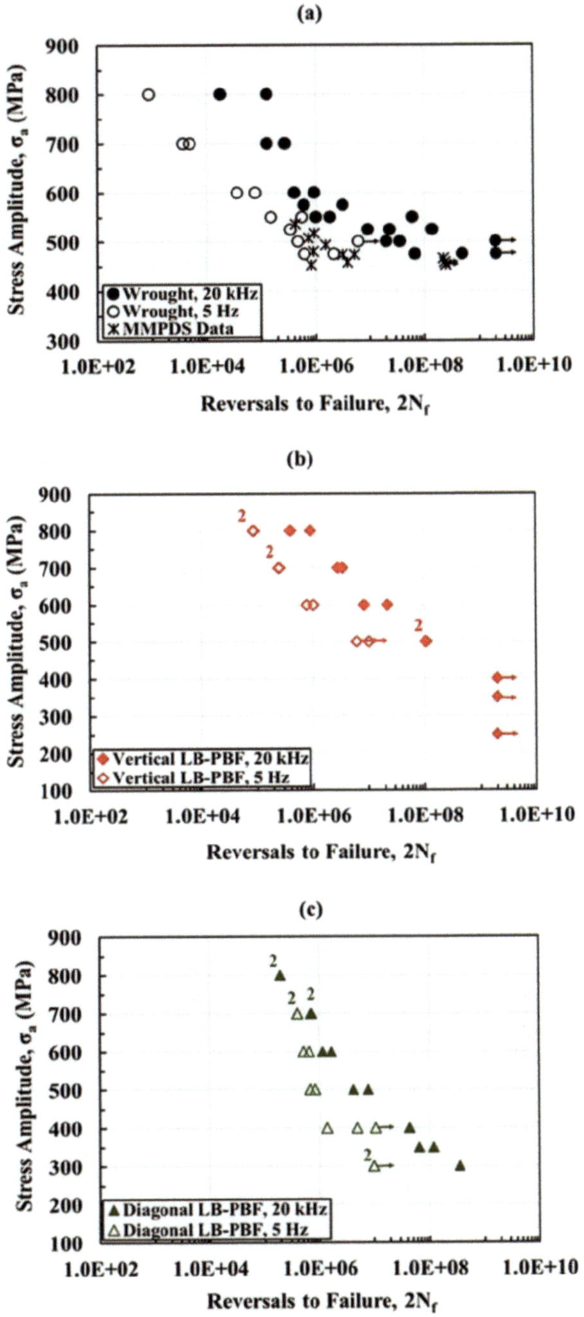

FIGURE 8.47 Stress-life data for (a) wrought IN718, (b) vertical SLM (LB-PBF) IN718, and (c) diagonal SLM IN718 tested at different loading frequency. MMPDS: Metal Material Performance Development and Standardization. (Reprinted with permission from Ref. [42], Elsevier.)

8.4.2 Variable Amplitude Loading

Amplitude refers to half of the difference between the maximum and minimum strains/stresses in a cyclic load, and variable amplitude loading refers to the variable amplitudes employed during the fatigue loading. In most cases, the load acting on a structure is not perfectly periodic and exhibits significant randomness with varying amplitudes. Higher amplitude loading can accelerate the evolution of dislocations and cracks and expedite the material damage, thereby reducing fatigue life.

The fatigue behaviors of 15-5 precipitation hardened stainless steel parts prepared by SLM were studied under variable amplitude loading at the zero minimum stress (tensile mean stress) and zero mean stress conditions [43]. The detailed variable amplitude loading process at zero minimum stress is shown in Figure 8.48. Two cyclic loads with different amplitudes are applied to the AM specimen, and the number of cycles is carried out in the order of first high and then low, and then the same load block is applied reversely to investigate the load sequence effect, as shown in Figure 8.48a and b. Secondly, the similar process is repeated for three test blocks with different load amplitudes as given in Figure 8.48c and d.

The number of cycles that the specimens could withstand under variable loading conditions with zero minimum stress and zero mean stress are shown in Figures 8.49 and 8.50, respectively [43]. The variation trend of calculated cumulative damage by Miner's rule, which is usually used to study the variable amplitude loading effect, is consistent with that of obtained fatigue life under zero minimum stress and zero mean stress conditions.

A severe effect of load sequence on fatigue life is detected for the two-block force under zero minimum stress in Figure 8.49. We can also observe that the endurance limit of AM materials is relatively low when subjected to higher loading stress first, whether it undergoes two or three amplitude changes. However, this difference is

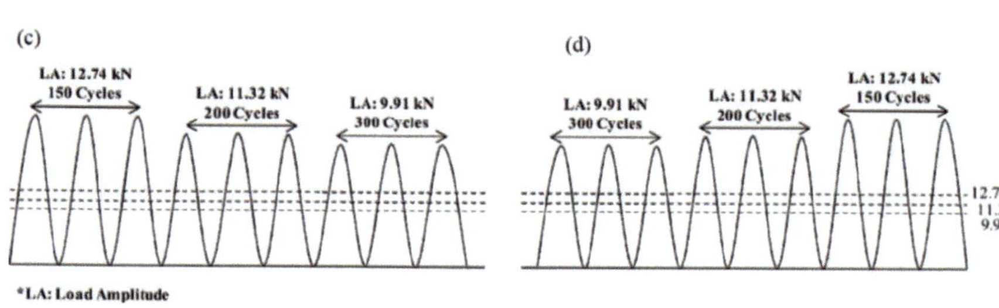

FIGURE 8.48 (a) Two blocks of cyclic forces, (b) two blocks of cyclic forces in reverse sequence, (c) three blocks of cyclic forces, (d) three blocks of cyclic forces in reverse sequence. (Reprinted with permission from Ref. [43], Elsevier.)

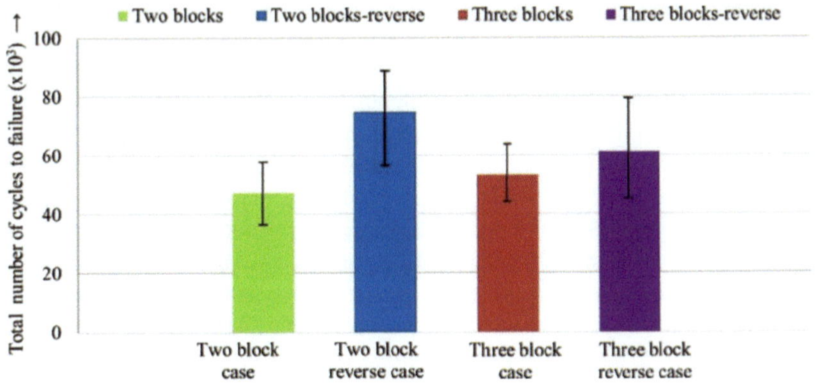

FIGURE 8.49 Fatigue life under variable amplitude loading in different sequences for zero minimum stress. (Reprinted with permission from Ref. [43], Elsevier.)

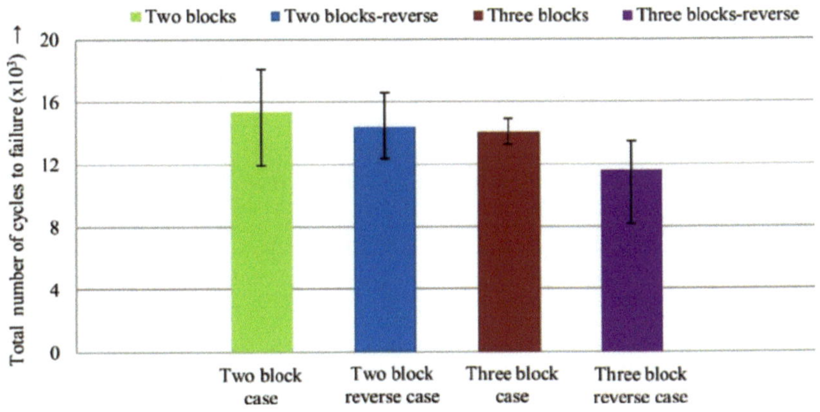

FIGURE 8.50 Fatigue life under variable amplitude loading in different sequences for zero mean stress. (Reprinted with permission from Ref. [43], Elsevier.)

getting smaller with the increase of force block. In Figure 8.50, for both two and three block forces, the effect of load sequence still exists under zero mean stress. In contrast to the previous results under zero minimum stress, when higher cyclic stress is applied first, a longer fatigue life is detected for this case. In summary, the load sequence effect (positive or negative) depends on the state of applied mean stress regardless of the number of force blocks.

8.4.3 Notch Effect

Notch is a common and necessary geometric feature of engineering load-bearing components, which can induce local stress gradient, and then lead to the change in fatigue performance and fatigue failure mode [44]. The surface quality of AM materials based on the powder bed fusion technology is relatively poor, especially when constructing a downward facing surface. A schematic illustration of the manufacturing

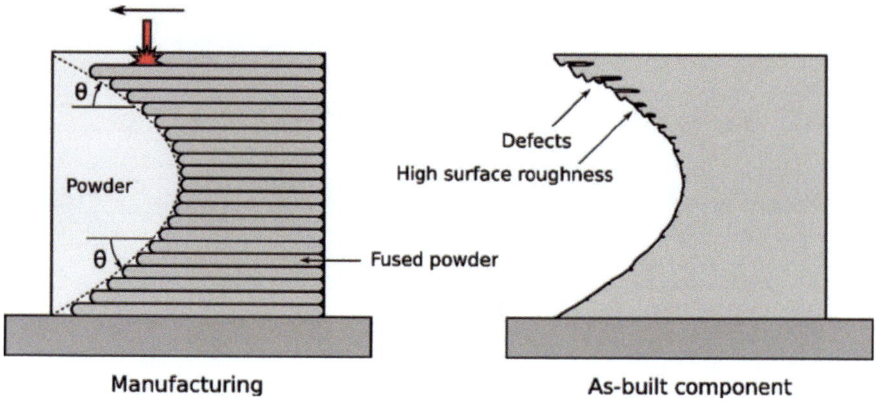

FIGURE 8.51 Schematic illustrations of a manufacturing and as-built AM notch component. The as-built notch has defects and high surface roughness on the downward facing surface. (Reprinted with permission from Ref. [45], Elsevier.)

process and as-built surface of an AM notch is shown in Figure 8.51 [45]. Many AM defects will appear on these surfaces to produce fatigue crack sources. A more obvious defect will be detected with the increase in building height. These defects also depend on the notch geometry and the orientation between building direction and the notch. Therefore, the notched fatigue property of AM materials will be affected by local surface defects near notches, the effect can be analyzed by changing the notch geometry and the orientation between the building direction and the notch. With increasing interest in using AM technology to fabricate engineering applications, it is necessary to evaluate the structural integrity of the AM notch components.

To study the effect of notch geometry on fatigue behaviors of the SLM IN718 alloy, four types of specimen geometries: (1) unnotched (30.3 mm); (2) semi-circular notch with notch radius is 5 mm; (3) V-shaped notch with notch radius is 1 mm; (4) V-shaped notch with notch radius is 0.1 mm were employed [45]. Both the notch acuity and the elastic stress concentration factor increase from the notch type 1 to the notch type 4. Figure 8.52a presents fatigue life data for all geometries where the marker also gives the size of the surface defect for crack initiation. As seen, with the increase of the notch acuity or the decrease of the notch radius, a larger size of the surface defect for crack initiation is observed, and the fatigue strength of AM metals decreases. The fatigue notch factor decreases with the increase of notch radius.

For AM metals, both the notch geometry and the local defects can act as the driving mechanism for fatigue. This special driving mechanism leads to the fatigue crack initiation site in a notched region depending on the notch acuity as shown in Figure 8.52b [45]. The scatter in fatigue crack initiation site increases for a decreasing notch acuity. For the unnotched specimens, the local defects within the gauge length dominate the fatigue crack initiation site, resulting in an obvious scatter. For the specimen with blunt notch, the local defects on the downward facing surface control the fatigue crack initiation site. However, for the specimen with a very sharp notch (very high stress concentration), the fatigue crack initiation site is controlled by the notch geometry, causing a low scatter.

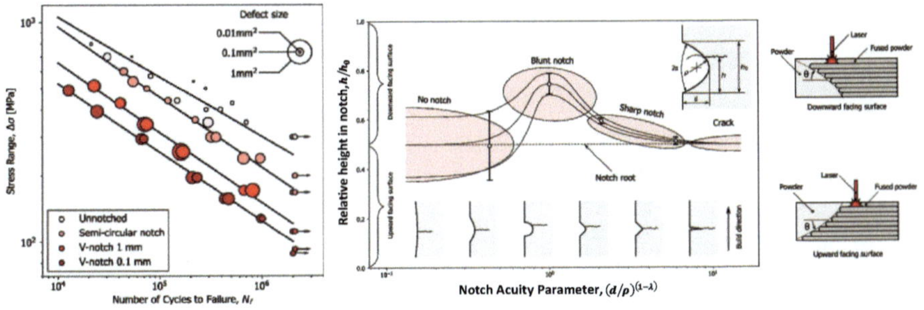

FIGURE 8.52 (a) Fatigue life data for all geometries where the marker also gives the size of the surface defect for crack initiation, (b) schematic illustration of the relationship between fatigue crack initiation site in a notched region and the notch acuity. (Reprinted with permission from Ref. [45], Elsevier.)

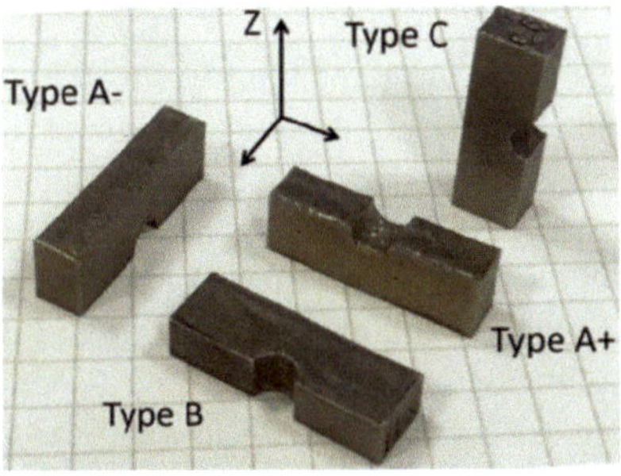

FIGURE 8.53 Specimen notch type depending on their orientations with respect to build direction Z. (Reprinted with permission from Ref. [47], Elsevier.)

The miniature fatigue specimens can be manufactured with their long axis oriented perpendicular and parallel to the build direction, and then are tested with surfaces in the as-built state as no support removal or surface machining is required [46]. Four types of notched fatigue specimens (Type A−, Type A+, Type B, and Type C) are constructed by SLM as shown in Figure 8.53 Specimen notch type depending on their orientations with respect to build direction *Z* [47]. These specimens have the same geometry but different orientations with respect to build direction *Z* to study how the orientation between the building direction and the notch affects the fatigue behavior of AM metals. The notch orientation with respect to the substrate plate can affect the quality of as-built SLM surfaces.

Figures 8.54–8.56 show the fatigue behavior of these four notch types [47]. It is evident that the fatigue life of notched specimens is consistently lower than that of unnotched specimens owing to the stress concentration effect of notches. The downskin fabrication condition (Type A−) is associated to a shorter fatigue life than the upskin condition (Type A+) as shown in Figure 8.56. The notched fatigue strength of SLM Inconel 718 with the as-built surface is highly directional due to different surface quality. Figures 8.54–8.56 are used to calculate the notch fatigue factor, which

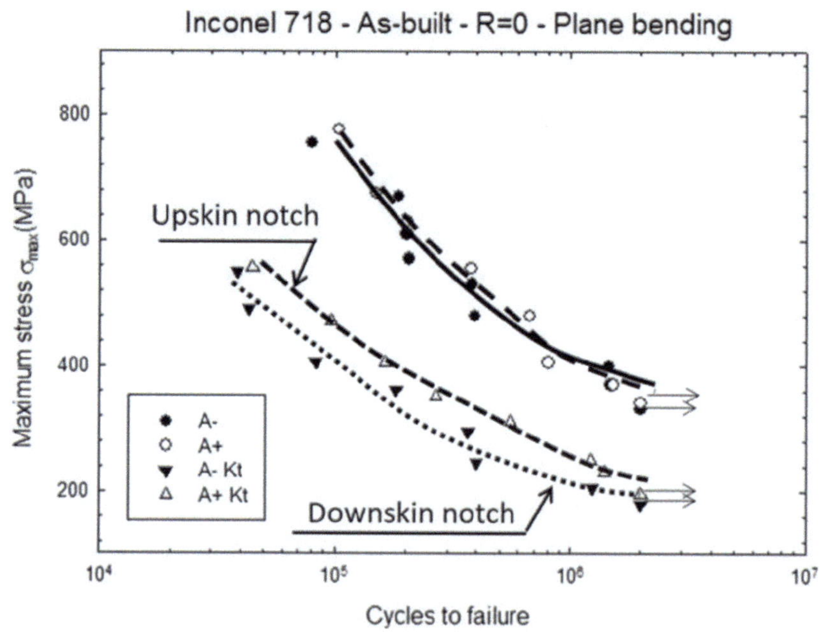

FIGURE 8.54 Notched fatigue behavior of Type A+ and Type A− specimens. (Reprinted with permission from Ref. [47], Elsevier.)

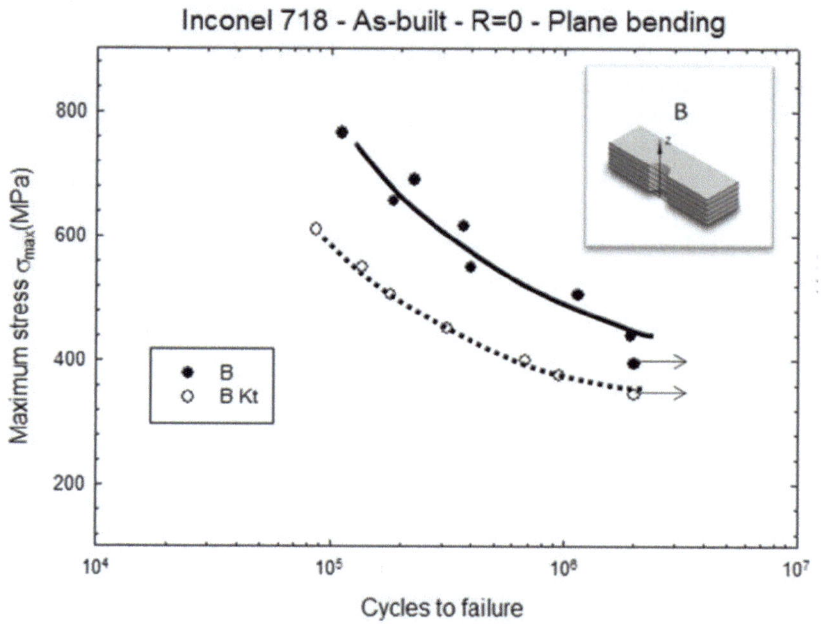

FIGURE 8.55 Notched fatigue behavior of Type B specimens. (Reprinted with permission from Ref. [47], Elsevier.)

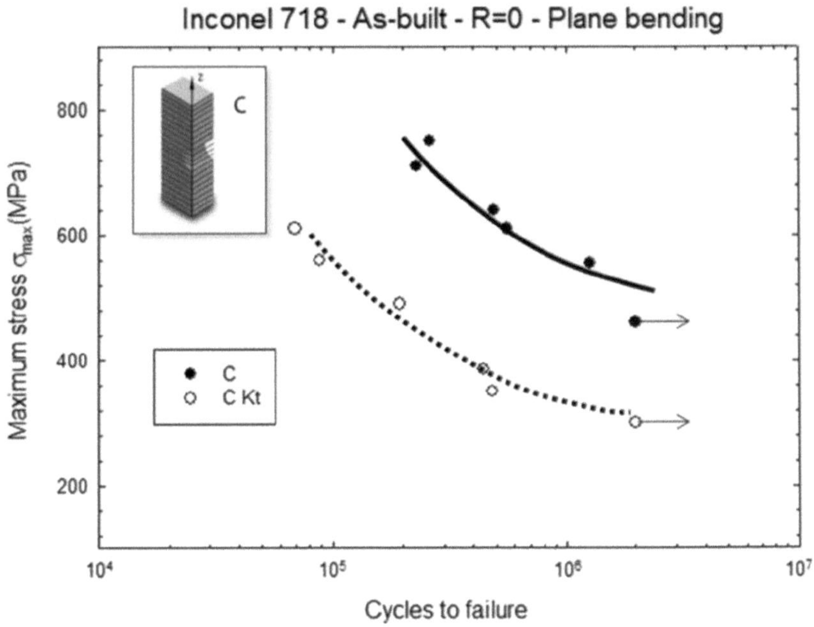

FIGURE 8.56 Notched fatigue behavior of Type C specimens. (Reprinted with permission from Ref. [47], Elsevier.)

refers to the ratio of the smooth fatigue strength and notched fatigue strength. The SLM-dependent notch fatigue factors gradually decrease with the change in notch types of Type A–, Type A+, Type C, and Type B. Namely, the AM technological impact is minimum for the case of Type B.

The fatigue behavior of metals manufactured by AM techniques is highly dependent on the notch geometry and the orientation of the notch construction. Moreover, the applied cyclic stress level also impacts the notched fatigue behavior of AM metals. In the high cycle fatigue regime (relative low stress level), the notched fatigue specimens failed from local defects in the downskin region rather than from the notch roots, while fatigue failure occurs in the notch roots in the low cycle fatigue regime (relative high stress level) [48].

8.4.4 Loading Ratio

The loading ratio (R) refers to the ratio of minimum load to maximum load in a loading cycle, including stress ratio and strain ratio. The loading ratio is a common load parameter and has a significant influence on the fatigue property and fatigue failure behavior of metals, especially in the very high cycle fatigue regime (fatigue life not lower than 1E7 cycles) [49,50].

The comprehensive studies [51] were performed on investigating the impact of various stress ratios ($R = -1$, -0.5, 0.1, and 0.5) on the fatigue properties and crack initiation characteristics of the Ti-6Al-4V alloy prepared through SLM in the high cycle fatigue and very high cycle fatigue regimes. The ultrasonic fatigue tests with a

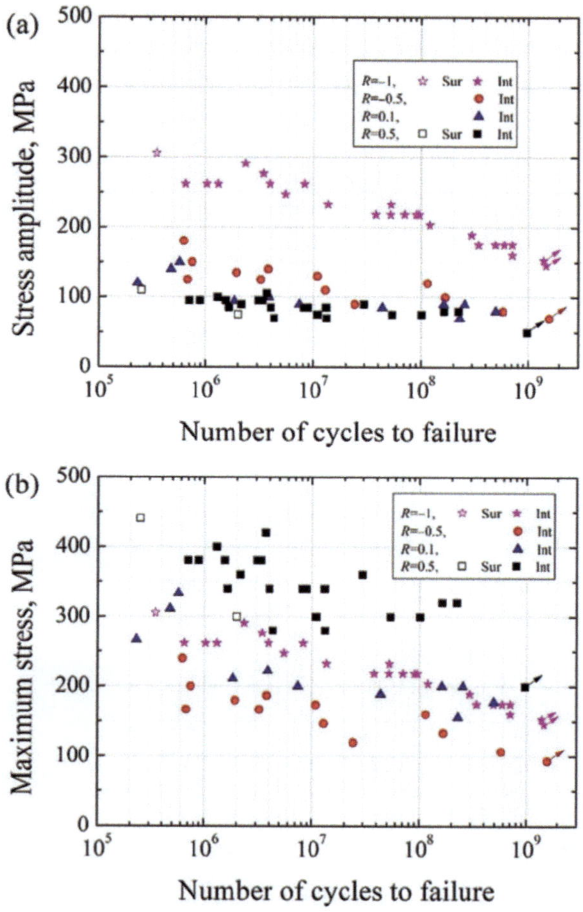

FIGURE 8.57 Fatigue life data of SLM Ti-6Al-4V presented by (a) stress amplitude and (b) maximum stress. (Reprinted with permission from Ref. [51], Elsevier.)

loading frequency of 20 kHz were performed on the AM specimen to accelerate the fatigue tests under these four stress ratios. The fatigue testing results are shown in Figure 8.57.

It is observed that both the stress amplitude and the maximum stress decrease as the number of cycles increases up to 1E9 cycles for the Ti-6Al-4V samples prepared by SLM, that is, the AM metal does not exhibit a traditional fatigue limit at all the stress ratios. Overall, with the increase of stress ratios, the fatigue strength expressed by stress amplitude decreases, while the fatigue strength expressed by maximum stress increases except for the case of $R = -1$. Haigh diagrams [51] (the relationship between stress amplitude and mean stress) show that the fatigue performances of the SLM Ti-6Al-4V alloy with a large number of defects under a certain positive mean stress are seriously degraded than what happened in conventionally titanium alloys in the long-life fatigue range. It implies that the effect of AM defects on the fatigue strength degradation is more severe for the positive mean stress ($R = -0.5$, 0.1, and 0.5).

Fatigue crack initiation can occur through two main modes for $R=-1$: surface crack initiation and interior crack initiation. Only one case of surface crack initiation is observed in the high cycle fatigue regime, fatigue cracks of all specimens are initiated internally in the very high cycle fatigue regime. Hence, the interior crack tends to dominate the crack initiation process at a longer fatigue life. A similar fatigue crack initiation behavior is detected under the stress ratio of $R=0.5$. Meanwhile, only interior fatigue crack initiation can be observed in the whole fatigue life range for stress ratios of $R=0.1$ and -0.5. In summary, fatigue cracks tend to start from the interior of the specimen in the long-life fatigue range, and this phenomenon is more obvious in the ultra-long-life fatigue range.

Figure 8.58 shows the SEM images of fatigue fracture surfaces under these four stress ratios [51]. The lack-of-fusion defects are the main reason for the fatigue fracture

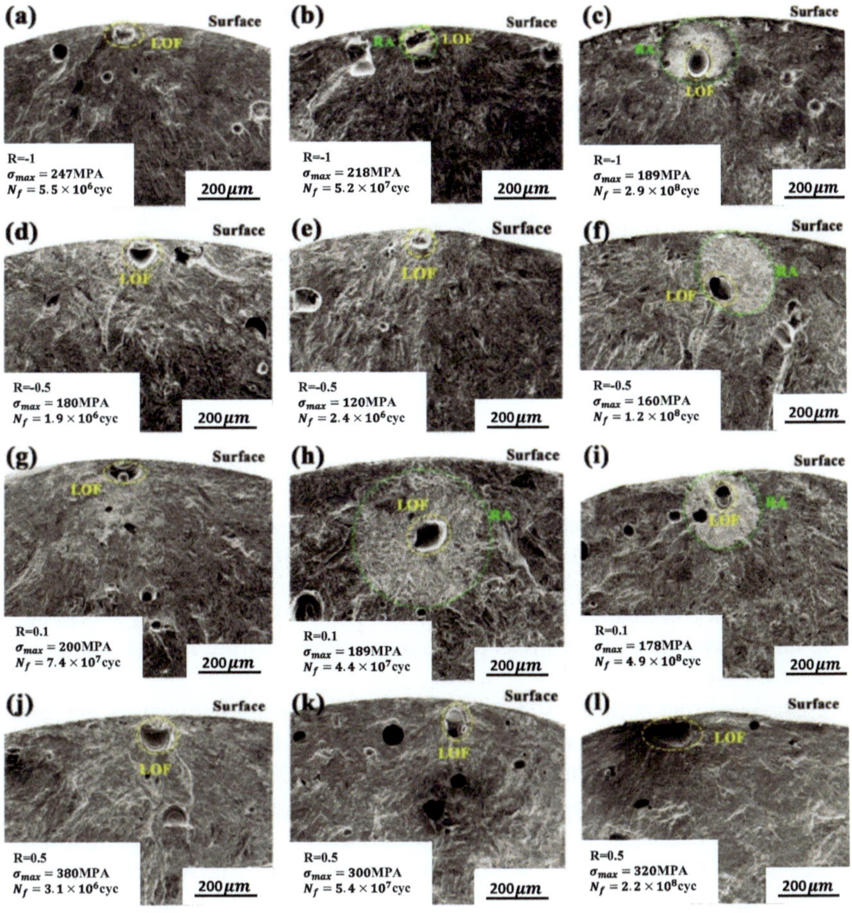

FIGURE 8.58 Fractographic morphologies of SLM Ti-6Al-4V specimens at stress ratios of (a–c) −1, (d–f) −0.5, (g–i) 0.1, and (j–l) 0.5. LOF: lack-of-fusion, RA: rough area. (Reprinted with permission from Ref. [51], Elsevier.)

of the SLM Ti-6Al-4V alloy. It is known that lack-of-fusion defects are generated mainly due to non-optimized AM processing parameters [2]. The rough area appears in the crack initiation region surrounding the fatal lack-of-fusion defects, and it seems that the location of lack-of-fusion defects has a significant effect on the size of rough area. The rough area is the characteristic region of fatigue crack initiation in very high cycle fatigue regime [52] for the SLM Ti-6Al-4V alloy. There is obvious rough area at $R = -1$, -0.5, and 0.1, especially at higher number of cycles. However, there is no rough area for $R = 0.5$ even if an ultra-long fatigue life is reached. Therefore, it is inferred that the rough area can form for the SLM Ti-6Al-4V alloy in very high cycle fatigue regime when the applied stress ratio is less than or close to 0 [51]. Hence, it seems that the formation of the rough area depends on the compressive stress. Overall, the location of defects, stress ratio, and fatigue life have effects on the formation of the rough area for the SLM alloy.

For the AM metals, the inhomogeneous microstructure with specific orientation and the manufacturing defect with 3D features usually appear, finally resulting in the mechanical properties of AM materials with anisotropic characteristics. This anisotropy in the microstructure and defect has a considerable impact on the fatigue behavior of AM materials, and this effect may be exacerbated under different stress ratios.

8.4.5 Size Effect (Risk Volume)

The design of engineering components for static and fatigue failure is based on mechanical properties, which are typically evaluated through lab-based tests on small specimens. The small specimens differ significantly in size from the final component. The size effect of materials refers to the mechanical properties of materials that may vary with the size or risk volume of the specimens, such as diameter or volume. For instance, the static strength decreases as the size of the specimen increases, indicating the size effect of static strength. The fatigue responses of engineering components usually are predicted according to the experimental result of the standard specimen with small size. Despite its critical effects on the structural integrity of engineering components, the effect of size on the fatigue response of AM components have been rarely investigated in the literature so far. Generally, size effect is experimentally verified by testing specimens with different size.

AlSi10Mg samples with two different risk volumes were prepared by SLM along horizontal and vertical directions to study the size effect of AM metals [53]. Two specimen geometries with different risk volumes, hourglass, and Gaussian were employed for the ultrasonic fatigue tests. According to the literature [54], almost all fatigue failures result from the limited material region, where its stress is larger than the 90% of the applied maximum stress (V_{90}). Accordingly, the risk volume can be assumed as the V_{90} ($V_{risk} = V_{90}$). As a result, the value of V_{90} for hourglass specimen is $250\,mm^3$, and the value of V_{90} for Gaussian specimen is $2330\,mm^3$.

Figure 8.59 is the fatigue life data of SLM AlSi10Mg samples along horizontal and vertical directions with two different risk volumes [53]. A significant size effect is experimentally detected for both building directions. A large risk-volume results in an obvious degeneration of fatigue property for AM metals regardless of building directions. The fatigue life of vertically constructed specimens shows a larger scatter both in Gaussian and hourglass specimens. Fatigue failures are mainly attributed

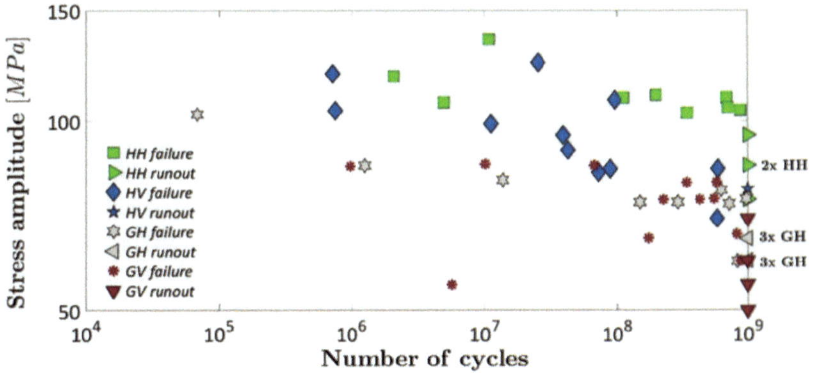

FIGURE 8.59 Fatigue life data of SLM AlSi10Mg samples. HH and HV are the hourglass specimens along horizontal and vertical directions, respectively; GH and GV are the Gaussian specimens along horizontal and vertical directions, respectively. (Reprinted with permission from Ref. [53], Elsevier.)

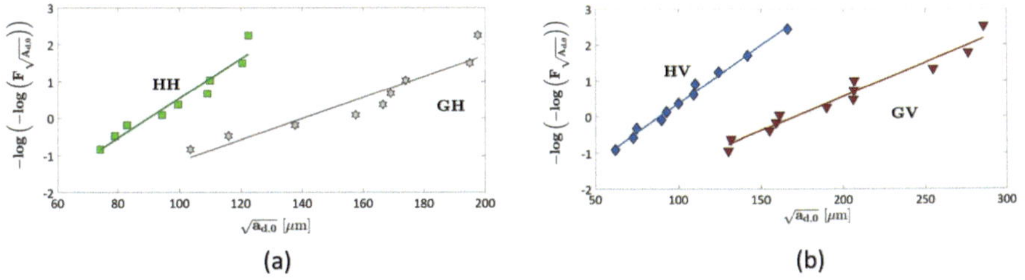

FIGURE 8.60 Critical defect size and estimated Largest Extreme Value Distribution on a Gumbel plot. (Reprinted with permission from Ref. [53], Elsevier.)

to manufacturing defects within the risk volume. Therefore, the characteristics of fatigue property should mainly associate with the features of the defect.

Pores is the dominant crack initiation defect in horizontally constructed specimens, causing about 70% of fatigue failures. For samples constructed vertically, the most critical defects are pores in the hourglass sample and lack-of-fusion defects in the Gaussian sample. Figure 8.60 gives the distributions of critical defect size to further evaluate the size effect associated with the defect size [53]. The largest defect in hourglass samples is just about 60% of that in the corresponding Gaussian sample for both building directions. This huge difference in largest defects, rather than the defect type, accounts for the significant size effect on fatigue property of AM metals.

8.5 ENVIRONMENTAL EFFECT ON AM FATIGUE

The fatigue behavior of a material is influenced by various environmental conditions, such as temperature, corrosion medium, etc. Temperature variations have a significant impact on material properties, and exposure to high temperatures generally has a detrimental effect on the fatigue behavior of the material.

Since metal materials are predominantly exposed to the surrounding liquid or gas environment in service, corrosion is also an important factor for consideration. The corrosion medium itself can act as an active medium and may significantly degrade the fatigue performance of metal materials. Fatigue failure in the low-cycle range usually results from surface-initiated cracks. However, as the fatigue life increases to the high-cycle or very-high-cycle range, cracks tend to initiate internally. Internal cracks are generally not affected by corrosion medium, hence surface and internal fatigue crack failure behavior will differ when metals are exposed in the surrounding corrosion environment.

When evaluating the fatigue performance of AM metals, it is important to study the effects of different environmental factors, with particular attention given to temperature and corrosion effects in this part. This section discusses the fatigue failure behavior of AM materials under different environmental conditions, expecting to provide guidance for inhibiting the initiation of fatigue cracks and reducing the fatigue crack propagation rate.

8.5.1 Temperature

To ensure the service safety of AM metals in different temperature environment, it is important to understand the relationship between temperature and the fatigue behavior of AM materials.

K536 nickel-based superalloy specimens were prepared by SLM in horizontal and vertical construction directions [55]. The residual stress introduced in the preparation process is eliminated by post-heat treatments. The specimens were subjected to fatigue tests at the high temperature of 400°C and 600°C. Figure 8.61 [55] is the fatigue testing results of SLM K536 alloys in the horizontal and vertical directions under the corresponding high-temperature conditions. As seen, the fatigue property of the AM metal decreases with the increase of the temperature regardless of building directions. A more obvious decrease in fatigue resistance is detected in the horizontal orientation specimen than in the vertical orientation specimen, when the applied temperature increases.

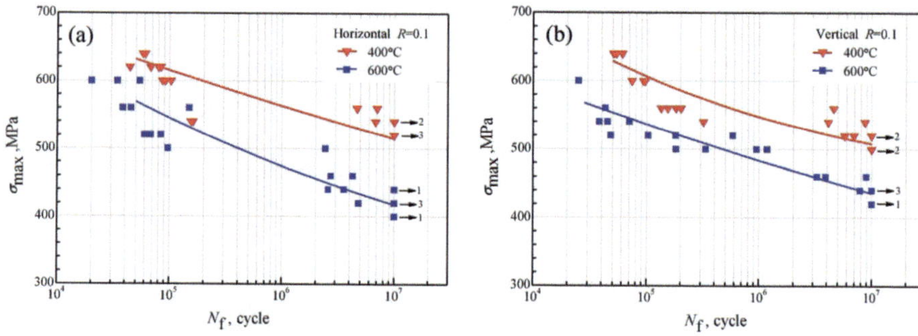

FIGURE 8.61 S-N curves of SLM K536 alloys in (a) horizontal orientation and (b) vertical orientation at 400°C and 600°C. (Reprinted with permission from Ref. [55], Elsevier.)

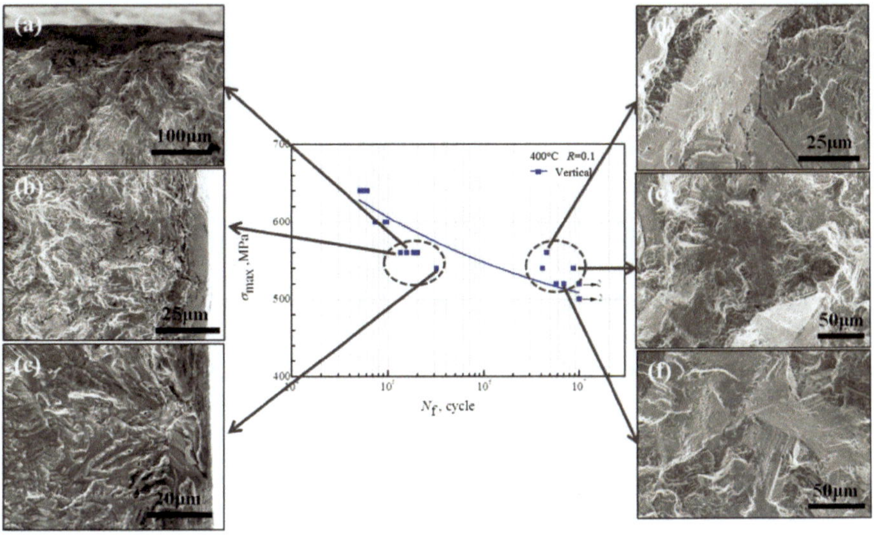

FIGURE 8.62 Crack initiation morphologies of SLM K536 specimens at 400°C. (a–c) Shorter failure life specimens; (d–f) longer failure life specimens. (Reprinted with permission from Ref. [55], Elsevier.)

For AM samples, cracks generally initiate from manufacturing defects (i.e., gas pores, LOF defects, etc.) under cyclic loading. However, in the case of SLM K536 specimen at elevated temperature, the surface slip and internal crystallography facets are found as the sources of crack initiation as shown in Figure 8.62 [55]. This may be due to that additional post-treatment was applied during specimen preparation. For specimens with short fatigue life, the fatigue failure is induced by the surface slip, while internal crystallography facets dominate the fatigue failure for specimens with long fatigue life even if the same stress level is applied. That is to say, the fatigue performance in high temperature severely depends on the location of the crack initiation. The high-temperature oxidation behavior is absent for interior crack initiation, this may account for the severe location sensitivity. For shorter life specimens, the surface slip initiation with a quasi-cleavage fracture mode appears in the fractography at both 400°C and 600°C. The longer life specimen exhibits internal crystallography facet characteristics with a cleavage fracture mode at both 400°C and 600°C. In conclusion, the high temperature has a significant effect on the fatigue resistance but plays little role in the crack initiation and propagation morphologies.

Figure 8.63 gives the other case about the effect of high temperature on fatigue behavior of AM metals, is the S-N curves of IN718 specimens prepared by SLM at 25°C, 450°C, and 650°C [56].

Overall, the fatigue strength decreases with the increase of temperature. However, a similar fatigue strength at 1E7 cycles is observed at 25°C and 450°C, this indicates that the SLM IN718 alloy exhibits little temperature sensitivity at moderate temperature in the long-life range. The fracture surface observation results [56] indicate that the fatigue crack sources are manufacturing defects (gas pore or LOF) located in

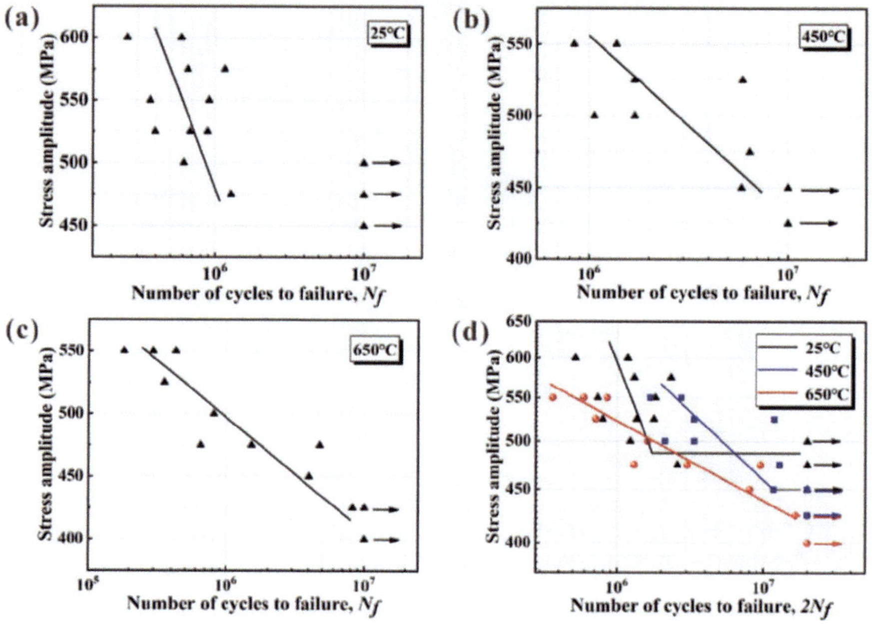

FIGURE 8.63 S–N curves of SLM IN718 alloy at different temperatures. (Reprinted with permission from Ref. [56], Elsevier.)

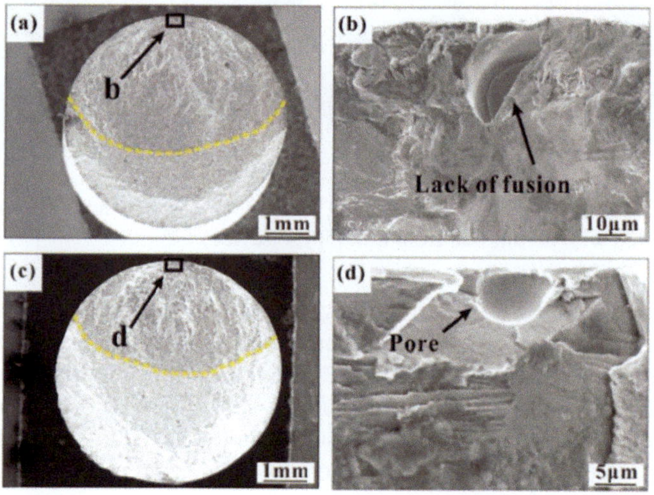

FIGURE 8.64 Fractographs of SLM IN718 fatigue specimens at 25°C. (Reprinted with permission from Ref. [56], Elsevier.)

the surface and subsurface at 25°C as shown in Figure 8.64. It is interesting that the fatigue crack sources are transformed into subsurface and internal facets at 450°C and 650°C as shown in Figure 8.65 [56]. This feature seems to contradict what we know.

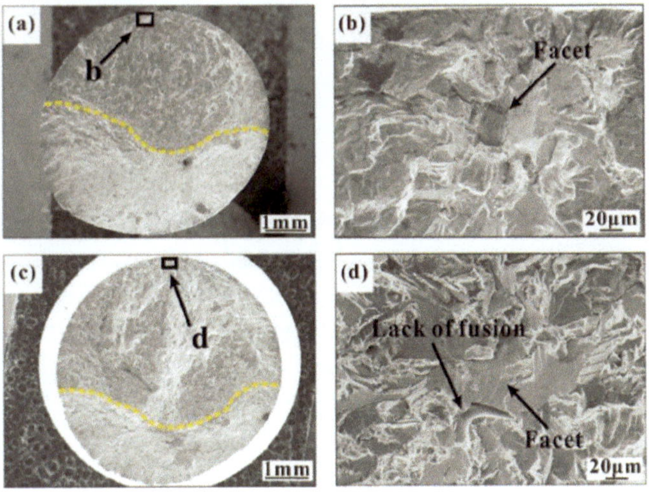

FIGURE 8.65 Fractographs of SLM IN718 fatigue specimens at (a, b) 650°C and (c, d) 450°C. (Reprinted with permission from Ref. [56], Elsevier.)

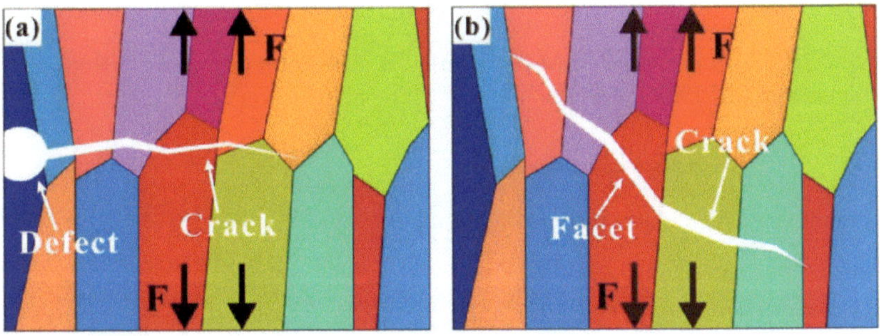

FIGURE 8.66 Schematic of fatigue cracking behavior at (a) room and (b) elevated temperatures. (Reprinted with permission from Ref. [56], Elsevier.)

At high temperatures, a dense oxide layer is formed on the surface of the sample, which can restrain the propagation of surface cracks. The decrease of matrix strength will blunt crack tips and reduce the notch sensitivity of the material. Moreover, dislocations are more likely to be activated at the weakest large grains along the preferential slip plane at high temperature, resulting in the formation of facets. For these reasons, the fatigue crack initiation are more likely to occur at subsurface and internal facets at 450°C and 650°C.

Figure 8.66 [56] summarizes the fatigue cracking mechanism at room temperature and high temperature. Surface and subsurface defects mainly act as the fatigue crack initiation sites at room temperature. Therefore, the fatigue property can be enhanced by reducing the defect number and size, which may be achieved by the optimization of AM parameters and even hot isostatic pressing. At elevated temperatures, the fatigue crack initiation generally results from subsurface and internal crystallography facets. At this time, the fatigue resistance can be enhanced mainly by reducing the size of coarse crystals and introducing the strengthened precipitate phase, which may be achieved by a faster cooling rate during the AM process and proper aging post-heat treatment.

8.5.2 Corrosion

Corrosion fatigue causes the gradual degradation of materials induced by the combined actions of cyclic loading and corrosive environments, finally leading to fatigue failure at a lower stress level than that determined by mechanical fatigue alone. In general, there are three stages for corrosion fatigue in aqueous media: active solvent, electrochemical passivation, and local passivation damage. During the active solvent stage, corrosion rates are high and material degradation is rapid. In the electrochemical passivation stage, a protective oxide layer forms on the metal surface, reducing corrosion rates and increasing resistance to further corrosion. However, at the local passivation damage stage, the protective oxide layer is locally breached, leading to rapid corrosion, and finally degenerating the fatigue property. In atmospheric environments, the fatigue life of materials is lower than that in vacuum. This may be because the oxygen atoms adsorbed on the fresh metal surface reduce the plasticity of crack tips, further preventing the crack closure. To mitigate the corrosion fatigue effect, surface coating and the use of corrosion-resistant alloys can be employed.

Ti-6Al-4V fatigue specimens were prepared by SLM, and the specimens were placed in air, artificial saliva, Ringer's solution, and 3.5%(wt) NaCl solution for studying its corrosion fatigue effects [57]. Figure 8.67 [57] shows the obtained fatigue crack growth rate curves in these four environments for comparison. As seen, a significant corrosion effect for the AM metal is detected in these three solutions. Compared with the air environment, the crack propagation rates increase 52%, 290%, and more than

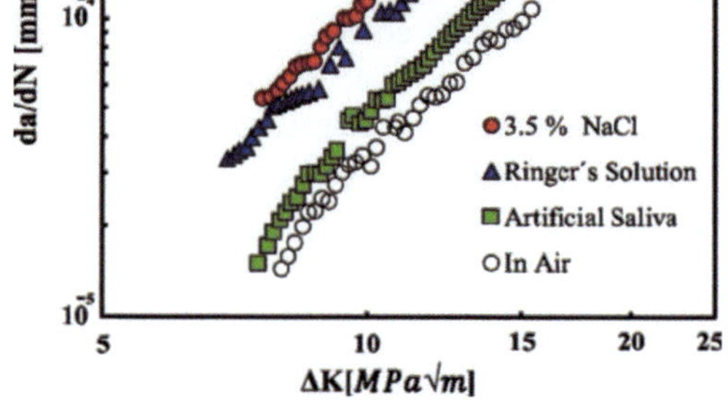

FIGURE 8.67 *da/dN-ΔK* curves in different corrosive solutions. (Reprinted with permission from Ref. [57], Elsevier.)

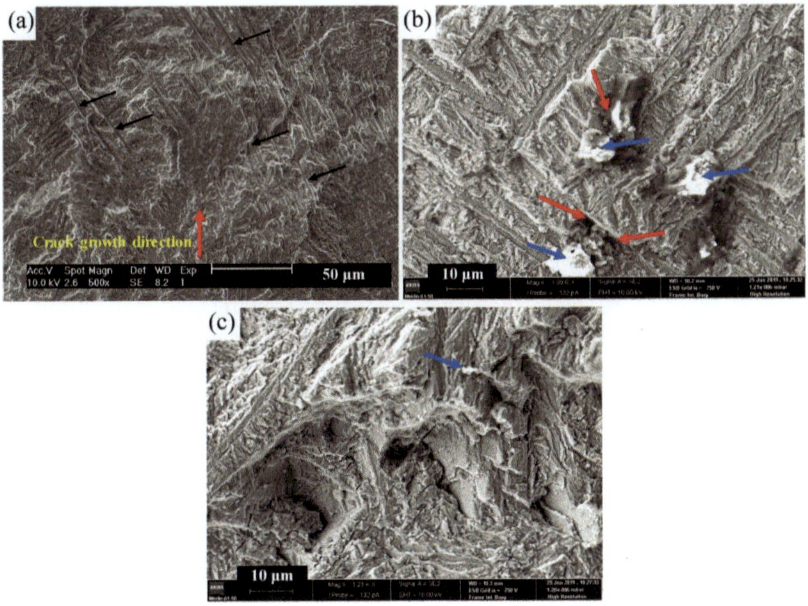

FIGURE 8.68 Fatigue fracture surfaces in (a) air, (b) NaCl solution, (c) artificial saliva. (Reprinted with permission from Ref. [57], Elsevier.)

330% in artificial saliva, Ringer's solution, and 3.5% NaCl solution, respectively [57]. The highest corrosive ability on the AM metal is observed in the 3.5% NaCl solution, while the artificial saliva presents the weakest corrosive ability on the AM metal. The passivation current density presents the following order: artificial saliva solution < Ringer's solution < 3.5% NaCl solution. This demonstrates that the protective oxide (TiO_2) film can be created more easily in artificial saliva but more difficult in 3.5% NaCl solution for the AM Ti-6Al-4V alloy. This tendency is consistent with the results of fatigue crack growth rate.

Figure 8.68 compares the fracture surfaces of the fatigue specimens tested in different environments [57]. Several characteristic regions with corrosion-related phenomena are obviously visible in the 3.5% NaCl solution, while it is not significant in the artificial saliva solution. The characteristic region is the result of the joint action of pitting corrosion and crevice corrosion. Pitting corrosion leads to surface depressions. Crevice corrosion generally occurs near pitting areas, resulting in the presence of light-colored agglomerates. These corrosion phenomena (pitting and crevices) conform that the NaCl solution severely accelerates fatigue crack growth of the AM metal.

8.6 SUMMARY

In order to realize the evolution from AM functional parts to AM structural parts, the fatigue behavior of AM metal materials must be paid great attention. At present, due to the void defects and inhomogeneous microstructure of AM metal materials, their fatigue strength is relatively low, and fatigue life dispersion is large. Hence, AM metal materials are difficult to be widely used in structural parts subjected to complex loads and environments at present. By optimizing AM process parameters and applying suitable post-fabrication treatments, fatigue properties of AM metal materials can be improved significantly. AM technology is increasingly used in key equipment innovations, such as advanced aeroengines (GE9X, GE) and aerospace engines (Raptor 3, SpaceX). It is expected that with the maturity of AM technology, it will be widely used in the future because of its significant advantages in near-net forming, etc.

ACKNOWLEDGMENTS

This work was financially supported by the National Natural Science Foundation of China (No.12202080 and No.12332012). Moreover, the authors warmly acknowledge Yuehui Jiang for his assistance in draft preparation.

REFERENCES

1. Sanaei, N., A. Fatemi, and N. Phan, Defect characteristics and analysis of their variability in metal L-PBF additive manufacturing. *Materials & Design*, 2019; 182:108091.
2. Yang, K., et al., Competing crack initiation behaviors of a laser additively manufactured nickel-based superalloy in high and very high cycle fatigue regimes. *International Journal of Fatigue*, 2020; 136:105580.
3. Parimi, L.L., et al., Microstructural and texture development in direct laser fabricated IN718. *Materials Characterization*, 2014; 89:102–111.
4. Biswal, R., et al., Interrupted fatigue testing with periodic tomography to monitor porosity defects in wire + arc additive manufactured Ti-6Al-4V. *Additive Manufacturing*, 2019; 28:517–527.
5. Sanaei, N. and A. Fatemi, Defects in additive manufactured metals and their effect on fatigue performance: A state-of-the-art review. *Progress in Materials Science*, 2020; 117:100724.
6. Liu, F., et al., Effects of defects on tensile and fatigue behaviors of selective laser melted titanium alloy in very high cycle regime. *International Journal of Fatigue*, 2020; 140:105795.
7. Wang, Z., et al., The microstructure and mechanical properties of deposited-IN718 by selective laser melting. *Journal of Alloys and Compounds*, 2012; 513:518–523.
8. Alabort, E., et al., Alloys-by-design: A low-modulus titanium alloy for additively manufactured biomedical implants. *Acta Materialia*, 2022; 229:117749.
9. Pistor, J. and C. Korner, A novel mechanism to generate metallic single crystals. *Scientific Reports*, 2021; 11(1):24482.
10. Zhao, Q., et al., Ultra-high cycle fatigue and ultra-slow crack growth behavior of additively manufactured AlSi7Mg alloy. *International Journal of Structural Integrity*, 2024; 15(2):382–407.
11. Li, W., et al., Subsurface faceted cracking behavior of selective laser melting Ni-based superalloy under very high cycle fatigue. *Scripta Materialia*, 2021; 194:113613.
12. Tusher, M.M.H. and A. Ince, Effect of stress-relieved heat treatment on very high cycle fatigue performance of additive manufactured Ti-6Al-4V alloy. *Fatigue & Fracture of Engineering Materials & Structures*, 2023; 46(10):3982–4000.
13. Yamashita, Y., et al., Defect analysis and fatigue design basis for Ni-based superalloy 718 manufactured by selective laser melting. *International Journal of Fatigue*, 2018; 117:485–495.

14. Dan, C., et al., Achieving ultrahigh fatigue resistance in AlSi10Mg alloy by additive manufacturing. *Nature Materials*, 2023; 22(10):1182–1188.
15. Rhein, R.K., et al., Effect of direct metal laser sintering build parameters on defects and ultrasonic fatigue performance of additively manufactured AlSi10Mg. *Fatigue & Fracture of Engineering Materials & Structures*, 2020; 44(2):295–305.
16. Guenther, J., et al., Fatigue life of additively manufactured Ti-6Al-4V in the very high cycle fatigue regime. *International Journal of Fatigue*, 2017; 94:236–245.
17. Chinnapat, P., Tang Y.T., and Reed R.C., Metal 3D printing as a disruptive technology for superalloys. *Nature Communications*, 2020; 11(1):2327.
18. Galati, M. and L. Iuliano, A literature review of powder-based electron beam melting focusing on numerical simulations. *Additive Manufacturing*, 2018; 19:1–20.
19. Zhang, M., et al., AlCoCuFeNi high-entropy alloy with tailored microstructure and outstanding compressive properties fabricated via selective laser melting with heat treatment. *Materials Science and Engineering a-Structural Materials Properties Microstructure and Processing*, 2019; 743:773–784.
20. Chew, Y., et al., Microstructure and enhanced strength of laser aided additive manufactured CoCrFeNiMn high entropy alloy. *Materials Science and Engineering A-Structural Materials Properties Microstructure and Processing*, 2019; 744:137–144.
21. Chastand, V., et al., Comparative study of fatigue properties of Ti-6Al-4V specimens built by electron beam melting (EBM) and selective laser melting (SLM). *Materials Characterization*, 2018; 143:76–81.
22. Soltani-Tehrani, A., et al., Ti-6Al-4V powder characteristics in laser powder bed fusion: The effect on tensile and fatigue behavior. *Additive Manufacturing*, 2022; 51:17.
23. Zhou, L.B., et al., Study of substrate preheating on the microstructure and mechanical performance of Ti-15Mo alloy processed by selective laser melting. *Journal of Alloys and Compounds*, 2022; 928:11.
24. Ali, H., et al., In-situ residual stress reduction, martensitic decomposition and mechanical properties enhancement through high temperature powder bed pre-heating of Selective Laser Melted Ti6Al4V. *Materials Science and Engineering A-Structural Materials Properties Microstructure and Processing*, 2017; 695:211–220.
25. Brandl, E., et al., Additive manufactured AlSi10Mg samples using selective laser melting (SLM): Microstructure, high cycle fatigue, and fracture behavior. *Materials & Design*, 2012; 34:159–169.
26. Jia, H.L., et al., Scanning strategy in selective laser melting (SLM): A review. *International Journal of Advanced Manufacturing Technology*, 2021; 113(9–10):2413–2435.
27. Wits, W.W., et al., The role of scan strategies in fatigue performance for laser powder bed fusion. Cirp Annals-Manufacturing Technology, 2022; 71(1):185–188.
28. Li, R.D., et al., Selective laser melting of an equiatomic CoCrFeMnNi high-entropy alloy: Processability, non-equilibrium microstructure and mechanical property. *Journal of Alloys and Compounds*, 2018; 746:125–134.
29. Zhang, M., et al., Fatigue and fracture behaviour of laser powder bed fusion stainless steel 316L: Influence of processing parameters. *Materials Science and Engineering A-Structural Materials Properties Microstructure and Processing,* 2017; 703:251–261.
30. Riemer, A., et al., On the fatigue crack growth behavior in 316L stainless steel manufactured by selective laser melting. *Engineering Fracture Mechanics*, 2014; 120:15–25.
31. Hatami, S., et al., Fatigue strength of 316L stainless steel manufactured by selective laser melting. *Journal of Materials Engineering and Performance*, 2020; 29(5): 3183–3194.
32. Gong, J.Q., et al., Microstructure and mechanical properties of AlSi10Mg alloy built by laser powder bed fusion/direct energy deposition hybrid laser additive manufacturing. *Additive Manufacturing*, 2022; 59:16.

33. Tang, M. and P.C. Pistorius, Oxides, porosity and fatigue performance of AlSi10Mg parts produced by selective laser melting. *International Journal of Fatigue*, 2017; 94:192–201.
34. Kaufmann, N., et al., Influence of process parameters on the quality of aluminium alloy EN AW 7075 using selective laser melting (SLM). *Physics Procedia*, 2016; 83(C):918–926.
35. Esmaeilizadeh, R., et al., On the effect of laser powder-bed fusion process parameters on quasi-static and fatigue behaviour of Hastelloy X: A microstructure/defect interaction study. *Additive Manufacturing*, 2021; 38:12.
36. Molaei, R., A. Fatemi, and N. Phan, Significance of hot isostatic pressing (HIP) on multiaxial deformation and fatigue behaviors of additive manufactured Ti-6Al-4V including build orientation and surface roughness effects. *International Journal of Fatigue*, 2018; 117:352–370.
37. Zhu Z.G., et al., Hierarchical microstructure and strengthening mechanisms of a CoCrFeNiMn high entropy alloy additively manufactured by selective laser melting. *Scripta Materialia*, 2018; 154:20–24.
38. Zhang, C., et al., Effect of heat treatments on fatigue property of selective laser melting AlSi10Mg. *International Journal of Fatigue*, 2018; 116:513–522.
39. Erfan, M., et al., Fatigue behaviour of notched laser powder bed fusion AlSi10Mg after thermal and mechanical surface post-processing. *Materials Science & Engineering A*, 2022; 829:142145.
40. Soyama, H. and Y. Okura, The use of various peening methods to improve the fatigue strength of titanium alloy Ti6Al4V manufactured by electron beam melting. *AIMS Materials Science*, 2018; 5(5):1000–1015.
41. Kahlin, M., et al., Improved fatigue strength of additively manufactured Ti6Al4V by surface post processing. *International Journal of Fatigue*, 2020. 134:105497.
42. Muhammad, M., et al., An investigation into the effects of cyclic strain rate on the high cycle and very high cycle fatigue behaviors of wrought and additively manufactured Inconel 718. *International Journal of Fatigue*, 2021. 144:106038.
43. Sarkar, S., C.S. Kumar, and A.K. Nath, Investigation on the mode of failures and fatigue life of laser-based powder bed fusion produced stainless steel parts under variable amplitude loading conditions. *Additive Manufacturing*, 2019; 25:71–83.
44. Yang, K., et al., Stress ratio and notch effects on the very high cycle fatigue properties of a near-alpha titanium alloy. *Materials*, 2018; 11(9):1778.
45. Solberg, K. and F. Berto, Notch-defect interaction in additively manufactured Inconel 718. *International Journal of Fatigue*, 2019; 122:35–45.
46. Nicoletto, G., Anisotropic high cycle fatigue behavior of Ti-6Al-4V obtained by powder bed laser fusion. *International Journal of Fatigue*, 2017; 94:255–262.
47. Nicoletto, G., Smooth and notch fatigue behavior of selectively laser melted Inconel 718 with as-built surfaces. *International Journal of Fatigue*, 2019; 128:105211.
48. Solberg, K. and F. Berto, The effect of defects and notches in quasi-static and fatigue loading of Inconel 718 specimens produced by selective laser melting. *International Journal of Fatigue*, 2020; 137:105637.
49. Yang, K., et al., Very high cycle fatigue behaviors of a turbine engine blade alloy at various stress ratios. *International Journal of Fatigue*, 2017; 99(Part 1):35–43.
50. Yang, K., et al., Stress ratio effect on notched fatigue behavior of a Ti-8Al-1Mo-1V alloy in the very high cycle fatigue regime. *International Journal of Fatigue*, 2018; 116:80–89.
51. Fu, R., et al., An experimental investigation of fatigue performance and crack initiation characteristics for an SLMed Ti-6Al-4V under different stress ratios up to very-high-cycle regime. *International Journal of Fatigue*, 2022; 164:14.

52. Yang, K., et al., Influence of the volume content of $\alpha+\beta$ colonies on the very high cycle fatigue behavior of a titanium alloy. *Fatigue & Fracture of Engineering Materials & Structures*, 2021; 44(10):2643–2658.
53. Tridello, A., et al., Size-effects affecting the fatigue response up to 10(9) cycles (VHCF) of SLM AlSi10Mg specimens produced in horizontal and vertical directions. *International Journal of Fatigue*, 2022; 160:11.
54. Y. Murakami, *Metal Fatigue: Effects If Small Defects and Nonmetallic Inclusions.* Elsevier, 2002.
55. Jiao, Z.H., et al., Experimental evaluation on elevated temperature fatigue and tensile properties of one selective laser melted nickel based superalloy. *International Journal of Fatigue*, 2019; 121:172–180.
56. Li, W.B., et al., The high-cycle fatigue properties of selective laser melted Inconel 718 at room and elevated temperatures. *Materials Science and Engineering A-Structural Materials Properties Microstructure and Processing*, 2022; 836:11.
57. Jesus, J.S., et al., Fatigue crack growth under corrosive environments of Ti-6Al-4V specimens produced by SLM. *Engineering Failure Analysis*, 2020; 118:104852.

9 Fatigue of Joints and Bridges

Cláudio S. Horas, Abílio M.P. De Jesus, José A.F.O. Correia, and Rui Calçada

9.1 INTRODUCTION TO STRUCTURAL INTEGRITY OF BRIDGES AND JOINTS

Metallic structures can suffer structural degradation over time due to environmental factors and loads that were not properly considered in the design. In this context, it is important to recognize fatigue as a significant threat to safety, causing localized damage at connections that can progress to failures of components and total collapses in the absence of redundancy. Concerns regarding these phenomena extend to all existing structures, but ageing bridges are particularly susceptible, due to the ongoing changes in traffic demand driven by socioeconomic development and environmental issues, with the railway ones being the most fatigue-prone.

Political authorities have shaped mobility policies to meet the need for reducing CO_2 emissions from the transportation of people and goods, assigning a critical role to railways and anticipating an increase in their tonnage in the coming years [1]. Numerous countries are investing in high-speed corridors that are integrated with complementary lines [2], necessitating infrastructure that can accommodate growing traffic demands. However, the development of railways cannot only rely on new structures, and the full capacity of existing ones should be explored, in line with a philosophy of maximizing material capacity utilization.

Railway bridges are subjected to complex cyclic loadings resulting from operating trains, which leads to progressive structural degradation from accumulated damage. Over time, a given metallic connection under repetitive loading may initiate fatigue cracks that then growth, compromising the load-bearing capacity. The awareness for the structural consequences of fatigue on bridges emerged only a few decades ago, significantly later than the initial era of metallic construction. This results in a notable number of such structures still in operation that were designed without explicit fatigue evaluations, with some constructed using outdated materials (Figure 9.1).

The historical progression of bridge construction technologies is confronted with developments in understanding fatigue in Figure 9.1, revealing an important gap to the first normative guidelines covering this issue in the 1950s. After an early period of metallic construction fostered by the Industrial Revolution, materials with properties similar to modern mild steel became prevalent in the early 20th century. However, many bridges still in service can be found made from puddle

DOI: 10.1201/9781003405122-9

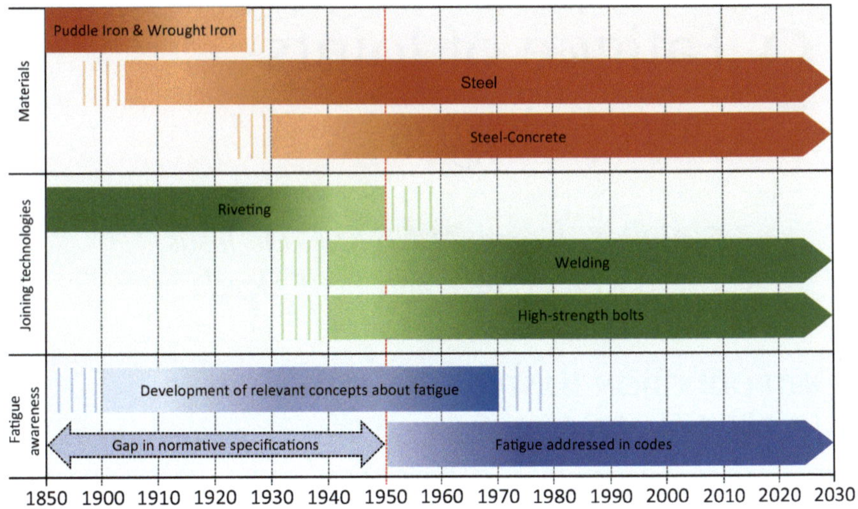

FIGURE 9.1 Metallic bridge construction and fatigue knowledge. (Reprinted with permission from Ref. [3], Elsevier.)

iron or wrought iron. Riveting was crucial for the evolution of structural systems and remained the main joining technique until the 1950s, when welding and high-strength bolts became predominant.

With the first catastrophic fatigue failures raising concern in several industries, pioneering research has been performed, but the localized nature of fatigue has posed challenges in researching focused on large structures like railway bridges. From an engineering perspective, fatigue consists of crack initiation and crack propagation, with the latter further divided into stable and unstable crack growth [4]. The full crack progression process develops from the atomic level up to the millimetre or centimetre scale [5,6], which is still smaller than the global scale of railway bridges, typically measured in metres or kilometres. This difference of scales creates a multi-scale problem that has been influencing decisively fatigue assessment methods.

As summarized in Figure 9.2, two main research lines have been explored for the fatigue assessment of large structures, namely: (1) global methods, covering, in general, the full fatigue life (including both crack initiation and crack propagation) with nominal stresses as input; and (2) local methods, subdivided into those focusing on crack initiation and those dealing with crack growth, as well as alternative methods that cover both stages, with all these models requiring fatigue local parameters.

From a scale compatible with structures and engineering components, the foundations of global fatigue methods were established at the end of the 19th century by Wöhler, relating nominal stresses to the total fatigue life expressed in the number of cycles [7], $\Delta\sigma_{n,i} - N_i$. Currently, these S-N relationships are catalogued and were derived mainly from wide experimental testing on certain geometries, inherently incorporating the respective local stress raisers that cause stress concentrations. Such an approach overcomes the need to assess local parameters at the millimetre scale, thus raising the level of analysis, as nominal stresses can be computed at the global

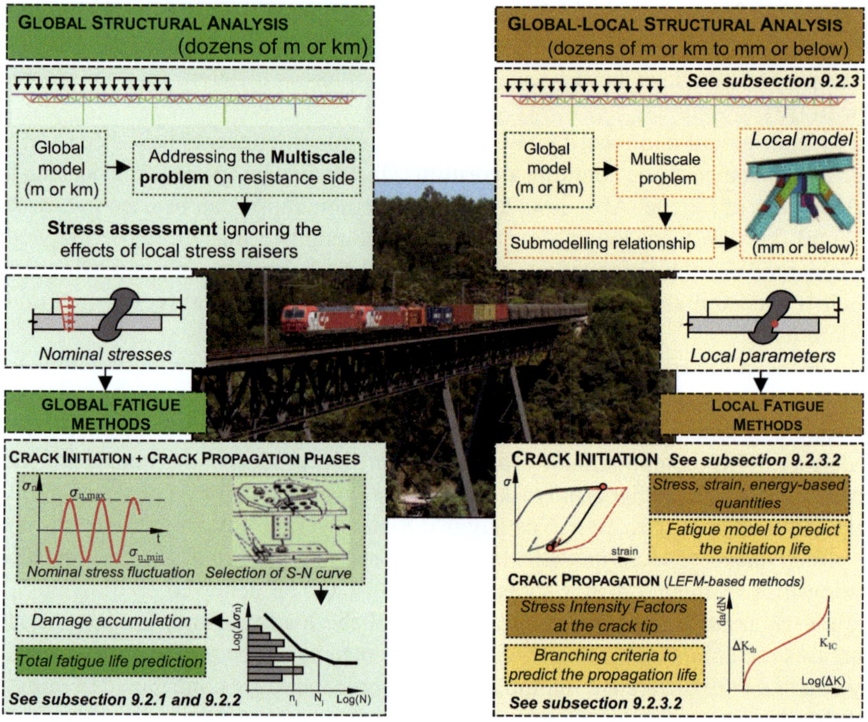

FIGURE 9.2 Research lines for fatigue assessment. (Reprinted with permission from Ref. [3], Elsevier.)

modelling scale. However, the application of such methods to existing details may be questionable, since a given real detail has to be approximated to a catalogued S-N curve, which is limited and does not reflect the diversity of connections found in existing structures or their loading. This inherent approximation introduces a degree of uncertainty, typically resulting in conservative estimates, excessively limiting the remaining fatigue life predictions, but unsafe fatigue results may be estimated in the case of relevant second-order effects.

Based on a scale more aligned with material behaviour, different local methods have been developed for crack initiation analysis, beginning with the Basquin model [8], which uses local elastic stress ranges to predict the number of cycles required for a crack to initiate ($\Delta\sigma^E - N_{f,i}$). In 1954, Coffin [9] and Manson [10] proposed a similar model for the plastic regime, relating strains with the number of cycles ($\Delta\varepsilon^P - N_{f,i}$). Combining these, the Basquin-Coffin-Manson model [11] was proposed to analyse the crack initiation in the elastoplastic regime ($\Delta\sigma^{EP} - \Delta\varepsilon^{EP} - N_{f,i}$). Other models such as those proposed by Morrow [12,13] or Smith-Watson-Topper [14] have been applied, with the latter as an example of those covering also multiaxiality responses.

The initiation phase is typically dominant in riveted and bolted details, but less relevant compared to propagation in welded ones. After a crack initiates, one can cover the entire fatigue life by performing the propagation assessment. Paris, Gomez, and Anderson introduced the Paris' law in 1961 [15], building on the principles of

Fracture Mechanics established previously in 1920 by Griffith [16], which relates the stress intensity factor range with the crack growth rate per cycle ($\Delta K - da/dN$). Other propagation laws have been suggested to address different regimes of crack growth [17].

For both global and local fatigue methods, common concepts have been formulated. For instance, cycle counting algorithms [18,19] may be applied, considering the cyclic nature of fatigue. Also, laws for damage accumulation can be used to estimate the total damage from loading events. In bridge engineering, the most frequently implemented accumulation rule is that proposed by Palmgren [20] and Miner [21], which is accepted as appropriate for fatigue events in the elastic regime and under random loads.

In terms of modelling, global methods incorporate local properties of details into the S-N curves, which allows stress inputs evaluated from a global model conceived with linear or planar Finite Elements (FE), eliminating the need for millimetre-scale modelling. This computational efficiency contributed to the dominance of such methods in regular engineering tasks, but it is associated with loss of reliability due to the inherent approximation to catalogued S-N curves, as noted. On the other hand, the effectiveness of local methods heavily depends on the accuracy of stress-strain relationships evaluated at hot spots, which can only be achieved by fully modelling the geometrical and material properties of the load transmission length. This decreases the computational efficiency, but advancements in computational technology coupled with different dynamic analysis approaches have contributed to mitigate such limitations.

Regarding the solving of the dynamic problem, direct time-integration has been dominant to evaluate loading cycles in small structures, applying algorithms like Newmark [22] or Hilber-Hughes-Taylor (HHT) [23]. However, for large structures, this approach typically requires the analysis of thousands of mutually dependent load steps within a single numerical model. Also, in general, global-scale modelling fails to capture local geometric detailing, resulting in an unreliable numerical representation of the response at hot spots, influencing decisively fatigue damage prediction. In order to overcome these challenges, the modal superposition of local stresses and strains, and stress intensity factors can be used. As stated by Horas et al. [24], localized material or contact nonlinearities do not impact the global linear behaviour, allowing analysis through the superposition of a limited number of vibration modes.

Given this context, submodelling techniques leveraged by modal superposition concepts allow the implementation of local fatigue methods to evaluate the response at specific hot spots both accurately and efficiently. Horas et al. [25] have developed two approaches to address fatigue issues using local parameters. However, applying these to complex structures, such as railway bridges, requires a realistic framework aimed at enhancing the benefits of local fatigue analysis without compromising computational efficiency, which must align with the pace of engineering tasks. This should enable infrastructure managers to conduct generalized safety evaluations to keep metallic railway bridge in service for as long as possible.

Considering that these structures may have hundreds of details, it is necessary to combine the advantages of global and local fatigue methods. Kuhn [26] proposed a multistep framework, establishing a probabilistic approach related to local fatigue concepts to complement global calculations, but the articulation between global and local fatigue methods was unclear, and no workflows were provided to implement fatigue assessment at lower scales. Later, Horas et al. [27] have developed an integrated methodology for predicting the fatigue life of existing metallic railway bridges, using global and local fatigue models with complementary but distinct ranges of application. Sequential phases of analysis were proposed to address details with different geometries and materials, aiming to reliably identify threats to structural safety. In the next Section 9.2, this methodology is presented before being applied to a real case study in Section 9.3. After, other relevant case studies related to the addressed topics are presented briefly in Section 9.4. Finally, in Section 9.5, the conclusions of the work are summarized, clearly highlighting the main advantages of the proposed fatigue assessment based on a global-local philosophy.

9.2 INTEGRATED METHODOLOGY FOR FATIGUE LIFE PREDICTION

Regarding the railway bridges in a specific network, it is recommended to classify the structural details according to the susceptibility to fatigue before important management choices. This methodology is organized into four successively linked phases: (1) Initial Assessment; (2) Standard Evaluation; (3) Detailed Analysis; and (4) Critical Decision. In Figure 9.3, a detailed workflow outlines the steps of fatigue analysis necessary to determine whether further interventions are required to maintain safety throughout the remaining planned service life. For reference, the normative guidelines proposed in the Eurocodes are used, but other relevant standards can be adopted, as the workflow of the proposed integrated methodology is generalizable.

As depicted in Figure 9.3, in order to characterize a certain investigated bridge, it is essential to progressively collect information using different sources, namely: (1) structural drawings to define geometrical and material properties; (2) inspection reports and visual inspections to identify undergoing damage; (3) experimental results to characterize dynamic characteristics through *in situ* tests or to define fatigue models from laboratory tests; and (4) traffic data to define the railway loading from historic information or weighting tests. Figure 9.3 also characterizes in general the calculations at each phase, with the Eurocodes used as a reference, as noted. Subsequent subsections provide a detailed workflow of each phase.

9.2.1 Phase I-Initial Assessment

The first phase is implemented to screen details without infinite fatigue life using a global normative method based on nominal stress, regardless of the age of the bridge. Details over the safety limit should be classified as fatigue-prone (refer to Figure 9.4).

As depicted in Figure 9.4, the process is logically divided into seven subtasks, part of two subphases: (i) analysis and updating of available information (subphase I.1, represented by the bluish top box); and (ii) initial fatigue assessment (subphase I.2,

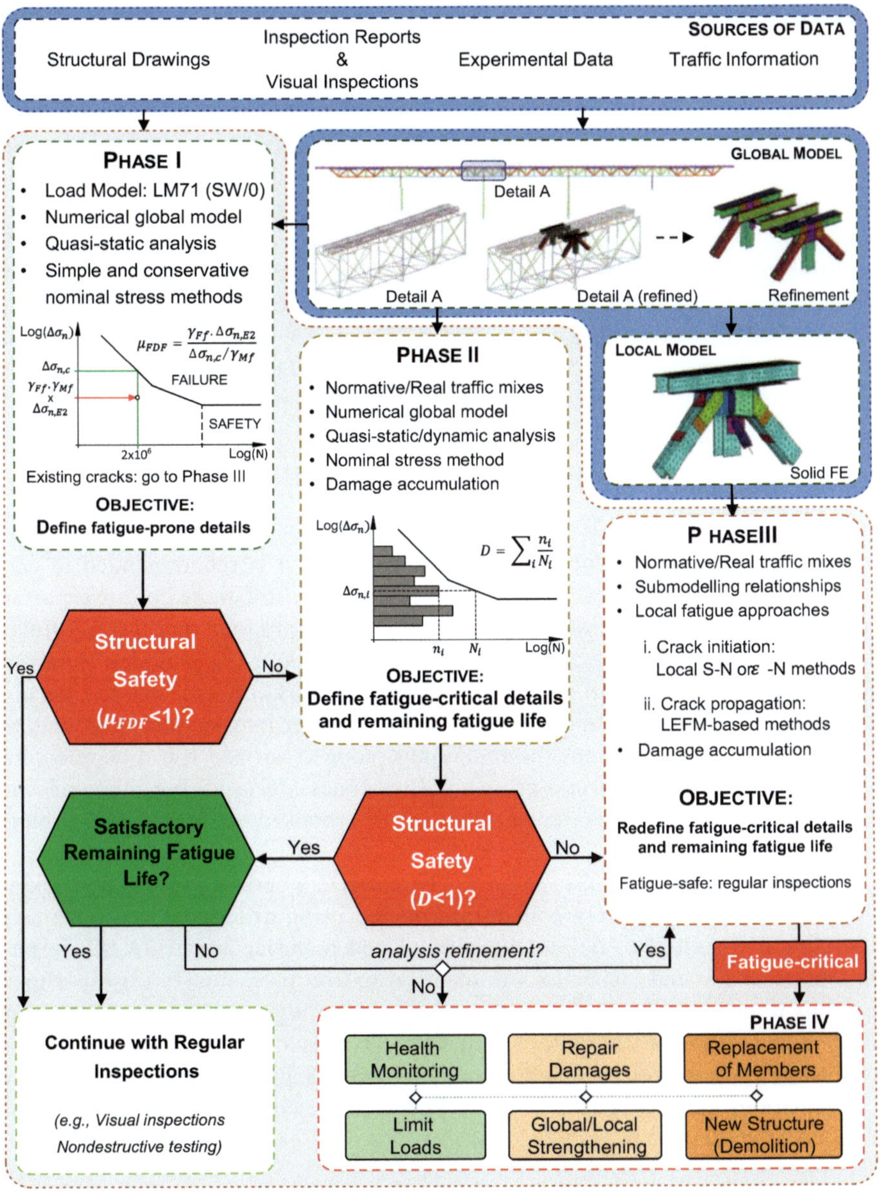

FIGURE 9.3 Integrated methodology for fatigue life prediction of existing metallic railway bridges-workflow. (Reprinted with permission from Ref. [27], Elsevier.)

shown in the greyish bottom box). Three subtasks are defined in subphase I.1, focusing on data collection and the development of global numerical modelling (subtasks i to iii). Regarding subphase I.2, four subtasks are established in relation to calculations for fatigue assessment and subsequent decision-making processes (iv to vii).

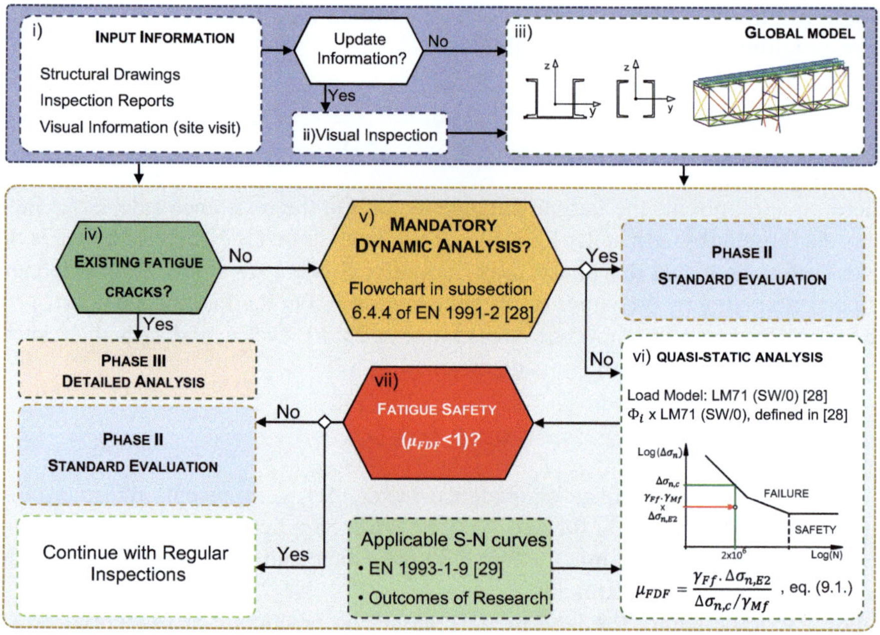

FIGURE 9.4 Phase I: Initial assessment-workflow. (Reprinted with permission from Ref. [27], Elsevier.)

9.2.1.1 Subphase I.1-Analysis and Updating of Available Information

In subtask (i), the information accessible about the bridge should be gathered from design drawings, inspection reports, and visual supports (e.g., photos or videos). Such results help in planning visual inspection to consolidate data, especially if the available information may be outdated, subtask (ii). These data are critical for developing the global numerical model, selecting appropriate modelling techniques and types of finite elements based on the size and type of structural system, subtask (iii).

9.2.1.2 Subphase I.2-Initial Fatigue Assessment

Based on the information gathered, subphase I.2 should begin with subtask (iv), which involves assessing whether propagating cracks were noticed in the inspections. If cracked details are found, the focus should shift to assessing the remaining life using local fatigue models in Phase III, Detailed Analysis (refer to Figure 9.3). Otherwise, the need for dynamic calculations should be verified, according to the workflow provided in subsection 6.4.4 of EN1991-2 [28], as in subtask (v). When dynamic analyses are proved necessary, Phase II, Standard Evaluation, must be started (see Figure 9.3). If quasi-static analyses are considered representative, in subtask (vi), the equivalent constant amplitude stress range method (λ-coefficient method) should be implemented as in EN1993-1-9 [29]. The load model LM71 (and SW/0, if applicable) proposed in EN1991-2 [28] is used to evaluate the infinite fatigue life through the comparison of an amplified nominal stress range with the corresponding allowable value, as follows:

$$\mu_{\text{FDF}} = \frac{\gamma_{\text{Ff}} \cdot \Delta\sigma_{n,E2}}{\Delta\sigma_{n,c}/\gamma_{\text{Mf}}} = \frac{\gamma_{\text{Mf}} \cdot \gamma_{\text{Ff}} \cdot \Delta\sigma_{n,E2}}{\Delta\sigma_{n,c}} \tag{9.1}$$

where, μ_{FDF} represents the fatigue damage factor. On the resistance side, $\Delta\sigma_{n,c}$ indicates the fatigue strength at 2×10^6 cycles from the assumed S-N curve, and γ_{Mf} is the partial safety factor. At this phase, only normative S-N curves are recommended and properly validated by outcomes of previous research. On the loading side, $\Delta\sigma_{n,E2}$ is the equivalent constant amplitude stress range at 2×10^6 cycles, and γ_{Ff} is the respective partial safety factor. $\Delta\sigma_{n,E2}$ is determined as follows:

$$\Delta\sigma_{n,E2} = \lambda \cdot \Delta\sigma_{n,71} \tag{9.2}$$

In Eq. (9.2), λ is the damage equivalence factor, $\Delta\sigma_{n,71}$ represents the maximum stress range related to LM71 (or $\Delta\sigma_{n,\text{SW0}}$ for SW/0, if applicable), and Φ is the amplification factor for the dynamic loading effects, vehicle and track imperfections. The coefficients λ and Φ are explicitly established in EN1991-2 [28].

For a specific detail, the load models should be positioned at sequential points along the track to generate a reliable stress spectrum, following a quasi-static analysis. When LM71 and SW/0 do not accurately represent the traffic characteristics, the responsible management authority should provide an adequate load model. Otherwise, Phase II should be initiated to accommodate alternative traffic scenarios. For shear loads, the respective stress range $\Delta\tau_{n,E2}$ should represent the loading side, with this calculated similarly to $\Delta\sigma_{n,E2}$, and compared against the resistance value set by $\Delta\tau_{n,C}$.

Once μ_{FDF} is calculated, a fatigue classification should be defined for each detail in subtask (vi), with the safety assessed based on this fatigue damage factor. If $\mu_{\text{FDF}} \geq 1.0$, the respective detail has limited fatigue life and is classified as fatigue-prone, with further investigation required in Phase II, Standard Evaluation. Otherwise, if $\mu_{\text{FDF}} < 1.0$, the fatigue life is theoretically infinite, provided that *in situ* visual inspections confirm the absence of initial defects. Regular inspection cycles should be upheld, as local damage may be hidden from observation, e.g., under the heads of rivets and bolts.

9.2.2 Phase II-Standard Evaluation

After identifying the connections prone to fatigue, Phase II requires further analyses. Improvements in numerical modelling are necessary, along with both quasi-static and dynamic calculations, adopting normative or real traffic mixes (refer to Figure 9.5).

As depicted in Figure 9.5, the process is logically divided into seven subtasks, part of two subphases: (i) the collection of experimental data (subphase II.1, represented by the bluish top box); and (ii) the standard fatigue evaluation (subphase II.2, shown as the greyish bottom box). Three subtasks are established in subphase II.1, focusing

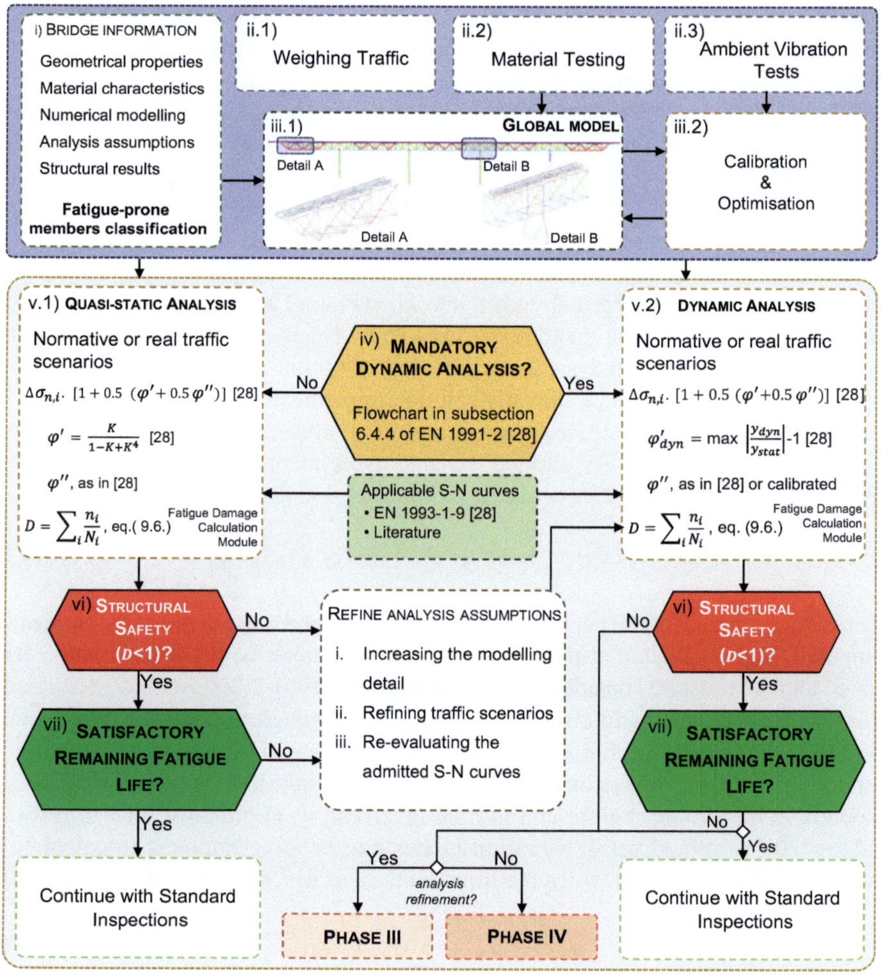

FIGURE 9.5 Phase II: Standard evaluation-workflow. (Reprinted with permission from Ref. [27], Elsevier.)

on acquiring additional information and enhancing numerical models (i to iii). Regarding subphase II.2, four subtasks are established in relation to calculations for fatigue assessment and subsequent decision-making processes (iv to vii).

9.2.2.1 Subphase II.1-Gathering Experimental Information

Logically, the data from Phase I serves as the foundation for gathering additional information. Identifying the limitations of the data in subtask (i) is crucial to plan new testing campaigns and subsequent modelling efforts. Next, in subtask (ii), three types of experimental tests can be planned, namely: (ii.1) weighing campaigns to quantify the real railway traffic load; (ii.2) material testing to define the elastic mechanical characteristics; and (ii.3) Ambient Vibration Tests (AVT) to assess modal values related to the dynamic behaviour. Depending on the details prone to fatigue identified, the data gathered in these two subtasks should contribute to enhancements in the global numerical model, with subtask (iii) divided into modelling improvements based on the updated structural information (iii.1) and those associated with calibration or optimization processes (iii.2), which can be interdependent.

9.2.2.2 Subphase II.2-Standard Fatigue Evaluation

In subtask (iv), the need for dynamic analyses is verified as per subsection 6.4.4 of EN1991-2 [28], with this check already made when Phase I is performed. Accordingly, if dynamic analyses are not mandatory, quasi-static calculations are acceptable (subtask v.1), and dynamic analysis are a refinement option (subtask v.2). Otherwise, dynamic calculations must be the only solution implemented (subtask v.2).

Regarding the quasi-static assessment, nominal stresses are calculated based on the load position, $\sigma_n(x)$. For fatigue damage assessment, the amplification factor may consider the average effects over a 100-year period [28], as follows:

$$\sigma_n(x) = \sigma_{n,\text{sw}} + \sigma_{n,\text{sta}}(x) \times \left[0.5(\varphi' + 0.5\varphi'')\right] \tag{9.3}$$

where, $\sigma_{n,\text{sw}}$ represents the permanent nominal stress and $\sigma_{n,\text{sta}}$ is the stress statically computed, both evaluated at the same location. Additionally, the amplification factors φ' and φ'' must be considered as proposed in EN1991-2 [28], with φ' associated with the effects of trains in circulation, while φ'' is related to the impact of vehicle and track imperfections. For dynamic assessment, modal superposition is recommended to increase the calculation efficiency. Assuming that the structure globally responds in the elastic regime and is time-invariant, as commonly seen in railway bridges [30], a limited set of vibration modes can be superimposed to calculate a given structural parameter, ψ, in the time domain, as described below:

$$\psi(t) = \psi_{\text{sw}} + \sum_j \psi_j \cdot Y_j\ (t) \tag{9.4}$$

where, ψ_{sw} represents the permanent part of ψ and ψ_j is the modal parameter calculated based on the modal shape of the j_{th} vibration mode. In the context of Phase II, the generic quantity ψ is a nominal stress, σ_n, which should be properly amplified according to Eq. (9.5):

$$\sigma_{n(\varphi)}(t) = \sigma_{n,\text{sw}} + \sigma_{n,\text{dyn}}(t) / [1 + \varphi'] \times \left[0.5(\varphi' + 0.5\varphi'')\right] \tag{9.5}$$

where, $\sigma_{n,\text{dyn}}$ is the dynamic nominal stress, calculated from $\sum_j \sigma_{nj} \cdot Y_j\ (t)$ as in Eq. (9.4). In dynamic analyses, φ' is inherently given by $\sigma_{n,\text{dyn}} / \sigma_{n,\text{sta}}$. For φ'', when moving loads are considered, normative guidelines similar to those used in quasi-static calculations can be used. Alternatively, a calibration process can be conducted, involving a two-step comparison: (i) first, benchmark results from a representative bridge-train interaction analysis, accounting for the profile of track irregularities, are compared with outcomes from the moving loads approach; then, (ii) these results are doubly compared with the corresponding static outcomes.

After quasi-static or dynamic analysis, nominal stresses are considered to implement the linear damage summation, which is commonly recommended in guidelines, such as EN1993-1-9 [29]. For a specific connection, several analysis steps should be performed for the stress response associated with each train loading. A cycle counting algorithm may be applied to quantify the relationship between a given nominal stress range and the corresponding number of cycles, $\Delta\sigma_{n,i} - n_i$. For each $\Delta\sigma_{n,i}$, the fatigue resistance may be evaluated in the number of cycles, derived from the adopted S-N curve, $\Delta\sigma_{n,i} - N_i$. In this step, S-N curves proposed in the state of the art may be used, such as those proposed by Taras and Greiner [31,32] for riveted connections, if the modelling accounts for all secondary effects. For j loading events, the relationship $\Delta\sigma_{n,i} - n_i - N_i$ allows for the linear accumulation of fatigue damage for a given traffic scenario, as follows:

$$D = \sum \frac{n_i}{N_i} = \frac{n_1}{N_1} + \frac{n_2}{N_2} + \cdots + \frac{n_j}{N_j} \leq D_L \tag{9.6}$$

where, D represents the total damage, calculated as the sum of damage of all j cycles or half-cycles, and D_L is the damage threshold limit, typically set at 1.00.

For each fatigue-prone detail, a fatigue classification is established based on D. In subtask (vi), safety is assessed by comparing D with D_L. If $D \geq 1.00$, the fatigue safety requirement is not met and the detail is classified as fatigue-critical. At this step, the analysis assumptions must be refined if possible. This may involve enhancing the accuracy of loading scenarios, improving the modelling (e.g., using shell FE), or conducting dynamic calculations if quasi-static ones were performed. Following such refinements, the fatigue damage should be reassessed, and the classification of details should be updated accordingly.

In subtask (vii), predictions for the fatigue life should be weighted depending on the past, present, and planned future traffic. The responsible management authority and railway companies must provide reliable traffic information. Alternatively, or as a complement, a weighing campaign may be performed based on Weigh-In-Motion (WIM) or Bridge-Weigh-In-Motion (B-WIM) algorithms. Depending on the safety criteria, an engineering decision is expected on the magnitude of the remaining fatigue life in comparison with the limit of service life. A primary risk analysis is recommended, considering aspects such as structural redundancy, potential consequences of failure or accessibility for inspection tasks.

Structural details not deemed safe to fatigue must undergo further investigation in Phase III, Detailed Analysis. Nevertheless, for a specific critical connection, the decision to proceed directly to Phase IV should be considered if the structural analyses performed in Phase II are judged accurate and if the connections and linked structural members can be easily repaired or replaced. Otherwise, details that demonstrate an acceptable remaining fatigue life should be classified as fatigue-safe, and the regular inspection cycles should be upheld.

9.2.3 Phase III-Detailed Analysis

Once fatigue-critical connections have been identified, the reliability of fatigue life predictions should be enhanced. For crack initiation and crack propagation, local notch models and Linear Elastic Fracture Mechanics (LEFM) principles should be applied, respectively. Modelling refinements to the global model can be made and related to the necessary submodels conceived for critical details at the mm scale (see Figure 9.6).

As depicted in Figure 9.6, two local fatigue approaches are suggested for implementing local fatigue methods, as suggested by Horas et al. [25], which benchmarked these against a reference "Approach 1" based on direct time-integration. As input, local responses obtained from elastic numerical models are considered, after establishing submodelling relationships leveraged by modal superposition to boost calculation efficiency, foreseeing analytical elastoplastic post-processing if necessary. These two routes differ in their approach to existing local contact nonlinearities, which affect the structural parameters considered for modal superposition. When local nonlinearities related to normal or friction contacts from rivets or bolts do not impact a specific hot spot, the so-called "Approach 2" can be applied by superimposing modal local parameters, taking the generic quantity outlined in Eq. (9.4) as a fatigue input of interest. As an alternative, the impact of nonlinear contacts at the early stages of the fatigue phenomenon can be addressed in "Approach 3" by

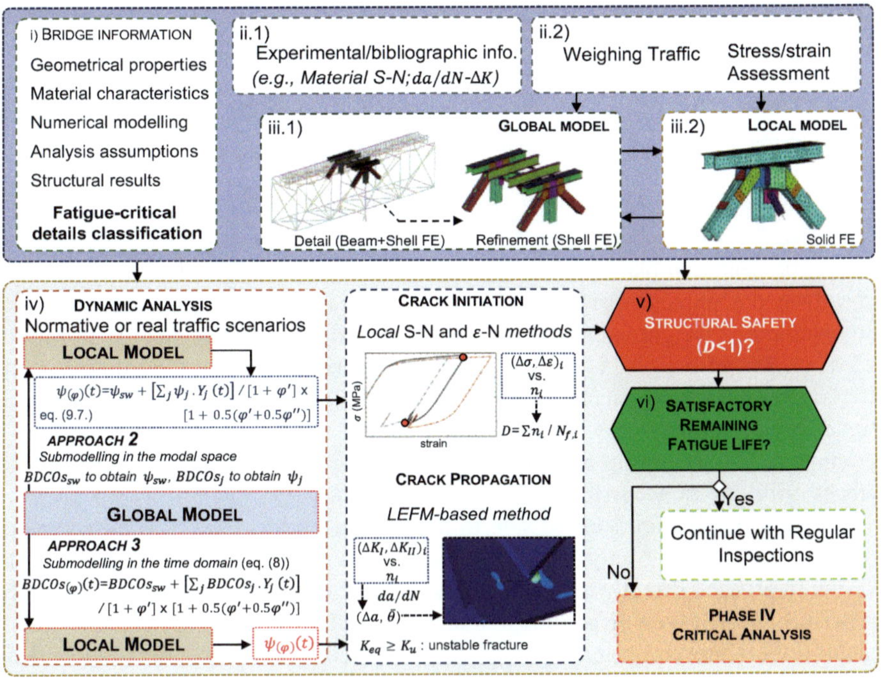

FIGURE 9.6 Phase III: Detailed analysis-workflow. (Reprinted with permission from Ref. [27], Elsevier.)

superimposing modal boundary conditions related to a specific submodelling relationship ($BDCOs_j$), modifying Eq. (9.4) accordingly. Such displacement fields are derived in the time domain, establishing the non-amplified dynamic component of BDCOs, which is part of those to be imposed on the limits of the related submodel, allowing the required local fatigue parameters to be subsequently assessed.

As shown in Figure 9.6, the workflow is logically organized into six subtasks, part of two subphases: (1) updating experimental information (subphase III.1, depicted by the bluish top box); and (2) detailed fatigue assessment (subphase III.2, shown in the greyish bottom box). Three subtasks are proposed in subphase III.1, focusing on gathering additional information and enhancing the modelling (i to iii), while subphase III.2 consists of three subtasks associated with predicting the remaining life (iv to vi).

9.2.3.1 Subphase III.1-Updating Experimental Information

The information gathered from Phase II should be reviewed to address limitations, as outlined in subtask (i). Given the complexity of the proposed local analyses and the advanced modelling required, new experimental data should be collected in subtask (ii). Laboratory tests may be performed to characterize the cyclic behaviour and fatigue properties of the material that composes the detail, but comparable material data may also be collected from the literature (ii.1). If Phase II did not provide sufficient information, a traffic weighting campaign may be conducted. This should be combined with an assessment of the stresses and strains near critical points under specific train loads, which will help validate the established submodelling relationship (ii.2).

For subtask (iii), improvements to the global model should be made in areas identified as being critical for fatigue. The nature of these enhancements depends on the type of FE used in the global model. For larger bridges modelled with line elements, a transition to a more refined shell FE should be done (beam-to-shell link) (iii.1). This refinement aims to increase the precision of the submodelling relationship with a specific local volumetric model (shell-to-solid coupling), which fully considers local geometrical, material, and contact properties (iii.2).

9.2.3.2 Subphase III.2-Detailed Fatigue Assessment

The fatigue assessment at the global scale is based on S-N curves for nominal stresses, which implicitly include the effects of local stress concentrators, addressing the challenge of multiscale modelling. However, this process has limitations that may reduce the reliability of the results, making it crucial to apply local methods.

In subtask (iv), dynamic calculations should be conducted, assuming submodelling relationships. Similar to Phase II, traffic scenarios representative of past, present, and planned future demands should be considered. Moving loads may be applied combined with modal superposition concepts, with values for φ'' calibrated as referred in subphase II.2, following detailed bridge-train interaction calculations. If such calibration has not been previously undertaken, it must be performed at this step. It is important to recognize that while bridge-train interaction methods are more robust and accurate, they increase greatly computational demands. Once defined the algorithm for dynamic calculations, "Approach 2" and "Approach 3" should be considered based on the local properties of the screen fatigue-critical connections [24,25].

For "Approach 2", the submodelling relationship is established within the modal space, where the modal BDCOs$_j$ are imposed on the submodel to compute the local modal parameters, which are then superimposed. From Eq. (9.4), amplification factors are applied to multiply the results locally, as per the following equation:

$$\psi_{(\varphi)}(t) = \psi_{\text{sw}} + \left[\sum_j \psi_j \cdot Y_j(t)\right] / \left[1+\varphi'\right] \times \left[1+0.5(\varphi' + 0.5\varphi'')\right] \quad (9.7)$$

where, $\psi_{(\varphi)}$ represents a specific local quantity that is amplified, and can be interpreted as local stress (σ), local strain (ε), or stress intensity factor (K), depending on the fatigue analysis being conducted. For either uncracked or cracked geometries, the local values are calculated in the time domain from a single set of numerical analysis, with the modal outcomes being a structural property and applicable to all railway loadings, avoiding repeated calculations. For "Approach 3", the submodelling relationship is defined within the time domain, based on the superposition of the modal boundary displacement conditions, which are initially evaluated in the global numerical model, as follows:

$$\text{BDCOs}_{(\varphi)}(t) = \text{BDCOs}_{\text{sw}} + \left[\sum_j \text{BDCOs}_j \cdot Y_j(t)\right] / \left[1+\varphi'\right] \times \left[1+0.5(\varphi' + 0.5\varphi'')\right] \quad (9.8)$$

where, BDCOs$_{(\varphi)}$ represent the boundary displacement fields that have been amplified to be imposed on the submodel, and the term $\sum_j \text{BDCOs}_j \cdot Y_j(t)$ is derived from Eq. (9.4). After applying BDCOs$_{(\varphi)}$, each time step t requires a static numerical analysis, assuming that the inertia effects in the submodel are insignificant in relation to the elastic responses resulting from the imposed boundary fields. To enhance computational efficiency, parallel computing algorithms can be implemented. From these calculations, local quantities are obtained in the time domain, with $\psi_{(\varphi)}(t)$ being σ, ε or K, allowing for the implementation of the appropriate local fatigue model.

9.2.3.2.1 *Crack Initiation Assessment*

From elastic numerical results for computational efficiency, the principal stress and strain components are calculated at the hot spot ($\sigma_1, \sigma_2, \sigma_3$ and $\varepsilon_1, \varepsilon_2, \varepsilon_3$, respectively). The equivalent quantities are then computed based on the theories of von Mises, Tresca, or Absolute Maximum Principal stress (AMP). For AMP, a sign correction must be applied, i.e., $\sigma_{1,\text{ AMP}} = \sigma_1$, if $|\sigma_1| > |\sigma_3|$, or $\sigma_{1,\text{ AMP}} = \sigma_3$, if $|\sigma_3| > |\sigma_1|$. For von Mises or Tresca stresses, the correspondent elastic values should be affected by the sign of $\sigma_{1,\text{ AMP}}$, since these are always positive, implementing an approximation to a multiaxial state defined mainly by σ_1 or σ_3. The elastic stress-strain analysis is finished by obtaining principal biaxial ratios that establish the proportionality between components, i.e., $\lambda_{i,\text{el}} = \sigma_{i,\text{el}}/\sigma_{1,\text{ AMP}}$ and $\phi_{i,\text{el}} = \varepsilon_{i,\text{el}}/\varepsilon_{1,\text{AMP}}$, which are considered to derive multiaxial elastoplastic local responses, if necessary.

When the elastically computed stresses at the hot spot are greater than the yield limit of the material, an elastoplastic post-processing should be incrementally applied. This analytical procedure is based on the Neuber [29] or Glinka criteria [33–35], combined with the material model of Ramberg-Osgood [36], Eqs. (9.9)–(9.11), respectively, in order to calculate the equivalent elastoplastic values.

$$\Delta\sigma^{E} \cdot \Delta\varepsilon^{E} = \frac{\Delta\sigma^{\mathrm{EP}^2}}{E} + 2\Delta\sigma^{\mathrm{EP}}\left(\frac{\Delta\sigma^{\mathrm{EP}}}{2K'}\right)^{1/n'} \tag{9.9}$$

$$\Delta\sigma^{E} \cdot \Delta\varepsilon^{E} = \frac{\Delta\sigma^{\mathrm{EP}^2}}{E} + \frac{4\Delta\sigma^{\mathrm{EP}}}{1+n'}\left(\frac{\Delta\sigma^{\mathrm{EP}}}{2K'}\right)^{1/n'} \tag{9.10}$$

$$\Delta\varepsilon^{\mathrm{EP}} = \frac{\Delta\sigma^{\mathrm{EP}}}{E} + 2\left(\frac{\Delta\sigma^{\mathrm{EP}}}{2K'}\right)^{1/n'} \tag{9.11}$$

where, K' and n' are, respectively, the cyclic strain hardening coefficient and exponent, as in the relationship of Ramberg and Osgood; $\Delta\sigma^{E}$ and $\Delta\varepsilon^{E}$ are, respectively, the local elastic stress and strain ranges related by the modulus of Young (E), with $\Delta\sigma^{\mathrm{EP}}$ and $\Delta\varepsilon^{\mathrm{EP}}$ as the corresponding elastoplastic values.

After establishing the required stress-strain relationship, several local S-N or ε-N models can be implemented, with the former addressing only the elastic regime, while the strain-based ones take elastoplastic responses into account. Local S-N models are related to the material properties, with the stress inputs accounting for the local geometrical and contact characteristics. This differs from S-N curves for nominal stresses, which depend on the tested and analysed connections to derive them, integrating the respective geometrical and material properties. In the elastic regime, the initiation of a certain crack is governed by the local stress range and associated mean stress ($\Delta\sigma$ and σ_m, respectively). In this case, the local S-N model by Basquin [8] can be implemented as follows:

$$\frac{\Delta\sigma^{E}}{2} = \sigma'_f\left(2N_{f,i}\right)^{b} \tag{9.12}$$

where, σ'_f and b are, respectively, the cyclic fatigue strength coefficient and exponent, and $\Delta\sigma$ is the local elastic stress range related to the number of cycles required for crack initiation, $N_{f,i}$. This model excludes the mean stress effect, but it can be accounted for by using the term $\sigma'_f - \sigma_m$ instead of σ'_f. Alternatively, the Goodman and Gerber models [37] can be used to correct $\Delta\sigma$ considering σ_m.

Local fatigue ε-N models cover confined local plasticity, and the load sequence must be considered. These models should then be implemented for each $\Delta\sigma^{\mathrm{EP}} - \Delta\varepsilon^{\mathrm{EP}}$ relationship, with the damage accumulated as in Eq. (9.6). After increment i, the elastoplastic post-processing is reapplied for the next increment $i + 1$, progressively defining the hysteresis cycles. Among the possibilities, one can implement local fatigue models such as Basquin-Coffin-Manson [11] (Eq. 9.13), Morrow [12,13] (Eq. 9.14), or Smith-Watson-Topper (SWT) [14] (Eq. 9.15).

$$\frac{\Delta\varepsilon^{\text{EP}}}{2} = \frac{\sigma'_f}{E}\left(2N_{f,i}\right)^b + \varepsilon'_f\left(2N_{f,i}\right)^c \tag{9.13}$$

$$\frac{\Delta\varepsilon^{\text{EP}}}{2} = \frac{\sigma'_f - \sigma_m}{E}\left(2N_{f,i}\right)^b + \varepsilon'_f\left(2N_{f,i}\right)^c \tag{9.14}$$

$$\frac{\Delta\varepsilon^{\text{EP}}}{2}\sigma_{\max} = \frac{\sigma'^2_f}{E}\left(2N_{f,i}\right)^{2b} + \sigma'_f\varepsilon'_f\left(2N_{f,i}\right)^{b+c} \tag{9.15}$$

where, σ'_f and b are associated with the elastic part as mentioned; ε'_f and c are, respectively, the fatigue ductility coefficient and exponent; σ_m and $\sigma_{\max}$ are, respectively, the local mean and maximum stresses, both for the relationship $\Delta\sigma^{\text{EP}} - \Delta\varepsilon^{\text{EP}}$. The SWT model can also be applied in its multiaxial form [14], considering as input the equivalent elastoplastic stress-strain relationship of von Mises, Tresca, or AMP, influenced by the biaxial ratios earlier defined to compute the principal components.

9.2.3.2.2 *Crack Propagation Assessment*

Once a macroscopic crack is formed, the propagation process must be assessed to compute the remaining fatigue life. The principles of LEFM have been considered in various fields and implemented in large structures with complex details susceptible to fatigue after recent computational developments.

Three pure crack loading modes may be developed: (i) opening mode (mode I, K_{I}); (ii) sliding mode (mode II, K_{II}); and (iii) tearing mode (mode III, K_{III}). Two or all of the defined pure modes may coexist, leading to mixed-mode crack propagation. In the case of railway bridges, complex loading results in intricate crack propagation processes. For such geometries submitted to relevant mixed-mode conditions (I+II), Tanaka [38] suggested the equivalent stress intensity factor, K_{eq}, as follows:

$$K_{\text{eq}}(t) = \sqrt[4]{K_{\text{I}}^4(t) + 8K_{\text{II}}^4(t)} \tag{9.16}$$

where, K_{I} and K_{II} are the stress intensity factors related to the local response. Considering K_{eq}, the Paris' law can be written following the relationship $\Delta K_{\text{eq}} - n_i$, relating these values with da/dN as below:

$$\frac{da}{dN} = C \cdot \Delta K_{\text{eq}}{}^m \tag{9.17}$$

where, C and m are constants associated with the material. Through the integration of the previous Eq. (9.17), the crack increment Δa can be computed:

$$\Delta a = \sum_i n_i \cdot C \cdot \Delta K_{\text{eq},i}{}^m \tag{9.18}$$

The crack path orientation may be evaluated by a crack branching criterion. For example, the Maximum Tangential Stress criterion (MTS) [39] can be implemented, stating that the maximum tangential stress value at the crack tip defines the direction of crack progression. From the values of the stress intensity factors, the growth angle may be established by the kink angle time history, $\theta(t)$, according to:

$$\theta(t) = \cos^{-1}\left(\frac{3 \cdot K_{\mathrm{II}}^2(t) + \sqrt{K_{\mathrm{I}}^4(t) + 8 \cdot K_{\mathrm{I}}^2 \cdot K_{\mathrm{II}}^2(t)}}{K_{\mathrm{I}}^2(t) + 9 \cdot K_{\mathrm{II}}^2(t)}\right) \tag{9.19}$$

In Eq. (9.19), at a time instant t, θ is the angle related to the original plane of the crack, being positive when K_{II} is negative, following the local coordinate system of Fracture Mechanics. For a complex loading, such as that derived from a moving train, an equivalent kink angle, $\bar{\theta}$, may be obtained from a weighted average function [40]:

$$\bar{\theta} = \frac{\sum_{i}^{n} \frac{da}{dN}\left(\Delta K_{\mathrm{eq},i}\right) \cdot \theta_i\left(K_{\mathrm{I}}, K_{\mathrm{II}}\right)}{\sum_{i}^{n} \frac{da}{dN}\left(\Delta K_{\mathrm{eq},i}\right)} \tag{9.20}$$

Considering Eqs. (9.16)–(9.20), when a reasonable set of trains cross a certain bridge, Δa and $\bar{\theta}$ should be computed to defined crack growth under mixed-mode loading (I+II). The modelling of the cracked geometry associated with each train would be unfeasible due to computational limitations for complex connections. The crack propagation ends when the unstable propagation regime is reached, i.e., when the equivalent stress intensity factor is equal to or greater than the material toughness, $K_{\mathrm{eq}} \geq K_u$. At this point, the fatigue life associated with crack propagation is determined, with the predicted crack path reflecting the results of the crack initiation assessment or crack characteristics detected *in situ*, if applicable.

Finally, the fatigue classification for each detail should be updated from the recalculated damage in subtask (v), where safety is evaluated. In subtask (vi), the remaining fatigue life of the structure is compared to the safety standards required for the bridge. Connections that fail to meet the fatigue safety criteria require further analysis in Phase IV. Otherwise, details that demonstrate an adequate remaining fatigue life should be deemed fatigue-safe, with regular inspection cycles upheld as planned.

9.2.4 Phase IV-Critical Decision

The previous three phases provide a consolidated fatigue classification of details. If any connection is still identified as critical, appropriate corrective measures must be planned to extend the fatigue life of the structure (see Figure 9.7).

As depicted in Figure 9.7, the process is logically organized into four subtasks, part of two subphases: (1) refinement of management information (subphase IV.1, represented by a bluish box on the left); and (2) quantification of the safety level and critical analysis (subphase IV.2, shown as a greyish box on the right). Subphase IV.1

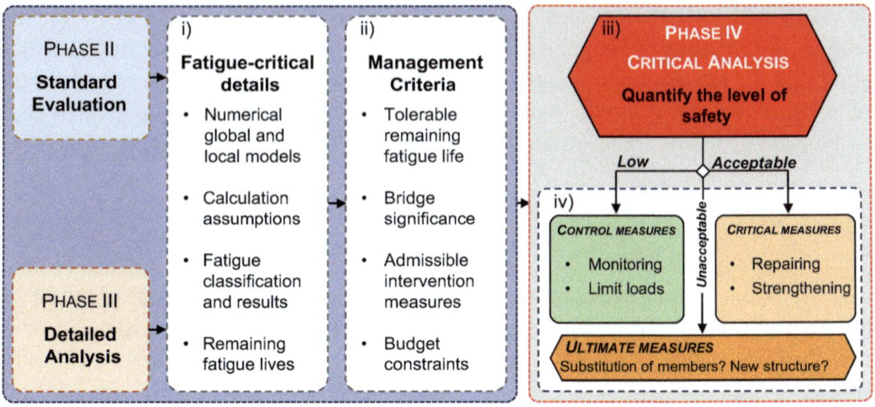

FIGURE 9.7 Phase IV: Critical decision-workflow. (Reprinted with permission from Ref. [27], Elsevier.)

focuses on two key tasks: analysing the findings from previous phases and incorporating additional input from the responsible management authority (i and ii). In subphase IV.2, the emphasis is on the decision-making process for implementing corrective measures (iii and iv).

9.2.4.1 Subphase IV.1-Refinement of Management Information

In general, Phase IV follows the Detailed Analysis (Phase III) to critically evaluate the results calculated, prioritizing the operational and engineering focus. Nevertheless, as noted in Phase II, if it is determined that no relevant improvements can be made through the third step of calculations, Phase IV may begin directly for certain details, such as secondary members with simple replacement or repair.

In subtask (i), the information gathered from the calculations performed should be analysed, including geometrical and material properties of the investigated details, as well as the locations of stress concentrations (hot spots), the nature of the damage, and the predictions for fatigue life. Also, the previously developed numerical models should serve as the foundation to investigate corrective measures.

Regarding non-structural restrictions, the criteria set by the responsible management authority must be accounted for, including acceptable remaining fatigue life, significance of the bridge, permissible interventions, and financial constraints. In subtask (ii), this input should already be partially available from the previous phases but needs further refinement to address the risk of global or local failures.

9.2.4.2 Subphase IV.2-Quantification of the Safety Level and Critical Analysis

In subtask (iii), the information from subtasks (i) and (ii) should be combined to set the input data to start a risk analysis for defining the safety level. The properties of a given detail critical to fatigue and the impact of its failure should be prudently weighed, namely the relevance of the related structural members for the global and local behaviour. The acceptable level of damage for a non-redundant connection must be significantly lower than for one with redundancy. Three different safety levels can be established: (i) acceptable; (ii) low; and (iii) unacceptable. In subtask (iv), corrective measures may be planned based on this classification:

a. control measures for an acceptable level of safety;
b. critical measures for a low level of safety; and,
c. ultimate measures for an unacceptable level of safety.

In severe situations, the necessary corrective actions may be integrated with those specified for higher safety levels in order to optimize the remaining fatigue life. A general overview of these corrective measures is provided for context.

9.2.4.2.1 Control Measures

For mild fatigue issues, short or long-term monitoring can be implemented to collect data on the global or local dynamic responses that impact the progression of fatigue damage. This information can be used to refine or validate numerical models [41] and provide valuable insights into railway loading, such as traffic weight distribution [42–44]. Alternatively, or as complementary action, periodic inspections can be scheduled to obtain discrete information on the structural integrity [45,46]. Also, imposing load limitations can be considered a temporary measure to address immediate concerns and slow the development of damage, for example.

9.2.4.2.2 Critical Measures

For a low safety level, two corrective approaches can be planned to enhance the fatigue strength of metallic bridges: (1) localized interventions, targeting the critical area; and (2) broad measures, impacting on the response of multiple structural members or the entire structure. About the first type, such localized interventions include reducing stress concentration at hot spots by making repair cuts in welded connections [47], eliminating mechanical fasteners near critical points and replacing the others with high-strength bolts [48,49], or drilling stop holes at the crack front [50,51]. In this latter case, steel cover plates can be welded or bolted to help reduce the crack propagation rate. For older metallic bridges composed of non-weldable materials, CFRP patches can be glued as an alternative [52,53], if geometrical constraints allow it.

Regarding the broad measures, new solutions based on CFRPs have been proposed for improving metallic bridges, according to concepts analogous to those of external prestressing with unbonded cables in concrete members. Prestressed CFRP bands may be applied, impacting the structural behaviour by significantly reducing effective stress ranges and mean stress values, thereby influencing the prediction of fatigue life [54,55].

9.2.4.2.3 Critical Measures

Reduced remaining fatigue lives require critical measures to ensure safe circulation. Two main scenarios can occur, namely: (1) unacceptable damage in a few connections; or (2) extended damage. In the first scenario, replacing structural elements may be necessary, depending on the consequences of this measure for the structural system. Additionally, the roots of the identified fatigue damage must be fully investigated and understood to prevent further degradation. For generalized damage, replacing structural members may not be satisfactory, and it may be prudent to neutralize the affected bridge.

9.2.5 Summary of the Integrated Methodology for Fatigue Life Prediction

Key characteristics of the multiphase analysis process should be stressed, namely:

i. Four successively dependent phases are integrated to study all connections of a given bridge, gradually limiting the number of details to be investigated;
ii. Phase I and Phase II rely on normative global fatigue methods, assuming a certain conservatism level to identify connections critical to fatigue;
iii. Phase III is defined based on two local approaches to fully account for the local geometrical and material properties of connections with unacceptable remaining fatigue life;
iv. The local analyses have the goal of fully exploring the installed fatigue capacity of critical connections characterized by numerous geometries of diverse materials;
v. The first three phases offer valuable information and analysis tools for managing the investigated bridge, including the design of corrective interventions if required.

This integrated methodology is applied to a real case study. Representative results are presented to demonstrate the contributions of this multiphase strategy.

9.3 CASE STUDY: VÁRZEAS BRIDGE

Located on the international railway line of Beira Alta in Portugal, the Várzeas Bridge is a 281-m-long riveted structure, built in 1958 without specific fatigue checks, which is now nearing the end of its service life. Due to mobility policies that address environmental issues, the traffic demand is foreseen to increase in the Portuguese railway lines. This context makes the Várzeas Bridge a relevant concern for the responsible management authority and an important case study.

9.3.1 Phase I-Initial Assessment of the Várzeas Bridge

9.3.1.1 Subphase I.1-Analysis and Updating of Available Information

In the initial subtasks of the integrated methodology, information about the structure was gathered, with a focus on the original drawings. A visual inspection was planned and carried out (Figure 9.8), revealing only minor corrosion pits and no significant signs of degradation, without deviations from the initial structural configuration. Also, based on the identified data gaps, a full geometric survey was performed (refer to Figure 9.9).

(a) (b)

FIGURE 9.8 Partial views of the Várzeas Bridge-site photos: (a) trussed structure; (b) ballastless track. (Reprinted with permission from Ref. [56–58], Elsevier.)

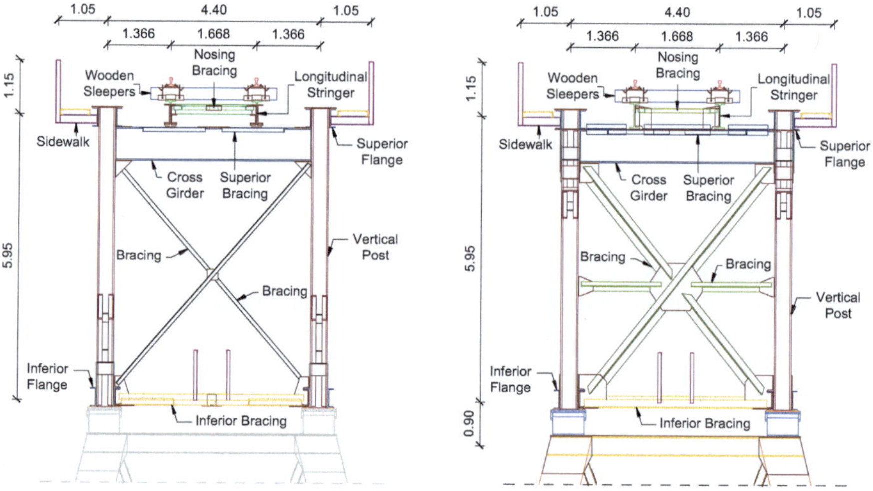

(a) (b)

FIGURE 9.9 Cross-sections of the Várzeas Bridge (m): (a) general cross-section; (b) cross-section at the columns. (Reprinted with permission from Ref. [58], Elsevier.)

The structure supports a single ballastless railway track and has a total length of 281 m, divided into three central spans of 60 m and two extreme spans of 50.50 m. The structural system can be categorized into several groups of elements: (i) two inverted Warren trusses are linked by cross-girders, with the latter supporting the ballastless railway track, which gives origin to a relevant subsystem; (ii) three levels of longitudinal bracing linking the longitudinal members (inferior, superior, and nosing bracings); (iii) cross-sectional bracings regularly spaced; (iv) four trapezoidal shape trusses serve as columns; and (v) one granite masonry column on the west side and one granite masonry abutment on the east side also support the railway loadings.

Concerning the primary dimensions for the numerical modelling, the two inverted Warren trusses are 5.95 m high and 4.40 m wide. The initial and final five girder panels are 6.50 m in length, while the others measure 6.00 m. The metallic columns vary in height from 10.10 to 30.20 m, and the granite masonry column is 12.20 m tall.

The superior flanges of the inverted Warren trusses consist of two U-shaped sections, each 400mm in height, linked by a plate that is 650mm wide and 12mm thick. The inferior flanges have a similar configuration, but the plate is 13mm thick, increasing to 20mm near the columns. The diagonals have a cross-section made up of two U-shaped sections, each 350mm in height, which is adjusted near the columns to a cross-section with two U-shaped sections of 400mm in height. The vertical-posts have an I-shaped cross-section, 348mm in height and 297mm in width, changed at the truss-to-column connections to an I-shaped section, 348mm in height and 560mm in width. The longitudinal stringers under the track are supported by the cross-girders, both composed of I-shaped sections: the former measuring 550mm in height and 200mm in width, and the latter being 688mm in height and 297mm in width. The cross-sectional bracings and the three levels of longitudinal bracings (inferior, superior, and nosing bracings) are made up of L-shaped, half I-shaped, or U-shaped sections (see Figure 9.9).

Concerning the nodes, the confluent members form connections with intricate geometry. Angles and filling plates are used to level the unions between line members. At this scale, the geometry of the gusset plates was not explicitly represented due to the limitations of the model type. However, the geometrical properties of the converging structural members were increased to obtain a calibrated equivalent stiffness in the area of influence of these nodal connections. Also not modelled in this phase, but assumed as additional mass, a large number of rivets were found in the joints, with two nominal diameter ranges: (i) 22–32mm for connections between the main structural members (truss flanges, diagonals, vertical-posts, cross-girders and longitudinal stringers); and (ii) 9–20mm for the remaining details involving bracing members.

In the longitudinal plane, rotations are considered free due to the existing metallic bearing supports. On the west side, the longitudinal displacement is also free, while it is restricted at the remaining supports. The east abutment received most of the horizontal forces due to the flexibility of the metallic columns.

After defining the geometrical characteristics of the structure, a global model using beam finite elements was conceived, given the structural complexity and length of the Várzeas Bridge. This modelling aims to reproduce the global dynamic behaviour and was built in the ANSYS software [59] (see Figure 9.10).

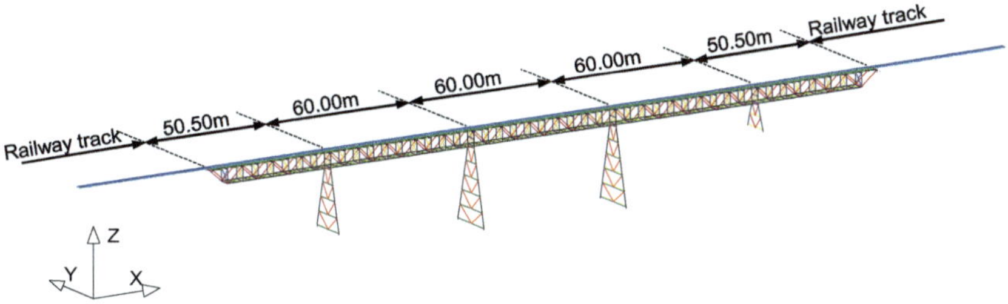

FIGURE 9.10 Global model of the Várzeas Bridge. (Reprinted with permission from Ref. [57,58], Elsevier.)

Beam elements (BEAM188 [60]) were used to model the metallic components of the deck and columns. The rails and wooden sleepers were also modelled by BEAM188 elements, composing part of the ballastless railway track. Spring-damper elements (COMBIN14 [60]) were applied to represent the elastic stiffness of the metallic pads between the rails and wooden sleepers, as well as between the wooden sleepers and the longitudinal stringers. COMBIN14 elements were also used to model the stiffness of the foundation of the metallic columns. Nodal mass elements (MASS21 [60]) accounted for the mass of rivets, filling plates, and non-structural elements, such as the sidewalk and deck inspection access. Lastly, rigid connections (MPC184 [60]) were applied to link the centre of gravity of the beam elements modelling the railway track.

Despite the bridge's age, the steel materials are expected to be similar to modern steel grades. For example, the St52 grade specified in the original design for the main structural elements corresponds to the modern S355 steel grade [61]. Thus, the following mechanical properties were adopted: a Young's modulus (E) of 210 GPa; a density (ρ) equal to 7850 kg/m³; a yielding stress (f_{yd}) of 355 MPa; and a Poisson's ratio (ν) equal to 0.27. Those values are in line with material tests conducted on samples from a similar bridge built by the same contractor during the same period [62].

9.3.1.2 Subphase I.2-Initial Fatigue Assessment

Based on the characteristics of the structure, the most important connections between the main metallic members were identified, taking into account their geometrical characteristics and expected structural behaviour (see Figure 9.11), i.e.: (i) D.1, related to the inferior flanges at the nodes with and without confluent diagonals; (ii) D.2, associated with the superior flanges at the nodes with and without confluent diagonals; (iii) D.3, related to the vertical-posts at the inferior nodes, (iv) D.4, associated with the vertical-posts at the superior nodes; (v) D.5, related to the diagonals at the superior and inferior nodes; (vi) D.6, associated with the flanges of the cross-girders; and (vii) D.7, related to the web of the cross-girders. Different structural responses are expected for the nodal connections with and without confluent diagonals, but the concept of nominal stress is limited in reflecting these differences, and the global structural analysis may lack reliability in modelling secondary effects. In addition, eight detail types were identified for the secondary connections related to the bracings (D.8 to D.13).

The structural analysis varies based on the type of structural system and the speed of moving trains. Given that the bridge is continuous and the railway traffic does not exceed 200 km/h, dynamic calculations are not mandatory, according to the guidelines outlined in subsection 6.4.4 of EN1991-2 [28]. Thus, the λ-coefficient method can be implemented, as recommended in EN1993-1-9 [29] and foreseen in Phase I.

On the resistance side, the S-N curve detail category 71 proposed in EN1993-1-9 [29] for welded details was adopted as stated in previous research [3]. From Eq. (9.1), $\Delta\sigma_{n,c}$ was then defined as 71 MPa at 2×10^6 cycles. The value of γ_{Mf} also impacts the fatigue strength, but in a semi-deterministic calculation, its effect can be analysed when combined with the dynamic amplification factor applied to the normative load model. For the loading, the SW/0 load model has to be applied in addition to LM71, since the bridge is continuous, resulting in the calculation of the μ_{FDF} from

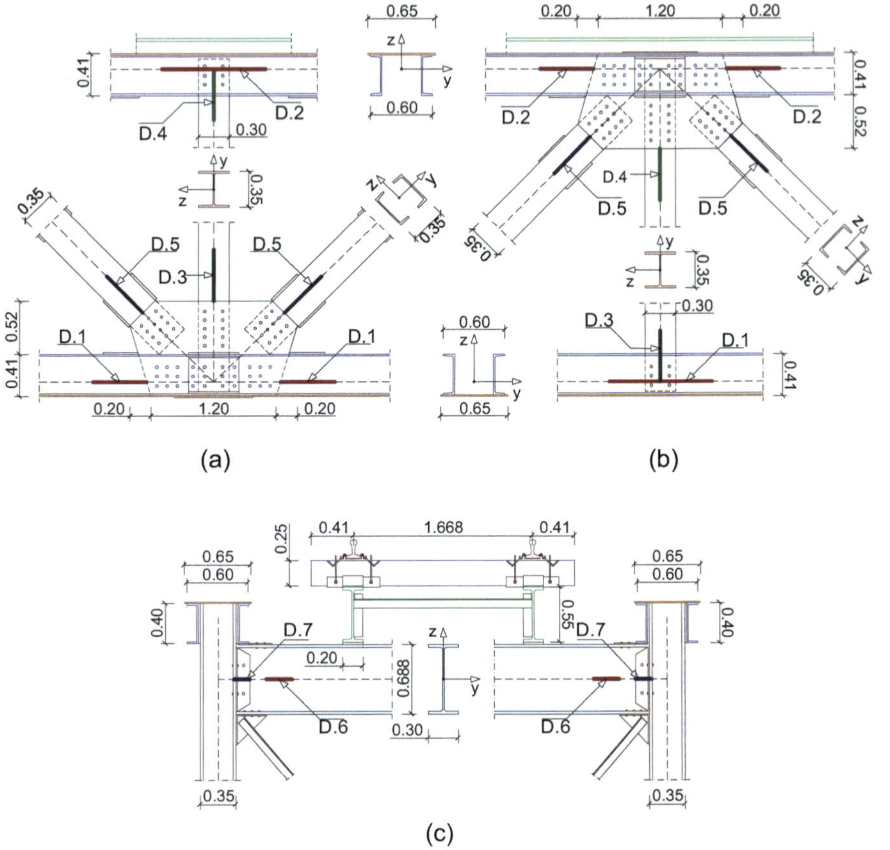

FIGURE 9.11 Main riveted connections of the Várzeas Bridge (m): (a) inferior node with confluent diagonals and superior node without confluent diagonals; (b) inferior node without confluent diagonals and superior node with confluent diagonals; (c) cross-girder-to-vertical-post. (Reprinted with permission from Ref. [58], Elsevier.)

the maximum stress range. $\Delta\sigma_{n,71}$ and $\Delta\sigma_{n,\mathrm{SW0}}$ were evaluated at the critical points defined based on the expected behaviour for this type of connection. In details with multiple lines of rivets, the hot spot is located at the holes of the first line, defining the longitudinal position of the stress assessment points. For a first approximation of the group effect, the nominal net stress values were adjusted in accordance with that proposed in prEN1993-1-9 [63] for non-preloaded bolts, considering that the rivets have reasonably low clamping forces. Thus, four points were defined for each cross-section to evaluate the nominal stresses, with a fifth one added to assess the shear connection related to D.6 (Figure 9.12).

Effective nominal stress ranges were calculated taking into account 60% of the compressive elastic part, as suggested in EN1993-1-9 [29], considering the permanent loads in a characteristic load combination. Such static stress ranges were

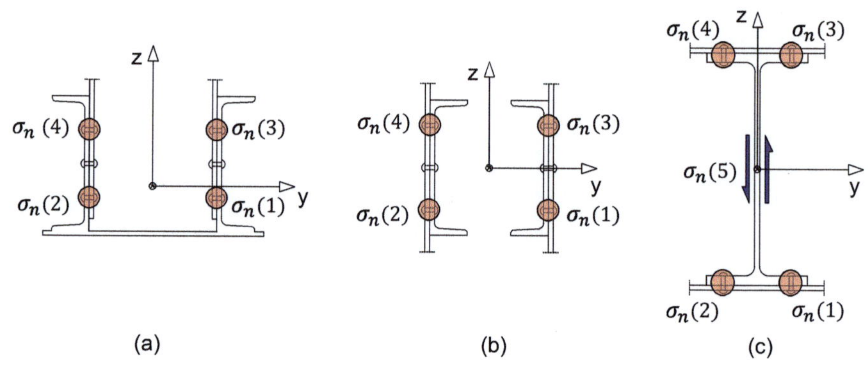

FIGURE 9.12 Assessment points for nominal stresses: (a) inferior flange; (b) diagonal; (c) cross-girder. (Reprinted with permission from Ref. [58], Elsevier.)

amplified as in Eq. (9.2), requiring the evaluation of λ and Φ. These parameters depend on the determinant length of the structural members related to the riveted connections (L_Φ), with conservative values being assumed, e.g., L_Φ (D.5, general cross-section) = 18.50 m, L_Φ (D.5, at the columns) = 17.25 m.

For the bridge characteristics and respective loading, the applicable load models represent the real traffic when affected by λ_1 for trains with 25 t per axle [64], assuming a fatigue verification for a service life of 100 years, which results in $\lambda = \lambda_1$. Regarding the amplification factor Φ, the λ-coefficient method is related to Φ_2, which is associated with the "careful maintenance" of the track [28]. However, it may be reasonable to conservatively consider Φ_3 for "standard maintenance" when verifying structures such as the existing ones with a history of cyclic loading. As in Eq. (9.1), such a value combined with γ_{Mf} amplifies the static results, with the latter depending on the maintenance philosophy [29]. Thus, $\Delta\sigma_{n,71}$ and $\Delta\sigma_{n,\text{SW0}}$ were amplified by $1.35 \times \Phi_3$ for the primary connections, ensuring a "safe life" approach since their failure has severe consequences, and by $1.15 \times \Phi_3$ for the secondary connections, guaranteeing a "damage tolerant" approach due to the lesser importance of these members.

Details types D.1 to D.13 were categorized as noted, but presenting all the results would not bring scientific gains. Thus, only the fatigue analysis of D.5, between the diagonals and gusset plates, is presented, as higher μ_{FDF} values were found (Figure 9.13).

As previously referred, the cross-section of diagonals is reinforced to 2UPN400 near the columns with two strengthening plates 14 mm thick on each side, which contrasts with the general cross-section of 2UPN350, since these members have high nominal stresses. As shown in Figure 9.13, out of 92 connections (half of the existing 184, due to the structural symmetry in the $0xz$ plane, see Figure 9.10), 66 riveted details were classified as prone to fatigue, with a μ_{FDF} above 1.00.

Overall, for both principal and secondary connections, 252 out of 1143 details were identified as fatigue-prone, assuming structural symmetry (504/2286). Therefore, Phase II was implemented based on these results.

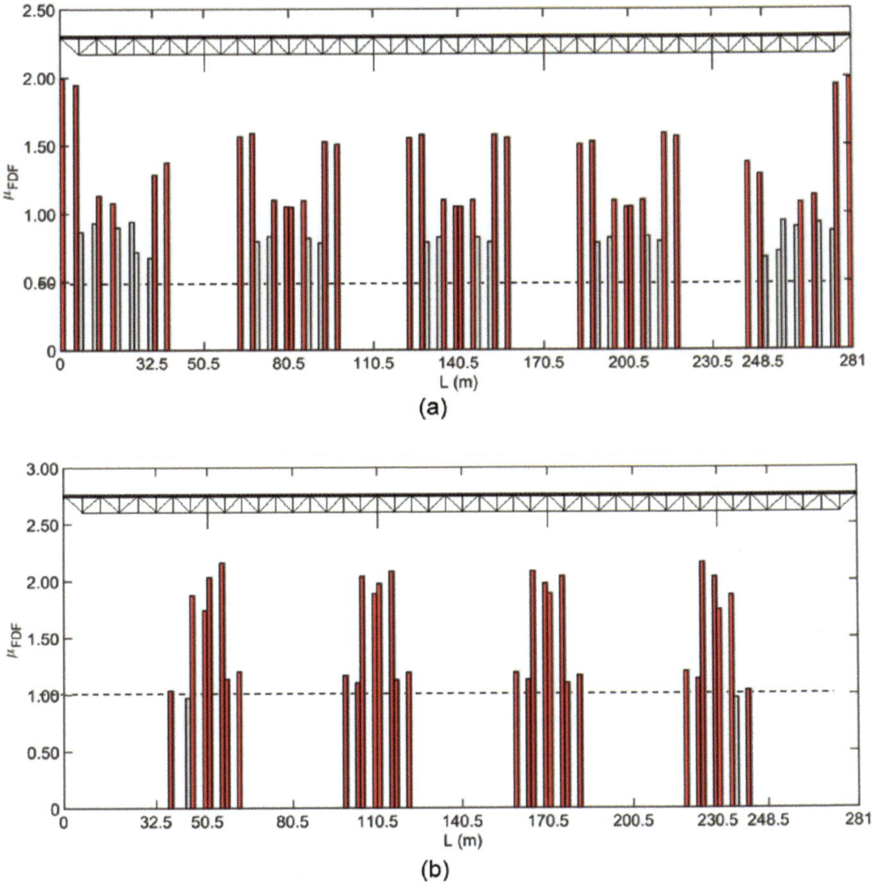

FIGURE 9.13 D.5, Fatigue damage factor (μ_{FDF}): diagonals (a) general cross-section; (b) cross-section at the columns. (Reprinted with permission from Ref. [27], Elsevier.)

9.3.2 Phase II-Standard Evaluation of the Várzeas Bridge

9.3.2.1 Subphase II.1-Gathering Experimental Information

From the data of Phase I, two ambient vibration tests were planned considering the characteristic of the bridge: (i) a global AVT to assess modal values associated with the global system; and (ii) a localized AVT to evaluate the modal properties of the subsystem related to the ballastless track. Respectively, 18 and 10 vibration modes were identified. After, the frequencies and modal shapes were used as inputs to optimize the global model by employing a genetic algorithm, which aimed to approximate the numerical properties to the real values obtained experimentally (see Figures 9.14 and 9.15).

The final Modal Assurance Coefficient (MAC) was 0.948 on average, while an average error of 2.464% was achieved between numerical and experimental frequencies. At this step, evaluating the real railway loading to establish past, present, and

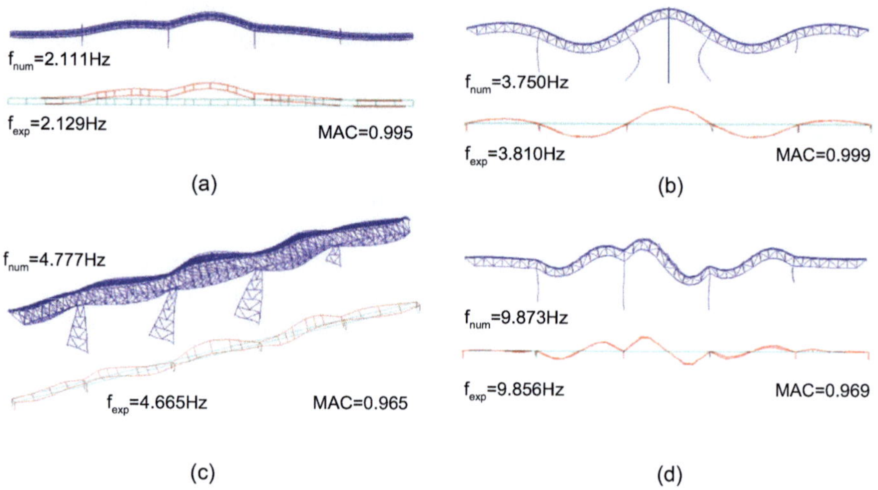

FIGURE 9.14 Global modes – experimental vs. numerical values: (a) global mode 1 (lateral, *yy*-direction); (b) global mode 5 (vertical, *zz*-direction); (c) global mode 7 (torsion, *zz* and *yy*-direction); (d) global mode 16 (vertical, *zz*-direction). (Reprinted with permission from Ref. [57].)

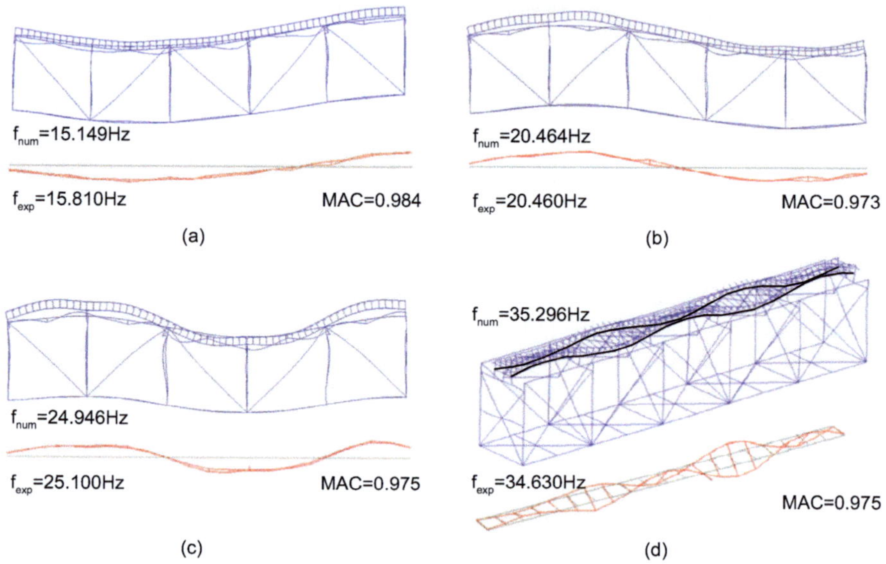

FIGURE 9.15 Local modes – experimental vs. numerical values: (a) local mode 1 (vertical, *zz*-direction); (b) local mode 3 (vertical, *zz*-direction); (c) local mode 5 (vertical, *zz*-direction); (d) local mode 9 (torsion, *zz*-direction). (Reprinted with permission from Ref. [57].)

future traffic scenarios was hindered by logistical difficulties, reflecting problems in analysing existing structures. Therefore, the available information is assumed to be sufficient, with the focus on the resistance side in the analyses to be conducted.

9.3.2.2 Subphase II.2-Standard Fatigue Evaluation

Nominal stresses were calculated from the optimized global model after quasi-static analyses, with these values being amplified as in Eq. (9.3), computing φ' and φ'' as recommended in EN1991-2 [28]. The normative heavy and standard traffic scenarios also defined in this guideline were assumed, given the lack of data on the real traffic. The nominal stresses were then used to compute the relationship between stress ranges and their frequency values (i.e., $\Delta\sigma_{n,i} - n_i$), necessary for the linear damage accumulation following Eq. (9.6). As in Phase I, the S-N curve detail category 71 defined the fatigue resistance, which has been validated by historical research [3].

Following the reasoning behind the integrated methodology, only the fatigue-prone connections should be investigated here, but the 2286 details were checked to evaluate the consistency between phases. Similar to subphase I.2, the fatigue damage was accumulated for a service life of 100 years (D_{100}). For detail D.5, related to the diagonals, the fatigue results are emphasized due to the higher damage (Figure 9.16).

Comparing Figures 9.13 and 9.16, the connections identified as critical to fatigue in Phase II were also screened as fatigue-prone in Phase I, confirming consistency between the different steps of fatigue analysis under similar calculation assumptions. In addition, the second Phase II proved that some prone details are not fatigue-critical. For the general cross-section, the higher fatigue damage value obtained was 1.634 at $x = 6.50$ m and $x = 274.50$ m. Near the columns, the maximum D_{100} was equal to 2.256 at $x = 116.50$ and $x = 164.50$ m.

Subsequently, dynamic calculations were conducted to improve the accuracy of amplification factors associated with the railway loading (φ') and those related to vehicle and track imperfections (φ''), as outlined in Figure 9.5. The former was inherently accounted for by considering moving loads, which ensured computational efficiency. The latter factor was calibrated by accounting for the real track irregularities profile in benchmark analyses based on bridge-train interaction, as described in Section 9.2.2.2. The fatigue damage was then revaluated, providing a measure of the improvement in results achieved through this refinement.

In Table 9.1, the maximum D_{100} for the detail types with at least one detail critical to fatigue are provided for the normative heavy traffic scenario. For these loading, reduced D_{100} values were calculated after dynamic analyses. However, if this traffic mix had been loading the bridge since 1958, various riveted connections would now be approaching their fatigue limit, with short-term concerns regarding the details associated with the diagonals and those related to the cross-girders. As foreseen in Figure 9.5, three analysis refinements may be conducted alongside dynamic analysis. Among these, the quantification of the real traffic through weighing campaigns would lead to significant enhancements in the reliability of fatigue life prediction.

In total, 50 connections were classified as fatigue-critical from the 1143 evaluated for the heavy traffic mix, assuming structural symmetry (100/2286), while only 11 details did not meet the safety criterion for the normative standard traffic (22/2286). Twenty of these critical details to fatigue are associated with

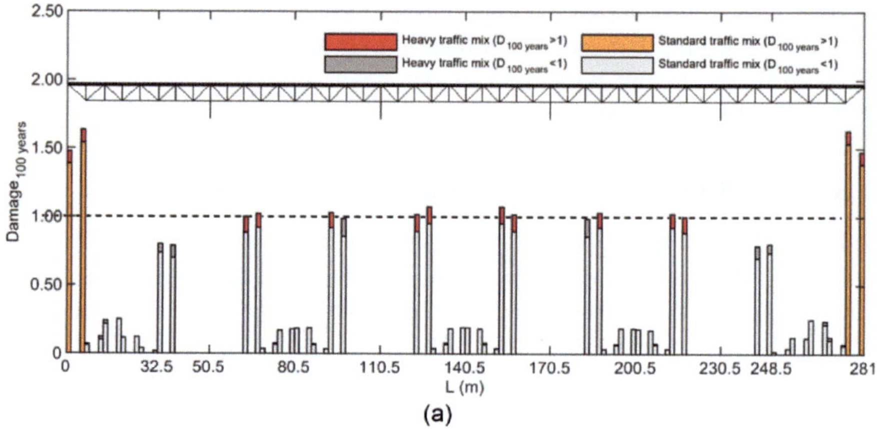

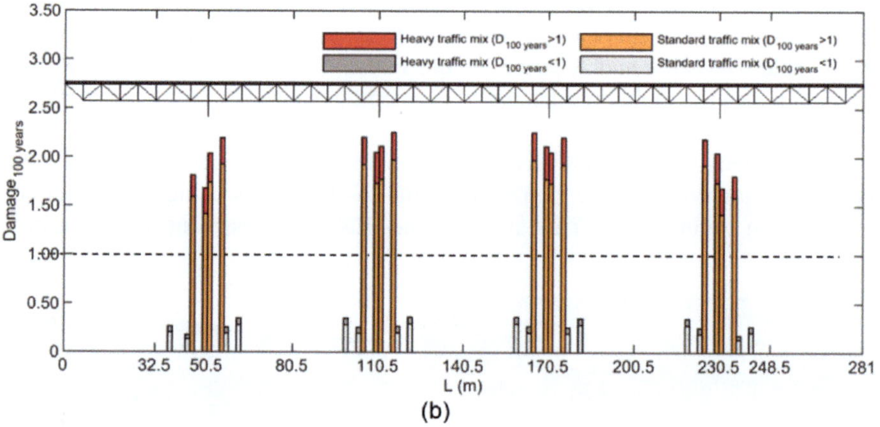

FIGURE 9.16 D.5, Damage for 100 years (D_{100})– diagonals: (a) general cross-section; (b) cross-section at the columns. (Reprinted with permission from Ref. [27,58], Elsevier.)

TABLE 9.1
Phase II: Principal Connections (Maximum Results for the Heavy Traffic Mix)

		i. Quasi-Static Analysis (φ' and φ'' as in [28])		ii. Dynamic Analysis (Inherent φ' and Calibrated φ'')	
Detail ID.	**Structural Element**	D_{100} **(Max.)**	**Fatigue Life (Years, Months)**	D_{100} **(Max.)**	**Fatigue Life (Years, Months)**
D.1	Inferior flange	1.225	81, 7	1.167	85, 7
D.4	Vertical-post (sup. nodes)	1.716	58, 3	1.090	91, 8
D.5	Diagonal	2.256	44, 3	1.772	56, 6
D.6	Cross-girder	2.103	47, 6	1.502	66, 5

diagonals (D.5), highlighting the importance of further analyses focused on this detail type. Given the damage index values, the riveted connection at $x = 109.30$ m ($D_{100} = 1.623$, see Figure 9.16), part of the superior node at $x = 110.50$ m, was investigated at a lower scale.

9.3.3 Phase III-Detailed Analysis of the Várzeas Bridge

9.3.3.1 Subphase III.1-Updating Experimental Information

Phases I and II were based on global assessment, but local fatigue methods require modelling the real characteristics of the riveted length. Given the localized nonlinear contacts associated with the mechanical fasteners, local fatigue "Approach 3" was applied (see Section 9.2.3.2). This involved conceiving the appropriate submodelling relationship in the time domain and calculating BDCOs from the modal quantities of interest evaluated from the global model ($BDCOs_j$), following Eq. (9.8).

To lower the scale from the global to the local one, more consistent with fatigue, a detailed geometrical survey was conducted to assess the local dimensions, including the characteristics and location of the rivets. This updated information was used to enhance the global model and to develop the submodel at the mm scale (see Figure 9.17).

On a 15 m length, shell finite elements (SHEL181 [60]) were included in the global model and related to the line elements by line-to-shell links using multipoint constraints, as in Figure 9.17a. This smooths the BDCOs assessment at the locations

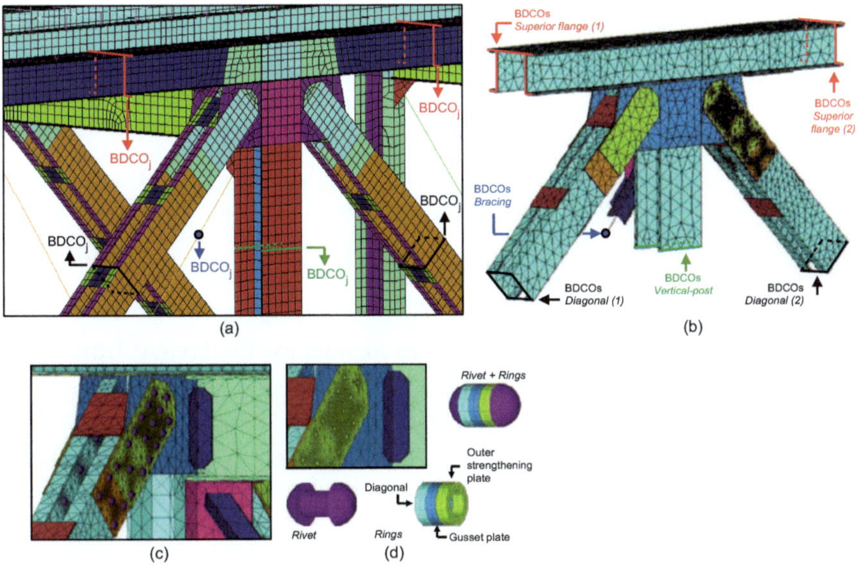

FIGURE 9.17 Submodelling relationship between the refined global model and the local model at $x = 109.30$ m: (a) global model (partial perspective); (b) submodel (perspective); (c) submodel (partial perspective); (d) submodel, rivets and rings (details). (Reprinted with permission from Ref. [27,57], Elsevier.)

associated with the limits of the conceived submodel presented in Figure 9.17b. Solid finite elements with 20 nodes (SOLID186 [60]) modelled the metallic members as in Figure 9.17b, including the 44 rivets (22 per side) related to the diagonal-to-gusset plates connection and the two strengthening plates that materializes alternative load paths (see Figure 9.17c and d). The exception was the bracing members, which were modelled by beam elements (BEAM188 [60]) due to their lesser relevance. Between the interfaces of the solid elements, the normal and friction contacts were simulated by the contact pair elements CONTA174-TARG170 from ANSYS [60] to prevent finite penetration and permit limit sliding, admitting respectively "hard" contact and the Coulomb model (friction coefficient $\mu = 0.15$). A clamping stress of 60 MPa was applied to the rivets by an equivalent negative temperature, assuming an initial adjustment approach that allows the preload to vary with the internal response of the load transfer mechanism.

Aiming to validate the submodelling relationship, an *in situ* test was performed to evaluate strains near the critical detail, using 14 strain gauges to obtain stress values from the linear elastic material behaviour. These experimental results were compared to the numerical ones, with such local stress values derived from the submodel related to the optimized and refined global model as previously described. Moving loads were applied to obtain φ' from the equivalent static calculations, also considering the calibrated φ'' shown in Table 9.1.

For a passenger train with known wheelbases and axle weights, moving at 53.34 km/h, a strong agreement was found for stresses at two locations near the riveted length as outlined in Figure 9.18, thereby validating the established submodelling relationship.

9.3.3.2 Subphase III.2-Detailed Fatigue Assessment

The validated models allowed the complete implementation of the concepts underlying the local fatigue "Approach 3", as described in Section 9.2.3.2 [57,58]. The submodelling relationship helped identify the critical locations for crack initiation based on a preliminary stress analysis. Without initial defects, the hot spots were found at

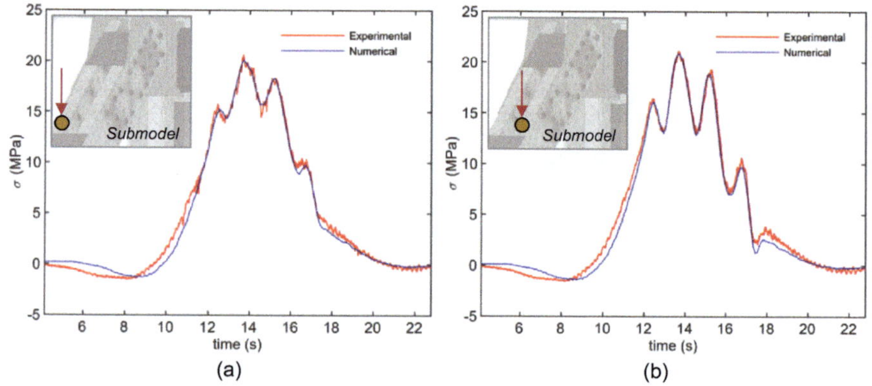

FIGURE 9.18 Experimental vs. numerical stress results: (a) strain gauge 9; (b) strain gauge 10. (Reprinted with permission from Ref. [27,58], Elsevier.)

the holes related to the rivets, as expected. Regarding the loading side, the normative heavy traffic mix was maintained as a reference due to the ongoing absence of reliable data about the real traffic loading.

Following Eq. (9.8), the amplified boundary conditions were derived in the time domain, as the normal and tangential contacts affect the local stress-strain relationship at the hot spots. Quasi-static numerical analyses for all steps of train loadings were then required. However, the boundary displacements associated with the instant of time t are not dependent on those associated with $t-1$, which allows independent calculations. Thus, parallel computing was applied on two different setups: (1) two calculations on a personal computer (3.40 GHz Intel® Core(TM) i7-6700CPU processor and 16 GB RAM memory); and (2) six calculations on a workstation (2.20 GHz Intel® Xeon® E5-2630CPU processor and 64 GB RAM memory).

Reduced multiaxiality was observed in the local results, with the components σ_1 or σ_3 dominating the response [57]. Thus, von Mises stress results were adopted (σ_{VM}) to evaluate the local stresses, considering the through-thickness average values of the most stressed member. This approach was assumed as an approximation for estimating the fatigue life of an edge crack that reaches the entire width. Among the 44 rivets of the mechanism of loading transfer, the local results of the greatest magnitude are highlighted in this work, corresponding to the most limited fatigue life.

Figure 9.19 presents the stresses at the hot spot for initiation of a crack at the marked rivet hole, considering the normative trains of the heavy traffic scenario, with FT5 standing for Fatigue Train type 5, for example [28]. To evaluate the efficiency of calculations, it took 4 days and 12 hours to analyse FT5 (the longest time), while FT11 required 1 day and 21 hours (the shortest time). For FT5, the von Mises stress exceeds the yield strength ($f_{yd} = 355$ MPa), resulting in localized plasticity. This was handled using an elastoplastic post-processing, as proposed by Horas et al. [24,25] for riveted details, considering the analytical model suggested by Glinka [33–35].

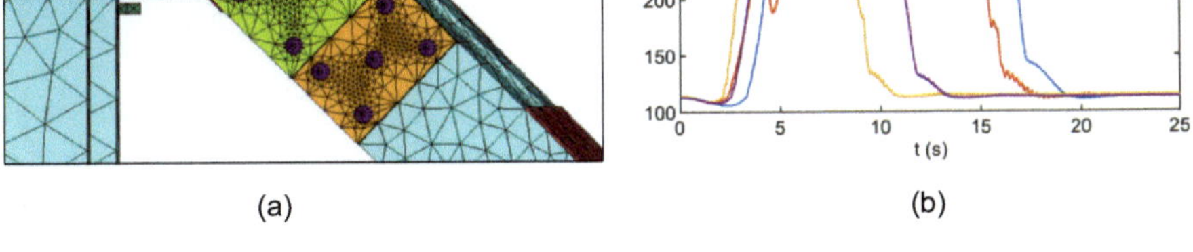

FIGURE 9.19 $\sigma_{VM}(t)$ for the most stressed rivet hole (FT5): (a) rivet and rivet hole analysed (local view, perspective); (b) time histories of the local von Mises stresses. (Reprinted with permission from Ref. [27], Elsevier.)

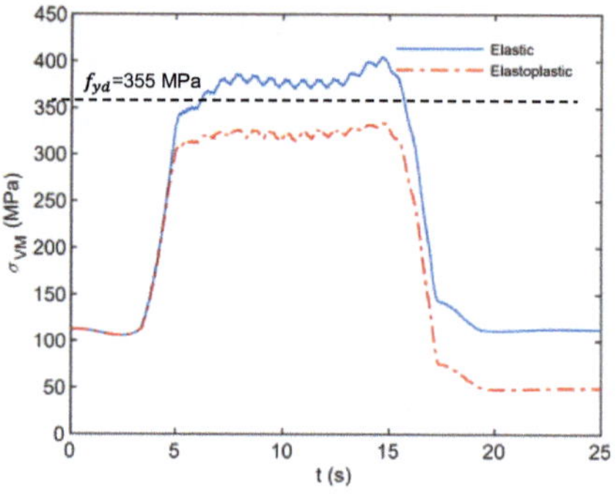

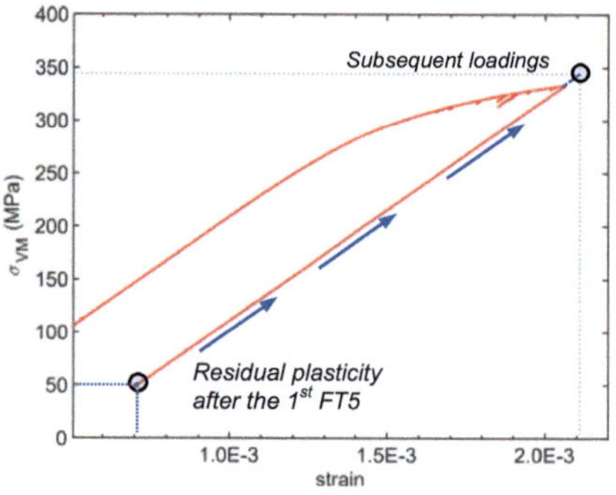

FIGURE 9.20 $\sigma_{VM}(t)-\varepsilon_{VM}(t)$ for the most stressed rivet hole (FT5): (a) time histories of local stresses; (b) stress-strain diagrams (elastoplastic). (Reprinted with permission from Ref. [27], Elsevier.)

Figure 9.20a compares the elastic stress with the corresponding elastoplastic values for the rivet hole exhibiting the highest local stresses. Additionally, Figure 9.20b outlines the stress-strain relationship of von Mises for the elastoplastic results ($\sigma_{VM}(t)-\varepsilon_{VM}(t)$), including the resulting confined plasticity. The findings revealed that the induced residual plasticity lowers the permanent local stress state, indicating that the initial stress for subsequent events is 48.04 MPa, rather than the earlier 112.82 MPa. Considering that the maximum von Mises stress range for FT5 is 298.71 MPa, this leads to a cyclic loading in the linear elastic regime, as $48.04+298.71=346.75$ MPa. Consequently, only the first passage of FT5 induces an elastoplastic response, which will only be repeated if an unexpected overload occurs.

After calculating $\sigma_{VM}(t)-\varepsilon_{VM}(t)$, the rainflow algorithm was implemented to determine the fatigue life expressed as the number of cycles ($N_{f,i}$). The Basquin model was implemented, including the mean stress effect, as in the elastic part of the Morrow proposal [12,13], relating all the quantities for a given loading cycle i ($\sigma_{m,i}$, $\Delta\sigma_i$ and n_i, respectively), as in Eq. (9.21).

$$\frac{\Delta\sigma_i}{2}=(\sigma'_f-\sigma_{m,i})\cdot(2N_{f,i})^b \tag{9.21}$$

From reference [48], the cyclic fatigue strength coefficient and exponent were adopted as $\sigma'_f=952.52$ MPa and $b=-0.089$, respectively. Such material parameters define the fatigue model limit and can be affected by surface finishing, size effect, and type of loading, which were not considered in this work but are highlighted for future parametric analysis.

For all loadings, the linear damage accumulation was calculated following Eq. (9.6) modified for the relationship $n_i-N_{f,i}$, also accounting for the configuration and tonnage of the heavy traffic scenario. A fatigue life prediction equal to 8421 years was calculated [58], which can then be assumed as a theoretical infinite fatigue life for crack initiation, in contrast to the limited total fatigue life obtained in

Phase II ($D_{100} = 1.623$, after dynamic analysis). This demonstrates the importance of correctly addressing the local characteristics of the riveted length, instead of relying on conservative approximations given by S-N curves for nominal stresses. In total, only 4 out of 44 rivet holes showed local stress values that exceeded the yielding stress, with all showing local stress-strain values lower than that presented above at the hot spot for crack initiation.

A comparable integrated analysis may be conducted for other connections that presented relevant fatigue damage in Phase II. However, the findings for the diagonal-to-gusset plate detail at $x = 109.30$ m suggest that fatigue problems will not arise on this bridge in the absence of initial failures.

9.3.4 Critical Analysis of the Methodology for Fatigue Life Prediction

After applying the progressive global-local modelling for multiphase fatigue assessment, some key conclusions can be stated:

i. Phase I and Phase II efficiently identify the details critical to fatigue, thereby logically reducing the number of local fatigue analyses required;
ii. Phase III allowed performing local fatigue analysis based on numerical and analytical tools, opening possibilities to study multiple mechanisms of loading transfer in competitive computational times;
iii. On the resistance side, the limitations of existing catalogues of S-N curves for nominal stresses were clearly stated by confronting the drastically different results from the global and the local fatigue assessments under the same railway loading;
iv. The accuracy of local fatigue analysis depends on the material properties that influence the nonlinearity of fatigue models, which should be considered;
v. Real load scenarios were not considered due to limitations in the available data, but these must be quantified to increase the reliability of the fatigue assessment.

As shown, this methodology integrates global and local fatigue methods and approaches to create innovative lines of analysis for existing geometries and materials, using data from the existing bridges to increase the reliability of fatigue assessment. Maximizing the fatigue life of bridges leverages the development of railways as an environmentally friendly transport.

9.4 OTHER RELEVANT CASE STUDIES

After the presentation of the integrated methodology, and respective implementation to analyse a real case study, this section discusses similar applications, which implemented the same concepts. Two case studies related to different stages of fatigue are summarily presented to complement the previous discussion, offering additional results that exemplify the potential of the developed tools.

9.4.1 Crack Initiation Assessment-Fundão River Bridge

On the Vitória-Minas railway line (EFVM) in Brazil, fatigue damage was detected on a riveted structural system common to several bridges. Teixeira et al. [65] implemented a multiphase global-local analysis, applying local fatigue models at the mm scale to study a bridge that represents such a structural system susceptible to damage.

Figure 9.21 presents the investigated Bridge over the Fundão River which is 41 m long and was constructed in 1944, alongside the damage found on the Bridge over the Correntes River, which has a similar trussed structure and is subjected to similar cyclic loads. For the global fatigue assessment, the S-N curves for riveted details proposed in AREMA [66] were considered for the resistance side, along with the detail category 71 defined in EN1993-1-9 [29]. Additionally, historical data allowed to establish past and present traffic scenarios. After calculations related to Phase I and Phase II of the integrated methodology, the connection of the vertical-post to the gusset plates was identified as critical at the extreme and intermediate nodes (respectively, U1 and U3). The observed real damage confirmed U1 as critical (refer to Figure 9.20b), while conservative numerical results were found for U3, since no real damage was found. Thus, Phase III was implemented, establishing appropriate submodelling relationships between the refined global model and the submodels (see Figure 9.22, for example).

FIGURE 9.21 Riveted structural system investigated concerning fatigue damage: (a) Fundão River Bridge; (b) damage in the River Correntes Bridge. (Reprinted with permission from Ref. [65], Elsevier.)

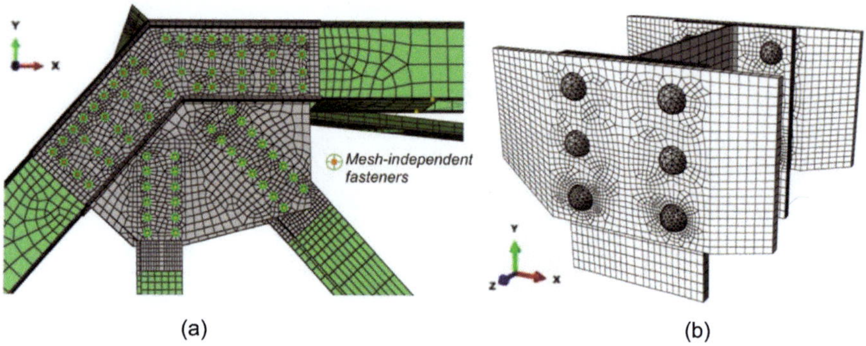

FIGURE 9.22 Modelling for the extreme node U1: (a) refinement in the global model; (b) volumetric submodel. (Reprinted with permission from Ref. [65], Elsevier.)

Due to the length of the bridge and the reduced dynamic effects caused by low-speed heavy traffic, the refinement of the global model included shell and solid finite elements, properly related to the volumetric submodel, both conceived in the ABAQUS software [67]. In the global model, solid elements were used in the riveted length, allowing the rivets to be modelled by mesh-independent fasteners with stiffness derived from elastic short-beam behaviour, as well as accounting for the normal and tangential contacts between the remaining plates. This may improve the BCDOs to be imposed to the submodel in the time domain but is limited to simple cases as an alternative to the so-called "Approach 3".

At the hot spots, local elastic stresses were calculated, with an incremental elastoplastic post-processing being performed for those above the yield limit by implementing the Neuber and Ramberg-Osgood models (Eqs. 9.9 and 9.11). After confined plasticity, residual stress led to further loadings in the elastic regime, as in the results presented in Section 9.3.3.2. From these calculations, the Basquin model was implemented, as in Eq. (9.12), taking into account the effect of mean stress through the correction derived from the Goodman model.

The local fatigue analysis confirmed the riveted detail at the extreme node U1 as critical, while the middle node U3 was found to be fatigue-safe with a theoretical infinite life. Both results align with the real damage observed. Regarding the critical detail related to U1, a fatigue life of 68 years was computed, which compares with the 63 years observed. The influence on the fatigue results of the surface finishing and size effect was investigated, and the strong influence of material parameters was stated. In addition, the location of the hot spot was compared to the real damage observed on the Bridge over the Correntes River, showing remarkable agreement (see Figure 9.23).

Overall, the integrated methodology enabled the fatigue damage to be analysed on a scale more aligned with this phenomenon, yielding results distinct from those achieved at the global assessment (Phases I and II). These outcomes confirm that the bridge can continue operations after the repair measures carried out in 2007 on the extreme node U1, providing critical data to the responsible management authority.

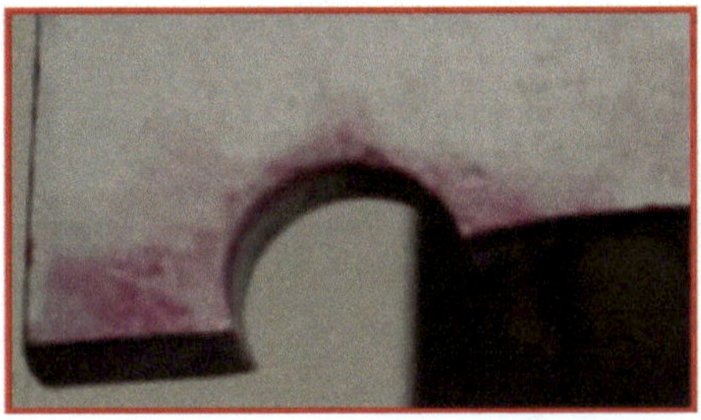

(a)

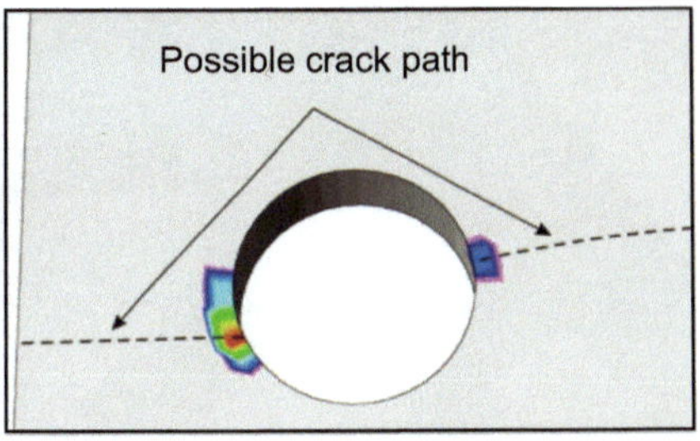

(b)

FIGURE 9.23 Real vs. numerical damage – extreme vertical-post: (a) damage in the Correntes River Bridge; (b) damage magnitude (submodel). (Reprinted with permission from Ref. [65], Elsevier.)

9.4.2 Crack Propagation Assessment-Sado River Bridge

Located on the southern line in Portugal, the Bridge over the Sado River was built in 2010, consisting of three 160-m arches that suspend the box girder deck through hangers. To evaluate the impact of fatigue, Albuquerque et al. [68] and Horas et al. [69] implemented a multiphase analysis to identify critical details.

Figure 9.24 presents an overview of the bridge and the detail type related to the welded connection of the diagonals of the diaphragm under the hangers, located inside the box girder. Calculations related to Phase I and Phase II of the integrated methodology identified the detail under hanger 51, located 36 m from one end of the bridge, as the most prone to fatigue, even though it is safe in the absence of initial defects. For the resistance side, the detail category 45 was considered as proposed in EN1993-1-9 [29], while real traffic was evaluated by a weighing campaign and assumed as loading. Despite this result of the global assessment, the analysis of the impact of an initial crack derived from an unexpected construction defect was considered relevant. The global model was then refined close to the screened detail, allowing a submodelling relationship with a local model conceived by solid finite elements in the ANSYS software (see Figure 9.25).

Given the absence of local contacts, the submodelling relationship was established in the modal space by implementing "Approach 2" described in Section 9.2.3.2. In order to study the propagation phase, an initial crack of 1 mm was assumed, modelling a crack box at its front to calculate K_{I} and K_{II} using the Virtual Crack Closure

(a) (b)

FIGURE 9.24 Welded structural system investigated concerning fatigue damage: (a) Sado River Bridge; (b) detail type prone to fatigue. (Reprinted with permission from Ref. [68], Elsevier.)

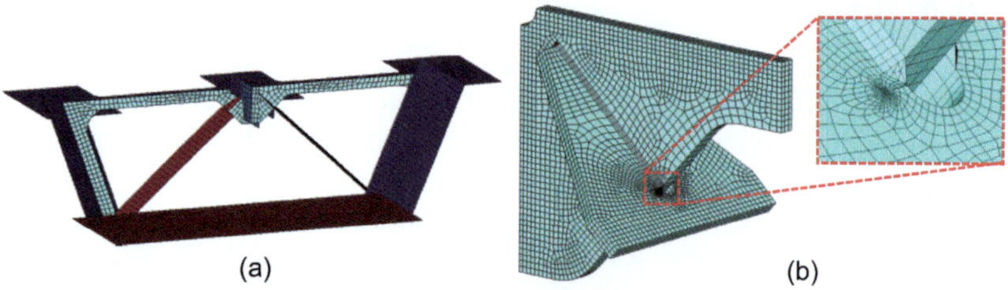

(a) (b)

FIGURE 9.25 Submodelling refinement for diaphragm 51: (a) shell refinement in the global model; (b) volumetric submodel. (Reprinted with permission from Ref. [69], Elsevier.)

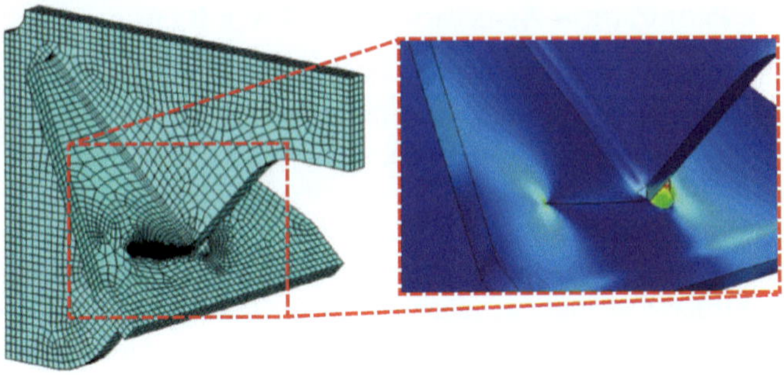

FIGURE 9.26 Numerical crack propagation before failure. (Reprinted with permission from Ref. [69], Elsevier.)

Technique (VCCT). To implement modal superposition, the vibration modes impacting the local response were identified, while irrelevant modes were excluded to reduce the number of numerical analyses of the submodel for each crack increment. By adapting Eq. (9.7), K_I and K_{II} were calculated in the time domain, and K_{eq} was obtained by applying the mixed-mode criterion (I+II) proposed by Tanaka, as in Eq. (9.16).

After material characterization, crack propagation was numerically simulated under real traffic conditions and the normative heavy traffic [28], using the concepts of equivalent kink angle ($\bar{\theta}$) and fixed crack increment (Δa) for a given traffic block (see Figure 9.26, for example). When the maximum value for K_{eq} reached the material toughness (K_u), the propagation was stopped. For the heavy traffic mix, remaining fatigue life of 1 year and 10 months was calculated, while 44 years were obtained for the real traffic. The difference in results highlighted the importance of accurately defining the real loading, since the discrepancy with the normative one can be critical.

Such results were obtained in time ranges between 11 and 14 hours, assuming parallel processing to compute the modal quantities for each crack increment in a personal computer with a 3.40 GHz i7-6700CPU processor and 16 GB RAM memory. Overall, from these results, even in the case of an initial defect in the most prone detail, the propagation life is sufficiently long for the crack to be detected in a timely manner.

9.5 SUMMARY

Presently, normative global fatigue methods dominate the design of new structures, based on a trade-off between computational efficiency and safety. However, these methods may not adequately consider the local geometrical and material properties of existing details in ageing bridges. This drawback can result in unnecessary restrictions in management, leading to significant direct and indirect financial and economic costs.

In this context, an integrated methodology was outlined to sequentially implement global and local fatigue methods, following a progressive global-local modelling approach to evaluate the consequences of extending the fatigue life.

The presented case studies allowed for the conclusion on the potential of applying local fatigue methods to large structures such as bridges.

Overall, the provided tools and the logical organization of the multiphase methodology allow for the full exploration of the capacity of materials under cyclic railway loadings, in accordance with a sustainable philosophy, while also addressing the increasing traffic demands towards net-zero emission targets.

ABBREVIATIONS

AMP	Absolute Maximum Principal stress
AVT	Ambient Vibration Tests
B-WIM	Bridge-Weigh-In-Motion
EFVM	Vitória-Minas railway line
FE	Finite Elements
HHT	Hilber-Hughes-Taylor
LEFM	Linear Elastic Fracture Mechanics
MAC	Modal Assurance Coefficient
MTS	Maximum Tangential Stress criterion
SWT	Smith-Watson-Topper
VCCT	Virtual Crack Closure Technique
WIM	Weigh-In-Motion

ACKNOWLEDGEMENTS

This work was financially supported by: Base Funding-UIDB/04708/2020 with DOI 10.54499/UIDB/04708/2020 (https://doi.org/10.54499/UIDB/04708/2020) of the CONSTRUCT-Instituto de I&D em Estruturas e Construções-funded by national funds through the FCT/MCTES (PIDDAC).

REFERENCES

1. European Commission, Climate Action. Transport emissions, (2019). https://climate.ec.europa.eu/eu-action/transport_en (accessed April 27, 2023).
2. UIC, *High-Speed around the World: Historical, Geographical, and Technological Development*, UIC, Paris, France, 2023.
3. C.S. Horas, J.N. Silva, J.A.F.O. Correia, A.M.P. De Jesus, Fatigue damage assessment on aging riveted metallic railway bridges: a literature review, *Structures* 58 (2023) 105664. https://doi.org/10.1016/j.istruc.2023.105664
4. J. Schijve, Fatigue of structures and materials in the 20th century and the state of the art, *Int J Fatigue* 25 (2003) 679–702. https://doi.org/10.1016/S0142-1123(03)00051-3
5. D. Broek, *The Practical Use of Fracture Mechanics*, Kluwer Academic Publishers, Dordrecht/Boston/London, 1988.
6. D. Radaj, Review of fatigue strength assessment of nonwelded and welded structures based on local parameters, *Int J Fatigue* 18 (1996) 153–170. https://doi.org/10.1016/0142-1123(95)00117-4
7. W. Schütz, A history of fatigue, *Eng Fract Mech* 54 (1996) 263–300. https://doi.org/10.1016/0013-7944(95)00178-6

8. O.H. Basquin, The exponential law of endurance tests, *Am Soc Testing Mater Proc* 10 (1910) 625–630.
9. L.F. Coffin, A study of the effects of cyclic thermal stresses on a ductile metal, *Trans ASME* 76 (1954) 931–949. https://doi.org/10.1115/1.4015020
10. S.S. Manson, Behaviour of materials under conditions of thermal stress, Report 1170, NACA (1954) 37.
11. D. Radaj, C.M. Sonsino, W. Fricke, *Fatigue Assessment of Welded Joints by Local Approaches*, Woodhead Publishing, Cambridge, MA, 2006.
12. J. Morrow, Cyclic plastic strain energy and fatigue of metals, in: *Internal Friction, Damping, and Cyclic Plasticity*, ASTM International, Philadelphia, PA, 1965: pp. 45–87. https://doi.org/10.1520/STP43764S
13. J. Morrow, Fatigue properties of metals, in: *Fatigue Design Handbook*, No. AE-4, Society of Automotive Engineers, Warrendale, PA, 1968: pp. 1–132.
14. K.N. Smith, P. Watson, T.H. Topper, A stress-strain function for the fatigue of metals, *J Mater* 5 (1970) 767–778.
15. P. Paris, M. Gomez, W. Anderson, A rational theory of fatigue, *Trend Eng* 13 (1961) 9–14.
16. A.A. Griffith, VI. The phenomena of rupture and flow in solids, *Proc R Soc Lond A Math Phys Sci* 221 (1921) 163–198. https://doi.org/10.1098/rsta.1921.0006
17. J.T.P. de Castro, M.A. Meggiolaro, T. Topper, *Fatigue Design Techniques: Vol. I - High-Cycle Fatigue*, CreateSpace Independent Publishing Platform, South Carolina, USA, 2016.
18. S. Downing, D. Socie, Simple rainflow counting algorithms, *Int J Fatigue* 4 (1982) 31–40. https://doi.org/10.1016/0142-1123(82)90018-4
19. ASTM E 1049–85, Standard Practices for Cycle Counting in Fatigue Analysis, (2017). https://doi.org/10.1520/E1049-85R17
20. A.Z. Palmgren, Die Lebensdauer Von Kugellagern (Life Length of Roller Bearings or Durability of Ball Bearings), *Zeitschriyt Des Vereines Der Deutschen Ingenioeren* 68 (1924) 339–341.
21. M.A. Miner, Cumulative damage in fatigue, *J Appl Mech* 12 (1945) A159–A164. https://doi.org/10.1115/1.4009458
22. K.-J. Bathe, *Finite Element Procedures*, Prentice Hall Inc, Englewood Cliffs, NJ, 1996.
23. J. Chung, G.M. Hulbert, A time integration algorithm for structural dynamics with improved numerical dissipation: the generalized-α method, *J Appl Mech* 60 (1993) 371–375.
24. C.S. Horas, J.A.F.O. Correia, A.M.P. De Jesus, P. Kripakaran, R. Calçada, Application of the modal superposition technique combined with analytical elastoplastic approaches to assess the fatigue crack initiation on structural components, *Eng Fract Mech* 185 (2017) 271–283. https://doi.org/10.1016/j.engfracmech.2017.06.001
25. C.S. Horas, A.M.P. De Jesus, R. Calçada, Efficient computational approach for fatigue assessment of riveted connections, *J Constr Steel Res* 153 (2018) 1–18. https://doi.org/10.1016/j.jcsr.2018.10.005
26. B. Kühn, Assessment of existing steel structures - recommendations for estimation of the remaining fatigue life, *Procedia Eng* 66 (2013) 3–11. https://doi.org/10.1016/j.proeng.2013.12.057
27. C.S. Horas, A.M.P. de Jesus, R. Calçada, New multiphase methodology for fatigue damage screening of aging metallic railway bridges, *Int J Fatigue* (2023) 107498. https://doi.org/10.1016/j.ijfatigue.2022.107498
28. CEN, Eurocode 1: Actions on structures, Part 2: Traffic loads on bridges (2017).
29. CEN, Eurocode 3: Design of steel structures, Part 1–9: Fatigue (2010).
30. C. Albuquerque, P.M.S.T. de Castro, R. Calçada, Efficient crack analysis of dynamically loaded structures using a modal superposition of stress intensity factors, *Eng Fract Mech* 93 (2012) 75–91. https://doi.org/10.1016/j.engfracmech.2012.06.009

31. A. Taras, R. Greiner, Statistical background to the proposed fatigue class catalogue for riveted components, ECCS TC6, Lausanne, Switzerland, (2010).
32. A. Taras, R. Greiner, Development and application of a fatigue class catalogue for riveted bridge components, *Struct Eng Int* 20 (2010) 91–103. https://doi.org/https://doi.org/10.2749/101686610791555810
33. G. Glinka, Calculation of inelastic notch-tip strain-stress histories under cyclic loading, *Eng Fract Mech* 22 (1985) 839–854. https://doi.org/10.1016/0013-7944(85)90112-2
34. G. Glinka, Energy density approach to calculation of inelastic strain-stress near notches and cracks, *Eng Fract Mech* 22 (1985) 485–508. https://doi.org/10.1016/0013-7944(85)90148-1
35. G. Glinka, Relations between the strain energy density distribution and elastic-plastic stress-strain fields near cracks and notches and fatigue life calculation, *ASTM STP* 942 (1988) 1022–1047. https://doi.org/10.1520/STP24538S
36. W. Ramberg, W.R. Osgood, Description of stress-strain curves by three parameters, Technical Note No. 902, NACA (1943) 32.
37. G. Sendeckyj, Constant life diagrams — a historical review, *Int J Fatigue* 23 (2001) 347–353. https://doi.org/10.1016/S0142-1123(00)00077-3
38. K. Tanaka, Fatigue crack propagation from a crack inclined to the cyclic tensile axis, *Eng Fract Mech* 6 (1974) 493–507. https://doi.org/10.1016/0013-7944(74)90007-1
39. F. Erdogan, G.C. Sih, On the crack extension in plates under plane loading and transverse shear, *J Basic Eng* 85 (1963) 519. https://doi.org/10.1115/1.3656897
40. I. Fracture Analysis Consultants, Franc3D Reference Manual, 2017.
41. F. Magalhães, Á. Cunha, E. Caetano, Dynamic monitoring of a long span arch bridge, *Eng Struct* 30 (2008) 3034–3044. https://doi.org/10.1016/j.engstruct.2008.04.020
42. F. Marques, C. Moutinho, W.H. Hu, Á. Cunha, E. Caetano, Weigh-in-motion implementation in an old metallic railway bridge, *Eng Struct* 123 (2016) 15–29. https://doi.org/10.1016/j.engstruct.2016.05.016
43. R. Pimentel, D. Ribeiro, L. Matos, A. Mosleh, R. Calçada, Bridge weigh-in-motion system for the identification of train loads using fiber-optic technology, *Structures* 30 (2021) 1056–1070. https://doi.org/10.1016/j.istruc.2021.01.070
44. X. Xiao, D. Pi, Q. Zhu, A bridge weigh-in-motion algorithm for fast-passing railway freight vehicles considering bridge-vehicle interaction, *Mech Syst Signal Process* 181 (2022) 109493. https://doi.org/10.1016/j.ymssp.2022.109493
45. T.D. Righiniotis, A comparative study of fatigue inspection methods, *J Constr Steel Res* 62 (2006) 352–358. https://dx.doi.org/10.1016/j.jcsr.2005.08.008
46. J. Herter, *Inspection Strategies to Prevent Fatigue Failure of Gusset Plates in Steel Truss Bridges*, CRC Press, Taylor & Francis Group, Boca Raton, FL, 2012. <Go to ISI>://WOS:000328502802017.
47. F.V. Lippi, M. Orlando, W. Salvatore, Assessment of the dynamic and fatigue behaviour of the Panaro railway steel bridge, *Struct Infrastruct Eng* 9 (2013) 834–848. https://doi.org/10.1080/15732479.2011.625955
48. C.W. Roeder, G.A. MacRae, A.Y. Kalogiros, A. Leland, *Fatigue Cracking of Riveted, Coped, Stringer-to-Floorbeam Connections*, National Technical Information Service, Washington, DC, 2001.
49. J.A.F.D.O. Correia, B.A.S. Pedrosa, P.C. Raposo, A.M.P. De Jesus, H.M. dos Santos Gervásio, G.S. Lesiuk, C.A. da Silva Rebelo, R.A.B. Calçada, L.A.P.S. da Silva, Fatigue strength evaluation of resin-injected bolted connections using statistical analysis, *Engineering* 3 (2017) 795–805. https://doi.org/10.1016/j.eng.2017.12.001
50. M. Al-Emrani, R. Crocetti, B. Akesson, B. Edlund, Fatigue damage retrofitting of riveted steel bridges using stop-holes, in: P. Mäkeläinen, P. Hassinen (Eds.), *4th International Conference on Light-Weight Steel Aluminium Strcutures*, Espoo, Finland, 1999: pp. 223–230.
51. FHWA, Manual for repair and retrofit of fatigue in steel bridges, Tech Report, FHWA-IF-13-020, (2013) 134.

52. A. Bassetti, P. Liechti, A. Nussbaumer, Fatigue resistance and repairs of riveted bridge members, *Eur Struct Integr Soc* 22 (1999) 207–218. https://doi.org/10.1016/S1566-1369(99)80043-0
53. A. Mohabeddine, J.A.F.O. Correia, P.A. Montenegro, J.M. Castro, Fatigue crack growth modelling for cracked small-scale structural details repaired with CFRP, *Thin-Walled Struct* 161 (2021) 107525. https://doi.org/10.1016/j.tws.2021.107525
54. E. Ghafoori, G. Prinz, E. Mayor, A. Nussbaumer, M. Motavalli, A. Herwig, M. Fontana, Finite element analysis for fatigue damage reduction in metallic riveted bridges using pre-stressed CFRP plates, *Polymers (Basel)* 6 (2014) 1096–1118. https://doi.org/10.3390/polym6041096
55. E. Ghafoori, M. Motavalli, Innovative CFRP-prestressing system for strengthening metallic structures, *J Comp Construct* 19 (2015). https://doi.org/10.1061/(ASCE)CC.1943-5614.0000559
56. C.S. Horas, Integrated methodology for fatigue life prediction of existing metallic railway bridges, PhD Thesis, Faculty of Engineering of the University of Porto, 2021.
57. C.S. Horas, A.M.P. De Jesus, D. Ribeiro, R. Calçada, Efficient multiscale methodology for local stress analysis of metallic railway bridges based on modal superposition principles, *Eng Fail Anal* 138 (2022) 106391. https://doi.org/10.1016/j.engfailanal.2022.106391
58. C.S. Horas, A.M.P. De Jesus, R. Calçada, Efficient progressive global-local fatigue assessment methodology for existing metallic railway bridges, *J Constr Steel Res* 196 (2022). https://doi.org/10.1016/j.jcsr.2022.107431
59. ANSYS, Academic Research, Release 17.1, ANSYS Inc., 2017.
60. ANSYS. ANSYS Mechanical APDL Element Reference, Release 17 Documentation for ANSYS, Academic Research, ANSYS Inc., 2017.
61. CEN, EN10025, Hot rolled products of structural steels, 2005.
62. A.M.P. Jesus, J.A.F.O. Correia, Fatigue assessment of riveted railway bridge connections. Part I: Numerical investigations, in: *7th European Conference on Steel Bridges*, Guimarães, Portugal, 2008.
63. CEN, Eurocode 3: Design of steel structures, Part 1–9: Fatigue, (2020).
64. CEN, Eurocode 3: Design of steel structures, Part 2: Steel Bridges, (2006).
65. R. Teixeira, C.S. Horas, A.M.P. De Jesus, R. Calçada, T.N. Bittencourt, Innovative hierarchical fatigue analysis of critical riveted railway bridges: A case study, *Eng Struct* 317 (2024) 118629. https://doi.org/10.1016/j.engstruct.2024.118629
66. AREMA, *Manual for Railway Engineering*, American Railway Engineering and Maintenance-of-way Association, Maryland, USA, 2021.
67. ABAQUS, *Analysis User's Manual, Version 6.12*, Dassault Systèmes Simulia, Inc., Rhode Island, USA, 2012.
68. C. Albuquerque, A.L.L. Silva, A.M.P. de Jesus, R. Calçada, An efficient methodology for fatigue damage assessment of bridge details using modal superposition of stress intensity factors, *Int J Fatigue* 81 (2015) 61–77. https://doi.org/10.1016/j.ijfatigue.2015.07.002
69. C.S. Horas, G. Alencar, A.M.P. De Jesus, R. Calçada, Development of an efficient approach for fatigue crack initiation and propagation analysis of bridge critical details using the modal superposition technique, *Eng Fail Anal* 89 (2018) 118–137. https://doi.org/10.1016/J.ENGFAILANAL.2018.03.009

10 Fatigue Modeling with Artificial Intelligence and Machine Learning

Changqi Luo, Shun-Peng Zhu, and Qingyuan Wang

10.1 INTRODUCTION

Fatigue is a major factor in mechanical failure, with 80%–90% of engineering components fractures resulting from fatigue [1]. For example, 80% of accidents in welding structures are caused by fatigue damage [2]. Clarifying the fatigue mechanism and completing fatigue modeling are of great significance for economy and safety. Preliminary research on fatigue can be traced back to the Industrial Revolution in the 19th century. William Albert carried out relevant research work and published a paper on the relationship between cyclic load and durability of the metal in 1837. Two years later, Jean-Victor Poncelet first officially used the word "fatigue" in a book on mechanics, which originated from the Latin expression with the meaning of "to tire". After that, great amounts of studies have been conducted on fatigue modeling, which can be summarized into four stages as shown in Figure 10.1.

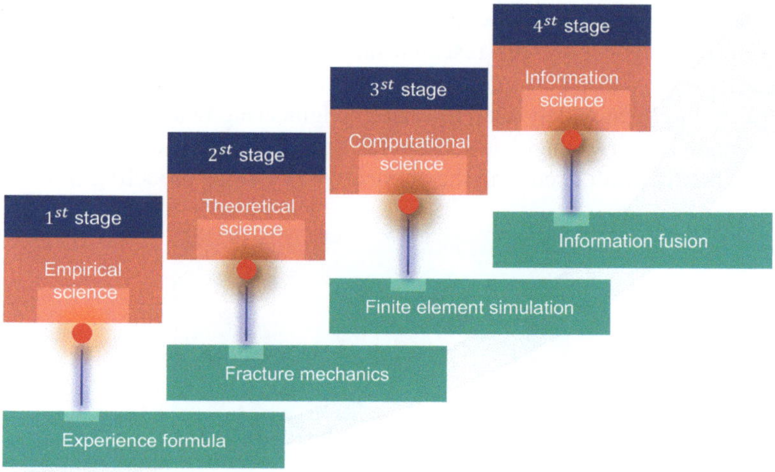

FIGURE 10.1 The development of fatigue modeling. (Reprinted with permission from Ref. [1], Elsevier.)

DOI: 10.1201/9781003405122-10

The first stage is empirical science where fatigue modeling mainly relies on experiments and experiences. For example, the famous S-N curve proposed by August Wöhler was summarized by lots of experiments. In the theoretical science stage, Alan A. Griffith who investigated the cracks in brittle glass led to the birth of fracture mechanics. This promoted the understanding of fatigue, and several theorems were developed. In the computational science stage, with the development of computer technology and finite element (FE) theory, the fatigue process can be easily explained, and the experimental costs have been greatly saved. The last stage is information science stage. Driven by data and theory, fatigue modeling based on artificial intelligence (AI) and machine learning (ML) has become a new paradigm. ML is a powerful tool that can dig deep into data information and extract key inputs for efficient and accurate fatigue modeling. A great deal of work has been done in this field, and fatigue modeling with AI and ML will continue to evolve as data augmentation and ML theory develops. Therefore, this chapter is valuable as a primer on fatigue and ML for a quick overview of related research.

This chapter is organized as follows. First, the basic concepts of ML and the main applications of fatigue modeling are reviewed. Then, recent advances in fatigue modeling with AI and ML are summarized. Next, three engineering applications are presented and discussed. Finally, summary and prospects are developed.

10.2 BASIC CONCEPTS

AI emerged as a branch of computer science in the 1950s [2]. It is a novel subject that studies and develops theories, methods, technologies, and application systems for simulating, extending, and expanding human intelligence. ML is the mainstream method to solve several AI problems at this stage. As an independent direction, it is in the process of rapid development and has achieved good results. For various reasons, research on ML has stagnated many times. However, with the improvement of computer computing power, the accumulation of data and the introduction of deep learning (DL) algorithms, the research on ML has become the focus of attention again. The relationship between AI, ML, and DL is shown in Figure 10.2. Recently, ML theory has been widely used in various engineering fields, and its application in fatigue modeling has brought new vitality to fatigue. In this section, the basic concept of ML is briefly introduced, and the application in fatigue modeling is discussed.

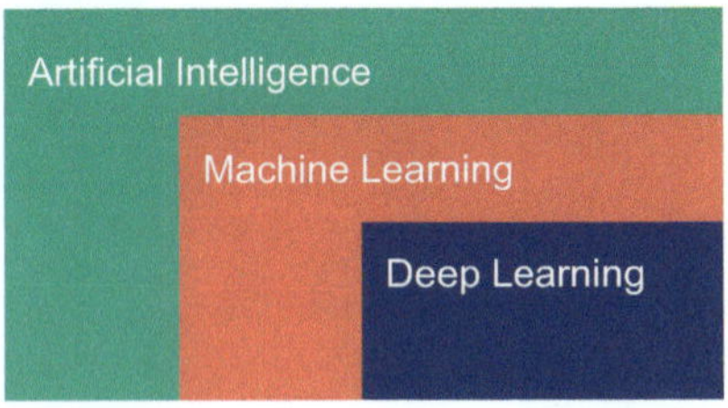

FIGURE 10.2 The relationship of artificial intelligence, machine learning and deep learning.

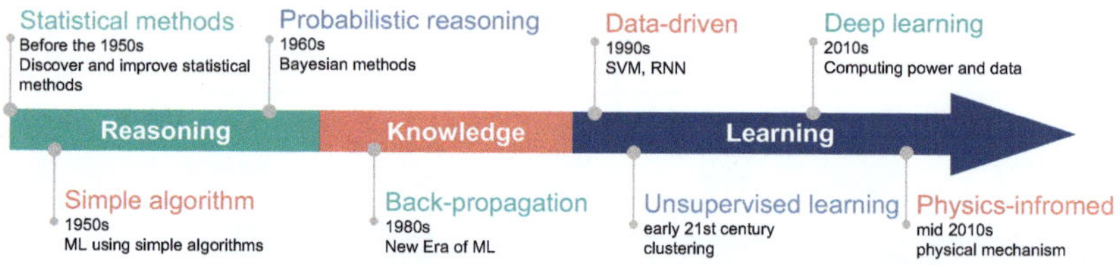

FIGURE 10.3 The development history of ML.

10.2.1 Introduction to ML

The earliest ML algorithms can be traced back to the early 20th century, and as of today, more than 100 years have passed. It has been nearly 40 years since ML was called an independent direction in 1980. The research history of ML has a natural and clear vein from focusing on "reasoning", to focusing on "knowledge", and then focusing on "learning" as shown in Figure 10.3. Reasoning Period: The time is from 1950s to 1970s. People think that if the machine is endowed with logical reasoning ability, the machine can be intelligent. However, as research moved forward, it was gradually realized that logical reasoning alone was far from achieving AI. Knowledge period: The time is from 1970s to 1980s. During this period, a large number of expert systems came out and made great contributions in many fields. However, the expert system faces the "knowledge engineering bottleneck," so some scholars thought that it would be great if the machine could learn knowledge by itself. Learning Period: From 1980s to the present, ML began to receive attention, became an independent subject field and began to develop rapidly, and a period when various ML technologies flourished.

ML models can take advantage of high-dimensional and heterogeneous data and automatically and continuously update themselves by learning from the data. These characteristics mean that ML has great advantages in fatigue modeling. Before the introduction of ML models based on physical information, data-driven methods and ML methods were almost equivalent in the field of fatigue. As presented in Figure 10.4, according to the theory and network structure of ML algorithms, ML can be divided into supervised learning, unsupervised learning, semi-supervised learning, and reinforcement learning.

10.2.1.1 Supervised Learning

Data information is a treasure house that needs to be deepened, and supervised learning has realized the direct use of data information. Supervised learning is the most widely used ML model. Common supervision learning algorithms include regression analysis and statistical classification such as support vector machine (SVM), Gaussian process regression (GPR), Logical regression (LR), Random forest (RF), Decision tree (DT), Bayesian network (BN), K-nearest neighbor (KNN), and neural network (NN). It should be pointed out that related research shows that NNs account for 57% of related ML models in fatigue life estimation [2]. Among them, NN can be further divided into artificial NNs (ANN), deep NN (DNN), convolutional NN

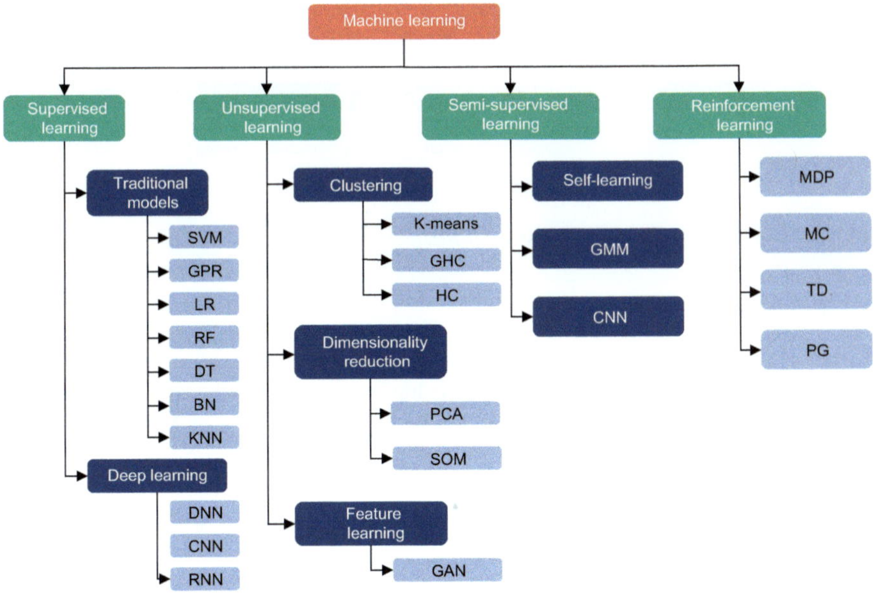

FIGURE 10.4 The categories of ML models. (Reprinted with permission from Ref. [1], Elsevier.)

FIGURE 10.5 Supervised learning. (Reprinted with permission from Ref. [1], Elsevier.)

(CNN), and recurrent NN (RNN). In addition, there are several variants in the above-mentioned ML models, which are no longer listed here.

Supervised learning algorithms build corresponding mathematical models based on existing input-output correspondences. This data set is often called training data and consists of a set of training examples. Each training set consists of one or more inputs and corresponding one or more outputs. In the mathematical model, the input training data is represented by an array or vector, also known as a feature vector. Meanwhile, the whole data sets are represented by a matrix. Through iterative optimization of an objective function, supervised learning algorithms learn a function that can be used to predict the output associated with new inputs. Supervised learning is essentially learning the statistical law of nonlinear mapping from input to output. Its schematic diagram is displayed in Figure 10.5. Undoubtedly, based on the

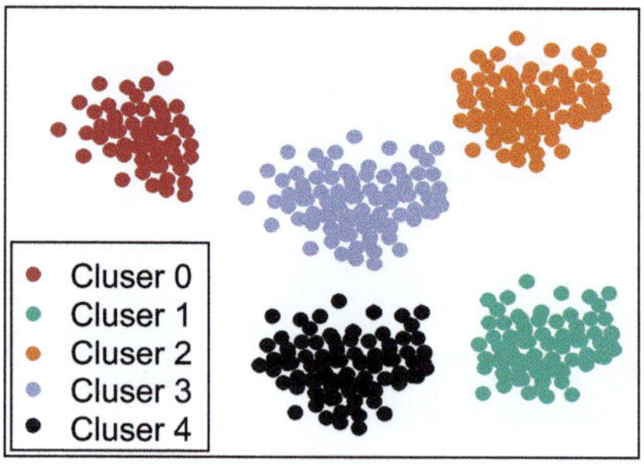

FIGURE 10.6 The schematic diagram of clustering.

above characteristics, supervised learning is the most widely used type of ML model in fatigue modeling.

10.2.1.2 Unsupervised Learning

Unsupervised learning algorithms only contain input data sets and use some strategies to find commonalities in the data, like grouping or clustering of data points. The training set has not been labeled, classified, or categorized. Similarly, unsupervised learning algorithms learn from each new test set looking for and reacting to such commonalities. Unsupervised strategies have significant advantages in visualizing and identifying trends in massive data sets, breaking out irrelevant inputs, and finding manufacturing conditions that lead to similar material properties, especially when expert knowledge is unavailable or in big data situations. Common unsupervised learning includes clustering, dimensionality reduction, and feature learning. As seen in Figure 10.6, the clustering algorithm clarifies data to clarify its data relationship, such as K-means clustering, Gaussian hybrid clustering (GHC), and hierarchical clustering (HC). The dimension reduction is used to extract key input information in fatigue modeling, stripped out high correlation input, related models include the principal component analysis (PCA), and self-organizing map (SOM). As a feature learning algorithm, it generates data that obeys real features during the network training process, thereby achieving feature enhancement.

10.2.1.3 Semi-Supervised Learning

Semi-supervised learning is somewhere in between the first two, that is, the labels of the training data may be incomplete, noisy, inaccurate, and limited. Data sets obtained in this way are cheaper than supervised learning, but the trained models are more accurate and more effective in training than unsupervised learning.

Semi-supervised learning is divided into pure semi-supervised and transductive learning. In pure semi-supervised learning, the test data is not in the training set, and the trained model is desired to have strong generalization performance. Transductive learning was introduced by Vladimir Vapnik [3]. It was motivated by the fact that it is easier than inductive learning, given the fact that inductive learning tries to learn a general function to solve a specific problem, while transductive learning tries to learn a specific function for the problem at hand. The flow of pure semi-

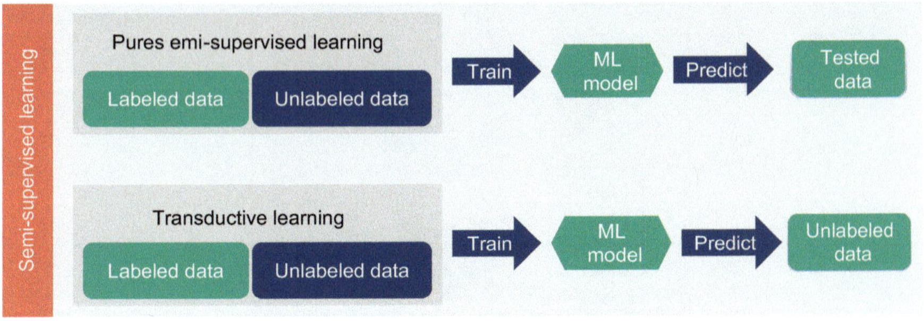

FIGURE 10.7 The comparison of pure semi-supervised and transductive learning. (Reprinted with permission from Ref. [1], Elsevier.)

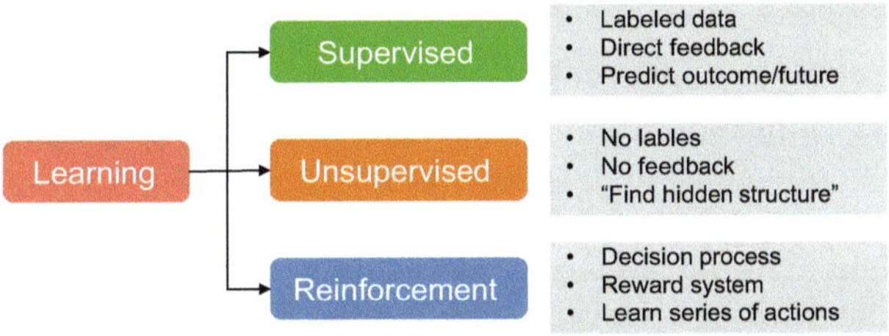

FIGURE 10.8 The comparison of different ML categories.

supervised and transductive learning is exhibited in Figure 10.7. The commonly used semi-supervised learning models are self-training, Gaussian mixture model (GMM), and semi-supervised CNN.

10.2.1.4 Reinforcement Learning

Reinforcement learning is an area of ML concerned with how software agents ought to take actions in an environment to maximize some notion of cumulative reward, as seen in Figure 10.8. Based on this characteristic, reinforcement learning is widely used in game theory, control theory, operations research, information theory, simulation-based optimization, multi-agent systems, swarm intelligence, statistics, and genetic algorithms. In ML, the Markov decision process is often considered and reinforcement learning algorithms do not assume knowledge of an exact mathematical model. The commonly used reinforcement learning algorithms are Monte Carlo learning (MCL), temporal-difference learning (TD), Q-learning, policy gradient (PG), etc.

10.2.2 Fatigue Modeling with ML

The fatigue modeling process based on ML mainly includes the following features, i.e. input, output, model selection, model training, and model evaluation. This section briefly discusses the above features, as seen in Figure 10.9. In addition, some limitations of ML are also discussed.

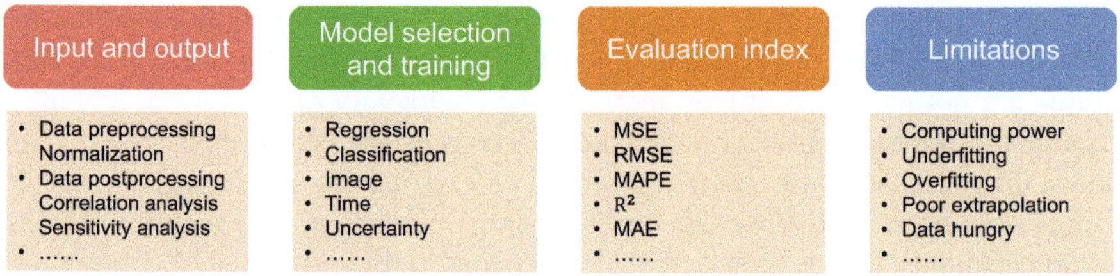

FIGURE 10.9 The typical features of fatigue modeling with ML. (Reprinted with permission from Ref. [1], Elsevier.)

10.2.2.1 Input and Output

Input is the cornerstone of ML modeling, and its basic content includes data establishment, data preprocessing, and data postprocessing. Firstly, establishing datasets is a critical part of fatigue modeling, which is used for ML to learn patterns from it. The commonly used input of fatigue includes structural features, mechanical properties, experimental conditions, microstructure, and defect features [4,5]. Specifically, the input data may contain feature vectors, image matrices, time, etc.

Data preprocessing is an important part for ML to build an accurate model. In the process of data establishment, different types of inputs are collected. In this case, the orders of these inputs are different in scales, for example, the modulus of elasticity may reach 10^9, while the size may only be 10^{-3}. This difference can lead to an increase in model training time or the failure of ML model converges. To eliminate the dimensional influence among features, prevent gradient explosion, and improve prediction accuracy and generalization ability of the ML models, the normalization should be considered [6]. The commonly used normalization methods usually map the data to [0,1], [–1,1] or centered on 0. For example, the linear normalization method as be expressed as,

$$x^* = \frac{x - x_{\min}}{x_{\max} - x_{\min}} \tag{10.1}$$

where x denote the original data, $x^* \in [0,1]$ denotes the normalized data. $x_{\min}$ and $x_{\max}$ are the maximum and minimum of original data, respectively. Also, to normalize the data to [–1,1],

$$x^* = \left(\frac{x - x_{\min}}{x_{\max} - x_{\min}}\right) \times 2 - 1 \tag{10.2}$$

In fatigue modeling, data means information. The effectiveness of the data-driven approach is highly dependent on input data obtained in the previous process. In cases where input data are limited or not enough, the resulting predictive error may be significant.

Data postprocessing is also important for fatigue modeling with ML. Data postprocessing usually refers to extracting the key input features of the ML model established by raw data through correlation analysis, sensitivity analysis, etc.

If the information within the features is weakly correlated, it will cause the ML model to learn wrong information, resulting in large errors. Besides, If the input features are not identified, it may increase the complexity of the model, leading to the curse of dimensionality. Although data postprocessing is not considered in many studies, it is because many studies are based on existing research work, and the key features have been relatively clear. But for a large number of data features obtained from experiments, data postprocessing is very important.

Accurate output is the goal of fatigue modeling. The inputs and outputs should include necessary information related to the underlying physics of the fatigue modeling [7]. As mentioned in the previous section, the output can be regression (like fatigue life prediction) or classification (like damage monitoring). Appropriate output helps to elucidate fatigue mechanisms.

10.2.2.2 Model Selection and Training

The choice of ML model is not unique. On the one hand, for the same fatigue modeling problem, there are many ML algorithms with similar performance. On the other hand, different ML algorithms have different characteristics, and choosing the correct algorithm will help improve modeling accuracy and analysis efficiency. There are many factors that affect model selection, common ones include input type, output type, model complexity, input dimension, data size, etc. As shown in Table 10.1, several typical fatigue modeling problems are provided, along with recommended ML models. The influencing factors of fatigue modeling greatly limit the selection of ML algorithms, and improper models will lead to large errors and even fail to solve

TABLE 10.1
Model Selection

Influencing Factors	Recommended Model	Description
General regression problem	ANN, Support Vector Regression (SVR), etc.	For general regression/classification problems, almost all machine learning models can achieve good results. At this point, the simplicity of the model needs to be considered.
General classification Problems	SVM, DT, etc.	
Image	CNN, DNN, etc.	Image data is usually complex and requires a reasonable machine learning structure to achieve better performance.
Time	RNN, Long Short-Term Memory (LSTM)	Due to the particularity of temporal data, correlations between outputs should be considered.
Uncertainty	Probabilistic Neural Network (PNN), GPR, etc.	How to quantify the uncertainty of data requires special machine learning models.
Small samples	physics inform-based neural network (PINN), GAN et al.	Insufficient data will cause machine learning to fail to learn generalized information accurately. Therefore, it is necessary to introduce machine learning models that can supplement relevant information.

specific fatigue modeling problems. Meanwhile, there is still not commonly accepted criteria for selecting the optimal ML model for fatigue modeling. Model selection is an open research problem in fatigue modeling.

Model training is crucial for fatigue modeling with ML. Selecting appropriate training algorithm and setting appropriate training parameters can effectively improve the accuracy of the model. Take the typical Back propagation neural network (BPNN) as an example, setting a suitable learning rate can not only improve the efficiency of algorithm convergence but also improve the accuracy of the model. In addition, the core of ML algorithms is actually an optimization problem, that is, how to minimize the training target error function. Choosing a suitable target function can affect the convergence performance of the model. On the other hand, optimization algorithm of ML (such as Gradient Descent, Stochastic Gradient Descent, Adam) can significantly improve the accuracy of the model. Therefore, these methods should be chosen carefully, and further improvement is needed for specific problems.

10.2.2.3 Evaluation of ML Model

The evaluation of the ML model intuitively reflects the quality of the model performance. In the process of Model training, some indicators are usually used as optimized target functions, such as mean squared error (MSE). And, they may directly represent a certain algorithm. For example, in linear regression as seen in Figure 10.10, by adding different regularization items to the target function, different regression algorithms are formed [8,9].

$$f(\boldsymbol{x}) = \boldsymbol{w}^T \boldsymbol{x} + b \tag{10.3}$$

$$\text{MSE} = \frac{1}{n}\sum_{i=1}^{n}(f(x_i) - y_i)^2 \tag{10.4}$$

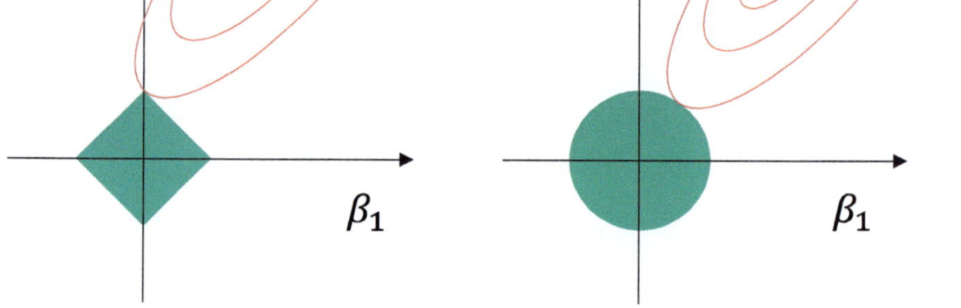

FIGURE 10.10 Different regularization items.

$$\text{cost}_{\text{Lasso}} = \frac{1}{n}\sum_{i=1}^{n}(f(x_i) - y_i)^2 + \lambda\|\boldsymbol{w}\|_1 \tag{10.5}$$

$$\text{cos}\,t_r = \frac{1}{n}\sum_{i=1}^{n}(f(x_i) - y_i)^2 + \lambda\|\boldsymbol{w}\|_2 \tag{10.6}$$

where $\boldsymbol{x}$ denote the input data, and y denote the output; $f(\boldsymbol{x})$ is the linear regression model and $\boldsymbol{w}$ represents the weights and b is a constant value; MSE is the MSE indicator. $\text{cost}_{\text{Lasso}}$ and cost_r is Lasso (least absolute shrinkage and selection operator) and ridge regression, respectively. $\|\cdot\|_1$ and $\|\cdot\|_2$ represents the L_1 regularization and L_2 regularization, respectively. It is worth mentioning that the linear combination of Eqs. (10.3) and (10.4) is the elastic network model [10] in ML.

In addition to MSE (Eq. 10.4), common evaluation indicators of the ML model also include mean absolute error (MAE, Eq. 10.7), root mean squared error (RMSE, Eq. 10.8), mean percentage error (MPE, Eq. 10.9), mean absolute percentage error (MAPE, Eq. 10.10) and the coefficient of determination (R^2, Eq. 10.11). For classification issues, common indicators are: Accuracy (Eq. 10.12), Precision (Eq. 10.13), Recall (Eq. 10.14), and so on.

$$\text{MAE} = \frac{1}{n}\sum_{i=1}^{n}|f(x_i) - y_i| \tag{10.7}$$

$$\text{RMSE} = \sqrt{\frac{1}{n}\sum_{i=1}^{n}(f(x_i) - y_i)^2} \tag{10.8}$$

$$\text{MPE} = \frac{1}{n}\sum_{i=1}^{n}\frac{y_i - f(x_i)}{y_i} \tag{10.9}$$

$$\text{MAPE} = \frac{100\%}{n}\sum_{i=1}^{n}\left|\frac{y_i - f(x_i)}{y_i}\right| \tag{10.10}$$

$$R^2 = \frac{\frac{1}{n}\sum_{i=1}^{n}(f(x_i) - \overline{y})^2}{\frac{1}{n}\sum_{i=1}^{n}(y_i - \overline{y})^2} \tag{10.11}$$

$$\text{Accuracy} = \frac{\text{TP} + \text{TN}}{\text{TP} + \text{TN} + \text{FP} + \text{FN}} \tag{10.12}$$

$$\text{Precision} = \frac{\text{TP}}{\text{TP} + \text{FP}} \tag{10.13}$$

$$\text{Recall} = \frac{\text{TP}}{\text{TP} + \text{FN}} \tag{10.14}$$

where TP, FP, FN, TN is True Positives, False Positives, False Negatives and True Negatives, respectively [10].

In addition, the efficiency of ML training, such as the time required for training models, and computer computing power, are often used as evaluation indicators. For different indicators, the performance of each ML model may not always be the best. Therefore, the performance of ML models also needs to be discussed further according to the problem of actual fatigue modeling.

10.2.2.4 Limitations of ML

Computing power: ML is a technology that is extremely thirsty for computing power. Although related ML theories were proposed in the 1950s to 1970s, the real application of ML can be traced back to the 1990s. With the development of computer computing power, especially the development of Graphic Process Units (GPUs) that are good at image computing, DL can be truly applied. However, for complex ML models, large data input, and high-dimensional problems, the training time of the ML model may be as long as more than a week. Therefore, computer calculation examples also become one of the limitations of ML. How to use some specific technologies to reduce the dependence of ML algorithms on computing power remains to be studied.

Underfitting: Underfitting means that ML model has not yet learned the general properties of the training samples, as seen in Figure 10.11. With the improvement of ML theory and algorithms, this problem is relatively rare. Usually, this problem can be better solved by increasing the complexity of the model or increasing the number of iterations.

Overfitting: Overfitting is a major difficulty in Fatigue modeling with ML as seen in Figure 10.11. Since the selected ML may be too complex, there are too many sample input features (the actual effective features are less than the input features), the data noise is too large, and ML learns unnecessary information, resulting in a model that performs well on the training set, but shows a large error on the test set. For the above problems, potential solutions include reducing the complexity of the model, reducing the dimension of the input, and increasing enough training data.

Poor extrapolation: Poor extrapolation may be overfitting or underfitting in some cases, as seen in Figure 10.11. However, what needs to be mentioned here is that for fatigue modeling, the ML model cannot obtain accurate results due to insufficient information in the interval beyond the training area. Thus, conventional ML

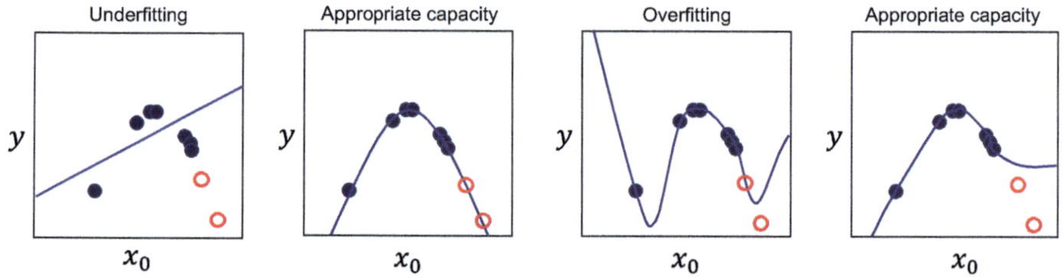

FIGURE 10.11 Different fitting case: Appropriate capacity; underfitting, overfitting, poor extrapolation.

algorithms generally perform poorly at extrapolation. Interestingly, in recent years, ML models based on physical information may be a potential way to solve the above problems.

Data hungry: Data hunger is a pain point in ML. ML is well known as data-driven which requires a large amount of data for training. The lack of data extremely limits the performance of ML. However, in fatigue modeling, data usually means a huge cost [11]. The main methods used for data hungry include data augmentation by data processing and physics models [12–14].

Poor interpretability: Poor interpretability has always been a criticism of ML. As a black-box model, ML establishes the relationship between input and output through a complex function or structure. However, how does ML works have always been unknown. Therefore, research on interpretable ML is also developing. Recently, PINN performs well in interpretable ability.

In this section, some limitations of ML are briefly discussed. In fact, the causes and solutions of the above-mentioned limitations in fatigue modeling are still worth exploring in depth. In this way, a ML model with good performance can be obtained to help with subsequent analysis and research.

10.2.3 Optimization in ML

In fatigue modeling with ML, whether it is ML training or ML application, optimization is an inseparable topic. In this section, the optimization problem of ML is specifically mentioned. In fact, optimization widely exists in all aspects of ML from modeling to application. It can be divided into three aspects: error optimization, model optimization, and output optimization as seen in Figure 10.12. There are a lot of studies that can be referred to for the relevant theories of optimization, and a specialized discipline has been formed, so we will not discuss it here. It should be noted that the optimization algorithms are generally divided into two categories which are deterministic algorithms and heuristic algorithms, respectively. Meanwhile, the more information the optimization algorithm needs (such as gradient information), the faster its efficiency will be, but in many cases, it is very difficult to obtain such information. This also determines some basic guidelines for choosing an optimization algorithm.

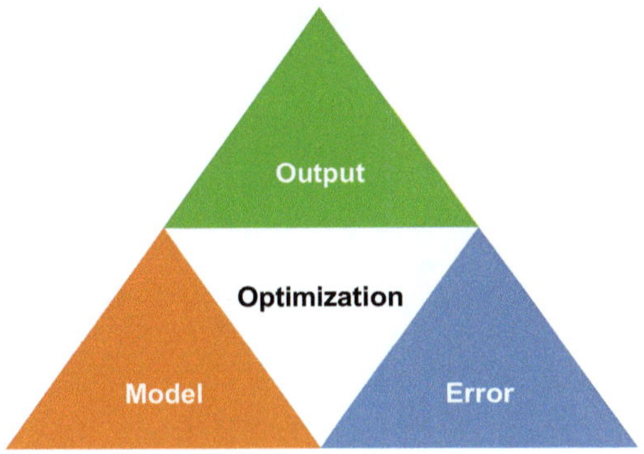

FIGURE 10.12 The optimization type of ML.

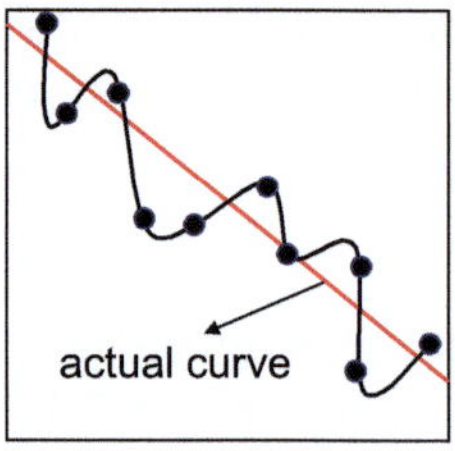

FIGURE 10.13 The schematic diagram of error.

10.2.3.1 Error Optimization

Error optimization can be seen as a subset of model training in the previous section. How to reduce the error between the model prediction value $f(x_i)$ and the actual output y_i as much as possible has been widely discussed in ML. On the one hand, different objective functions may imply different ML fitting performance. As shown in Figure 10.13, Eqs. (10.15) and (10.16) provides the L_1 loss function and L_2 loss function, respectively. Obviously, since the L_1 loss function squares the error, if the loss is used for general regression problems, the outliers have a greater impact on the loss. This further leads to a difference in model fitting performance.

$$L_1 = \sum_{i=1}^{n} (f(x_i) - y_i)^2 \tag{10.15}$$

$$L_2 = \sum_{i=1}^{n} |f(x_i) - y_i| \tag{10.16}$$

In optimization, the above can be viewed as the objective function. On the other hand, the choice of optimization algorithm is also very important in ML. The optimization algorithm will directly affect the convergence ability and convergence speed of the model. Meanwhile, error optimization also needs to consider how to transform the objective function into a solvable optimization problem. Taking the SVM [15] as an example, the optimization algorithm of this model first considers converting the objective function into a dual problem and then adopts the Lagrange method to solve it.

10.2.3.2 Model Optimization

The model optimization of ML is a broader concept than error optimization. In many studies, this content accounts for the majority of fatigue modeling. Besides, for different ML algorithms, the content that needs to be optimized is also different. For example, for SVM, to obtain a model with stronger generalization ability, researchers may need to optimize its three hyperparameters [16]. But for NNs, in addition to selecting appropriate hyperparameters, due to the randomness of neural network weights, some intelligent optimization algorithms (such as genetic algorithms [17], particle swarm optimization [18], etc.) are usually used to optimize their initial weights to achieve good performance. In addition, for specific problems, the structure of the model needs to be further optimized, such as RNN, LSTM, and its variants [19]. In a word, the purpose of model optimization is to make the ML model achieve better performance, such as prediction accuracy and generalization ability.

10.2.3.3 Output Optimization

Output optimization, that is, to optimize the research object. It can be classified into the application of ML. In fatigue modeling with ML, the output is usually fatigue life (can be other targets as well). Considering the high cost of material design or structural optimization through experiments or FE simulation, ML algorithms are widely used in surrogate modeling to enable efficient optimal design [7,20].

10.3 FATIGUE MODELING WITH AI

Since the birth of fatigue, research on fatigue has been improving. With the development pf related fatigue theories, experiments, and FE simulation, it lays the foundation for fatigue modeling with ML and develops at high speed. For example, between 1990 and 2021, more than 251 fatigue-related publications related to neural networks have been presented, involving all aspects of fatigue applications. Meanwhile, in recent years, ML modeling based on physical information has told development that it has injected new vitality into fatigue modeling. To enable the readers to have the basic understanding of fatigue modeling with ML, this section briefly reviews the recent advance.

10.3.1 ML-Based Fatigue Applications

After several generations of efforts, several classical methods have been formed and a series of applications for fatigue modeling have been developed. In this section, five classical applications are briefly introduced which is fatigue life prediction, fatigue crack, fatigue damage, fatigue strength, and fatigue load as presented in Figure 10.14.

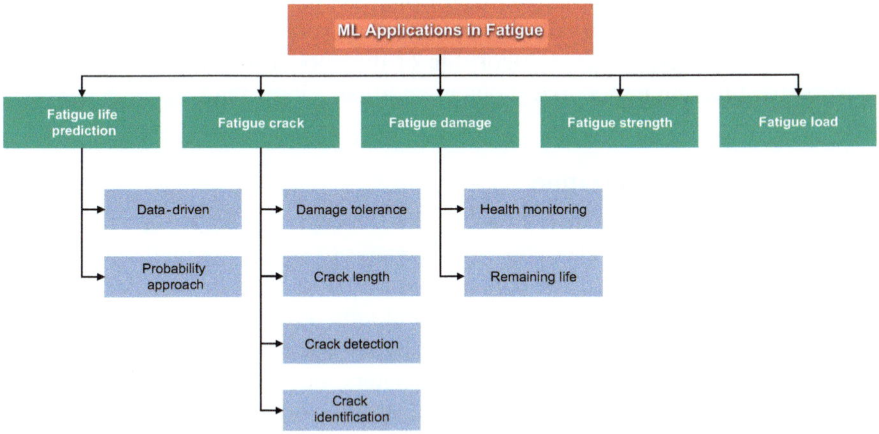

FIGURE 10.14 The categories of ML applications in fatigue.

10.3.1.1 Fatigue Life Prediction

Among all ML applications in fatigue modeling, fatigue life prediction is the most widely studied topic. Data-driven methods based on ML usually take material properties, fatigue loads (such as stress, strain), geometric dimensions, pressure, temperature, processing technology, etc. as input features, and fatigue life as output. In most cases, ML models matched experimental data and outperformed traditional methods in accuracy. However, different ML models may behave differently depending on the dataset. Although they are all fatigue life predictions based on ML, the specific details are different for different objects. The specific research objects are uniaxial fatigue life prediction, multiaxial fatigue life prediction, residual fatigue life prediction, etc. In addition, there are fatigue life predictions for metals or alloys, and fatigue life predictions for specific structures.

In recent years, researchers have also developed probability-based fatigue life prediction methods. Deterministic methods only provide a definite result in fatigue life prediction, such as fatigue life prediction value. Whereas probability-based fatigue life prediction methods provide point estimates and uncertainty quantifications such as variance or confidence intervals for fatigue life predictions. Although such methods seem to be very close to the prediction results of GPR in terms of description, they are established from a statistical point of view and based on the uncertainty quantification of fatigue life.

A common practice for probabilistic fatigue life prediction using ML is to include the probability value as another input variable. First, a probabilistic model of the assumed distribution is used to fit the fatigue life data. Then the fatigue life under any probability of failure is calculated. The NN is then trained using fatigue life, the corresponding probability of failure, and other variables such as stress levels. On the other hand, by reasonably designing the structure of the ML model, researchers have developed results that can simultaneously output the predicted value of fatigue life and the predicted standard deviation, such as the probabilistic physics-guided NN. In addition, Bayesian-based ML methods have also been used to establish probabilistic fatigue life predictions. In addition to establishing the relationship between probabilistic fatigue life and influencing variables, NNs are used to model the distribution of fatigue life data.

10.3.1.2 Fatigue Crack

Damage tolerance design based on fatigue cracks is an important part of fatigue analysis. In damage tolerance design, the fatigue process is considered as a highly nonlinear crack growth process [1]. In general, fatigue crack growth describes the relationship between the fatigue crack growth rate (da/dN) and the stress intensity factor range (ΔK). Thus, the ML model can be built by the input ΔK and the output da/dN. Also, other influencing factors (including maximum stress intensity factor K_{max}, load ratio, crack length, overload ratio, and temperature) are considered to investigate different problems. In addition, the short and long cracks, stress ratio and overload effect, and crack driving force are also investigated.

Crack length prediction is considered in fatigue modeling where the relationship can be modeled between crack size (a) and number of cycles (N) without crack growth analysis. In one study, acoustic emission (AE) data and cycle count were utilized as inputs, with crack length as the output variable. Employing a cascade feedforward neural network (FNN), the results demonstrated strong performance, achieving prediction errors of no more than 3%. Another investigation delved into the influence of stress ratio and cycle count on crack length through various models leveraging intelligent optimization techniques. Additionally, FE analysis was employed to explore the correlation between morphological and material factors, stress levels, cycle count, and crack length. ML techniques were further utilized to enhance simulation efficiency. Furthermore, research has been dedicated to crack length monitoring over time, incorporating recurrent neural networks (RNN) and long short-term memory (LSTM) models for accurate modeling.

Fatigue crack detection is another application of the ML model in fatigue modeling. The crack length, shape, size and so forth is quantified by some indirect methods like acoustic, electromagnetic, and thermal signals. Crack identification is also an important application with ML which is closely related to computer vision. Although there are many overlaps with crack detection, crack identification focuses on the crack segmentation from an image with crack and background pixels.

10.3.1.3 Fatigue Damage

Fatigue damage can be defined as crack initiation and growth under fatigue loading which can be classified as structural health monitoring, and its main applications can be divided into health state diagnosis and remaining life prediction (prognosis). Due to the excellent information fusion and pattern recognition capabilities of ML, it is widely used in the field of structural health monitoring. In fatigue damage detection and monitoring, the state of fatigue damage can be classified as multiple damage levels from health to failure. Usually, fatigue damage manifests as residual stiffness degradation or the reduction of static strength.

10.3.1.4 Fatigue Strength

Fatigue strength denotes the stress value that a material can withstand at a particular life of interest which is also called fatigue limit or fatigue endurance. A common input-output combination for fatigue strength prediction often considers the microstructure, chemical composition, processing parameters, and work conditions as input. For particular problems, some of them may not be considered.

10.3.1.5 Fatigue Load

The fatigue modeling of load spectrum is also widely used. Various studies have explored the use of neural network (NN) models to predict fatigue loads and equivalent stress ranges for different applications. Neural network models have been developed for online health monitoring of helicopter flight-critical components, fatigue load prediction at wind farms, and estimation of equivalent stress ranges in steel bridge decks. These models have shown promising results, with prediction errors less than 5% compared to FE models. Furthermore, multi-level deep neural networks combining long short-term memory and fully connected layers have been used to estimate the strain response of acceleration data, showing strong consistency with real-time history.

This section discusses the basic application of AI and ML in fatigue modeling. In most cases, the ML model could improve the accuracy of prediction, reduce the computational cost, and enable further applications and developments. Further development based on the above research content, such as structural optimization, material design, etc., requires more in-depth research and discussion.

10.3.2 Fatigue Modeling with ML by Informed-Driven

Information is the key to ML. The core of ML modeling is to learn useful information from data as much as possible. As mentioned earlier, learning insufficient information can lead to underfitting, and learning wrong information can lead to overfitting. Meanwhile, good extrapolation can only be achieved if the correct information is learned. However, when the data-driven ML model is no longer able to further learn information from the data (such as insufficient data, noise), how to make the ML model obtain physical consistency has gradually become an urgent problem to be solved. To solve this problem, physics informed-driven machine learning (PIML) algorithms have been proposed and widely used in fatigue modeling. Several studies have shown that PIML has shown excellent performance in practical applications, including: transparency, interpretability, analysis ability, response speed, generalization ability, and accuracy. This section attempts to understand how PIML works from various aspects.

10.3.2.1 A General Perspective

In fact, when a large number of data-driven ML research gradually showed inconsistent results with physical knowledge, researchers gradually realized the importance of physical knowledge in building ML models. Aykol et al. [21] envision several ML structures for coupled physical models. Figure 10.15a and g show purely physics-driven and purely data-driven models. Obviously, they have some typical characteristics, such as physical models with strong generalization capabilities but large errors, and data-driven models with high training data accuracy but weak extrapolation capabilities. Figure 10.15b–d show the structure of several sequential integration physical models coupled with ML, and physical knowledge is not integrated into the ML structure. Figure 10.15e and f show the structure of hybrid PB and ML model. Obviously, Figure 10.15e has become the current mainstream PIML structure, and the most widely used one is PINN. On the other hand, considering the high costs of solving partial differential equations (PDEs), a large number of NN-based work coupled with physics laws have been developed. The related work greatly reduces the calculation amount of FE simulation and realizes high-fidelity calculation. And the relevant ideas, theories and algorithms have been extended to other fields and applications including fatigue modeling.

It should be noted that the characteristics of the PIML model determine the need for a specific function library to support it. Taking PINN as an example, in addition to constructing a specific loss function, the training of the model is the key to related algorithms. How to transfer its error and gradient is the key. Fortunately, there are a large number of ML libraries to solve this problem, including TensorFlow, Pytorch, Keras, and JAX. Meanwhile, there are some actively developed libraries including

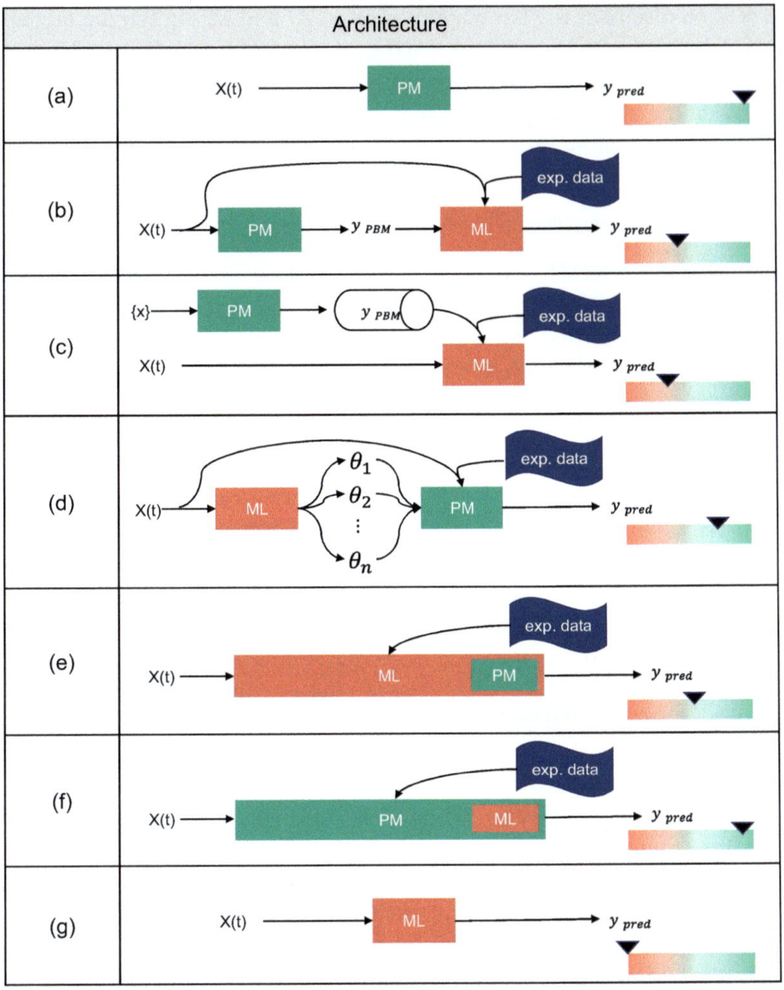

FIGURE 10.15 A map of integration strategies for physics-based model (PM) and ML models. (Reprinted with permission from Ref. [21].)

DeepXDE, SimNet, PyDEns, NeuroDiffEq, NeuralPDE, SciANN, and ADCME. Since Python is the dominant programing language for ML, it's convenient to use Python for PIML modeling. This preference stems from Python's dominance in the ML community and the availability of a large number of libraries customized for ML tasks. Many of these libraries take advantage of the automatic differentiation mechanisms provided by frameworks such as PyTorch. Using Python, users can easily define fatigue modeling problems, leaving the underlying implementation details to the solver. This abstraction simplifies the modeling process, allowing users to focus on defining the specific aspects of the fatigue modeling problem, while the solver can handle complex problems and solve them efficiently. Besides, there are still other software designed for fatigue modeling like MATLAB® [22].

10.3.2.2 Input Data

In the early exploration of ML modeling combined with physical information, the preprocessing of training data was one of the widely adopted solutions. The raw data is preprocessed to extract hidden physical information, thus transforming the weakly correlated input of fatigue modeling into strongly correlated input. Relevant research can be divided into two categories. The physical model calibrated with experimental data is used as the input of ML. To some extent, the ML model can be regarded as a PIML model. By combining the real plant, standard tribometers, and reference numerical model as input to train the RBFNN, the results showed that it's very close to the experimental data obtained in the production plant. On the other hand, for the complex data obtained by simulation or experiment, through the physical model, it is simplified to direct data related to fatigue modeling, which can also be regarded as a kind of PIML model. Wang et al. [23] considered the sensitive features and continuous damage mechanics as input, and the results demonstrate that the physical knowledge-guided ML model using sensitive features exhibits better performance of fatigue life prediction.

10.3.2.3 PIML Model

The PIML model uses the explainable reasoning mechanism of the physical model to guide ML to mine physical laws from the data. Similar to DL theory, by changing the topology of the model, the model can show superior performance in solving specific types of problems. The PIML model also greatly draws on this idea. As shown in Figure 10.16, through different topological structures, physical knowledge is introduced at the appropriate position, so as to achieve the consistency between the physical mechanism and the ML model. In addition, proper PIML also involves the information transfer (such as error and gradient) in the model training process, which is very important for the PIML model based on it. The current mainstream PIML algorithms (such as TensorFlow, Pytorch) all use the form of tensors to realize information transmission.

10.3.2.4 PIML Loss

Compared with introducing physical information through various topological structures, the introduction of physical information through loss functions has gradually become the current mainstream method. In fatigue modeling, constructing

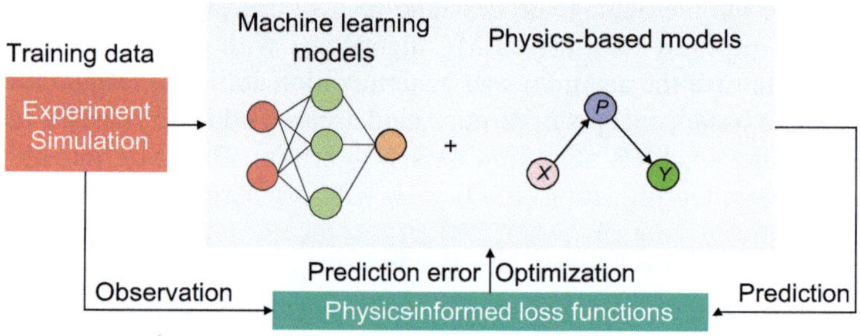

FIGURE 10.16 Structure of PIML. (Reprinted with permission from Ref. [1], Elsevier.)

the correct Loss function and realizing the consistency of physical laws have been greatly developed. Specifically, in life prediction, there are some broad regularities, such as fatigue life decreases with increasing stress. For the dispersion of fatigue life, especially in additive manufacturing materials (defects are the main factor affecting its fatigue life), the smaller the stress, the more obvious the analysis of fatigue life caused by defects. Or it is a physical model fitted through a large number of experiments, and its physical laws can guide the construction of PIML. In the FE simulation, by combining the constitutive equations of the structure, boundary conditions, etc., the constructed PIML is consistent with the above information in terms of physical laws. The PIML constructed based on the above information can realize a ML model with high generalization ability through a small sample input, so it is widely used. However, it should be noted that the construction of the loss function may not be straightforward, for example, researchers cannot obtain an accurate functional relationship between input and output. At this time, physical consistency can be maintained to some extent by considering the first-order and second-order derivative information between the input and output. Some studies have also pointed out that the results of PIML based on the loss function gradually approach the data-driven model when the amount of data is sufficient. This is because when the amount of data is large, the data contains a large amount of physical information that can be directly learned by ML; on the other hand, due to the increase in the amount of data, in the construction of the loss function, the loss of data fitting takes up The proportion of is much larger than that of physical loss, therefore, the two methods finally regress to the data-driven method.

10.4 CASE STUDY

In this section, three typical works of fatigue modeling with ML are presented including the purely data-driven ML model for fatigue life prediction, fatigue reliability analysis of specific structural component, small sample modeling fatigue modeling by PIML.

10.4.1 Fatigue Life Prediction Based on Data-Driven

Random defects are a significant contributor to the variability in fatigue performance observed in additive manufacturing. However, mechanics-based models often oversimplify the fatigue damage process and neglect crucial influencing factors. In this case, the two purely data-driven ML algorithms with optimization strategy is considered to improve the accuracy and generalization ability of fatigue life prediction. Thus, some basic concepts of physics model-based additive manufacture (AM) fatigue life prediction, SVR algorithm, ANN algorithm, Bayes optimization (BO), and Particle Swarm Optimization (PSO) are briefly discussed.

10.4.1.1 Physics-Based Fatigue Life Prediction

In AM materials, the defects will cause severe stress concentration, which leads to the production of fatigue cracks. Therefore, the defect-related parameter is defined by Murakami and Endo [24,25], which is which is defined as the square root of the projected defect area on the plane perpendicular to the maximum principal stress,

$$\Delta K_0 = Y\Delta\sigma\sqrt{\pi}\sqrt{\text{area}} \tag{10.17}$$

where ΔK_0 represents the original stress intensity factor, Y is 0.5 for internal defects and 0.65 for surface and sub-surface defects, $\Delta\sigma$ is the range of applied stress.

The fatigue crack growth life can be evaluated by the Paris law [26],

$$da/dN_f = C(\Delta K)^m \tag{10.18}$$

where da/dN_f denotes the fatigue crack propagation rate, C and m denote material constants. Noted that ΔK is a variable where $\Delta K = F\Delta\sigma\sqrt{\pi a}$. Due to the fatigue cracks predominantly initiated at large defects, the crack initiation life, N_i, could be ignored as most of the loading cycles are used for crack propagation. Taking $\sqrt{\text{area}}$ and a_c as the original and final crack length, the fatigue crack propagation life is given by the following equation [27],

$$N_f = \int_{\sqrt{\text{area}}}^{a_c} \frac{1}{C(\Delta K)^m} da = \frac{\left(\sqrt{\text{area}}\right)^{1-m/2} - a_c^{\,1-m/2}}{C(1-m/2)\left(F\Delta\sigma\sqrt{\pi}\right)^m} \tag{10.19}$$

When $\sqrt{\text{area}} \ll a_c$, N_f could be provided as follows:

$$N_f = \frac{\sqrt{\text{area}}}{C(1-m/2)(\Delta K_0)^m} \tag{10.20}$$

10.4.1.2 SVR Algorithm

The SVR algorithm is a supervised ML model based on the statistical learning theory, which is a variant of SVM. Based on statistical theory, SVR is widely used due to its excellent generalization ability and prediction accuracy. Suppose the training data D contains d number of instances each with an attribute x_i and its associated class y_i, i.e. $\{(x_1, y_1), (x_2, y_2), \dots, (x_d, y_d)\}$. Then, as seen in Figure 10.17, the linear function $f(x)$ on the dataset D can be defined as [28],

$$f(\boldsymbol{x}) = \boldsymbol{w}^T\boldsymbol{x} + b \tag{10.21}$$

where $\boldsymbol{w} = (w_1, w_2, \dots, w_d)$ is the normal vector of the hyperplane, b is the displacement term. The deviation of each error for each training data (x_i, y_i) is allowed to be within ε, that is $|y_i - f(x_i)| \le \varepsilon$. However, the assumption does not hold for all cases which in turn requires the slack variables ξ_i, ξ_i^* to cope with the cases that violate the assumption. Thus, the optimization problem can be defined as:

$$\min : \frac{1}{2}\|\boldsymbol{w}\|^2 + C\sum_{i=1}^{d}\left(\xi_i + \xi_i^*\right) \tag{10.22}$$

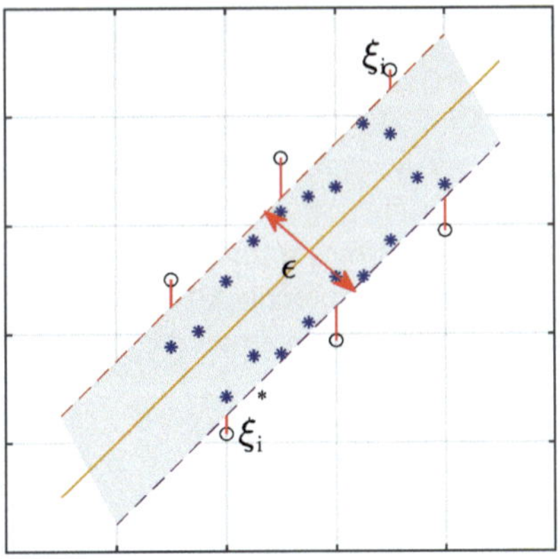

FIGURE 10.17 Principle of SVR. (Reprinted with permission from Ref. [33], Elsevier.)

$$\text{s.t. } \forall i: \begin{cases} f(x_i) - y_i \leq \varepsilon + \xi_i \\ y_i - f(x_i) \leq \varepsilon + \xi_i^* \\ \xi, \xi_i^* \geq 0 \end{cases}$$

where $\|\cdot\|$ represents the Euclidean norm and C is a constant penalty term, the optimal outcome can be achieved through the Lagrange function. To avoid computing the dot product on transformed data tuples, which are the training tuples x_i and x_j in the dual space, a kernel function defined over the original input data $K(x_i)$ can be utilized to replace each occurrence of the dot product between transformed tuples. This approach allows all calculations to be performed in the original input spaces. Specifically, the Gaussian kernel function has been chosen as:

$$K(x_i, x_j) = \exp\left(-\frac{\|x_i - x_j\|^2}{2\sigma^2}\right) \tag{10.23}$$

where σ is the radial basis of the Gaussian kernel function. The generalization ability of SVR is mainly determined by three hyperparameters C, σ, and ε. In this case, the Bayes optimization algorithm is used for the optimization of hyperparameters.

10.4.1.3 ANN Algorithm

The ANN model is used to map the complex nonlinear prediction regression issues by simulating the processing of complex information by the neural system of the human brain. In this case, the three-layer NN model is used for fatigue modeling as seen in Figure 10.18. The behavior of the neuron within each layer could be obtained as follows:

$$y_i = f\left(\sum \omega_{ij} x_i + \theta_i\right) \tag{10.24}$$

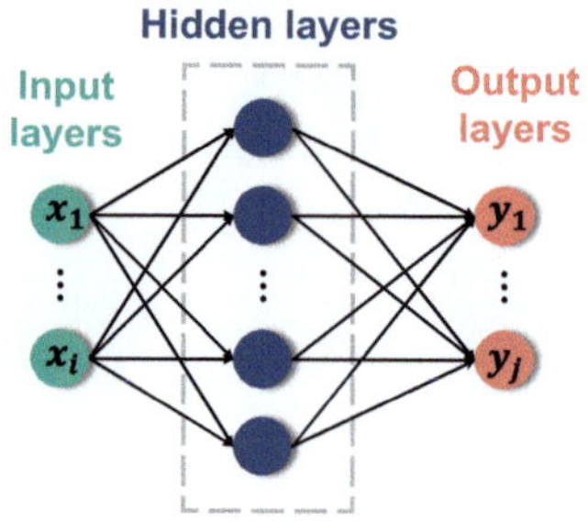

FIGURE 10.18 The schematic diagram of three layers ANN.

In each layer of the ANN, neurons receive input information which is multiplied by corresponding weights ω_{ij} and compared with the respective threshold θ_i. Subsequently, an activation function f_i is applied to determine the output of the neuron after receiving inputs from the previous layer. This process continues until the information is transferred to the output layer. For this particular ANN model, the sigmoid activation function is employed,

$$f_i(x_i) = \frac{1}{1 + \exp\{x_i\}} \tag{10.25}$$

For the ANN model, although there are many optimized contents. For example, the selection of the activation function, the hidden layer of the model, and the selection of error convergence algorithm. In this case, the above-mentioned content was determined through many attempts without further discussions. The randomness of the ANN model is a key factor affecting the performance of the model. For example, the initial weights of the ANN model will lead to unstable performance when the model converges. Therefore, the PSO algorithm is considered to optimize the initial weights of the ANN model.

10.4.1.4 Bayes Optimization

Bayes optimization is a black-box optimization algorithm that is used to solve the extreme value of the unknown function of the expression. The algorithm predicts the functional value of any functional point based on a set of functional values of a set of sampling points and gives its predictive difference, which is achieved through the GPR, as seen in Figure 10.19. According to the results of the GPR, the constructor acquisition function is used to measure the degree of exploration of each point and solve the extreme value of the acquisition function to determine the next sampling point. Finally, the extreme value of this set of sampling points is returned as the extreme value of the function.

In general, the performance of the SVR model can be measured by RMSE. To obtain better generalization ability of the SVR model, the BO is performed with objective function min min $Z = \text{RMSE}(\boldsymbol{X})$, where $\boldsymbol{X} = [C, \sigma, \varepsilon]$. The BO algorithm is divided into the following five steps:

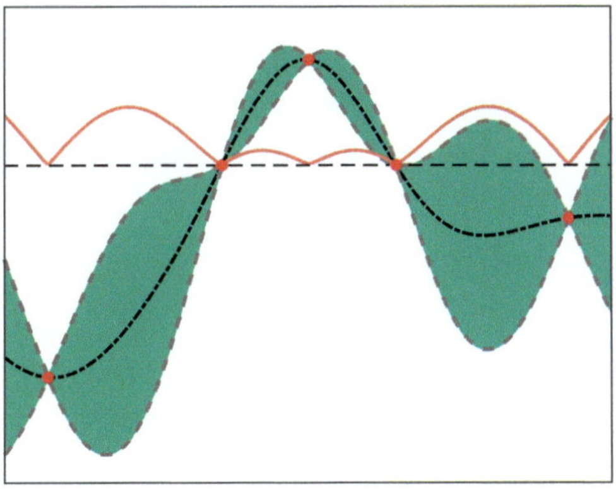

FIGURE 10.19 The schematic diagram of Bayes optimization. (Reprinted with permission from Ref. [29].)

TABLE 10.2
The Optimizaition Interval of the SVR Optimization Hyperparameters

C	ε	σ
$[10^2,10^6]$	$[10^{-4},10^{-1}]$	$[0.1,30]$

1. Initialize the initial distribution of the surrogate function $\alpha(\boldsymbol{X})$;
2. Choose the data point $\boldsymbol{X}_i$ so that the surrogate function takes the maximum value;
3. Evaluate the objective function results Z_i associated with the data point $\boldsymbol{X}_i$;
4. Update the surrogate function with the new data $\boldsymbol{X}_i$ to obtain a successive distribution (i.e., the initial distribution at the next step);
5. Repeat (i)–(iv) until the stopping criterion is matched [29].

What's more, the interval of the optimization hyperparameters is limited as seen in Table 10.2.

10.4.1.5 Particle Swarm Optimization

The PSO has the advantages of fast convergence, few parameters, and simple and easy to achieve. As an intelligent optimization algorithm, the problem of high - dimensional optimization, PSO converges faster than genetic algorithms to the optimal solution.

In this case, the ANN model uses the sigmoid activation function, the number of hidden layers is 20 layers, and Adam is used as the training algorithm. Considering that the optimization of ANN initial rights is generally high, the PSO algorithm is selected in this case to optimize the initial authority of ANN model. Subsequently, implement model training through the Adam algorithm. Generally, the ANN model optimized through the PSO usually has a more robust model performance [30].

10.4.1.6 Data and K-Fold Cross Validation

For a set of data for ML models, to verify the performance of the model, the data set is usually divided into training sets, validation, and test sets. Usually, the training set is used to train the ML model. ML model obtained the information from the training data. Validation set does not participate in training and is used to validate the accuracy of the model. The test set also does not participate in training, it can be seen as the actual application of the ML model. However, the ML model often performs well on the training set, but the prediction error is relatively large in the test set. This is because, during the training process, the ML model may learn the noise in the training data, which leads to the overfitting. To obtain good generalization capabilities, the ML model performs well on training sets as well as the test sets. In this case, the K-fold cross-validation is considered to further optimize the performance of the model.

In this case, the AM material Ti-6.5Al-2Zr-Mo-V is considered [31]. The Ti-6.5Al-2Zr-Mo-V samples were manufactured via directed metal deposition and then subjected to a double annealing heat treatment before being machined into standard cylindrical specimens with a diameter of 5 mm at the smallest section. A series of fatigue tests were conducted on these specimens at three stress levels, using a sine-wave constant amplitude stress with a stress ratio (R) of 0.06 and a frequency (f) of 10 Hz, all performed at room temperature under atmospheric conditions. SEM was utilized to examine the fracture morphology, particularly focusing on features indicative of crack initiation sites [32]. The characteristic parameters of Ti-6.5Al-2Zr-Mo-V are shown in Table 10.3, where $\sigma_{\max}$ is the peak stress of fatigue experiments, h_c indicates the nearest distance from defect center to surface.

In this case, there are 29 experiment points which are divided into 24 training points and 5 test points. For the 24 training points, the K-fold cross-validation is used. Note that, the error of the ML model can be divided into the training error, validation error, and test error which can be represented as $\text{RMSE}_{\text{train}}$, $\text{RMSE}_{\text{validation}}$ and $\text{RMSE}_{\text{test}}$, respectively. Besides, as seen in Figure 10.20, the K-Fold cross-validation divides the training set data into the K part, using the K-1 portion as training, the remaining part as a test. Thus, the optimization objective function of BO-SVR and PSO-ANN is $\min\left(\sum_{i}^{K}(\text{RMSE}_{\text{train}}+\text{RMSE}_{\text{validation}})_i\right)$ where the validation sets are not used for model training. By doing so, the generalization ability of ML model will be improved and the prediction accuracy in test sets will increase.

10.4.1.7 A General Framework

In this case, the framework of purely data-driven ML model for AM fatigue life prediction is presented in Figure 10.21 and the specific steps can be summarized as:

1. Obtain input and output data through fatigue tests, and divide them into training sets and test sets randomly. In this case, the input data also includes defect information obtained by SEM;
2. Build the basic SVR or ANN model by the training sets;
3. Obtain the objective function of the BO and PSO through the K-fold cross-validation, where $K = 5$ in this case.

TABLE 10.3
Characteristic Parameters of Ti-6.5Al-2Zr-Mo-V [31]

σ_{max} / MPa	$\sqrt{area}$ / μm	h_c / μm	N_f
720	4	315	875,322
720	7	594	627,590
720	12	960	375,137
720	37	204	804,089
720	38	953	777,717
720	45	1018	714,969
720	47	122	232,542
720	54	1124	303,411
720	59	66	405,901
720	77	908	159,196
720	84	738	891,389
720	100	1631	527,998
720	114	443	252,952
760	10	526	452,019
760	24	224	639,705
760	25	293	563,154
760	38	412	381,525
760	51	230	270,852
760	52	663	461,094
760	61	996	242,092
760	66	872	298,456
760	92	281	211,880
760	118	1292	173,684
760	132	1458	51,418
800	22	1260	545,806
800	37	1000	401,829
800	51	266	212,017
800	52	1320	146,029
800	88	453	94,538
800	114	99	29,930

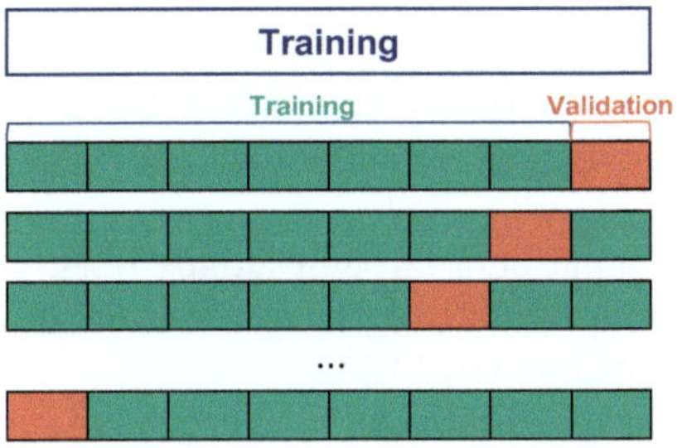

FIGURE 10.20 The schematic of K-fold cross validation. (Reprinted with permission from Ref. [29].)

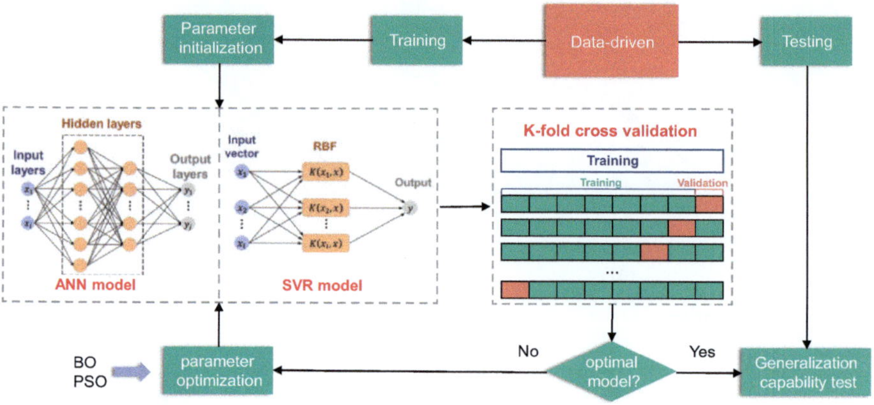

FIGURE 10.21 The framework of purely data-driven ML model for AM fatigue life prediction. (Reprinted with permission from Ref. [29].)

4. Use the BO algorithm to optimize the SVR's hyperparameters and use the PSO algorithm to optimize ANN's weights. Repeat this step until meet the convergence conditions (100 iterations in this case).
5. Use the established ML model to predict the fatigue life on the test set.

10.4.1.8 Results

The comparison of different data-driven and physics models is shown in Figure 10.22. To avoid the difference in model performance caused by data set differences, all methods use the same training set and test set. Figure 10.22a–c provides Paris Law, BO-SVR, and PSO-Ann's prediction results in the training set and test set with 2 band width. As seen in Figure 10.22d, for training set, the BO-SVR obtained the lowest RMSE which is 0.1751. The PSO-ANN is better than Paris law in training set with 0.2657 and 0.2703, respectively. For test set, the BO-SVR obtained the lowest RMSE which is 0.2126. However, The Paris law is better than PSO-ANN in training set with 0.2126 and 0.2306, respectively. Similarly, the data-driven BO-SVR obtained the highest value which is 0.6312, 0.6941, 0.5659 for Paris law, BO-SVR, and PSO-ANN, respectively. First of all, the BO-SVR provides a typical ML prediction result, that is, the error of the training set is lower than the error of the test set. It can be seen from Figure 10.22a that the fitting of the physical model is usually large even if the fitting data, which may be caused by incomplete understanding of physical mechanisms. For PSO-ANN, the poor model results may be caused by inappropriate model structures, such as the model is too complicated and the data is insufficient. Note that, the training set error in PSO-ANN is higher than the result of the test set, which may be caused by small samples of test data.

In this case, the following content is worth noting. First, the inaccuracy of physical models may be caused by incomplete physical mechanisms. Therefore, a more in-depth analysis of the fatigue process of fatigue may be of great help to predict fatigue life. Second, the ML model based on data-driven is not always better than physical models. The appropriate model structure and superstructure will greatly affect the performance of the model. It is also important to choose the ML model correctly.

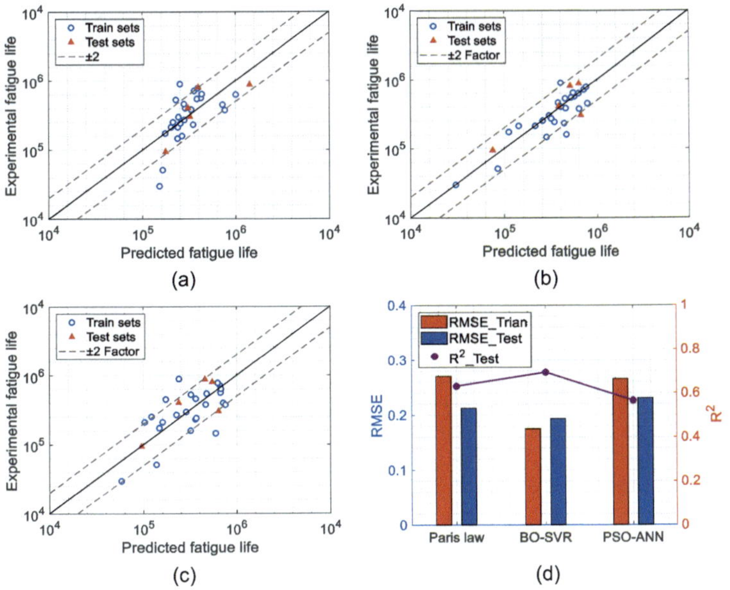

FIGURE 10.22 Comparison of different fatigue life prediction methods. (a) Paris law; (b) BO-SVR; (c) PSO-ANN; (d) Comparison of evaluation index.

Third, the fatigue life prediction method of ML coupling physical model may be a potential optimization strategy. Fourth, data is very important for ML modeling. Therefore, data sharing is of great significance to fatigue research.

10.4.2 Structural Reliability Analysis of Turbine-Bladed Disk

Uncertainties widely exist in material properties, loads, and geometrical parameters, resulting in the unsafety of structures in practical engineering. Therefore, it is of significant importance to perform the structural reliability analysis (SRA) [33]. SRA is a comprehensive estimation for considering the influence of various uncertainties of engineering structures, so as to guarantee the safe and reliable service of structures [34]. In this case, the FE model is established to evaluate the reliability index of turbine-bladed disk with multiple uncertainties. Considering the high computational cost of FE analysis, the active learning Kriging surrogate model is established with uniform importance sampling (UIS) for accurate and efficiency SRA [35].

10.4.2.1 Basic Concept of SRA

SRA problem can be seen as a high-dimensional conditional probability integration problem,

$$P_f = P\{G(\boldsymbol{X}) < 0\} = \int_{G(\boldsymbol{X})<0} f_X(\boldsymbol{x})d\boldsymbol{x} \tag{10.26}$$

where P_f indicates the failure probability; $\boldsymbol{X}$ denotes the random variables; $G(\boldsymbol{X})$ represents the limit state function (LSF or performance function, PF) and $f_X(\boldsymbol{x})$ is the joint probability density function. However, the direct integration is difficult. Thus, several SRA methods have been developed to deal with this integration problem, which can be mainly divided into simulation method and analytical method. Due to the low accuracy of analytical method, in this case, the simulation method is considered. Among the simulation method, C is most widely used. Based on the random variables $\boldsymbol{X}$, the samples the obtained by its statistical properties with sample size N_{MC}. Thus, the failure probability is,

$$\hat{p}_{f\text{-MCS}} = \frac{\sum_{i=1}^{N_{\mathrm{MC}}} I(\boldsymbol{X}_i)}{N_{\mathrm{MC}}} \tag{10.27}$$

where $\hat{p}_{f\text{-MCS}}$ denotes the estimated failure probability and $I(\boldsymbol{X}_i)$ is the indicator function defined as,

$$I(\boldsymbol{X}_i) = \begin{cases} 0 & G(\boldsymbol{X}_i) > 0 \\ 1 & G(\boldsymbol{X}_i) < 0 \end{cases} \tag{10.28}$$

where $G(\boldsymbol{X}_i) > 0$ denotes the structure safety and $G(\boldsymbol{X}_i) < 0$ denotes the structure failure. Besides, $G(\boldsymbol{X}_i) = 0$ is the limit state. Also, the coefficient of variation (CoV) of MCS can be estimated as,

$$\mathrm{CoV}_{\mathrm{MCS}} = \sqrt{\frac{1 - \hat{p}_f}{\hat{p}_f N_{\mathrm{MC}}}} \tag{10.29}$$

Note that, MCS is an accurate and robust but inefficient method. Therefore, in order to further improve the efficiency of SRA, this case adopts UIS. Besides, the reliability index (β) can be obtained,

$$\beta = \Phi^{-1}(P_f) \tag{10.30}$$

where $\Phi^{-1}(\cdot)$ is the inverse CDF of standard normal distribution.

10.4.2.2 Uniform Importance Sampling

UIS is a simulation method for computing weighted failure probabilities by uniform sampling, which can be regarded as a special case of importance sampling. Considering a limit state function $G(\boldsymbol{X})$, its variables are uniformly simulated by the following formula:

$$\begin{cases} L_j = F_j^{-1}(P_A) \\ U_j = F_j^{-1}(1 - P_A) \end{cases} \tag{10.31}$$

where L_j and U_j is the j-th lower bound and upper bound of random variable X_j, respectively. $F_j^{-1}(\cdot)$ is the inverse function of the cumulative distribution function

(CDF) of the random variable X_j. P_A is the sampling interval probability and its value can be obtained in [36]. Then, the failure probability can be estimated as,

$$\hat{p}_{f\text{-UIS}} = \frac{\sum_{i=1}^{N_{\text{UIS}}} I(X_i) w_i}{\sum_{i=1}^{N_{\text{UIS}}} w_i} \tag{10.32}$$

where $\hat{p}_{f\text{-UIS}}$ denotes the estimated failure probability by UIS, and N_{UIS} is the sample size of UIS; $w_i = \prod_{j=1}^{n} f_i\left(X_{i,j}\right)$ is the weight where $f_i\left(X_{i,j}\right)$ is the probability density function (PDF) of the i-th sample of variable $X_{i,j}$ and n is the total number of random variables. Also, the CoV of UIS can be estimated as:

$$\text{CoV}_{\text{UIS}} = \frac{\sqrt{\sum_{i=1}^{N_{\text{UIS}}} \frac{f_i^2(X_i)}{h^2(X_i)} I(X_i) - N_{\text{UIS}}(\hat{p}_{f\text{-UIS}})^2}}{N_{\text{UIS}} \hat{p}_{f\text{-UIS}}} \tag{10.33}$$

where $h(X_i)$ is the i-th probability density function of UIS.

$$h(X) = \prod_{j=1}^{n} \frac{1}{U_j - L_j} \tag{10.34}$$

Obviously, if the sampling interval probability P_A is fixed, $h(X)$ is constant value.

10.4.2.3 Problem Description

This case is considered a fatigue failure problem with a turbine-bladed disk under high-rotational load, as seen in Figure 10.23. The LSF of this component is presented as:

$$G = \log_{10}\left(N_f\right) - \log_{10}\left(N_c\right) \tag{10.35}$$

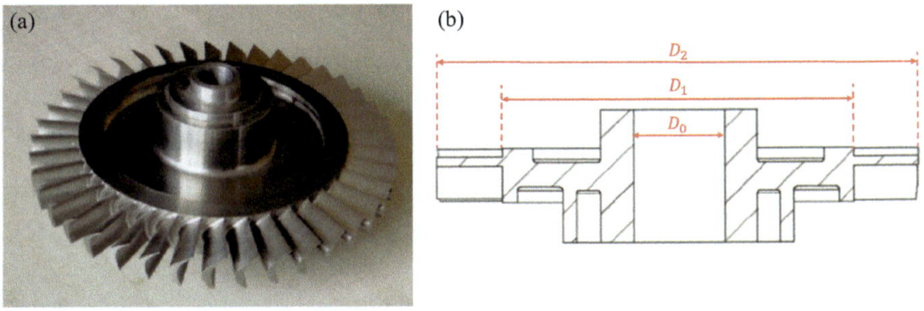

FIGURE 10.23 The turbine-bladed disk (a) structure, (b) cross-section. (Reprinted with permission from Ref. [33], Elsevier.)

where N_f is the actual failure life and $N_c = 10^7$ is the critical failure life [37,38]. Particularly, the N_f is obtained by the Fatemi-Socie damage criterion [39] with the equivalent shear strain as $\gamma_{a,\mathrm{eq}}$,

$$\gamma_{a,\mathrm{eq}} = \frac{\Delta\gamma_{\max}}{2}\left(1 + k\frac{\sigma_{n,\max}}{\sigma_y}\right) = \frac{\tau'_f}{G_s}(2N_f)^{b_\gamma} + \gamma'_f(2N_f)^{c_\gamma} \tag{10.36}$$

where $\Delta\gamma_{\max}$ represents maximum shear strain; $\sigma_{n,\max}$ represents the normal stress; $\tau'_f = \frac{\sqrt{3}}{3}\sigma'_f$ indicates the shear fatigue strength coefficient and σ'_f indicates the fatigue strength coefficient; b_γ denotes the fatigue exponent which equals to 0.07; $G_s = \frac{E}{2+2\nu}$ denotes the shear modulus where E represents Young's modulus and $\nu = 0.3$ represents the Poisson's ration. $\gamma'_f = \sqrt{3}\varepsilon'_f$ is the shear fatigue ductility coefficient and ε'_f is the fatigue ductility coefficient; c_γ is another fatigue exponent which is −0.96; k denotes the normal stress sensitivity coefficient which is determined as follows [40,41],

$$k = \left[\frac{\frac{\tau'_f}{G_s}\left(2N_f\right)^{b_\gamma} + \gamma'_f(2N_f)^{c_\gamma}}{(1+\upsilon)\frac{\sigma'_f}{E}(2N_f)^b + \left(1+\upsilon_p\right)\varepsilon'_f(2N_f)^c} - 1\right]\frac{2\sigma_y}{\sigma'_f(2N_f)^b} \tag{10.37}$$

where $c_\gamma \approx c$, $b_\gamma \approx b$; υ_p is plastic Poisson's ratio and σ_y is cyclic yield stress which is computed by Ramberg-Osgood equation as follows,

$$\sigma_y = K'\left(\varepsilon_{\mathrm{pa}}\right)^{n'} \tag{10.38}$$

where $\varepsilon_{\mathrm{pa}}$ is plastic strain as $\varepsilon_{\mathrm{pa}} = 0.0005$, K' is the cyclic strength coefficient with the hardening exponent $n' = \frac{b}{c}$.

In this case, the random variables are divided into four categories, which is load, material parameters, fatigue parameters and size parameters, as shown in Table 10.4. As seen in Figure 10.24, The FE model of the turbine-bladed disk has a total of

TABLE 10.4
Distribution Parameters of the Input Variables for Case 2

Uncertainties	Variables	Mean	Std	Distribution
Load	ω (rad/s)	2400	120	Lognormal
Materials	E (GPa)	108	2.16	Gumbel
Fatigue parameters	σ'_f (MPa)	1564	31.28	Weibull
	ε'_f	2.69	0.0538	Weibull
Size	SIZE I (mm)	9	0.225	Normal
	SIZE II (mm)	10	0.25	Normal

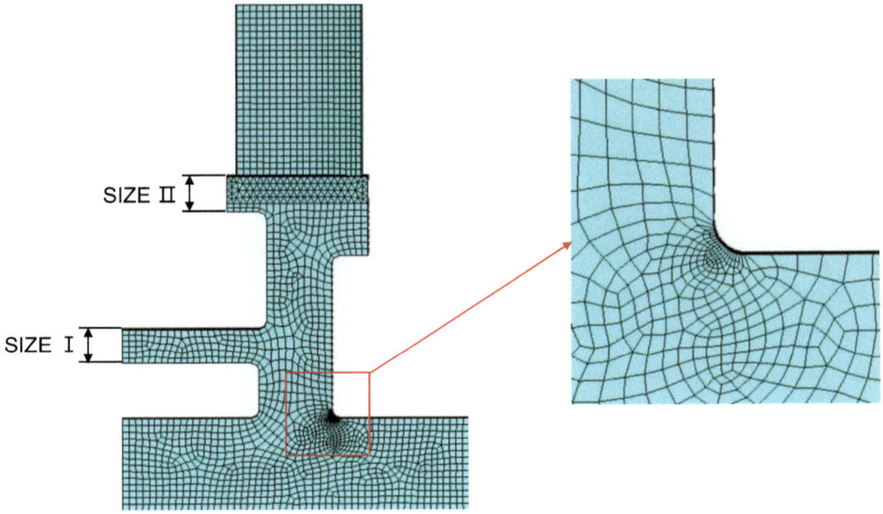

FIGURE 10.24 The schematic view of FE model for turbine-bladed disk.

20,994 nodes and 23,635 elements in a single random FE run with about 15 minutes of run time for a failure life of one group of random variables. Therefore, this case is a high computational cost and high dimension example.

10.4.2.4 Kriging Model and Active Learning

Kriging algorithm, as a variant of GPR, was first constructed by South African mining engineer Krige and used in geological statistics. Matheron [42] then refined this method and used it for nonlinear interpolation models. The LSF is assumed to be a realization of a stochastic process, including a regression term (mean) for global trend prediction and a Gaussian process term (variance) for local deviation measurement.

$$G(\boldsymbol{x}) = \boldsymbol{Q}^T q(\boldsymbol{x}) + Z(\boldsymbol{x}) \tag{10.39}$$

where $q(\boldsymbol{x}) = [q_1(\boldsymbol{x}), q_2(\boldsymbol{x}), \ldots, q_p(\boldsymbol{x})]^T$ is a vector of regression functions; $\boldsymbol{Q} = [Q_1, Q_2, \ldots, Q_p]^T$ is the corresponding regression coefficient; $Z(\boldsymbol{x})$ is a Gaussian process with mean 0 and its covariance function is obtained by [43].

Based on the predictive characteristics of the Kriging model, the active learning strategy can be used to build the accurate surrogate model by iterating. In this case, the U function [44] is considered for construct the Kriging model,

$$U(x) = \frac{|\mu_{\hat{G}}(x)|}{\sigma_{\hat{G}}(x)} \tag{10.40}$$

where $\mu_{\hat{G}}(x)$ is the prediction mean value of Kriging model and $\sigma_{\hat{G}}(x)$ is the prediction standard deviation.

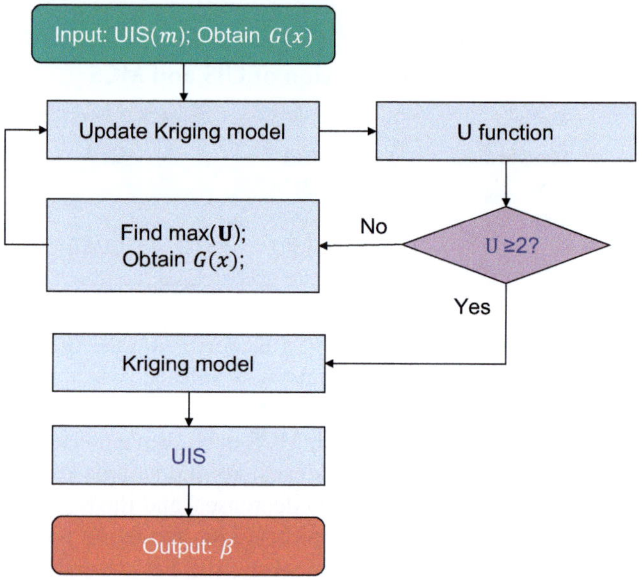

FIGURE 10.25 The framework of SRA for turbine-bladed disk under fatigue failure.

10.4.2.5 A General Framework

In this case, the framework of SRA for turbine-bladed disk under fatigue failure is presented in Figure 10.25 and the specific steps can be summarized as:

1. Determine the inputs distribution information of turbine blades disk;
2. Generate the uniform samples of inputs data x_{Train} based on Eq. (10.31) with sample size $m = 10$. Obtained the outputs (i.e. LSF) by FE analysis based on x_{Train};
3. Generate the uniform samples of inputs data x_{UIS} based on Eq. (10.31) with sample size $N = 1\text{e}6$;
4. Build the Kriging surrogate model based on the inputs and the FE outputs;
5. Obtained the Kriging output with x_{UIS}; Compute the U function based on Eq. (10.40);
6. Find $\min U(x_{\text{UIS}})$; If $\min U(x_{\text{UIS}}) > 2$, stop iteration and jump to step (8); Else, Obtained the output by FE analysis based on $x_{\min U}$, repeat step (4)–(6) until meets the convergence criterion;
7. Obtained the reliability index of turbine-bladed disk by UIS and Kriging model.

10.4.2.6 Results

Due to the particularity of the SRA problem, in the process of establishing a surrogate model, the accuracy of the specific function value, that is, the accuracy of the predictive function result is not important. It only needs to accurately determine the safety and failure of its nearly state function. Therefore, in this case, through the

TABLE 10.5
The Results Comparsion of UIS and MCS

Method	UIS		MCS	
Sample Size	β	CoV	β	CoV
1e5	4.7090	0.3174	–	–
1e6	4.6560	0.1679	–	–
1e7	4.6299	0.0505	4.6332	0.0556

active Kriging modeling strategy based on a uniform sample, the total of 110 FE simulation is conducted and the accurate limit state surface is established. As seen in Table 10.5, since this case is a high reliability problem, that is, the probability of low failure. When the sample capacity is small, MCS can't provide effective results, and UIS provides corresponding reliability indexes. As the sample capacity increases, the coefficient of variance of UIS gradually decreases, and the result is converging.

This case provides a typical application of fatigue modeling with ML, which is a SRA problem under fatigue failure. Obviously, the FE simulation is very time-consuming. Thus, an ML-based surrogate model is established to reduce the computational cost. In this case, it takes 10 minutes of FE simulation to calculate the result of a sample. This is almost impossible to complete the SRA based on the simulation method. However, based on the Kriging model, only several seconds are needed to finish the SRA. This shows the huge potential of the efficiency of ML methods in fatigue modeling. Besides, a method of improvement of a SRA is also considered, which shows higher efficiency than MCS.

10.4.3 Small Sample Modeling Based on PIML

Data hungry has greatly limited the application of ML in engineering, especially in fatigue modeling. Fatigue experiments are usually high cost and time-consuming, thus, the experimental data is relatively less and precious. It's difficult for purely data-driven ML model to learn enough information from these data. In this case, the PINN is considered for fatigue life prediction modeling which shows good performance in small samples.

10.4.3.1 Input Data

In this case, the experimental data of Ti-6Al-4V is available in [45], where the specimens were produced by electron beam melting without post-heat treatment. Fatigue tests were carried out under constant amplitude sinusoidal loading with a load ratio of 0.1 and a frequency f of 100 Hz. The SEM was used to detect the characteristics of critical defects. As seen in Table 10.6, the critical inputs of the Ti-6Al-4V fatigue life prediction are $\left[\Delta\sigma, \sqrt{\text{area}}, h, \text{Cir}\right]$, where $\Delta\sigma$ is the stress amplitude; $\sqrt{area}$ has the same definition in Case 1; h represents the location of critical defects and Cir is the circularity of defects,

$$h = d/\sqrt{\text{area}} \tag{10.41}$$

TABLE 10.6
Summary of Characteristic Parameters of Critical Defects and Observed Fatigue Lives

$\Delta\sigma$/MPa	$\sqrt{\text{area}}$/μm	h	Cir	N_f
766	331	0.91	0.96	42,543
676	57	0	0.55	72,700
766	43	0	0.64	96,550
652	43	0	0.51	112,000
652	213	0.47	0.97	158,600
766	28	0	0.69	236,006
698	192	7.55	0.59	310,347
676	61	9.02	0.97	388,510
676	207	5.56	0.98	657,700
676	140	0.71	0.99	902,800
720	91	1.65	0.76	1,497,869
676	8	31.25	0.39	1,623,600
676	49	15.31	0.72	2,525,897
676	31	12.90	0.98	4,736,800

$$\text{Cir} = \left(\sqrt{\text{area}}\right)^2 / \left(\pi \cdot r_{\max}{}^2\right) \tag{10.42}$$

where d is the nearest distance from the defect to the surface and $r_{\max}$ indicate the maximum distance from the center to the boundary in the projected area.

10.4.3.2 Physics Knowledge of Fatigue Modeling

Since the intricate nature of fatigue failure mechanisms in AM materials, several semi-empirical models oversimplify the impact of defects on fatigue life, often resulting in limited explanatory power when accounting for fatigue life variability. Recently, efforts have been made to integrate defect location and shape into fatigue life prediction models [45]. However, these models lack widespread validation and theoretical support. In response, a defect-driven Physics-Informed Neural Network (PINN) model is introduced for physically consistent fatigue life prediction. This model incorporates defect features to constrain the training process, effectively capturing the influence of defects on fatigue performance while maintaining physical consistency. Thus, the nonlinear mapping relationship between inputs $\left\{\Delta\sigma, \sqrt{area}, h, \text{Cir}\right\}$ and the output N_f,

$$N_f = f\left(\Delta\sigma, \sqrt{\text{area}}, h, \text{Cir}\right) \tag{10.43}$$

where $f(\cdot)$ is the nonlinear mapping established by PINN.

Note that, to obtain the partial derivatives of fatigue life with respect to defect characteristics and applied stress, the automatic differentiation technique in *Pytorch* is considered. Meanwhile, the mathematical description of physical knowledge in fatigue modeling can be summarized in Table 10.7.

TABLE 10.7
Mathematical Description of Physical Knowledge Under Fatigue

Mathematical Description	Physical Knowledge
$\frac{\partial N_f}{\partial \Delta\sigma} < 0$	N_f decreases with increasing applied stress
$\frac{\partial N_f}{\partial \sqrt{\text{area}}} < 0$	N_f decreases with increasing $\sqrt{\text{area}}$ of critical defect
$\frac{\partial N_f}{\partial h} > 0$	N_f decreases as the critical defect gets closer to the surface
$\frac{\partial N_f}{\partial \text{Cir}} > 0$	N_f decreases as the shape of critical defect becomes more complex
$\frac{\partial^2 N_f}{\partial \Delta\sigma^2} > 0$	$\frac{\partial N_f}{\partial \Delta\sigma}$ increases with increasing applied stress

To guarantee the physics consistence in Table 10.7, the PIML loss is modeled as a physical loss function embedded,

$$\text{loss} = (1-\lambda)\text{loss}_{\text{data-driven}} + \lambda \text{loss}_{\text{physics-informed}} \tag{10.44}$$

where loss is the loss function of defect-driven PINN; $\text{loss}_{\text{data-driven}}$ denotes the dara-driven loss function of defect-driven PINN and $\text{loss}_{\text{physics-informed}}$ denotes the physics information loss. λ is the scaling factor to control the influence factor of each loss. More specifically, The L_2 regularization is considered for $\text{loss}_{\text{data-driven}}$, and the physics information loss can be considered as,

$$\begin{cases} \text{loss}_{\text{physics-informed}} = \text{loss}_{1,\Delta\sigma} + \text{loss}_{\sqrt{\text{area}}} + \text{loss}_h + \text{loss}_{\text{Cir}} + \text{loss}_{2,\Delta\sigma} \\ \text{loss}_{1,\Delta\sigma} = \sqrt{\sum_{i=1}^{n} \max\left(0, \frac{\partial N_p^i}{\partial \Delta\sigma_i}\right)^2} \\ \text{loss}_{\sqrt{\text{area}}} = \sqrt{\sum_{i=1}^{n} \max\left(0, \frac{\partial N_p^i}{\partial \sqrt{\text{area}}_i}\right)^2} \\ \text{loss}_h = \sqrt{\sum_{i=1}^{n} \min\left(0, \frac{\partial N_p^i}{\partial D_i}\right)^2} \\ \text{loss}_{Cir} = \sqrt{\sum_{i=1}^{n} \min\left(0, \frac{\partial N_{\text{p}}^i}{\partial Cir_i}\right)^2} \\ loss_{2,\Delta\sigma} = \sqrt{\sum_{i=1}^{n} \min\left(0, \frac{\partial^2 N_p^i}{\partial \Delta\sigma_i^{\,2}}\right)^2} \end{cases} \tag{10.45}$$

where i is the i-th input with totally n inputs for training; N_p^i is the i-th prediction fatigue life of PINN.

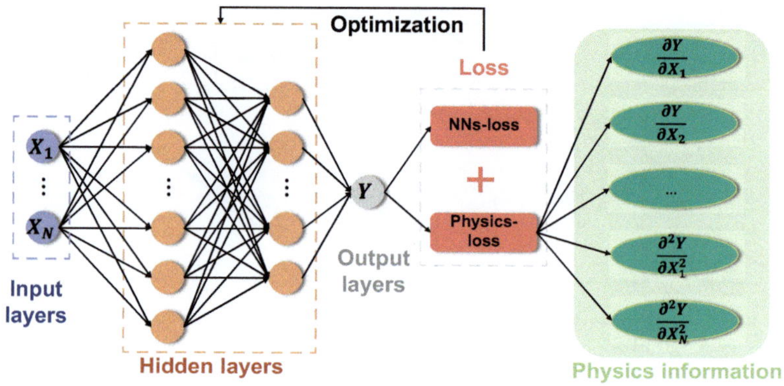

FIGURE 10.26 The framework of PINN for fatigue life prediction.

Obviously, if $\frac{\partial N_p^i}{\partial h_i}$, $\frac{\partial N_p^i}{\partial Cir_i}$ and $\frac{\partial^2 N_p^i}{\partial \Delta\sigma_i^{\ 2}}$ are less than zero, the ML modeling is contradictory to the physical knowledge in fatigue failure. Thus, to minimize the loss function of the PINN model, the training algorithm will continue to optimize model parameters and meet the physics consistence eventually.

10.4.3.3 The Framework of PINN for Fatigue Life Prediction

In this case, the framework of the PINN model for AM fatigue life prediction is presented in Figure 10.26 and the specific steps can be summarized as:

1. Obtain input and output data through fatigue tests, and divide them into training sets (75%) and test sets (25%) randomly. In this case, the input data also includes defect information obtained by SEM;
2. Build a loss function based on physical information, and establish a PINN model;
3. Train the PINN model and predict the fatigue life.

Note that, in this case, Except for different loss functions, the PINN model is no different from the ANN model. Besides, there are three hidden layers with 16 neurons, 32 neurons, and 24 neurons in the PINN model, respectively. The detail of activation function is Inputs-(Tanh)-First hidden layer-(Leaky-Relu with $\alpha = 0.2$)-Second hidden layer-(Leaky-Relu with $\alpha = 0.2$)-Third hidden layer-(Purelin)-Output. The scaling factor $\lambda = 0.52$ is determined based on several trails. Obviously, when $\lambda = 0$, the purely data-driven ANN model is obtained.

10.4.3.4 Results

After the 2000 of training epochs, the predicted lives by using Paris' law, ANN, fracture mechanics-based PINN and defect-driven PINN are shown in Figure 10.27 with training sets and test sets. As seen in Figure 10.27a–c, the Paris law obtained the worst predicting results where the test set is beyond the 2-band width. Although the purely data-driven ANN model obtained the highest predicting results in training set,

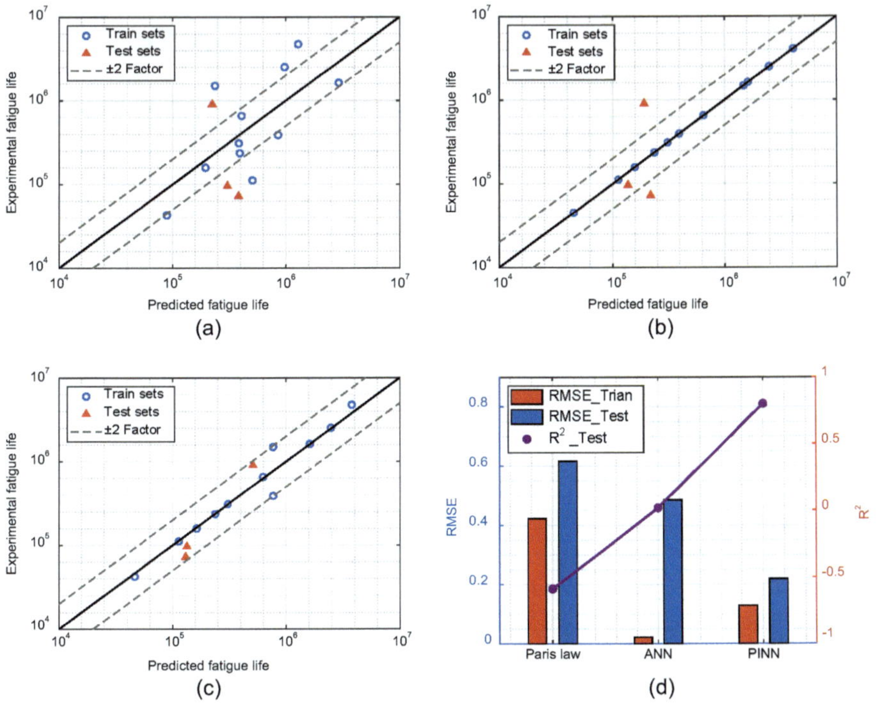

FIGURE 10.27 Predicted fatigue life of Ti-6Al-4V. (a) Paris law; (b) ANN; (c) defect-driven PINN; (d) Comparison of evaluation index.

the test set are also beyond the 2-band width. However, based on the defect-driven PINN model, the predicting results of training set and test set are both within the 2-band width. What's more, Figure 10.27d presents the RMSE and R^2 of different methods. Obviously, the PINN model obtained the lowest RMSE and highest R^2, which shows the good generalization ability of PINN model.

As seen in Figure 10.28, the different loss functions during the training process are provided. In Figure 10.28a, the data-driven loss of PINN is larger than the ANN model, thus, the predicting accuracy of PINN is also smaller than ANN. Obviously, due to no physical constraints being considered, the physical loss term of the ANN model does not converge during the training process, which in turn causes the purely data-driven model to learn the noise in the training data, thereby reducing its generalization ability. However, since the PINN model adds physical constraints, during the modeling process, by continuously adjusting the weights of the model, the relevant loss function gradually converges.

To further illustrate the effectiveness of the PINN model, Figure 10.29 shows the partial derivatives of input parameters and fatigue life. Note that, to improve the performance of the model, the data in this case are normalized to [−1,1]. Besides, when computing the partial derivative of each input, the other inputs are set as 0 and the numerical differentiation is considered. As shown in Figure 10.29, all the partial

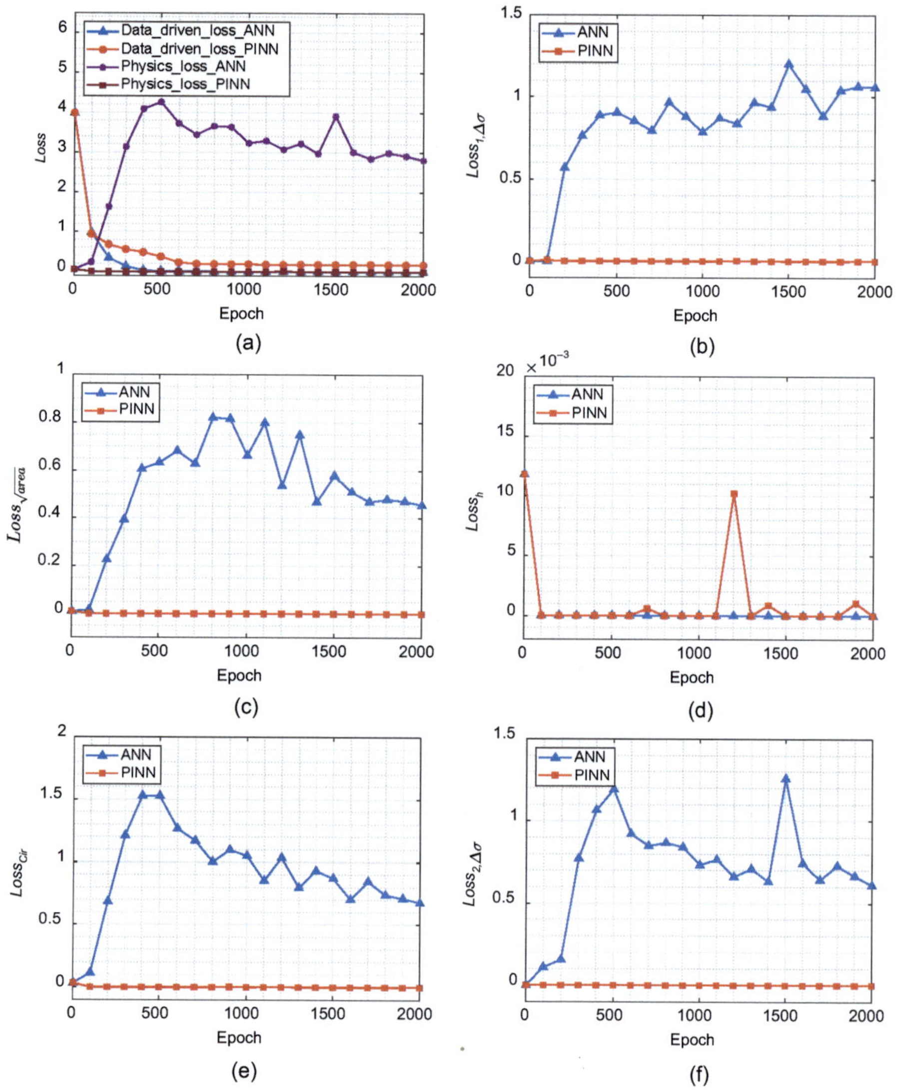

FIGURE 10.28 Different loss functions during training process. (a) the total loss; (b) $\text{loss}_{1,\Delta\sigma}$; (c) $\text{Loss}_{\sqrt{\text{area}}}$; (d) loss_h; (e) loss_{Cir}; (f) $\text{loss}_{2,\Delta\sigma}$

derivatives of PINN meet the physical mechanism. However, some partial derivatives of the ANN model don't meet the physical mechanism which leads the low generalization ability in test set.

In this case, the different characteristics of different fatigue life predicting models are discussed based on the AM materials of Ti-6Al-4V. First, due to incomplete understanding of the physical mechanisms, physical models often provide poor predictions. Second, purely data-driven ML models can obtain better prediction results in the training set. However, due to the possibility of learning noise information, the

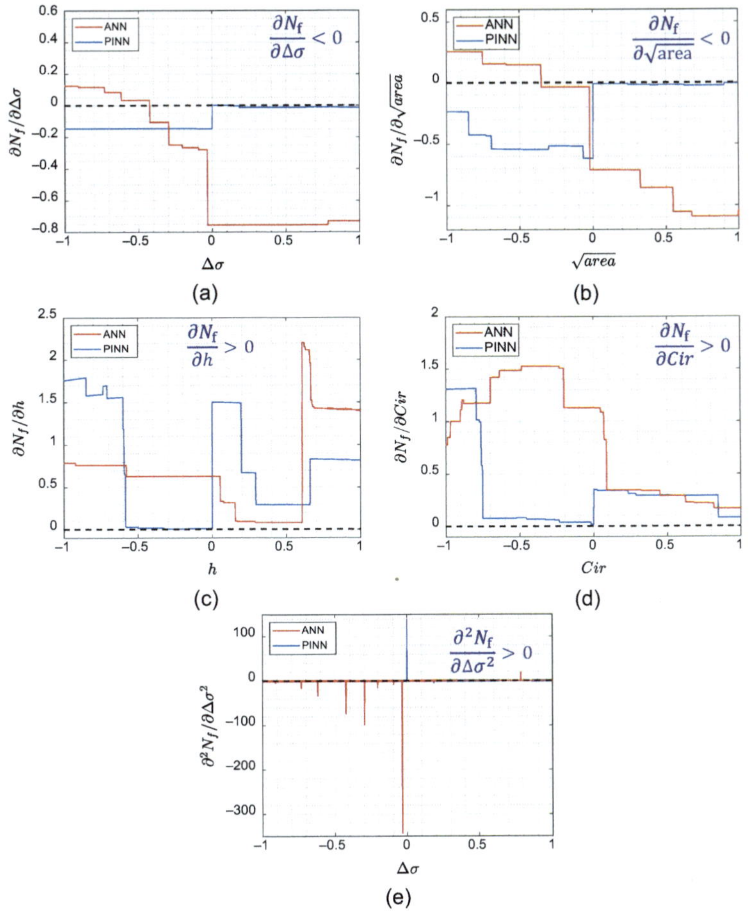

FIGURE 10.29 The partial derivatives of inputs and fatigue life. (a) $\frac{\partial N_f}{\partial \Delta\sigma}$; (b) $\frac{\partial N_f}{\partial \sqrt{\text{area}}}$; (c) $\frac{\partial N_f}{\partial h}$; (d) $\frac{\partial N_f}{\partial \text{Cir}}$; (e) $\frac{\partial^2 N_f}{\partial \Delta\sigma^2}$

model will appear overfitting, resulting in a large prediction error on the test set. Finally, since the PINN model considers both data-driven and physical information-driven, during the training process of the model, the relevant weights will be constrained to satisfy the physical laws. Therefore, the good generalization ability is obtained. This case shows the great potential of PIML for high-precision fatigue modeling.

10.5 PROSPECTS ON FATIGUE MODELING WITH AI

In this chapter, three typical case studies of fatigue modeling with ML are developed. In the first case, the data-driven models considering performance improvement strategy (including optimization strategy, normalization, K-fold cross-validation, etc.) are presented. The results showed that the appropriate ML model and effective improvement strategies can get better fatigue modeling results which requires professional knowledge for researchers. In the second case, high-precision surrogate calculation is one of the focus of fatigue research and fatigue failure analysis is often very complicated. Thus, based on some special ML algorithm and strategy, the computational cost will be greatly reduced. In the last case, fatigue life prediction with PIML is developed. The result is exciting that PIML gets rid of the restraint of the traditional data-driven algorithm. By introducing physical constraints, the generalization ability of the model has greatly improved and meets physical consistency. In general, to achieve the simple and common application of fatigue modeling with ML, there still needs a lot of development. Also, the main limitations and prospects should be discussed.

Uncertainty. One of the basic features of fatigue is its huge uncertainty. However, related ML modeling research for fatigue uncertainty is relatively small. Data-driven fatigue modeling is usually directly predicted as a certain result, which ignores its uncertainty. In the future, the application of ML in fatigue should focus more on the analysis of uncertainty quantification and nondeterministic (probabilistic and non-probabilistic) analysis.

Data Hungry and Data Augmentation. Whether it is fatigue test or fatigue FE simulation, its cost is huge, which leads to the data hungry of ML model and affects the performance of the model. On the one hand, to clarify the fatigue problems, a large number of test data is necessary. On the other hand, Data Augmentation also has several attempts in fatigue modeling, such as GAN, Bootstrap, etc. However, the development of reliable and persuasive data augmentation algorithms is still a current challenge.

ML Model Selection. Although there are currently several ML models, there is a lack of an ML model with excellent performance. Therefore, the model selection and application also increased the difficulty of researchers. Choose a suitable ML model for specific fatigue modeling problems, as well as develop new ML models to better deal with the problem of fatigue modeling. For complex fatigue modeling problems, develop advanced ML architectures to solve more challenging fatigue problems, such as image processing, time data processing, etc.

Performance Improvement. The characteristics of ML models are different, and their performance improvement needs to be considered. For data processing, training strategies, and model optimization, they currently lack a unified and practical framework, and related research needs to be further strengthened.

Benchmark. For some typical fatigue modeling issues, there is still a lack of relevant benchmarks. For example, based on the fatigue life prediction of test data, correlation analysis and sensitivity analysis need to be considered, and the ML model is established. Although there are many related research contents, a uniform and general benchmark remains to be developed to solve various typical fatigue modeling problems.

Online Application and Open Source. For beginners or researchers who do not understand ML, applying ML to solve fatigue modeling is usually difficult. Therefore, it is necessary to develop online applications based on ML fatigue modeling.

In addition, the open source of data and the open source of algorithms are also important for the development of fatigue research.

Application Development. The current fatigue modeling with ML is mainly concentrated in supervised learning (regression and classification). However, the functions and applications of ML are not limited to this. Developing more fatigue modeling methods based on ML is greatly helpful for the research of fatigue problems.

10.6 SUMMARY

In this chapter, the basic concept of fatigue modeling with ML are briefly introduced. First, ML can be mainly divided into supervised learning, unsupervised learning, semi-supervised learning and reinforcement learning. Next, the main parts of fatigue modeling with ML can be summarized as input, output, ML model, and evaluation index. Particularly, optimization is also a key part in fatigue modeling with ML. Thus, the different types of optimization application is discussed including error optimization, model optimization, and output optimization. What's more, the application of fatigue modeling with ML can be divided into fatigue life prediction, fatigue crack, fatigue damage, fatigue strength, and fatigue load. Among them, classification and regression are the most common applications. Note that, fatigue modeling with ML by informed-driven is the potential development direction in the future. Finally, three typical case studies and prospects are provided.

ABBREVIATIONS

AE	Acoustic Emission
AM	Additive Manufacture
AI	Artificial Intelligence
ANN	Artificial NN
BO	Bayes Optimization
BN	Bayesian Network
BPNN	Back Propagation Neural Network
CoV	Coefficient of Variation
CNN	Convolutional NN
CDF	Cumulative Distribution Function
DT	Decision Tree
DL	Deep Learning
DNN	Deep NN
FN	False Negative
FP	False Positive
FE	Finite Element
FNN	Feedforward NN
GHC	Gaussian Hybrid Clustering

GMM	Gaussian Mixture Model
GPR	Gaussian Process Regression
GAN	Generative Adversarial Network
GPU	Graphic Process Unit
HC	Hierarchical Clustering
KNN	K-Nearest Neighbor
Lasso	Least Absolute Shrinkage and Selection Operator
LSF	Limit State Function
LR	Logical Regression
LSTM	Long Short-Term Memory
ML	Machine Learning
MDP	Markov Decision Process
MAE	Mean Absolute Error
MAPE	Mean Absolute Percentage Error
MPE	Mean Percentage Error
MSE	Mean Squared Error
MCS	Monte Carlo Simulation
MC	Monte Carlo Learning
NN	Neural Network
PDE	Partial Differential Equation
PSO	Particle Swarm Optimization
PF	Performance Function
PINN	Physics-Informed Neural Network
PIML	Physics Informed-Driven ML
PG	Policy Gradient
PCA	Principal Component Analysis
PNN	Probabilistic Neural Network
PDF	Probability Density Function
POD	Proper Orthogonal Decomposition
RBFNN	Radial Basis Function NN
RF	Random Forest
RNN	Recurrent NN
RMSE	Root Mean Squared Error
SOM	Self-Organizing Map
SRA	Structural Reliability Analysis
SVM	Support Vector Machine
SVR	Support Vector Regression
TD	Temporal-Difference Learning
TN	True Negative
TP	True Positive
UIS	Uniform Importance Sampling

REFERENCES

1. H. Wang, B. Li, J. Gong, F.Z. Xuan, Machine learning-based fatigue life prediction of metal materials: Perspectives of physics-informed and data-driven hybrid methods, *Eng. Fract. Mech.* 284 (2023). https://doi.org/10.1016/j.engfracmech.2023.109242.
2. J. Chen, Y. Liu, Fatigue modeling using neural networks: A comprehensive review, *Fatigue Fract. Eng. Mater. Struct.* 45 (2022) 945–979. https://doi.org/10.1111/ffe.13640.
3. A. Gammerman, V. Vovk, V. Vapnik, Learning by transduction, *arXiv* (2013) 148–155. https://arxiv.org/abs/1301.7375.
4. X.C. Zhang, J.G. Gong, F.Z. Xuan, A deep learning based life prediction method for components under creep, fatigue and creep-fatigue conditions, *Int. J. Fatigue.* 148 (2021) 106236. https://doi.org/10.1016/j.ijfatigue.2021.106236.
5. E. Maleki, S. Bagherifard, S.M.J. Razavi, M. Bandini, A. du Plessis, F. Berto, M. Guagliano, On the efficiency of machine learning for fatigue assessment of post-processed additively manufactured AlSi10Mg, *Int. J. Fatigue.* 160 (2022) 106841. https://doi.org/10.1016/j.ijfatigue.2022.106841.
6. B. Li, W. Zhang, F. Xuan, Machine-learning prediction of selective laser melting additively manufactured part density by feature-dimension-ascended Bayesian network model for process optimisation, *Int. J. Adv. Manuf. Technol.* 121 (2022) 4023–4038. https://doi.org/10.1007/s00170-022-09555-9.
7. K. Dresia, G. Waxenegger-Wilfing, J. Riccius, J. Deeken, M. Oschwald, Numerically efficient fatigue life prediction of rocket combustion chambers using artificial neural networks, *8th European Conference for Aeronautics and Aerospace Sciences (EUCASS)*, Madrid, (2019). https://doi.org/10.13009/EUCASS2019-264.
8. H. Zou, The adaptive lasso and its oracle properties, *J. Am. Stat. Assoc.* 101 (2006) 1418–1429. https://doi.org/10.1198/016214506000000735.
9. G.C. McDonald, Ridge regression, *Wiley Interdiscip. Rev. Comput. Stat.* 1 (2009) 93–100. https://doi.org/10.1002/wics.14.
10. H. Zou, T. Hastie, Regularization and variable selection via the elastic net, *J. R. Stat. Soc. Ser. B Stat. Methodol.* 67 (2005) 301–320. https://doi.org/10.1111/j.1467-9868.2005.00503.x.
11. W. Solano-Alvarez, M.J. Peet, E.J. Pickering, J. Jaiswal, A. Bevan, H.K.D.H. Bhadeshia, Synchrotron and neural network analysis of the influence of composition and heat treatment on the rolling contact fatigue of hypereutectoid pearlitic steels, *Mater. Sci. Eng. A.* 707 (2017) 259–269. https://doi.org/10.1016/j.msea.2017.09.045.
12. H. Alqahtani, S. Bharadwaj, A. Ray, Classification of fatigue crack damage in polycrystalline alloy structures using convolutional neural networks, *Eng. Fail. Anal.* 119 (2021) 104908. https://doi.org/10.1016/j.engfailanal.2020.104908.
13. E. Yang, Y. Tang, L. Li, W. Yan, B. Huang, Y. Qiu, Research on the recurrent neural network-based fatigue damage model of asphalt binder and the finite element analysis development, *Constr. Build. Mater.* 267 (2021). https://doi.org/10.1016/j.conbuildmat.2020.121761.
14. Y.L. Han, X. Liu, S.H. Dai, Fatigue life calculation of flawed structures – Based on artificial neural network with special learning set, *Int. J. Press. Vessel. Pip.* 75 (1998) 263–269. https://doi.org/10.1016/S0308-0161(98)00040-4.
15. V. Vapnik, *The Nature of Statistical Learning Theory*, Springer Science & Business Media, 2013.
16. L. Wang, S.-P. Zhu, C. Luo, D. Liao, Q. Wang, Physics-guided machine learning frameworks for fatigue life prediction of AM materials, *Int. J. Fatigue.* 172 (2023) 107658. https://doi.org/10.1016/j.ijfatigue.2023.107658.
17. N. Wu, J. Sun, Fatigue detection of air traffic controllers based on radiotelephony communications and self-adaption quantum genetic algorithm optimization ensemble learning, *Appl. Sci.* 12 (2022). https://doi.org/10.3390/app122010252.

18. Y. Ma, S. Zhang, D. Qi, Z. Luo, R. Li, T. Potter, Y. Zhang, Driving drowsiness detection with EEG using a modified hierarchical extreme learning machine algorithm with particle swarm optimization: A pilot study, *Electronics.* 9 (2020). https://doi.org/10.3390/electronics9050775.
19. Z. Xiao, Z. Hu, L. Geng, F. Zhang, J. Wu, Y. Li, Fatigue driving recognition network: Fatigue driving recognition via convolutional neural network and long short-term memory units, *IET Intell. Transp. Syst.* 13 (2019) 1410–1416. https://doi.org/10.1049/iet-its.2018.5392.
20. B. Lotfi, P. Beiss, Application of neural networking for fatigue limit prediction of powder metallurgy steel parts, *Mater. Des.* 50 (2013) 440–445. https://doi.org/10.1016/j.matdes.2013.03.002.
21. M. Aykol, C.B. Gopal, A. Anapolsky, P.K. Herring, B. van Vlijmen, M.D. Berliner, M.Z. Bazant, R.D. Braatz, W.C. Chueh, B.D. Storey, Perspective—Combining physics and machine learning to predict battery lifetime, *J. Electrochem. Soc.* 168 (2021) 030525. https://doi.org/10.1149/1945-7111/abec55.
22. Y. Gu, C. Zhang, P. Zhang, M. V. Golub, B. Yu, Enriched physics-informed neural networks for 2D in-plane crack analysis: Theory and MATLAB code, *Int. J. Solids Struct.* 276 (2023). https://doi.org/10.1016/j.ijsolstr.2023.112321.
23. H. Wang, B. Li, F.Z. Xuan, Fatigue-life prediction of additively manufactured metals by continuous damage mechanics (CDM)-informed machine learning with sensitive features, *Int. J. Fatigue.* 164 (2022). https://doi.org/10.1016/j.ijfatigue.2022.107147.
24. Y. Murakami, M. Endo, Effects of defects, inclusions and inhomogeneities on fatigue strength, *Int. J. Fatigue.* 16 (1994) 163–182. https://doi.org/10.1016/0142-1123(94)90001-9.
25. Y. Murakami, S. Nemat-Nasser, Growth and stability of interacting surface flaws of arbitrary shape, *Eng. Fract. Mech.* 17 (1983) 193–210. https://doi.org/10.1016/0013-7944(83)90027-9.
26. P. Paris, F. Erdogan, A critical analysis of crack propagation laws, *J. Fluids Eng. Trans. ASME.* 85 (1963) 528–533. https://doi.org/10.1115/1.3656900.
27. A.S.H. Akebono, S. Matsuzaki, S. Mohd, Y. Hayashi, M. Kato, Effects of casting defects on fatigue strength of stainless cast steel SCS1 for the turbine runner of hydraulic power generation and fatigue life evaluation based on the defect size, *Japan Soc. Mech. Eng.* (2009) 1585–1590.
28. C. Luo, S. Zhu, B. Keshtegar, X. Niu, EUIS-SVR : An enhanced uniform simulation approach coupled with SVR for efficient structural reliability analysis, *Reliab. Eng. Syst. Safety.* 237 (2023) 109377.
29. X. Zhang, C. Luo, N. Menga, H. Zhang, Y. Li, S.-P. Zhu, Pressure and polymer selections for solid-state batteries investigated with high-throughput simulations, *Cell Reports Phys. Sci.* 4 (2023) 101328. https://doi.org/10.1016/j.xcrp.2023.101328.
30. K. Ziane, A. Ilinca, S.S. Karganroudi, M. Dimitrova, Neural network optimization algorithms to predict wind turbine blade fatigue life under variable hygrothermal conditions, *Eng.* 2 (2021) 278–295. https://doi.org/10.3390/eng2030018.
31. X. Wang, X. He, T. Wang, Y. Li, Internal pores in DED Ti-6.5Al-2Zr-Mo-V alloy and their influence on crack initiation and fatigue life in the mid-life regime, *Addit. Manuf.* 28 (2019) 373–393. https://doi.org/10.1016/j.addma.2019.05.007.
32. L. Wang, S.-P. Zhu, C. Luo, D. Liao, Q. Wang, Physics-guided machine learning frameworks for fatigue life prediction of AM materials, *Int. J. Fatigue.* 172 (2023) 107658. https://doi.org/10.1016/j.ijfatigue.2023.107658.
33. C. Luo, B. Keshtegar, S.P. Zhu, X. Niu, EMCS-SVR: Hybrid efficient and accurate enhanced simulation approach coupled with adaptive SVR for structural reliability analysis, *Comput. Methods Appl. Mech. Eng.* 400 (2022) 115499. https://doi.org/10.1016/j.cma.2022.115499.

34. C. Luo, B. Keshtegar, S.P. Zhu, O. Taylan, X.P. Niu, Hybrid enhanced Monte Carlo simulation coupled with advanced machine learning approach for accurate and efficient structural reliability analysis, *Comput. Methods Appl. Mech. Eng.* 388 (2022) 114218. https://doi.org/10.1016/j.cma.2021.114218.
35. C. Luo, S.P. Zhu, B. Keshtegar, X. Niu, O. Taylan, An enhanced uniform simulation approach coupled with SVR for efficient structural reliability analysis, *Reliab. Eng. Syst. Saf.* 237 (2023) 109377. https://doi.org/10.1016/j.ress.2023.109377.
36. M. Rashki, M. Miri, M. Azhdary Moghaddam, A new efficient simulation method to approximate the probability of failure and most probable point, *Struct. Saf.* 39 (2012) 22–29. https://doi.org/10.1016/j.strusafe.2012.06.003.
37. S.P. Zhu, Q. Liu, W. Peng, X.C. Zhang, Computational-experimental approaches for fatigue reliability assessment of turbine bladed disks, *Int. J. Mech. Sci.* 142–143 (2018) 502–517. https://doi.org/10.1016/j.ijmecsci.2018.04.050.
38. X.P. Niu, R.Z. Wang, D. Liao, S.P. Zhu, X.C. Zhang, B. Keshtegar, Probabilistic modeling of uncertainties in fatigue reliability analysis of turbine bladed disks, *Int. J. Fatigue.* 142 (2021) 105912. https://doi.org/10.1016/j.ijfatigue.2020.105912.
39. A. Fatemi, D.F. Socie, A critical plane approach to multiaxial fatigue damage including out-of-phase loading, *Fatigue Fract. Eng. Mater. Struct.* 11 (1988) 149–165. https://doi.org/10.1111/j.1460-2695.1988.tb01169.x.
40. S.P. Zhu, S. Foletti, S. Beretta, Probabilistic framework for multiaxial LCF assessment under material variability, *Int. J. Fatigue.* 103 (2017) 371–385. https://doi.org/10.1016/j.ijfatigue.2017.06.019.
41. S.P. Zhu, Z.Y. Yu, Q. Liu, A. Ince, Strain energy-based multiaxial fatigue life prediction under normal/shear stress interaction, *Int. J. Damage Mech.* 28 (2019) 708–739. https://doi.org/10.1177/1056789518786031.
42. Matherton, Principles of geo-statistics, *Econ. Geol.* 58 (1963) 1246–1266.
43. S Lophaven, H Nielsen, J Søndergaard. DACE: a Matlab kriging toolbox. Lyngby, Denmark: IMM, Informatics and Mathematical Modelling, The Technical University of Denmark, 2002.
44. B. Echard, N. Gayton, M. Lemaire, AK-MCS: An active learning reliability method combining Kriging and Monte Carlo Simulation, *Struct. Saf.* 33 (2011) 145–154. https://doi.org/10.1016/j.strusafe.2011.01.002.
45. Y. Hu, S. Wu, C. Xie, W. Wu, J. Zhang, Fatigue life evaluation of Ti-6Al-4V welded joints manufactured by electron beam melting, *Fatigue Fract. Eng. Mater. Struct.* 44 (2021) 2210–2221. https://doi.org/10.1111/ffe.13490.

11 Fatigue of Advanced Materials

Shaoxiong Xie, Wentao Huang, Yongjie Liu, and Qingyuan Wang

The development of the industrial field has posed further challenges to the efficiency and performance of the technology platform determined by materials, which has driven the exponential growth in developing advanced materials for industrial applications. As advanced materials have attracted increasing attention, ceramics and hydrogels have been explored for seamless integration into equipment and machines in the aerospace and biomedical fields. The anti-fatigue capability of ceramics and hydrogels is crucial to ensuring the structural integrity of these equipment and machines in service. Thus, in this chapter, advances in fatigue of ceramics and hydrogels are comprehensively explored and discussed, respectively.

DOI: 10.1201/9781003405122-11

11.1 ADVANCES IN FATIGUE OF CERAMICS

11.1.1 INTRODUCTION

Ceramics are widely known as one of the most perspective advanced materials for various applications, such as construction, cars, mobile phones, computers, electronics, nuclear power, aerospace, weapons, medicine, and so on, due to high chemical/heat stability, controllable multi-functionality, excellent mechanical hardness, and the like [1]. With the rapid development of science and technology, advanced functional ceramics have attracted considerable attention in consumer electronics industries. For instance, in the world of wireless communication, the 5G technique has become almost the most popular research topic, of which ceramic materials, such as microwave dielectric ceramics and polymer ceramic composites, play a significant role in frequency bands and filtering technology for 5G systems. With the increased concerns of energy crisis, energy storage materials and conversion systems have been significantly developed nowadays. As one of the most important energy storage materials, advanced ceramic materials, such as high entropy ceramics, ferroelectric ceramics, and antiferroelectric ceramics, could be used in high power or pulsed power devices for energy storage applications, thanking their excellent power density and fast charge-discharge rates as well as better frequency/temperature stability. In addition, industrial and scientific communities have also delivered the need for sensing techniques at high temperatures. The development of high-temperature piezoelectric ceramics, such as bismuth layer-structured ceramics [2], could guarantee adequate sensitivity of sensors in monitoring the health conditions of industrial facilities in high-temperature environments. Such high-temperature piezoelectric ceramics have been also more frequently used in aerospace and aircraft industries to evaluate and manage high-temperature components, such as engines. Undoubtedly, ceramic materials could significantly affect the daily life of human beings and accelerate the development of society. Moreover, ceramics materials, especially advanced functional ceramics, have become one of the most critical techniques in the increased competitiveness between different countries.

In spite of many functional advantages, ceramic materials are easily premature failure due to inherent brittleness when subjected to the applied mechanical loads in practical uses. For this reason, many researchers have focused on the investigation of mechanical fracture and crack propagation of ceramics by using different strategies. For instance, the fracture toughness and failure of ceramic materials have been studied by using a single-edge V-notched beam method [3], and the deformation behaviors of ceramics at different applied temperatures were revealed in detail. Xie et al. [4] analyzed the ferroelastic domains in piezoelectric ceramics by piezoresponse force microscope and explained the relationship between applied stress and ferroelastic deformation. *In-situ* crack propagation in ceramics was investigated by using scan electron microscope, showing three general categories for crack bridging elements including intact (elastic) bridges, frictionally sliding (plastic) bridges, and mechanically interlocked (elastic-plastic) bridges. Webber et al. [5] reviewed and discussed the current understanding of mechanical and fracture behaviors of large-strain lead-free ceramics for actuator applications and provided recommendations for further work in mechanics of ceramics. On the front of theory, researchers thought that a fracture criterion suitable for piezoelectric ceramics should consider material nonlinearity, and it is proposed that the energy release rate is strictly independent of the form of nonlinear electrical constitutive relation. In addition, the nonlinearity of ceramics is considered to stem from the ferroelectric domain switching. Thus, a small-scale polarization switching model was developed and the toughness mechanism under non-uniform electric and stress fields was well explained. Later, researchers developed two new fracture criteria for piezoelectric ceramics, where one is based on a generalized stress intensity factor affected by the stress field at the tip of crack, and the other is based on the crack opening displacement affected by the relative displacement of crack faces. The theoretical fracture criteria provide strong evidence and guidance for the experimental results of mechanical properties of ceramic materials and to some extent contribute to their structural and properties design, especially in anti-cracking and optimizing fracture failure.

However, as a significant branch of mechanical properties in materials, fatigue behaviors are often ignored in ceramics, since it is believed that the plastic deformation at the crack tip does not occur in brittle ceramic materials, unlike the metals. Actually, ceramic materials are often subjected to cyclic loads during services in many application situations, leading to the decrease in lifetime and even causing fatigue failure. For example, the piezoelectric ceramics used in piezoelectric accelerometer often suffer cyclic vibration loads when they are applied for structural health monitoring and/or nondestructive evaluation of turbines, jet engines, steam, and nuclear power plants. To ensure the safe operation of the applications, the ceramic materials need to maintain adequate sensitivity and reliability during long-term services, while a mechanical degeneration due to fatigue loads could significantly decrease the sensitivity and reliability, which could result in the distortion of vibration signal output and the inaccurate evaluation of health conditions of the industrial applications. In severe cases, fatigue degeneration even causes the occurrence of disaster accidents. Therefore, considering the reliability and stability in practical uses, one can see that the fatigue issues in ceramics should be paid more attention, and understanding the fatigue failure behaviors is significantly critical. To date, the investigation on fatigue behaviors in ceramics is very limited. Current existing researches on fatigue properties in ceramics mainly focus on the experimental observations of macroscopic fatigue degradation, fatigue crack propagation, fatigue properties prediction, and to some extent, empirical fatigue models. It is also indicated that the fatigue mechanisms in ceramics are attributed to the micro-pores, defects, domains, surface conditions, grain sizes, compositions, and so forth [6]. Despite these findings that could explain experimental observations of fatigue issues in ceramics in some cases, a deep understanding of fatigue mechanisms is still on the way. Therefore, more work on fatigue in ceramics is necessary, and the proposal of a novel fatigue mechanism is still challenging in our times.

This part mainly focuses on the findings of fatigue in advanced ceramic materials, such as piezoelectric or ferroelectric ceramics, which are often applied in smart sensors or actuators and are frequently subject to cyclic loads including electric loads and mechanical loads under long-term services, and thus the fatigue issues in these ceramics are typical. The fatigue in these ceramics could be divided into several groups, such as mechanical fatigue, dynamic fatigue, electrical fatigue, and electromechanical coupling fatigue. The relevant fatigue mechanics have been initially revealed in recent decades. As such, a detailed introduction to different fatigue failure behaviors and fatigue mechanisms in these ceramics will be presented. To show the fatigue in practical applications, the fatigue behaviors in some important ceramic devices, such as energy harvesters and MLCCs, are also elaborated. As there have been very few researches on the fatigue issues including fatigue degradation, fatigue crack, fatigue mechanism, and fatigue model of these ceramics, much of the presented research results are new and valuable. This knowledge of fatigue contributes to deeply understand the advances of fatigue in ceramics and provides a reference for the fatigue optimization and anti-fatigue design of advanced brittle ceramics in the future.

11.1.2 MECHANICAL FATIGUE IN CERAMICS

11.1.2.1 Mechanical Fatigue Characteristics and General Failure Mechanisms

Considering the inherent brittleness, the most common approach for mechanical fatigue failure characteristics of ceramic materials is cyclic bending test. As such, cyclic bending stress is generally employed to evaluate the fatigue strength and fatigue resistance of ceramics under mechanical loads. Generally, fatigue features are characterized by the relationship between bending stress and fatigue cyclic life (*S*-*N* curve) for ceramics. A general understanding of fatigue behaviors is that cyclic bending stress is decreased with the increase of the number of cycles to failure. For example, for fatigue characteristics of bismuth titanate ceramics [6], the cyclic bending stress is gradually decreased with increasing the cycles, suggesting the reduction of fatigue resistance at long-term loading. In solid materials, the fatigue failure process is often divided into three groups including crack initiation, crack propagation, and final fracture, respectively, of which crack initiation plays a significant role in fatigue failure process since most of fatigue lifetime will be consumed in this region. The fatigue damage accumulation in the crack initiation zone could be considered to determine the fatigue properties of materials. Therefore, crack initiation in fatigue failure process has increased considerable attention for scientific research. For brittle ceramics, the pre-existing defects such as pores, micro-cracks, or secondary phase impurities could be the origin of fatigue crack initiation. Due to locally structural heterogeneity around defects, significant strain mismatch comes up when applied mechanical loads. Under cyclic loads, macro cracks will be formed around these defects, significantly destroying the integrity of ceramics and decreasing the fatigue resistance. Moreover, in most of cases, the fatigue crack is prone to advance along grain boundaries of ceramics, which could substantially affect the functionality of grain boundary, leading to the formation of intergranular cracks during the long-term fatigue process. As an important consideration in mechanical strength design, grain boundary plays a significant role in mechanical properties of materials. However, grain boundary strength is weakened due to the intergranular fatigue crack propagation during cyclic stress, which will directly result in the decrease of

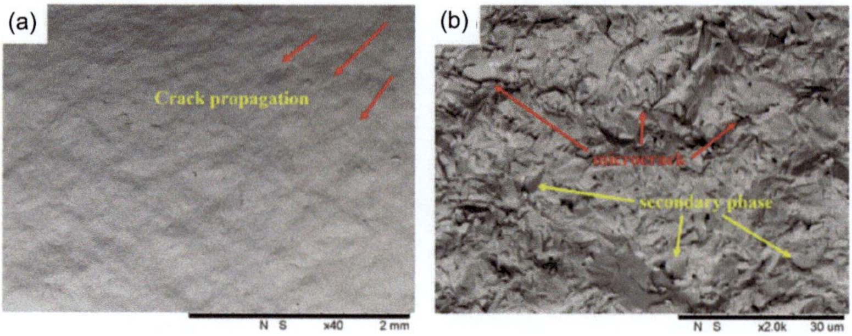

FIGURE 11.1 Fracture morphology of bismuth titanate-based ceramics. (a) macroscopic feature; (b) crack initiation zone after fatigue loading. (Reprinted with permission from Ref. [6], Elsevier.)

fatigue resistance and cause the final fatigue failure as the mechanical strength fails to bear the further fatigue crack propagation. For microstructural features of fatigue fracture in ceramics, it has been revealed that the cyclic fatigue fracture surface is much smoother than that under monotonic loading [6], as shown in Figure 11.1a. The fatigue striations in crack initiation and propagation zones are often observed in metals, while they cannot be detected in ceramic materials. In addition, the micro-cracks and secondary phase are found in ceramic matrix after cyclic loading, as shown in Figure 11.1b, which could be considered as the main fatigue crack initiation mechanism. It follows that reducing the amount of defects in ceramics will be an effective way to enhance fatigue strength and improve fatigue lifetime in practical applications.

The temperature dependence of fatigue properties is also an interesting research topic in the field of fatigue of materials, and the related research work on ceramic materials has also been revealed by cyclic bending test [7]. Interestingly, it shows that fatigue strength increases with the increase of applied temperature, of which the mean fatigue resistance at 10^5 cycles for samples tested at high temperature is twice as high as that at room temperature, as shown in Figure 11.2a. The intriguing behaviors are attributed to more distorted lattice structure at high temperature, since the tetragonal lattice structure of ceramics at room temperature will be changed to the cubic structure at high temperatures. For microstructural features of the ceramics, the fatigue fracture at room temperature is dominated by intergranular fracture model due to the high shear stresses concentrated at grain boundaries, while the fatigue fracture at high temperature is a combination of intergranular and transgranular fracture models, which is attributed to the formation of dislocations within grains due to the distorted grains induced by the change of the lattice structure system from room temperature to high temperature, with a schematic model of fracture mechanism illustrated in Figure 11.2b.

In addition, advances in bending fatigue characteristics of lead ceramics have demonstrated that fatigue degradation is directly related to the surface electrode status [8], such as nickel and silver electrodes. According to the typical *S-N* curves for the ceramics with nickel and silver electrodes, the stress amplitude for the nickel sample is located at a higher level compared to the silver one, and the fatigue strength for the nickel sample at 10^5 cycles is about 60 MPa, while the one for the silver

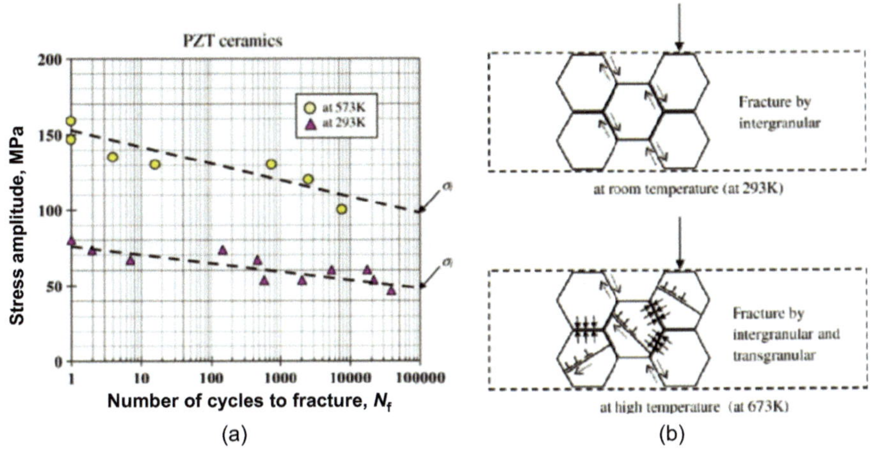

FIGURE 11.2 (a) *S-N* curves of lead ceramics at room temperature and high temperature; (b) Schematic illustration of the models of fracture mechanism at room temperature and high temperature, respectively. (Reprinted with permission from Ref. [7], Elsevier.)

sample is about 50 MPa, suggesting that the sample with nickel electrode has better fatigue resistance. For this fatigue mechanism, a general explanation is related to the electrode characteristics attached to the surface of ceramics. During the sintering process, lots of micro-pores are formed in the silver electrode, while the nickel electrode has higher density with less pores, leading to its better mechanical properties. Another interesting behavior for lead-based ceramics is that the hysteresis loop in stress-deflection curve is decreased with the increase of fatigue cycle, which could be attributed to the hardening behaviors in materials. To verify this, the values of Vickers hardness of lead-based ceramics after different fatigue cycles are characterized, and it is found that the hardness increases with the increase of fatigue cycle, further confirming the hardening behaviors in fatigue failure process.

In addition to the traditional bending fatigue test, novel fatigue test approaches, such as modified small punch (MSP) fatigue, compression fatigue, and compression-compression fatigue, are also applied to characterize the fatigue behaviors in ceramics. The MSP fatigue test approach is similar to the traditional bending fatigue test, while it keeps specific advantages in the mechanical test of materials especially for ceramics. For example, ceramics with big size and specific bulk shape are needed for the traditional fatigue test, while the MSP fatigue test just requires ceramic samples with small size and normal shape, which could not only save expensive raw materials used in ceramic synthesis but also improve the test efficiency in mechanical characteristics. Despite these advantages in MSP tests, the research on MSP fatigue test of ceramics is rare. As one of the few researchers, Xie et al. [9] have investigated fatigue behaviors and fracture mechanisms of unpoled and poled bismuth titanate ceramics by using the MSP fatigue test. It is found that the fatigue damage mainly occurs in the low cycle fatigue regime, which indicates that the corresponding fatigue mechanisms are related to the pre-existing defects and the extreme sensitivity of ceramic materials to high stress. In contrast, the nucleation and propagation of fatigue cracks at pre-existing defects are not observed at lower fatigue stress. It is proposed that the formation of fatigue cracks at lower fatigue stress or high cycle fatigue regime is due to slow crack growth induced by environmental corrosion. Moreover, the fatigue strength for poled ceramics is decreased due to the existence of internal stress induced in the poling process, reflecting that the fatigue resistance is weakened by the poling effect.

On compression fatigue, several studies have indicated that fatigue damage is related to strain accumulation under cyclic stress, and the corresponding failure mechanisms are attributed to domain switching deformation. As it is widely known, ceramic materials have excellent compression strength, and thus they have superior mechanical resistance under compression loading compared with tensile loading. However, for piezoelectric ceramics, the functionality such as ferroelectric or piezoelectric performance is easily changed under applied mechanical loads over coercive stress due to underlying polarization dynamic behaviors. The coercive stress in ceramics is the threshold value that domain structures could be oriented quickly under applied mechanical loads, which leads to the severe degeneration of the functionality of ceramics. Therefore, in spite of the ultra-high compression strength, the compression stress applied in ceramics is not allowed above coercive stress in practical uses. For this reason, the researches on compression fatigue of ceramics mainly focus on stress-induced domain switching behaviors, especially in ferroelastic domain deformation below coercive stress. It has been demonstrated that domain-switching behaviors could significantly affect the mechanical properties of ceramics. The significant roles of domain switching in the mechanical strength and fatigue properties of ceramics are also found in cyclic bending tests. To comprehensively understand the fatigue behaviors of ceramics, it is critical to elaborate the domain deformation characteristics and domain switching mechanisms under mechanical loads.

11.1.2.2 Stress-Induced Domain Switching Fatigue Behaviors

For more detailed analysis of mechanical failure behaviors in ceramics, the contribution of domain structure to mechanical properties needs to be considered. For ferroelectric materials, the plastic behaviors under applied mechanical stress could be characterized by nonlinear deformation or so-called ferroelastic deformation. For example, in the compression test [10], soft lead ceramics present typical nonlinear deformation in stress-strain curves, as shown in Figure 11.3. The nonlinear behaviors in the curves have a close relationship with ferroelastic domain switching. According to the stress-strain curves, the relative contributions from the nonlinear deformation and the linear elastic moduli could be determined. It can be seen from Figure 11.3a that the nonlinear ferroelastic deformation shows a strong dependence of temperature, i.e., a higher temperature could accelerate domain switching and thus decrease the ferroelastic deformation and nonlinear strain, which is attributed to the reduction in spontaneous strain related to a diminished tetragonal distortion. In addition, the poling state is also critical to the nonlinear deformation behaviors in ceramics, as shown in Figure 11.3b. The ceramics with poling treatment present increased nonlinearity due to the occurrence of more domain

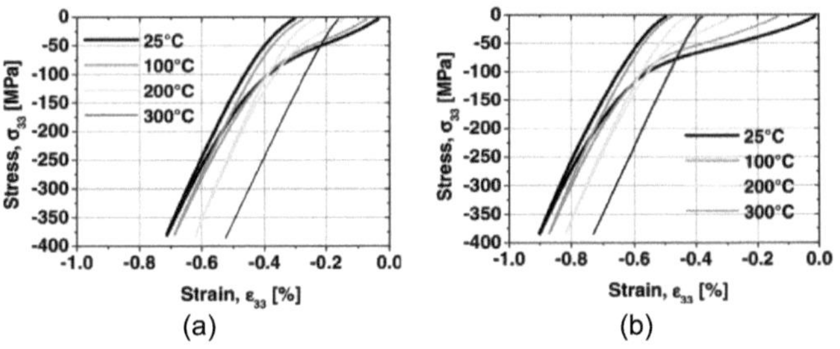

FIGURE 11.3 Ferroelastic deformation of soft lead ceramics with (a) unpoled and (b) poled states, respectively. (Reprinted with permission from Ref. [10], Elsevier.)

wall motion events during stress loading. In addition, the macroscopic ferroelastic response of antiferroelectric ceramics was investigated under compression stress at a wider temperature range from low temperature to high temperature. The ferroelastic hysteresis and domain switching strain could be observed in the weak ferroelectric, antiferroelectric, and paraelectric regions. Moreover, *in situ* stress- and temperature-dependent Raman spectroscopy measurements demonstrated that the compression stress could also induce changes in octahedra and cation displacement in the crystal structure of ceramics. These findings from the macroscopic view are widely recognized, which are solid references for the research on the mechanical properties of ceramics.

Obviously, the ferroelastic domain switching behaviors, to some extent, could largely determine the mechanical properties of ceramics. In compression fatigue events, neutron diffraction combined with a viscoelasticity model has revealed the domain switching fatigue behaviors and the levels of strain accumulated resulting from domain switching at different frequencies and different cyclic mechanical loads [11]. It shows that the magnitude of mechanical stress could affect the levels of ferroelastic strain, and the ferroelastic strain is strongly dependent on the original domain structures, while the frequency determines the time taken for the strain to reach saturation, and the frequency effects on domain orientation distribution and domain texture are more pronounced in unpoled ceramics. Ferroelastic strain is so-called irreversible strain in stress-strain curves. Under cyclic compression loading, every fatigue cycle will produce an increment in the cumulative irreversible strain due to ferroelastic deformation. Material hardening behaviors after cyclic fatigue stress could be observed and be indicated based on the narrowing of the hysteresis loop of stress-strain curves, which is attributed to the fact that the amount of switchable domains is decreased and the rest of domains are more hard to switch due to their specific microstructures and the existence of opposing internal stresses. The compression fatigue tests on hard ceramics have shown that the fatigue test rate has a significant effect on the domain switching fatigue behaviors [12]. In the faster rate tests, the maximum irreversible strain produced is smaller than that in the slower rate tests, along with smaller increments of irreversible strain every fatigue cycle and suppressed degree of hardening since more available domains keep no switching. Figure 11.4a shows the irreversible strain at different stress amplitudes under compression fatigue tests. Clearly, the irreversible

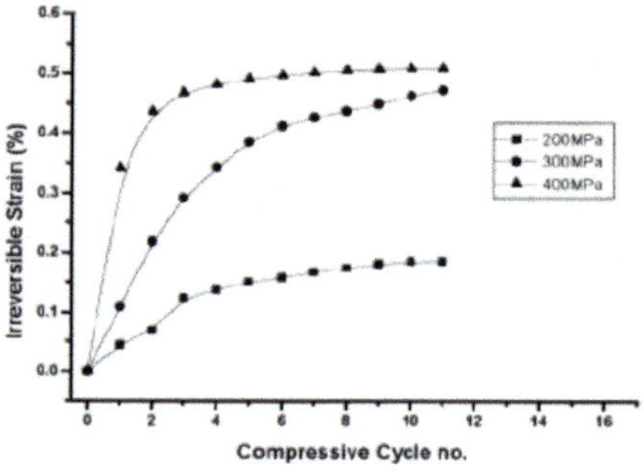

(a)

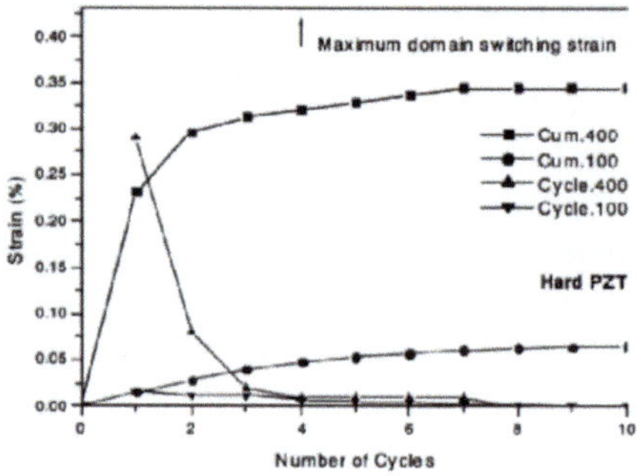

(b)

FIGURE 11.4 (a) Cumulative irreversible strain at different stress amplitudes; (b) Irreversible domain switching strain after cyclic compression tests. (Reprinted with permission from Refs. [12,13], Elsevier.)

strain at higher stress amplitude is much more severe than that at lower stress amplitude, and it seems like the ferroelastic domains at higher stress amplitude have quicker switching rate and are easier to achieve saturation state. Figure 11.4b further demonstrates the irreversible strain behaviors of a hard ceramic sample under compression fatigue tests [13]. Obviously, the irreversible strain is decreased with the increase of fatigue cycle, indicating the gradually decreased ferroelastic deformation. Moreover, most of irreversible strains are produced in the first few cycles, reflecting the ferroelastic domain switching events mainly take place in the earlier stage of fatigue.

On the ferroelastic deformation, another significant point is ferroelastic toughening behavior. Ferroelastic toughening is generally characterized by mechanical indentation test. Specifically, a crack will be introduced by an indenter, and the ferroelastic deformation event near the crack tip will be awake, along with the formation of compressive stress in a process zone due to the tetragonal anisotropy of ceramics. The compressive stress is generally considered as crack-closure stress, which could dissipate stress energy and shield crack tip as well as retard crack propagation. Therefore, the ferroelastic deformation, to some extent, toughens ceramics.

Based on the importance of ferroelastic deformation to mechanical properties, the ferroelastic domain switching characteristics have been extensively studied so far, and more novel mechanisms have been recently revealed in ceramics. For example, the ferroelastic domain switching during the unloading and loading process has been identified in lead ceramics by using X-ray and EBSD techniques, reflecting that the amount of domain switching strongly depends on the grains and different mechanical stress states, of which 90° domain switching and back-switching are simultaneously detected during unloading and loading process. Moreover, the 90° domain switching events increase with the increase of mechanical load and further increase even after the unloading process. Most importantly, two domain switching models are proposed that one is the 90° turn about the tetragonal *c*-axis, and the other is the 90° rotation of the tetragonal *a*-axis. A recent study has indicated that domain switching is easily switched when the domain is orientated close to the mechanical loading direction. From a structural analysis by EBSD, the domain switching takes place for ceramics mainly between 85.4° and 90.0°, and a domain switching model is proposed based on twin deformation theory, further verifying the domain switching angles between domain walls.

It is worth noting that most of the current existing work on ferroelastic deformation indicates that the domain switching events are energetically favorable, and relevant macroscopic switching models developed by Hwang et al. are often used to explain some common experimental results. Considering this point, *in-situ* observation of ferroelastic domain evolution in lead ceramics has been revealed by PFM, directly showing several domain switching evolution processes, such as broadening of lamellar domains, the motion of domain boundaries, and overwriting of differently oriented lamellar patterns, of which the overwriting process is related to a combination of energetically favorable and unfavorable transformations, which is very different with the macroscopic switching models that only consider energetically favorable behaviors in ceramics. Moreover, it is found that the cooperative active of favorable and unfavorable switching behaviors is attributed to the pre-existing domain structures in ceramics, and the favorable and unfavorable switching events could take place concurrently, rather than proceeding in order from most favorable to least favorable.

In addition, more intriguing microscopic structural behaviors of ferroelectric domains including stress-induced reversible and irreversible domain switching have been revealed by *in-situ* TEM experiments [14], as shown in Figure 11.5. In ferroelectrics, 90° domain could be ferroelectric or ferroelastic due to its large spontaneous strain, while 180° domain is only ferroelectric due to no spontaneous strain occurring across the corresponding domain wall. Therefore, we generally consider the stress-induced domain switching in ceramics is attributed to the motion of 90° ferroelastic domain and is not related to 180° ferroelectric domain. However, the kinetics of reversible and irreversible domain switching under applied stress have shown that in addition to the 90° ferroelastic domain switching, the 180° ferroelectric domain switching could be induced by high mechanical stresses, as indicated

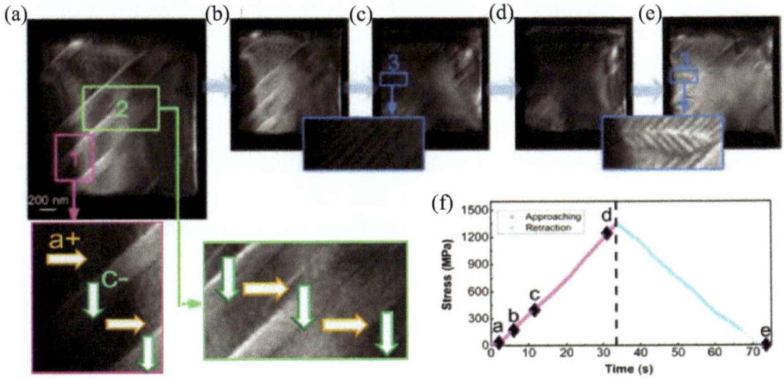

FIGURE 11.5 (a–e) A series of TEM images showing domain evolutions at different levels of tensile stress; (f) The corresponding stress-time plots during *in-situ* TEM observations [14].

by the domain pattern illustrated in Figure 11.5e, since the defect-induced pinning effect on the stable domain walls could be significantly overwhelmed or even be eliminated by applying high mechanical stress. The findings on stress-induced domain switching are so far valuable, which provide a way to diminish point or topological defects in ceramics.

In addition to the ferroelastic deformation, understanding the relationship between domain structure and fatigue crack is also necessary to rationalize fatigue failure behaviors. Previous research on fatigue failure characteristics of lead ceramics has shown that an acceleration of fatigue crack growth takes place when the crack propagates along the domain wall, while this conclusion is proposed based on the research on crack propagation of mechanical indentation test. Further study on fatigue properties of ceramics has demonstrated that the fatigue strength degradation and fatigue acceleration are attributed to the formation of microscopic internal stress at grain boundaries due to the domain switching deformation during long-term fatigue loading. The study on fatigue behaviors of titanate ceramics also indicated that the ferroelastic domain switching will produce internal stress and weaken the grain boundary, leading to the fatigue crack propagation along the grain boundary. However, the research on the relationship between domain switching and fatigue crack, so far, is not fruitful, thus a further research is urgently needed.

11.1.2.3 Mechanical Fatigue Behavior Assessments

The mechanical fatigue properties are generally accessed by the corresponding *S-N* curves. According to the fatigue test data, the relationship between applied stress amplitude and number of cycles could be fitted by linear method or polynomial method. Accordingly, the fatigue strength and fatigue life could be predicted. In addition, as we know, the material hardening behaviors could be observed in ceramics after fatigue loading. Therefore, the hardness value of ceramics after fatigue cycles could be a parameter to assess the fatigue behaviors. For brittle ceramics, one of the most efficient methods to assess the fatigue properties is the non-destructive apparent density test, which could evaluate internal damage and predict the fatigue life. The damage variable can be evaluated by variation of apparent density during fatigue test, and thus internal fatigue damage and fatigue life could be linked with each other. Another fatigue assessment approach is based on the evolution of the elastic coefficient of ceramics by resonance and anti-resonance frequencies. The variation of elastic coefficient as a scalar damage variable on the basis of the continuum damage mechanics could indirectly assess the internal damage in the samples subjected to mechanical loads. However, some lead ceramic materials show fatigue-free character in the elastic modulus after fatigue tests. For this reason, the evolutions of ferroelectric and piezoelectric properties could provide valuable assessment for fatigue behaviors of these ceramics, and deep understanding of the evolutions of crystal structure, internal stress, and domain wall dynamics under fatigue loading also sheds insight into damage mechanisms and fatigue behavior assessments for ceramic materials.

11.1.3 DYNAMIC FATIGUE IN CERAMICS

In complex environments, crack propagation in ceramic materials could take place even at lower stress during long-term loading, which is attributed to stress corrosion behaviors. Environmental effects, such as pure water, silicone oil, and air, could significantly affect the crack propagation and mechanical behaviors of ceramics. It is also indicated that the water in saliva could modify the energy that is required to break bonds between atoms in ceramic materials, thus resulting in subcritical crack growth (SCG) behaviors. Moreover, SCG could decrease the mechanical properties and shorten the fatigue life of ceramics, unfavorable for the long-term services of materials. Therefore, it is critical for ceramics operated in complex environments to understand the SCG behaviors induced by stress corrosion.

In ceramic materials, dynamic fatigue tests that reflect the relationship between mechanical strength and loading rate are generally employed to characterize the SCG behaviors. Specifically, the mechanical strength decreases with the decrease of loading rate, i.e., ceramic materials possess lower mechanical strength at lower loading rates, which is a good indication of slow crack growth due to the dynamic fatigue behaviors in ceramics [6]. Here, the crack growth velocity is relatively slow at lower loading rates, but crack nucleation still takes place in quantity during long-term loading due to the activation of environmentally assisted subcritical damage modes or stress corrosion induced SCG behaviors. With the reduction of fracture energy of crack growth in ceramics due to the crack nucleation, the crack growth resistance decreases, which leads to reaching the fracture threshold value in advance and premature failure. In addition, the dynamic fatigue behaviors could be elaborated by slow crack growth parameter n, which is also considered as the most efficient factor to evaluate the slow crack growth behaviors of ceramic materials. In brittle ceramics, a bigger slow crack growth parameter corresponds to a good dynamic fatigue behavior, reflecting a good slow crack growth resistance during long-term stress corrosion.

Dynamic fatigue behaviors of ceramics are affected by many factors. For example, the sintering temperature of ceramics could lead to different dynamic fatigue properties. Specifically, the ceramics with higher sintering temperatures have lower slow crack growth parameter values, indicating that the slow crack growth has a higher susceptibility to long-term stress corrosion and presents inferior dynamics fatigue properties. The dynamic fatigue properties are also related to the applied electric field. The ceramics under positive electric field has low dynamics fatigue and slow crack growth resistance, and the corresponding slow crack growth velocity could be evaluated by energy release rate combined with finite element analysis. The research on dynamic fatigue behaviors of Si_3N_4-based ceramics has demonstrated that the composition plays a significant role in slow crack growth behaviors [1]. The slow crack growth zone originates from the tensile surface of sample during three bending loading, and the slow crack growth generally results from a stress corrosion process involving the stable growth of pre-existing defects such as micro-pores. The porosity is detrimental to dynamic fatigue resistance, while the addition of an appropriate metal phase could significantly enhance the dynamic fatigue properties of ceramics by improving material density. The effect of metal phase such as cobalt on dynamic fatigue behaviors is also investigated in Al_2O_3/TiC ceramic, further confirming the metal phase has better resistance to slow crack growth and tends to increase dynamic fatigue properties. In addition, ambient temperature could also alter the dynamic fatigue behaviors in ceramics. The slow crack growth zone will become larger at high temperatures, and larger plastic deformation takes place at lower loading rates. Oxidative corrosion at high temperatures could break the surface smoothness and increase the fracture source, which leads to the reduction of slow crack growth resistance and the degeneration of dynamic fatigue with the unfavorable long-term behavior of ceramic materials specifically for the ceramics with metal phase.

11.1.4 ELECTRICAL FATIGUE IN CERAMICS

In piezoelectric applications, the preservation of electrical performance stability of ceramic materials under electric field is a significantly prerequisite. However, the electrical parameters, such as piezoelectric coefficient, remanent or maximum polarization, permittivity, and so forth are prone to be changed under cyclic electric field, causing electrical fatigue behaviors. Electrical fatigue is commonly characterized by the typical polarization versus electric field (*P-E*) hysteresis loops and unipolar or bipolar strain versus electric field (*S-E*) curves in ceramics [15], as shown in Figure 11.6. Before electrical fatigue loading, the *P-E* loops and *S-E* curves are saturated and symmetric, as shown in Figure 11.6a and b, while the ferroelectric properties decrease and the *S-E* curves become asymmetric after electric cycles, as indicated in Figure 11.6c and d. The degradation of electrical performance after electrical fatigue loading is related to the number of applied electric cycles, cyclic field amplitude, temperature, chemical doping, crystallographic structure, domain structures, and so forth, and relevant fatigue behaviors and mechanisms will be elaborated below.

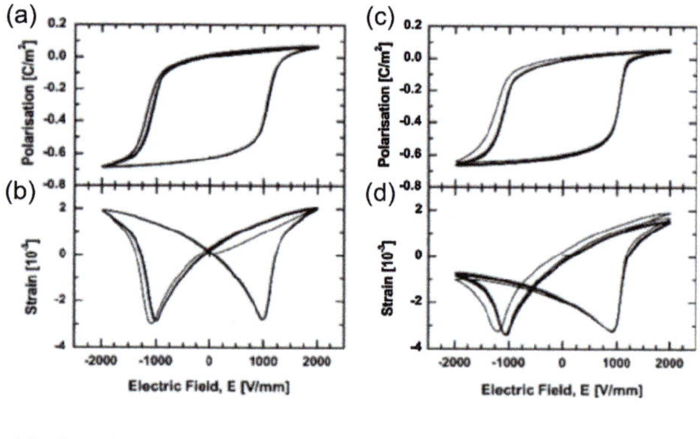

FIGURE 11.6 (a) *P-E* loops and (b) *S-E* curves of lead ceramics before electric cycles; (c) The corresponding *P-E* loops and (d) *S-E* curves of lead ceramics after electric cycles. (Reprinted with permission from Ref. [15].)

11.1.4.1 Electrical Fatigue Cracking

The formation of micro-cracks in ceramics is one of the main mechanisms for electrical fatigue failure. A previous study on the electromechanical and fatigue behaviors of piezoelectric ceramics has concluded that fatigue degradation is strongly affected by temperature and by the magnitude of applied electric field. Specifically, fatigue degradation gradually increases with the increase of electric field strength and with the decrease of temperature. The fatigue mechanisms are related to electrically induced stresses caused by internal anomalies and the formation of micro-cracks. However, the fatigue disappears above a critical temperature even at larger electric fields, which is attributed to the evolutions in electromechanical properties such as depolarization strain. The micro-cracks are generally induced by bipolar electric fields with enough large amplitude. Of course, unipolar electric loading could also lead to micro-crack growth. The electric cycle induced cracking often occurs close to the electrode interface and could cause the delamination of the electrodes, thus decreasing electrical functionality. The cracking at the electrode interface is thought to stem from stress concentration at the electrode edges or surface flaws. After removing about hundreds of micrometers from the electrode interface by polishing treatment, which means induced cracks are moved, the stress concentration is decreased and the electrical performance of ceramics is enhanced under cyclic electrical fatigue. In addition, pre-existing micro-cracks and pores as well as secondary phases are also the source of crack growth, since the stress concentration could be induced at these defects during electric cycles. The performance degeneration of ceramics under electric fatigue is very similar to the situations that occurred in mechanical fatigue. Therefore, the underlying mechanisms of such cracking behaviors during electric cycles actually originate from mechanical degeneration.

Considering the importance of cracking on electric fatigue degradation, the crack evolution behaviors during electric fatigue process have been investigated intensively. For example, Weitzing et al. [16] reported the evolution of pre-cracks produced by Vickers indentation under bipolar cyclic electric loading, as illustrated in Figure 11.7.

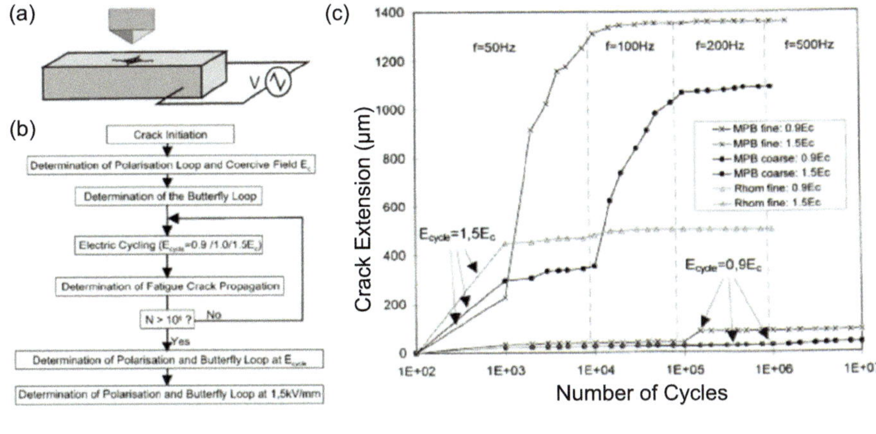

FIGURE 11.7 (a) Experimental set-up for the determination of fatigue-crack growth; (b) Experimental procedure for fatigue tests; (c) Crack extension as a function of fatigue cycles. (Reprinted with permission from Ref. [16], Elsevier.)

It can be seen from Figure 11.7c that a higher cycling field shows a quicker fatigue crack growth rate, while the fatigue crack growth rate decreases with the increase of fatigue cycles, and a saturation point is reached after 10^5–10^6 cycles. Besides, the increase of frequency of electric loading could cause abrupt crack growth, and the fatigue crack growth behaviors are related to ceramic compositions. It is proposed that the mechanical strain due to electromechanical coupling is the main crack driving force, while the elastic strain is not enough strong to advance crack growth after the saturation point. Fatigue crack growth has been also reported in the lead ceramics with a surface crack produced by Vickers indentation under electric cycles. Due to indentation residual stress and crack tip stress intensity, it is found that fatigue crack growth is also initiated at smaller electric fields. The theoretical work investigated by Zhu and Yang has shown that the fatigue crack growth is a step-by-step cleavage process caused by a domain switching sequence that produces a cyclic driving stress field near the fatigue crack tip, and the fatigue crack starts to grow as the stress intensity factor is larger than inherent fracture toughness of ceramics. Moreover, the stress intensity factor at the crack tip under a negative electric field is several times as large as that under a positive electric field, indicating easier fatigue crack growth and weak fatigue resistance under negative electric fields.

Some reports also revealed the fatigue crack growth in ferroelectric ceramics with a short notch. Interestingly, there exist two typical characteristics in the fatigue crack growth under electric loading. Under low electric loads, micro-cracks initiated near main crack are the major fatigue mechanism that could impede the main crack growth, whereas the main crack is the only fatigue cracking mode that dominates the fatigue mechanism under high electric loads. Moreover, the fatigue crack generally grows along the grain boundaries, and the crack growth rate is nonlinearly related to the applied electric fields, where relevant fatigue behaviors could be further elaborated by domain switching effect and fracture mechanics concepts. A similar experimental design on fatigue crack growth of ceramics with a pre-crack has been performed in other reports. It shows that three regimes can be observed during crack growth, i.e., a pop-in from the pre-crack, steady-state crack growth, and a decrease of the crack growth rate with the increase of electric cycles. Combined with a nonlinear finite element analysis, it reveals the driving forces for fatigue crack initiation and growth could be determined by the distribution of residual stresses caused by ferroelectric switching, and the inhomogeneous electric fields originated from notch or crack are mainly responsible for the fatigue behaviors. Recently, physical mechanisms of fatigue crack growth of ceramics with a U-shaped notch have been rationalized by employing an *in-situ* digital image correlation technique [17]. It shows the existence of incompatible deformation near the U-shaped notch during the electrical fatigue process, and the pop-in behaviors are attributed to the interaction of incompatible domain switching deformation zones near the main crack. Besides, the steady-state crack growth generally takes place at the stage of crack closure, as illustrated in Figure 11.8a–c, and the corresponding fatigue mechanism is related to three physical factors including arcing/bubble effect, wedging, and grain-to-grain interaction, as shown in Figure 11.8d–f. It is worth noting that fatigue cracking under cyclic electric fields is considered as permanent damage in ceramics, i.e., this type of fatigue failure behavior is irreversible, which significantly affects the reliability and stability of ceramic materials during long-term services.

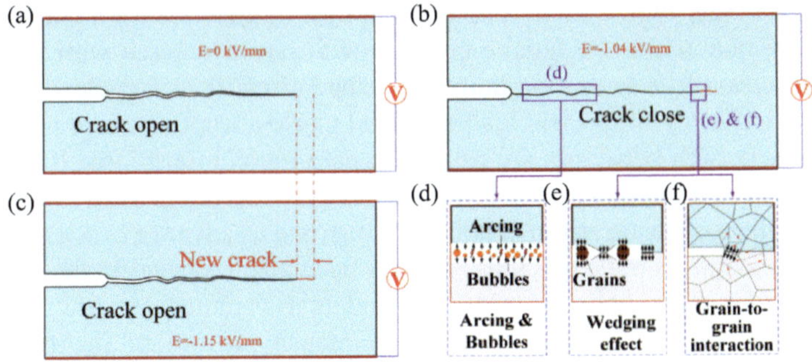

FIGURE 11.8 (a–c) Illustrations of steady-state crack growth in ceramics; (d–f) Illustrations of fatigue mechanisms. (Reprinted with permission from Ref. [17], Elsevier.)

11.1.4.2 Domain Wall Pinning and Space Charge Effects

It has been widely reported that the cyclic electric field could cause continuous reorientation of ferroelectric domains and also make the electronic and ionic charge carriers redistributed, and the interaction between domain walls and charge carriers takes place during cyclic electric fatigue. After several electric cycles, domain walls could be pinned by defects such as charge carriers, limiting the motion of domain walls under electric fields. With the increase of electric cycles, the pining effect is increased, and domain walls become immovable, leading to the degeneration of ferroelectric performance and the appearance of fatigue behaviors. Therefore, domain wall pining is a typical electrical fatigue mechanism in ceramics. *In-situ* transmission electron microscopy has indicated that the large ferroelectric domains are disrupted and are replaced with accumulated defect clusters and fragmented domains after electric cycles. The evolutions of domain structures under cyclic electric fields are profoundly connected with fatigue damage. Recent research has demonstrated that the electrical fatigue loading will cause a decrease in domain dynamics and piezoelectric responses in ceramics. Moreover, the fatigue mechanism is determined by the domain wall pinning effect in the early period of electric fatigue loading, while the microstructural damage such as cracking will dominate the fatigue process with the increase of fatigue cycles. Besides, unlike the permanent damage of fatigue cracking, the fatigue degradation dominated by domain wall pinning can be reversed by annealing at high temperatures due to the unstable domain wall features, as demonstrated in some relaxor ceramics.

In addition, electric bias field behaviors are always observed in ferroelectric hysteresis loops. In this case, the hysteresis loops shift along the coordinate system resulting in an asymmetry of coercive fields under applied electric field, and the phenomenon is more obvious when ceramic materials are subjected to electric cycles. The formation of electric bias fields is related to asymmetric pinning of ferroelectric polarization direction. The asymmetric of pinning process could be described by a kinetic imprint model, which shows the cyclic electrical fatigue originating from spatially variational bias fields, resulting in incomplete domain switching events occurred in ceramics. For this reason, lots of ferroelectric domains are frozen in order to adapt the asymmetric pinning of polarization orientation. The research based on a gradual domain switching model has shown that the nonhomogeneous distribution of space charges may give rise to the asymmetric variation of ferroelectric loops. In addition, the development of bias fields and asymmetry of coercive fields are mainly attributed to the migration of charge carriers to grain boundaries driven by unscreened depolarization field, and the fatigue degradation could also be restored by thermal annealing, indicating space charge-dominated fatigue mechanism is also not a permanent damage in ceramics.

A new finding on electrical fatigue degradation has been revealed by a direct observation of nanoscale dynamics of ferroelectric domain [18]. It has been indicated that *c* domains could be nucleated at the original domain walls with the increase of electric cycles. Most importantly, the change of charge distribution takes place during electrical fatigue. By using *in-situ* biasing transmission electron microscopy, it reveals the charges are mainly accumulated at domain walls during electric cycles, as shown in Figure 11.9, and the charge accumulation at domain walls contributes to the formation of *c* domains. Besides, these accumulated charges could serve as compensation charges to freeze the *c* domains and limit their polarization dynamics, making ferroelectric domains less responsive to the applied cyclic electric fields and thus leading to the ferroelectric degradation in materials. Another similar research has shown that 180° domain switching events during electric cycles in bismuth ferrite ferroelectrics are vulnerable to fatigue. It is proposed that non-neutral domain walls are expected to appear during 180° domain switching due to their abundant switching routes under applied electric fields based on a multistep switching process model. However, in order to decrease higher electrical energy, the non-neutral domain walls can be stabilized by trapping charge carriers, leading to fatigue degradation due to the domain wall pinning effect. Despite these findings based on ferroelectric single crystal or film materials, they shed new insights into the nature of ferroelectric fatigue and reveal the role of charge carriers in domain dynamics during electric cycles. It has great significance for us to understand novel electrical fatigue mechanisms in ceramics.

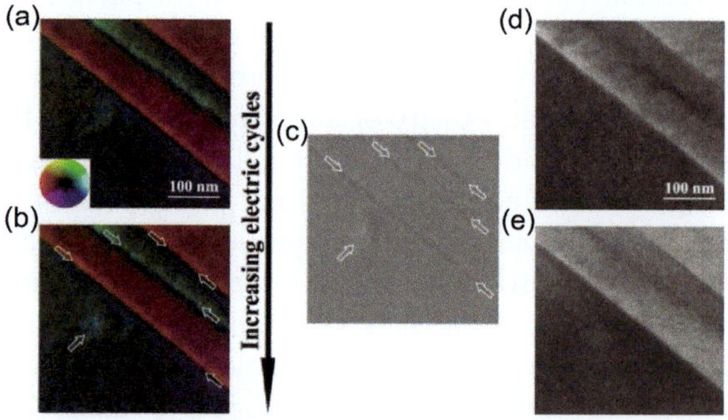

FIGURE 11.9 (a, b) Electric field mapping before and after cyclic elecric loading; (c) A graysclae image obtained from the subtraction of the image in (a); (d, e) TEM images before and after electric cycles. (Reprinted with permission from Ref. [18].)

11.1.4.3 General Strategies for the Enhancement of Fatigue Properties

In ceramics, defects such as oxygen vacancies play important roles in tailoring electrical fatigue behaviors, and it has been reported that decreasing the amount of oxygen vacancy by doping strategy is an effective way to enhance fatigue properties. Recent research on fatigue strain recoverability of lead-free piezoelectric ceramics has provided a novel strategy to enhance fatigue properties, i.e., defect engineering. The defect dipoles could be manipulated by introducing chemical ions into ceramics, and a giant strain response of ceramics is produced due to the contribution of defect dipoles at unipolar electric fields. During electrical fatigue tests, defect dipoles gradually deviate from their initial positions and migrate to new positions. As such, with the increase of the distance between defect dipoles, a stronger rebound force will be generated, causing an enhanced recoverability of defect dipoles and keeping the ceramic with good fatigue resistance. Some work has revealed that an ultrahigh strain response could be attained in lead-free ceramics with ferroelectric-nonpolar phase boundary, and a reversible transformation between the polar ferroelectric phase and nonpolar phase could also be achieved under cyclic electric field, leading to enhanced fatigue performance. Therefore, the stable nonpolar phase will be a significant consideration in designing ceramic materials with improved cycling reliability and fatigue resistance.

In addition, it is also an effective approach to enhance fatigue properties by reducing the domain wall pinning effect, improving domain wall activity, or optimizing domain structures in ceramics. For example, by controlling chemical compositions and tuning phase boundaries in lead-free ceramics, good fatigue properties such as stable up to 10^6 cycles could be achieved in ceramics with rhombohedral-tetragonal phase boundary. In this case, hierarchical nano-sized domain structures are formed, of which local structure inside nano-sized ferroelectric domains consists of rhombohedral and tetragonal twin variants. Therefore, the physical origin of enhanced good fatigue properties could be due to low domain wall energy and nearly vanishing polarization anisotropy of such optimized domain structures, resulting in a good polarization dynamic behavior and a quick response to cyclic electric fields. The investigation on the fatigue behaviors in lanthanum-substituted bismuth titanate ceramics has found that fatigue degradation takes place in bismuth titanate ceramics mainly due to single nucleation site at the electrode interface of new ferroelectric domains during the electrical fatigue process. However, after lanthanum modification, there exists a high density of antiphase boundaries in ceramics. Such typical domain structures could provide additional nucleation sites of ferroelectric domains in the fatigue switching process, and not just dependent on the electrode interface, thus leading to enhanced fatigue properties. A similar fatigue mechanism has also been reported in multi-ion co-doped bismuth titanate ceramics [2], where the introduction of high-density layered 180° domain walls contributes to the excellent electrical fatigue properties because the layered domain walls present superior polarization dynamics and easily nucleate at existing 90° domain walls under applied cyclic electric field. In addition, the decrease of oxygen vacancy concentration in this ceramics by donor doping strategy also benefits the improvement of electrical fatigue properties.

11.1.5 ELECTROMECHANICAL COUPLING FATIGUE IN CERAMICS

Understanding fatigue behaviors of ceramics under cyclic electrical and combined mechanical loading is also significant for us to design materials, since ceramic materials are often subjected to simultaneous mechanical and electrical loading in many practical applications. The mechanical- and electrical field coupling fatigue has a certain similarity with single mechanical fatigue and electrical fatigue, while they still present different fatigue mechanisms. Historically, the research studies on mechanical- and electrical field coupling fatigue are very limited, and thus only a few findings can be referred for us to understand corresponding fatigue behaviors. For example, the investigation of the fatigue crack growth behaviors under constant mechanical loading and cyclic electrical loading by experimental method and finite element analysis has shown that, unlike the single electrical fatigue, the fatigue crack growth with the help of mechanical loading that provides an additional Model-1 crack opening could be evoked even at smaller applied electric fields, indicating that the mechanical loads could decrease the threshold for fatigue crack growth and accelerate the crack growth rate, namely, the mechanical loading is unfavorable for the fatigue resistance. However, a contradictory point has been provided by the investigation on fatigue behaviors of lead ceramics under constant mechanical loading and cyclic electrical loading. It reveals that under no external mechanical loads, a common ferroelectric fatigue is observed, while a simultaneous enhancement of both ferroelectric switching and remnant polarization is achieved under external compressive loads, reflecting an improved fatigue resistance, of which the flexible non-180° domain switching and suppression of micro-cracking under mechanical loads are responsible for the enhanced fatigue resistance. Besides, some work has recently investigated the influence of the domain switching process on mechanical- and electrical field coupling fatigue behaviors of lead ceramics. It shows that the fatigue process is mainly associated with the quick domain switching taking place near the coercive electric field, and the underlying fatigue mechanisms are related to the interplay between defect redistribution, charge carrier injection, and crack initiation. Most importantly, as the spontaneous polarization and ferroelectric strain of ceramics are smaller than their critical values, the domain switching process will gradually change near the coercive electric field, and the ceramics tend to present a good fatigue resistance.

In addition to the scenario under constant mechanical loading and cyclic electrical loading, the investigation of fatigue behaviors under cyclic mechanical- and cyclic electric loading is also intriguing. Based on the investigation on the fatigue failure of unpoled and poled lead ceramics under constant mechanical loading and cyclic electrical loading, it can be found the fatigue life of unpoled ceramics decreases more rapidly than that of poled ceramics, which is due to a more profound increase of micro-cracks in the unpoled ceramics with the increase of electric cycles. In contrast, under cyclic mechanical- and cyclic electric loading, the fatigue life of unpoled and poled lead ceramics decreases with the increase of cyclic mechanical loads, where the fatigue life of poled ceramics reduces more significantly than that of unpoled ceramics since ferroelectric domains of poled ceramics are more uniform, which could undergo more remarkably changed under coupled cyclic loads. To further reveal the fatigue behaviors under cyclic mechanical- and cyclic electric loading, fatigue tests of single-edge pre-cracked beam ceramic samples could be performed by using three-point bending [19]. It indicates that the number of cycles to failure of the ceramics tends to become smaller and the slope of the maximum energy release rate versus fatigue cycles becomes steeper under coupled cyclic loads. Moreover, the ceramics under single cyclic mechanical loading show both intergranular and transgranular fatigue fracture model, while fatigue fracture is dominated by the transgranular model under coupled cyclic loading. So far, the fatigue analysis of ceramics under a couple of cyclic fields is still a challenging research work, and the fatigue degradation mechanism in such cases is complex, thus more researches are needed in the future.

11.1.6 FATIGUE IN CERAMIC DEVICES

Fatigue failure issues in ceramic devices have attracted great attention in both commercial and academic fields, while relevant studies are currently scarce. As one of the most potential ceramic devices, energy harvesters are widely applied to many applications, such as monitoring traffic-induced pavement vibrations and supporting vehicle networking and intelligent transportation. The degradation of energy harvesters under equivalent traffic stress condition has been investigated by compression tests. It reveals that parallel piezoelectric devices have better electrical output with a repetitive number of cycles, while they present a higher degradation rate, which is attributed to the occurrence of micro-crack on the surface of ceramic devices. The cracking-dominated fatigue degradation mechanism has also been revealed in piezoceramic patch transducers, and it shows tensile loads have the strongest influence on fatigue damage and fatigue lifetime of devices. To further reveal fatigue failure features in energy harvesters under equivalent traffic load, the corresponding degradation mechanism and fatigue life have been investigated under intermittent and persistent loading conditions. The results show that the electrical output and fatigue degradation rate of ceramic devices increase with the excitation load and frequency, and it is suggested that the ceramic devices should be applied at places where fewer heavy trucks pass through so that they can deliver greater electrical output and possess better service life. Moreover, several micro-cracks are also observed on the silver electrode of the surface of devices, whereas fatigue degradation of energy harvesters is mainly affected by leakage and depolarization is the dominant fatigue mechanism. Besides, Salazar et al. [20] reviewed the fatigue issues in energy harvesters, which provided more information on the cyclic fatigue behaviors of a vibration energy harvester affected by fabrication defects and transduction materials, and also offered several suggestions for prolonging service life and achieving better properties which could resist the depolarization behaviors.

The fatigue behaviors of multilayer ceramic capacitors (MLCCs) that are one of the most prospective ceramic devices for energy storage applications, such as pulsed systems, are also an open topic to investigate so far, since a strong fatigue resistance of MLCCs could ensure physical integrity of pulsed systems and enable energy storage performance. Nevertheless, MLCCs often suffer from resistance degradation after long-term service, making them susceptible to thermal breakdown under large inrush currents. It is believed that the use of dielectric ceramics with low permittivity or placing a low permittivity layer close to electrodes could improve the fatigue reliability of MLCCs. Moreover, when the applied field is smaller than a certain threshold value, the fatigue breakdown of MLCC is hard to take place, whereas leakage current is still observed after a very long time. For MLCCs, oxygen vacancy migration is often thought of as a dominant fatigue mechanism. It has been demonstrated that the decrease in oxygen vacancy concentration is the main reason for enhanced fatigue reliability in MLCCs. In addition, it has been found that the absence of larger strain is also a main physical mechanism for improving fatigue degradation of MLCCs, since the low strain in ceramic devices hinders large cyclic changes in dimensions, which could prevent the formation of thermo- and mechanical-related micro-cracks. However, the cyclic charge and discharge process and fatigue failure of MLCCs are affected by inner ceramic materials, temperature, frequency, voltage fluctuation, and so like. Therefore, the fatigue issues in MLCCs, to some extent, are totally complex. The research on fatigue failure of MLCCs is so far fewer, and it is significantly expected to shed more insights into fatigue failure behaviors and fatigue degradation mechanisms in the future.

11.1.7 SUMMARY

In this part, advances in fatigue issues of piezoelectric ceramics have been investigated and general strategies for enhancement of fatigue of ceramic materials have been provided. Generally, the types of fatigue of the ceramics could be basically divided into several groups, such as mechanical fatigue, dynamic fatigue, electrical fatigue, and electromechanical coupling fatigue, which correspond to different failure behaviors and denegation mechanisms, respectively. In practical uses, the ceramic devices present more complicated fatigue issues, affected by multiple macro/micro factors. This knowledge of fatigue sheds insights into future work on the development of the next generation of ceramic materials for complicated cyclic field applications.

REFERENCES

1. Xu W, Yuan J, Yin Z. Dynamic fatigue behavior of Si_3N_4-based ceramic tool materials at ambient and high temperatures. *Ceramics International* 2019;45:21572–8.
2. Xie S, Xu Q, Chen Q, Zhu J, Wang Q. Realizing super-high piezoelectricity and excellent fatigue resistance in domain-engineered bismuth titanate ferroelectrics. *Advanced Functional Materials* 2024; 34:2312645.
3. Gogotsi GA. Fracture toughness of ceramics and ceramic composites. *Ceramics International* 2003;29:777–84.
4. Xie S, Tan Z, Jiang L, Nie R, Xu Q, Chen Y, et al. Ferroelastic properties and compressive stress-strain response of bismuth titanate based ferroelectrics. *Ceramics International* 2020;46:1183–8.
5. Webber KG, Vögler M, Khansur NH, Kaeswurm B, Daniels JE, Schader FH. Review of the mechanical and fracture behavior of perovskite lead-free ferroelectrics for actuator applications. *Smart Materials and Structures* 2017;26:063001.
6. Xie S, Xu J, Chen Y, Tan Z, Nie R, Wang Q, et al. Flexural fracture mechanisms and fatigue behaviors of $Bi_4Ti_3O_{12}$-based high-temperature piezoceramics sintered at different temperatures. *Ceramics International* 2018;44:16758–65.

7. Okayasu M, Otake M, Bitoh T, Mizuno M. Temperature dependence of the fatigue and mechanical properties of lead zirconate titanate piezoelectric ceramics. *International Journal of Fatigue* 2009;31:1254–61.
8. Okayasu M, Ozeki G, Mizuno M. Fatigue failure characteristics of lead zirconate titanate piezoelectric ceramics. *Journal of the European Ceramic Society* 2010;30:713–25.
9. Xie S, Xu J, Chen Y, Jiang L, Tan Z, Nie R, et al. Poling effect and sintering temperature dependence on fracture strength and fatigue properties of bismuth titanate based piezoceramics. *Ceramics International* 2018;44:20432–40.
10. Webber K, Aulbach E, Key T, Marsilius M, Granzow T, Rödel J. Temperature-dependent ferroelastic switching of soft lead zirconate titanate. *Acta Materialia* 2009;57:4614–23.
11. Pojprapai S, Jones JL, Studer AJ, Russell J, Valanoor N, Hoffman M. Ferroelastic domain switching fatigue in lead zirconate titanate ceramics. *Acta Materialia* 2008;56:1577–87.
12. Moreno JC, Guiu F, Meredith M, Reece M, Alford NM, Penn S. Anisotropic and cyclic mechanical properties of piezoelectrics—compression testing. *Journal of the European Ceramic Society* 1999;19:1321–4.
13. Calderon-Moreno JM. Stress induced domain switching of PZT in compression tests. *Materials Science and Engineering: A* 2001;315:227–30.
14. Chen Z, Huang Q, Wang F, Ringer SP, Luo H, Liao X. Stress-induced reversible and irreversible ferroelectric domain switching. *Applied Physics Letters* 2018;112:152901.
15. Verdier C, Lupascu D, Rödel J. Unipolar fatigue of ferroelectric lead–zirconate–titanate. *Journal of the European Ceramic Society* 2003;23:1409–15.
16. Weitzing H, Schneider G, Steffens J, Hammer M, Hoffmann M. Cyclic fatigue due to electric loading in ferroelectric ceramics. *Journal of the European Ceramic Society* 1999;19:1333–7.
17. Li Y, Chen Z, Duan G. Fatigue crack growth mechanism in ferroelectric ceramics under alternating electric field: An investigation by digital image correlation technique. *Acta Materialia* 2022;235:118065.
18. Huang Q, Chen Z, Cabral MJ, Wang F, Zhang S, Li F, et al. Direct observation of nanoscale dynamics of ferroelectric degradation. *Nature Communications* 2021;12:2095.
19. Narita F, Shindo Y, Sato M. Fatigue behavior of cracked piezoelectric ceramics in cyclic three-point bending under AC electric fields. *Journal of Solid Mechanics and Materials Engineering* 2013;7:530–9.
20. Salazar R, Serrano M, Abdelkefi A. Fatigue in piezoelectric ceramic vibrational energy harvesting: A review. *Applied Energy* 2020;270:115161.

11.2 ADVANCES IN FATIGUE OF HYDROGELS

11.2.1 INTRODUCTION

Hydrogels are polymers composed of water and polymer networks. In comparison to other polymers, they possess a high water content [1], which means that hydrogels integrate the properties of both solids and liquids. They can maintain similar physicochemical properties to liquid water while exhibiting solid-like rheological behavior [2]. Due to their ability to form networks with different polymers, hydrogels exhibit a broad range of tunable physical and chemical properties [1–3]. These characteristics have made the hydrogel systems garner significant attention and have found applications in various fields, including biomedical engineering [4], soft robotics [5], triboelectric nanogenerators [6], sensors [7], and energy storage [8], as seen in Figure 11.10.

Traditional hydrogels, composed of simple polymer network structures, typically exhibit weak mechanical properties (strength < 50 kPa, stiffness < 10 kPa, and toughness < 10 J/m^2) [9]. Gong et al. [10] first discovered a tough double network (DN) hydrogel, which consists of a short-chain rigid network as the first network and a long-chain flexible network as the second network. When the DN hydrogel begins to bear a tensile load, the long-chain flexible network transmits stress throughout the entire hydrogel network, whereas the short-chain rigid network bears most of

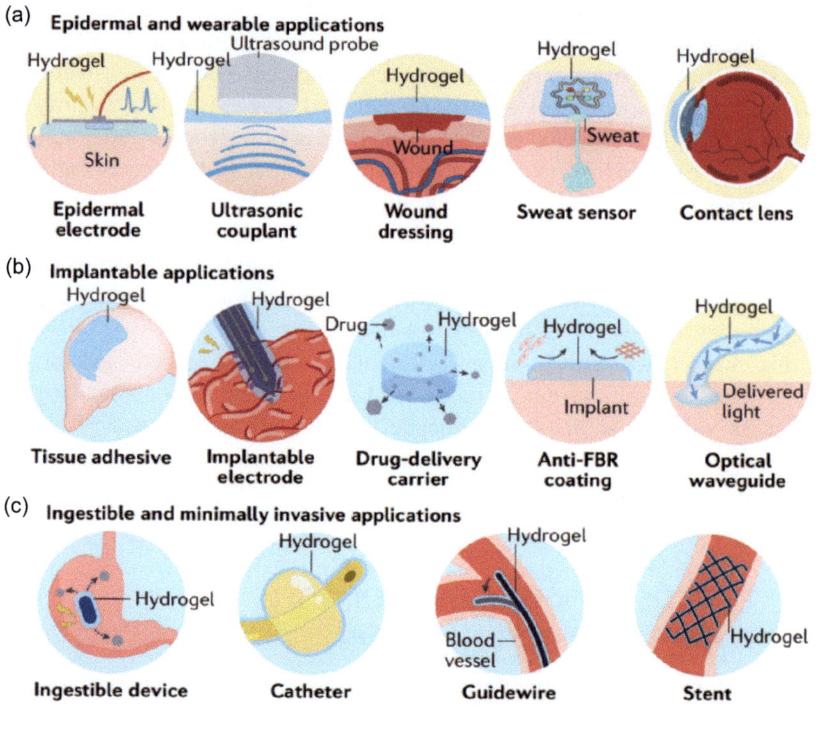

FIGURE 11.10 Applications of hydrogels at interfaces [3].

the load and provides stiffness. With increasing load, the short chain rigid network begins to break and dissipates energy to enhance the toughness of the hydrogel, acting as a toughener. The synergistic action of these two mechanisms achieves high toughness through an energy dissipation mechanism [11]. The introduction of DN hydrogels has sparked exploration among scholars worldwide for tougheners, leading to proposals utilizing hydrogen bonds [12], hydrophobic interactions [13], and ionic bonds [14]. In addition, novel processing technologies, such as freeze-casting [5,15] and compression annealing [5], have been employed to regulate the microstructure of hydrogels, resulting in the development of hydrogels as tough as tendons [15].

With the improvement of hydrogel performance, the range of applications for hydrogels has also expanded, requiring consideration of more usage conditions. Especially in certain working conditions, hydrogels need to endure prolonged static and cyclic loads, inevitably raising concerns about their susceptibility to fatigue damage [11]. For example, strain sensors [7], artificial skin [16], and artificial muscles [17] in practical applications necessitate repeated stretching, while soft robots [5] experience repeated bending of their mechanical arms. Likewise, materials used for artificial heart valves must withstand billions of cyclic loads within their expected lifespan [18]. Current research indicates that both prolonged static and cyclic loads can induce fatigue in hydrogels, manifested through changes in performance, nucleation, and propagation of fatigue cracks [11]. Fatigue symptoms exhibit variation depending on the chemistries of bonds and topologies of networks [11]. Therefore, to adapt to a wider range of application scenarios and facilitate the extensive use of hydrogels, it is urgent to investigate the factors influencing hydrogel fatigue and develop fatigue-resistant hydrogels. In comparison to materials such as metals, ceramics, plastics, elastomers, and composites, which have been extensively studied for fatigue, research specifically focused on hydrogel fatigue is still in its early stages. Tanaka et al. [19] first reported on crack propagation in hydrogels under static loading in 2000, while studies on fatigue crack propagation in hydrogels under cyclic loading were first reported by Tang et al. in 2017 [20]. Suo et al. [11] provided the first comprehensive review on hydrogel fatigue in 2019. Qu et al. [21] reviewed the anti-fatigue design of hydrogels in 2024. As a result, research methods and characterization techniques for hydrogel fatigue are not yet mature, and further research on the constitutive relationship of hydrogels is needed.

In this chapter, we provide an overview of the concept of hydrogel fatigue, its characteristics, and the current methods used to experimentally assess fatigue. We then discuss various factors that influence the fatigue performance of hydrogels. Subsequently, we examine the latest advancements in fatigue-resistant hydrogels, highlighting five representative strategies. Lastly, we emphasize the main challenges and potential research directions in the field of hydrogel fatigue.

11.2.2 FATIGUE CONCEPTS OF HYDROGELS AND FATIGUE TEST METHODS

Load-bearing materials, such as metals, are subjected to disturbance stress at a certain point or points. After enough cyclic disturbances, cracks or complete fractures form. The development process of this local permanent structural change is fatigue. Suo and colleagues [11] characterized fatigue in hydrogels as the conditions displayed under prolonged loading, sub-dividing it into two types as per the two archetypical load curves, namely static fatigue and cyclic fatigue. Static fatigue, analogous to creep fatigue in metals, is fatigue that occurs under extended static loading (Figure 11.11), while cyclic fatigue results from extended cyclic loading

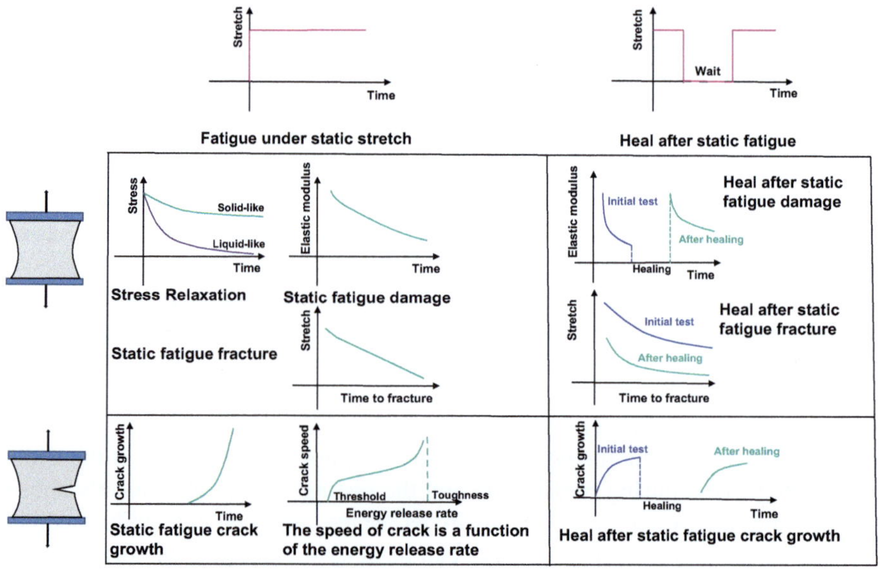

FIGURE 11.11 Static fatigue. The figure is created based on the work of [11].

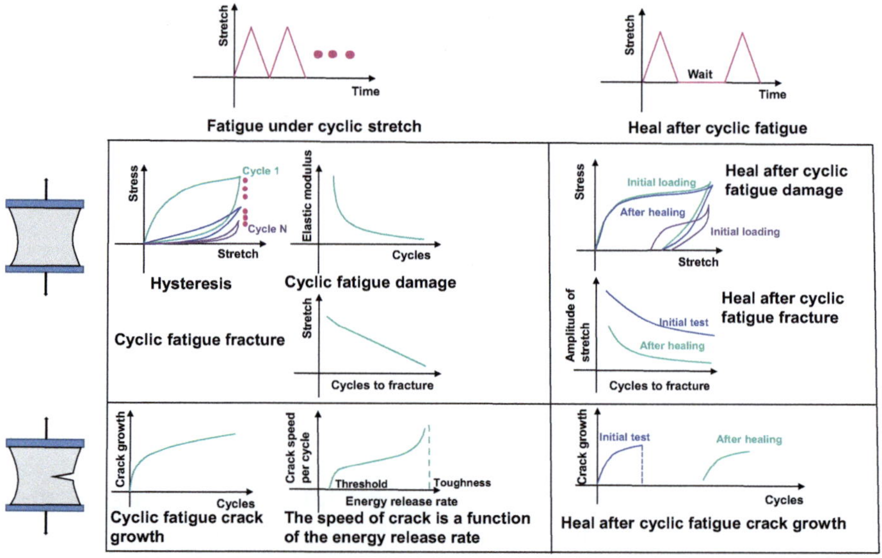

FIGURE 11.12 Cycle fatigue. The figure is created based on the work of [11].

(Figure 11.12). Persistent loading in hydrogels has the potential to induce a variety of symptoms, encompassing performance modifications, as well as the nucleation and propagation of cracks. Performance modifications primarily appear as diminished macroscopic responses—modulus, cyclic stress, electrical conductivity, and swelling ratio among others—which may or may not result in material fractures. Contrarily, crack nucleation (fatigue crack initiation) and propagation (fatigue crack expansion) culminate in material fractures (fatigue fractures), a phenomenon collectively known as fatigue damage. Contemporary studies suggest that fatigue damage in hydrogels commences at the molecular level, fundamentally due to molecular units of a hydrogel changing neighbors under prolonged loading [11].

This section summarizes the conventional fatigue testing methods and damage characterization techniques for hydrogels: monotonic stretch testing, pure shear testing, cyclic loading test, constant stretch test, constant stress test, digital image correlation (DIC) technology, photoelasticity, and mechanochemical damage visualization technology. All experimental materials mentioned above should be selected uniformly and without defects. Water can migrate within hydrogels, implying they can lose and gain water from their surroundings [11]. Notably, water content dramatically influences the mechanical properties of hydrogels [22]. For prolonged tests, such as fatigue tests, samples should be sealed in a humidity-controlled chamber [23] or submerged in oil or water baths [24]. The change in water content can then be determined by measuring the before-and-after weights of the samples. If the hydrogel is homogenized during preparation and the testing time is significantly shorter than the water migration time in the hydrogel, water migration can be deemed negligible [11]. The migration time of water can be determined using the formula L^2/D, where L is the representative length of the sample, and D is the effective diffusivity rate of water in the hydrogel [11]. For specimens without precut cracks, the relevant length is the minimum dimension (e.g., the thickness of the sample), and the effective diffusion rate of water in the hydrogel can be measured [25]. When water migration within the hydrogel can be considered negligible, the hydrogel is deemed incompressible [11]. This review primarily summarizes the research progress on hydrogel materials regarding cyclic fatigue, with the subsequent content referring to cyclic fatigue unless otherwise stated.

11.2.2.1 Static Fatigue Test

11.2.2.1.1 Static Fatigue Test Without Precut Crack

11.2.2.1.1.1 Monotonic Stretch Test

A monotonic stretch test was performed using dog-bone-shaped specimens (GB/T 528-2009-4). Nominal stress is defined as the applied force in the deformed state divided by the cross-sectional area in the undeformed state. Stretch is calculated by dividing the length of the sample in the deformed state by its length in the undeformed state, and the difference between this value and 1 is known as engineering strain. Due to the typically large elongations exhibited by hydrogels, the stress-stretch curve is best measured using video extensometers or photographic markers. Hydrogels are considered incompressible materials, and the initial slope of the stress-stretch curve under monotonic stretch or compressive loading reflects Young's modulus. The stress-stretch curve for some hydrogels depends on the loading rate, due to the time-dependent dissociation or reassociation of molecules.

11.2.2.1.1.2 Constant Stretch Test (Stress Relaxation Test)

Samples are rapidly stretched to a predetermined displacement before maintaining a constant displacement and recording their stress-time curves. These curves initially rise quickly before gradually falling over time, ultimately approaching zero or achieving a consistent stress level. If the final stress relaxes near zero, the hydrogel is liquid-like while a finite level of stress relaxation signifies solid-like. However, the sample may break before complete stress relaxation. As static stretch decreases, the time to fracture increases—a phenomenon described by Suo et al. as static fatigue fracture [23], also recognized as delayed fracture, subcritical fracture, stress relaxation fracture, or stress corrosion, with the latter two terms specifying the fracture mechanisms [11]. The mechanism of delayed fracture in hydrogels remains unclear: in physically crosslinked polymer networks, such delay may be due to the time required for polymer chains to dissociate physical crosslinks and possibly pull out across the plane of the crack [23]. Other possible causes include time-dependent solvent migration [23] and progressive damage to the polymer network [26].

11.2.2.1.1.3 Constant Stress Test (Creep Test)

Quickly stretch the sample to a specified stress, then keep the stress constant and record its stretch-time curve. Under constant stress, solid-like hydrogels stretch close to constant, whereas liquid-like hydrogels stretch infinitely. At the same time, hydrogels may suffer from delayed fracture, also known as creep rupture [11].

11.2.2.1.2 Static Fatigue Test with Precut Crack

To investigate the mechanisms of crack propagation in materials, a sample with a precut crack is typically used [27]. By inducing a precut crack, stress concentration often triggers the initiation and expansion of cracks from the tip of the precut crack, thereby facilitating the study of crack propagation behaviors in materials and their underlying mechanisms. Similarly, hydrogels utilize a sample with a precut crack to examine their fatigue crack propagation behavior and mechanism. Unlike the conventional monotonic stretch test where a dog-bone-shaped sample is used, precut crack experiments often employ rectangular hydrogels of dimensions: width L, height $2H$, thickness b, and precut crack length c, where $L \gg 2H \gg b$, $c \gg 2H$. This is often referred to as a "pure shear test."

11.2.2.1.2.1 Energy Release Rate

When comparing the fatigue performance of different materials, the crack propagation rate is a commonly used parameter, and it is a function of the energy release rate G and the stress intensity factor K [27,28]. In fatigue fracture studies of metals, aside from the small crack-tip area, the whole is typically considered to be under minor strain conditions, and linear elastic theory aptly describes the behavior outside the crack-tip region. Similarly, for fatigue fractures in high-elasticity bodies, nonlinear elastic theory accurately characterizes the behavior beyond the crack tip region. Therefore, in crack propagation studies of elastic bodies, regardless of whether their elasticity is linear or nonlinear, as long as the non-elastic region around the crack tip is small compared to the sample size, energy release rate G can serve as a measure of the applied load [29].

In an undeformed state, the height of the sample is H. In a deformed state, the sample's height becomes λH, with λ being the applied stretch. For pure shear tests, the form of energy release rate is:

$$G = HW(\lambda)$$

where $W(\lambda)$ is the energy density per unit volume of the sample without a precut crack, obtained by integrating the area under the stress-stretch curve of a sample without a precut crack. The energy release rate, irrespective of crack length, is a loading parameter, not a material attribute [11].

11.2.2.1.2.2 Toughness

The toughness Γ, defined as the critical energy release rate during crack propagation, is a material property related to the energy dissipation in crack extension, with a unit equivalent to the unit of energy per unit area [11,29]. The experimental methods for measuring the toughness of hydrogels include pure shear test, peel test, tear test, and notch test on a single sample. Existing research suggests that material toughness is independent of material size and test type [30].

Typically, toughness is determined through pure shear test, using two hydrogel samples of identical size—one without a precut crack, and the other with one [14]. As seen in Figure 11.13, the stress-stretch curve measured from the sample without a precut crack determines the function $W(\lambda)$. The sample with a precut crack is stretched to fracture, noting the critical stretch λ_c (the onset of crack propagation). The toughness of the hydrogel is then calculated by:

$$\Gamma = HW(\lambda_c)$$

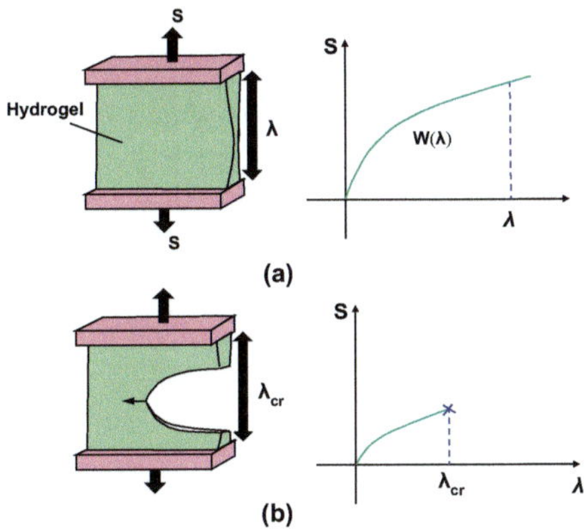

FIGURE 11.13 Pure shear test. The figure is created based on the work of [11].

This method applies to elastic materials with stress-stretch curves of any shape. For other experimental methods, please refer to the literature on measuring and interpreting hydrogel toughness [30].

11.2.2.1.2.3 Flaw Sensitivity

Traditional hydrogels are extremely sensitive to flaws. For instance, an intact polyacrylamide hydrogel can be stretched multiple times its original length, but with a few millimeters of the precut crack, a small stretch can quickly expand the crack and compromise the material. However, tough hydrogels mitigate this flaw sensitivity effectively. Suo et al. [31] defined a characteristic length for flaw sensitivity, the fractocohesive length - Γ/W_f as an indicator of flaw sensitivity in hydrogels. Here, Γ denotes material toughness and W_f represents work of fracture measured using a "perfect sample" containing no cracks, where $W_f = W(\lambda_f)$, and λ_f signifies the stretch that ruptures the sample with the precut crack. Under static loads, Γ represents the static-fatigue threshold while W_f stands for the fracture work quantified in a sample devoid of a precut crack and subjected to an infinitesimally low loading rate. When the crack length is less than this critical physical length, the ultimate material properties, such as maximum energy density, maximum stretch, and tensile strength, are independent of crack length, indicating defect insensitivity at this size. Since the fracture energy W_f is almost constant, the fractocohesive length follows the same trend as the toughness (see Figure 11.14).

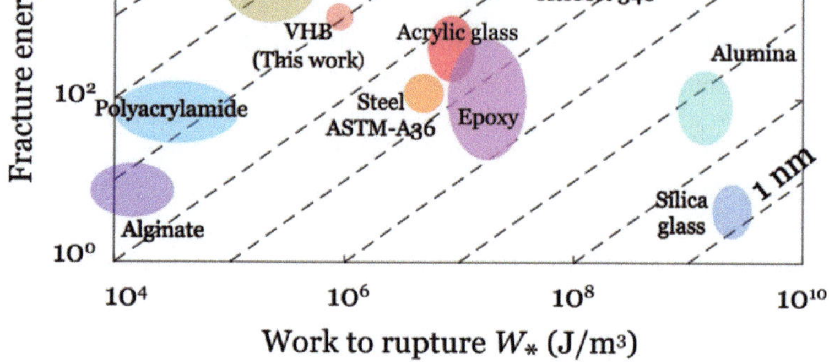

FIGURE 11.14 A diagram of toughness Γ and work of fracture W_f, along with the flaw-sensitivity length Γ/W_f. (Reprinted with permission from Ref. [31], Elsevier.)

11.2.2.1.2.4 Static Fatigue Crack Growth

Static fatigue crack extension is a common method to study complex rheological hydrogels. The energy release rate G approaches the toughness Γ when the crack is rapidly expanding and approaches the threshold when the crack is not expanding. The threshold is the toughness of the hydrogel after it undergoes complete relaxation, and the relaxation is inelastic to the loading rate. Static fatigue crack extension was performed using the same experimental approach as for toughness measurements, and a video camera was used to record the fatigue crack extension process [11].

11.2.2.2 Cyclic Fatigue Test

During the cyclic fatigue experiments of structural materials such as metals, fatigue life is an essential performance parameter. It is determined by the number of cyclic loadings a material can withstand before experiencing fatigue failure. The S-N curve is often utilized to depict fatigue life, where "S" denotes the cyclic stress and "N" signifies the fatigue life. According to the concept of fatigue life, fatigue can be categorized into three major types [32]: low-cycle fatigue (LCF), with a fatigue life below 10^5 cycles; high-cycle fatigue (HCF), with a fatigue life between 10^5 and 10^7 cycles; and very high-cycle fatigue (VHCF), with fatigue lives exceeding 10^7 cycles. In general, the loading stress in low-cycle fatigue is greater than the yield stress, whereas the loading stress in HCF and VHCF is lesser than the yield stress, thus resulting in distinct fatigue failure mechanisms for each. Research on the fatigue of soft materials like hydrogels is nascent, and fatigue testing relies on standard mechanical testing methods for elastomers or metallic materials. The fatigue mechanism of soft materials such as hydrogels has yet to be clearly understood, and there has been no specific categorization of their fatigue types. Furthermore, it remains unclear whether there exist different failure mechanisms in these materials due to varied loading stresses. As a result, this review refers to the hydrogel fatigue simply as below 10^5 cycles for LCF, within the range of 10^5–10^7 cycles for HCF, and above 10^7 cycles for VHCF. Restricted to test equipment and methods, most of the tests on cyclic fatigue of hydrogels are within 10^5 cycles, which can be categorized as LCF.

11.2.2.2.1 Cyclic Fatigue Test without Precut Crack

Fatigue tests without precut crack are commonly used to investigate crack initiation mechanisms and measure material hysteresis. The test samples, shaped like dog bones, experience cyclic tensile loading, with stretch or stress at λ or σ respectively, tracking the cycle count N to failure. The acquired data, plotted as strain or stress on the y-axis versus fatigue life on the x-axis, provide the material's fatigue strength under given conditions, as depicted in Figure 11.15. Fatigue life decreases with increasing loading stretch (λ) or stress (σ); there exists a certain value beneath which the material can withstand an infinite number of cycles without breaking. In metals and like materials, this value is known as the fatigue strength, while for hydrogel materials, it is referred to as the endurance stretch λ_e, as indicated by Lu et al. [33] (see Figure 11.16).

Cycle fatigue damage originates from irreversible changes occurring within materials under cyclic loading. Data from cyclic fatigue tests without precut cracks

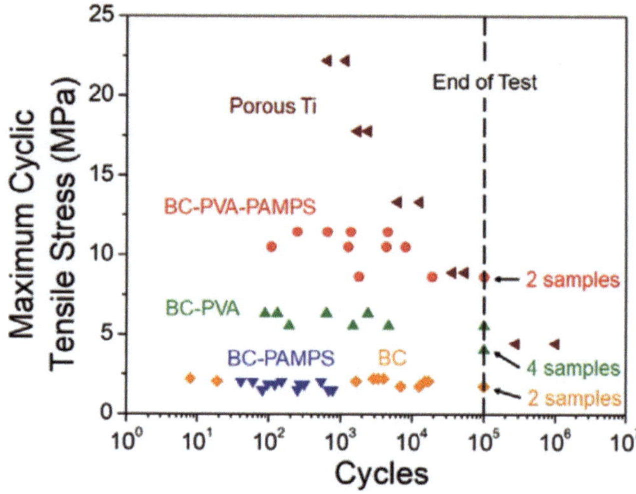

FIGURE 11.15 Maximum cyclic tensile stress-fatigue life diagram (arrows indicate samples that did not fail after 10^6 cycles) [34].

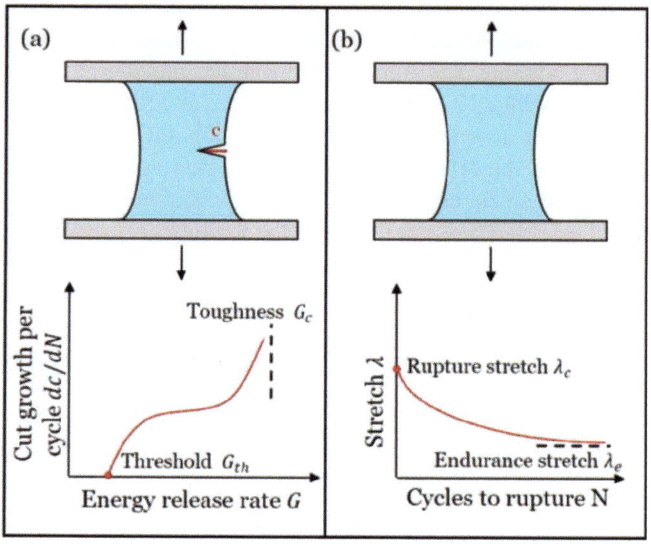

FIGURE 11.16 Two types of fatigue tests: (a) cyclic fatigue test with precut cracks, crack expansion rate (*dc*/*dN*)-energy release rate (*G*) plot; (b) cyclic fatigue test without precut cracks, strain-life plot. (Reprinted with permission from Ref. [33], Elsevier.)

can be collected and plotted as cyclic stress-response curves to observe the fatigue damage process. For nearly elastic hydrogels (e.g., polyacrylamide), the hysteresis is minimal. However, hydrogels with toughener exhibit significant hysteresis, which typically depends on the loading rate [11]. Consequently, the cyclic stress-response

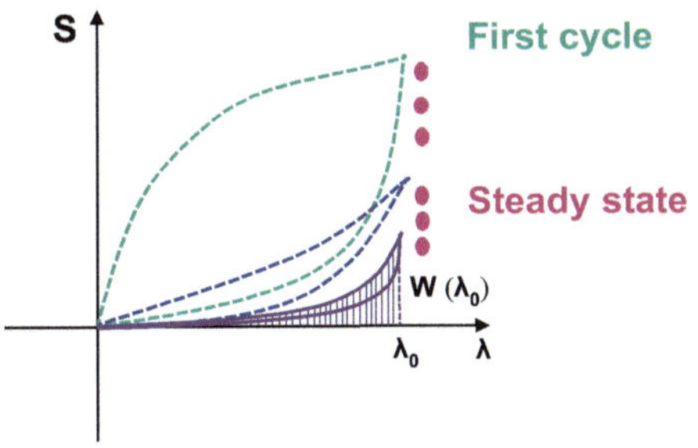

FIGURE 11.17 Cyclic stress response curve. The figure is created based on the work of [11].

curves of tough hydrogels vary with the load cycle count, with the hysteresis loop gradually diminishing, signifying a reduction in the energy consumed in each cycle (Figure 11.17). This occurs because most sacrificial bonds within the toughener irreversibly break during the relatively short cycle periods of approximately 1 second. The progressive damage due to the gradual rupture of chemical bonds over the loading cycles is referred to as shakedown [35]. After a certain number of cycles, the curves stabilize, and no further changes occur. This steady state may result either from the total damage of irreversible sacrificial bonds induced by loading or from a balance between the breaking and recovery of reversible bonds. For hydrogels with toughener, the entire sample undergoes an inelastic phase during fatigue, making elastic theories inapplicable, and it is unclear whether the energy release rate sufficiently represents the applied load [29]. Therefore, the effective energy release rate—the energy release rate at a steady state—replaces the initial energy release rate as the applied load. Calculate the effective energy release rate using the following procedure: Utilize the stress-strain curve from the Nth cycle of an uncut sample when the hydrogel approaches a steady state. Integrate the loading portion of the stress-strain curve to compute the energy density $W(\lambda_{max})$, where the effective energy release rate is $G = HW(\lambda_{max})$. Plot the steady-state crack growth per cycle, da/dN, as a function of the effective energy release rate G, and use the area under the stress-strain curve for the uncut sample in a steady state to calculate the energy release rate at steady state [35].

11.2.2.2.2 Cyclic Fatigue Experiments with Precut Crack

Pre-cracked pure shear specimens were used for cyclic fatigue experiments, with test setups similar to those employed for non-pre-cracked cyclic fatigue tests. The experimental process was recorded using a camera to calculate the fatigue crack growth rate (da/dN), and linear fitting was applied to the G and da/dN, as shown in Figure 11.16a. When the energy release rate approaches toughness, the crack growth rate increases sharply; however, when the energy release rate approaches a specific value, the crack growth rate is nearly zero. This value is known as the fatigue threshold (G_{th}) or the fatigue crack growth threshold [11]. When the applied energy release rate exceeds G_{th}, fatigue crack growth initiates, while crack growth does not occur when the applied energy release rate is below G_{th}. The water content of hydrogels significantly impacts their fatigue thresholds.

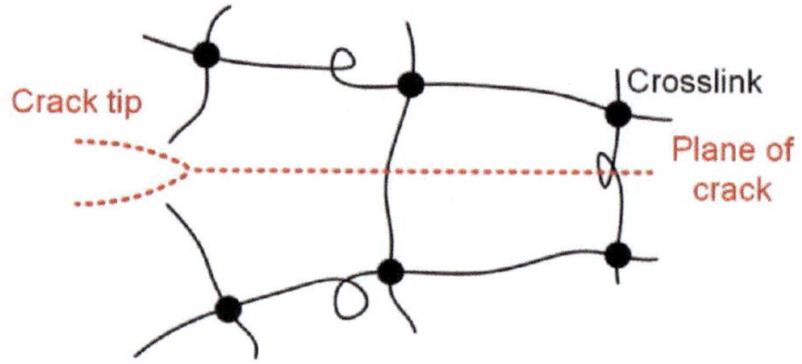

FIGURE 11.18 Lake-Thomas model [11].

For polyacrylamide hydrogels (PAAM) with water contents of 96%, 87%, 78%, and 69%, the measured fatigue thresholds are 4.3, 8.4, 20.5, and 64.5 J/m², respectively. Furthermore, the crack growth rate at a steady state depends on the energy release rate (Figure 11.18).

As with the monotonic stretch test, hydrogel materials exhibit a characteristic length scale, $G_{th}/W_e(0)$, in cyclic fatigue tests, known as the endurance fractocohesive length [33]. G_{th} represents the fatigue threshold, while $W_e(0)$ denotes the fracture energy of the uncut sample upon reaching a steady state after numerous cyclic loads.

11.2.2.3 The Lake-Thomas Model

The Lake-Thomas model is extensively utilized for the interpretation of fatigue fracture thresholds in elastomers [36]. As per the Lake-Thomas model [11,23], when the applied energy release rate approaches the fatigue threshold, only a single dissipation process is activated within the elastomer: polymer chains located at the fracture plane undergo damage, determining the fatigue threshold. In other words, the model suggests that most hysteresis within the elastomer contributes negligible effects toward the threshold. Further assumptions include that, chemical bonds along each polymer chain are much stronger than physical interactions between different polymer chains. Hence, as a chain is elongated to near rupture, all chemical bonds (C-C) along the entire chain are stretched to an identical condition. Upon the rupture of a bond within a chain, the energy within the entirety of the chain is dissipated. In this Lake-Thomas model, the threshold is essentially the bond energy of a single-layer polymer chain divided by the area of the elastomer in its undeformed state. The bond energy per unit volume of the elastomer is J/V , where J represents the C-C bond energy, and V denotes the unit volume of the individual monomers. The chain layer thickness of a single-layer polymer chain can be estimated as $l\sqrt{n}$, the average distance between the ends of the chain, where l symbolizes the monomer length, and n represents the number of monomers per chain. The Lake-Thomas model estimates the fatigue threshold of the elastomer as:

$$G_{th} = \alpha l \sqrt{n} J/V$$

where α is a pre-factor of order unity that is yet to be determined clearly.

Suo et al. [11,23] advanced the Lake-Thomas model, adapting it for gel materials and verifying it in single-network hydrogels. First, considering the formation of a network within a solvent, let's define φ as the volume fraction of the polymer chains in the gel. As such, the bond energy per unit volume reduces to $\varphi J/V$. Upon further consideration of the network in a dry state, which absorbs the solvent and forms a gel, the average distance between the chain ends can be estimated as $\varphi^{-\frac{1}{3}} l\sqrt{n}$. Thus, the resulting estimate for the fatigue threshold is $G_{th} = \alpha\varphi^{\frac{2}{3}} l\sqrt{n} J/V$. In another scenario, consider forming the network in a solvent—defining φ_0 as the volume fraction of the polymer chains in the gel. As the gel absorbs or loses the solvent to form the final gel, φ represents the volume fraction of the polymer chains in the current state of the gel, with the average distance being estimated as $(\varphi\varphi_0)^{\frac{1}{3}} l\sqrt{n}$. The resultant fatigue threshold estimate is $G_{\text{th}} = \alpha\varphi^{\frac{2}{3}}\varphi_0^{\frac{1}{3}} l\sqrt{n} J/V$. From the equation, it is apparent that the initial volume fraction of the polymer, the final volume fraction, and the number of monomers per chain influence the fatigue threshold. Lake-Thomas model's accuracy is verifiable by adjusting each of these values.

Additionally, in polymer physics, the relationship $\varphi \sim Nn$ exists, where φ represents the volume fraction of the polymer chain and is directly proportional to the product of the number of monomers in each chain (n) and the number of molecular chains per unit volume (N), leading to $\Gamma_{\text{th}} \sim Nn^{\frac{3}{2}} J/Vl$. The exponent of $n^{\frac{3}{2}}$ is more significant than that of N, implying that n has a greater impact on the polymer's fatigue threshold compared to N [37]. For double-network hydrogels, the fatigue threshold measured in the hydrogel aligns closely with the Lake-Thomas model estimates [29], with thresholds typically increasing in tandem with polymer chain length. However, no systematic characterization or report of the experimental findings is available yet.

11.2.2.4 Damage Observation and Characterization Methods

Due to the high water content and polymer properties of the hydrogel, most of the traditional microscopic characterization methods do not apply to the fatigue damage observation of hydrogels, and several damage observation methods suitable for hydrogels are introduced below.

11.2.2.4.1 DIC Technology

DIC is widely employed for surface and internal deformation measurements of materials and structures, encompassing applications in metals, polymers, composite materials, biomaterials, and biological tissues. DIC has extended from conventional sample sizes to macroscopic scales of tens of meters and micro- and nanoscales, as well as from static or quasi-static loading to high-speed dynamic loading, and from standard laboratory conditions to underwater and high-temperature extreme environments. Consequently, DIC techniques have been applied by researchers to soft materials such as hydrogels (Figure 11.19) [38], investigating deformation behavior and mechanisms, validating theoretical predictions, and ensuring the effectiveness

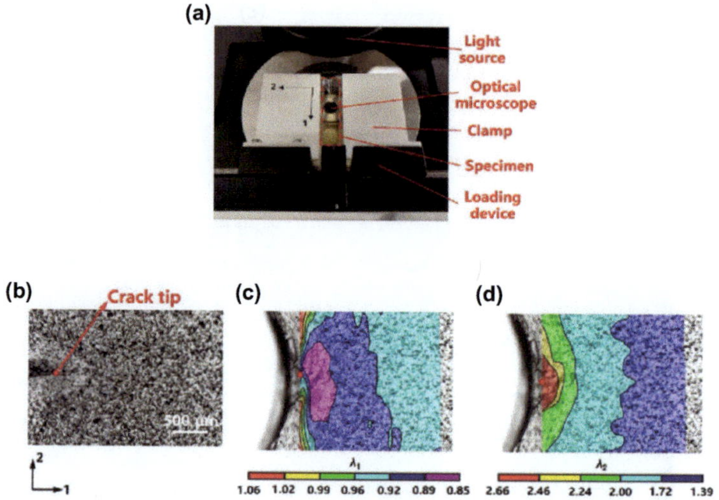

FIGURE 11.19 Application of DIC technology in hydrogels. (Reprinted with permission from Ref. [39], Elsevier.)

of finite element analyses. Although DIC possesses a broad range of applications, it still exhibits certain limitations when applied to materials with large deformations, like hydrogels. These limitations mainly arise from the artificial creation of speckle patterns using paint or marker dots, as the tensile properties of the paint coatings are weaker than the hydrogel materials, potentially leading to cracking during experiments and affecting the results. Careful consideration of the loading conditions and situations is necessary while employing DIC. Based on the TFM technique, the DIC technique can be further extended, and the observation of three-dimensional stress in hydrogels can be realized by using laser confocal as an observation tool and nanoscale fluorescent microspheres instead of scattering spots [39].

11.2.2.4.2 Photoelasticity

Photoelasticity is an effective experimental stress analysis method, derived from the principle of optical isotropy becoming anisotropic under load in glass plates, leading to birefringence in the polarized light field. It is a reliable solution for investigating the internal stress field distribution amidst complex mechanics issues. For hydrogels with good transparency, its internal stress distribution can be explored using a plane-polarized light field shown in Figure 11.20. The setup of a plane-polarized light field is the simplest and most fundamental optic layout in photoelasticity, as depicted in Figure 11.21. This fundamental setup consists of a light source and two polarizing plates. The polarizing plate near the light source, represented by "P" is known as the polarizer, while the other, marked as "A" is typically referred to as an analyzer. The planar photoelasticity specimen, labeled as "M" is placed between the two polarizing plates. Compared to DIC methods, it circumvents issues related to coating cracking and allows observation of the stress state throughout the loading process. However, photoelasticity is no longer applicable for opaque hydrogels.

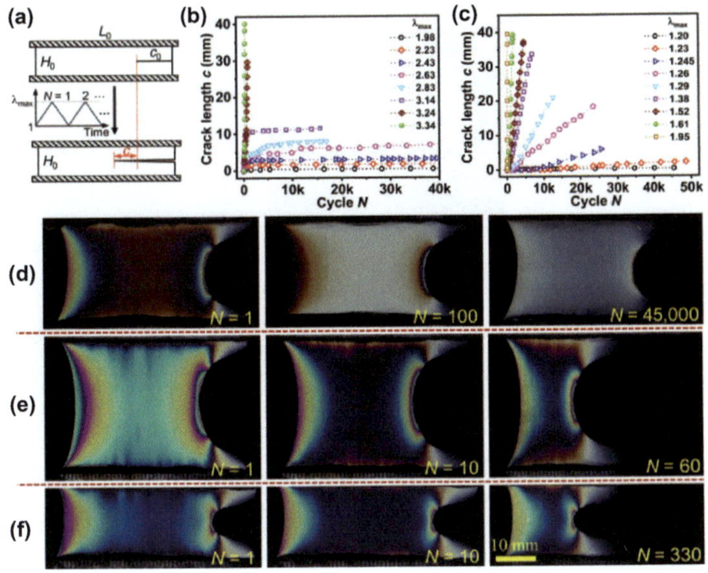

FIGURE 11.20 Application of photoelasticity in hydrogels. (Reprinted with permission from Ref. [40].)

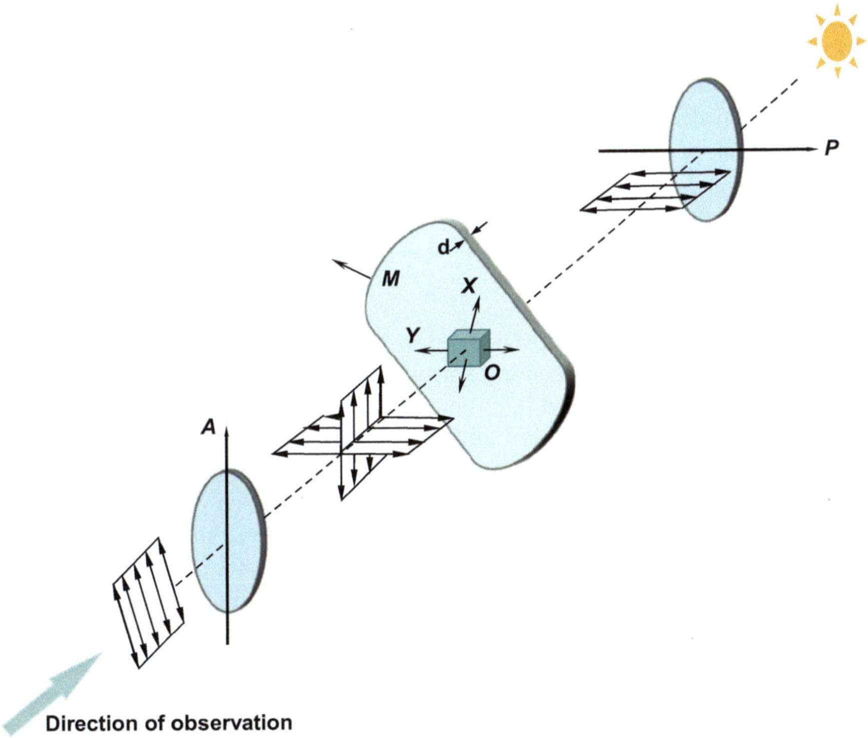

FIGURE 11.21 Schematic diagram of plane polarized light field.

11.2.2.4.3 Mechanochemical Damage Visualization Technology

The technique of mechanochemical damage visualization can be utilized for the visualization and quantification of stress, strain, and internal fractures around the crack tip in hydrogels and multi-network elastomers (Figure 11.22). This technique is particularly significant for characterizing the fracture damage at the molecular scale (i.e., bond breakage) [41]. For example, mechanophores embedded in the polymer backbone or crosslinker units emit fluorescence when broken due to the applied load on the polymer network. This process allows the visualization of mechanochemical damage within the materials. The limitation is that mechanophores may be activated under a lesser force than the original chemical bonds in the polymer, potentially altering the inherent fracture behavior of materials to a certain extent.

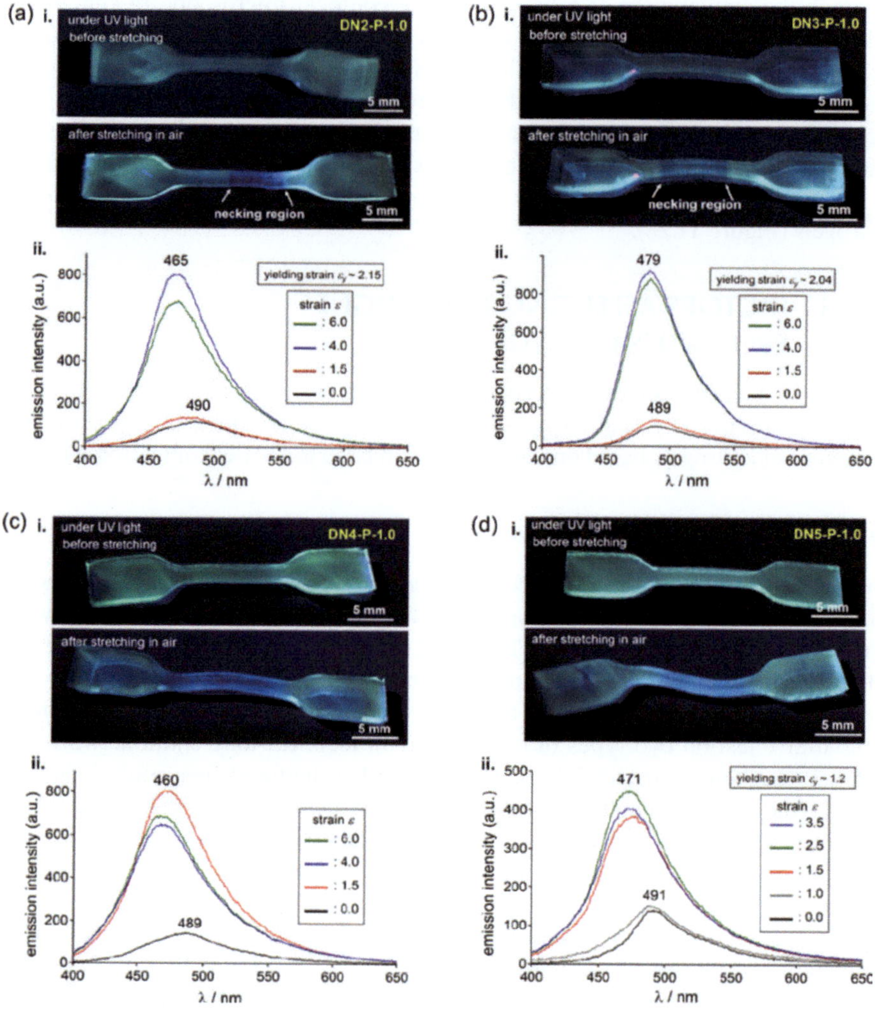

FIGURE 11.22 Application of mechanochemical damage visualization technology in hydrogels. (Reprinted with permission from Ref. [41].)

Additionally, the requirement of specific chemical synthesis can limit its wide application. An alternative strategy to reveal the mechanochemical damage is to detect mechanoradicals formed from the polymer network. Gong et al. [41] reported a method using radical-trapping prefluorescent probe P to visualize mechanochemical damage in double-network (DN) hydrogels, which allows in-situ observation of damage in the hydrogel network under ambient conditions. Techniques utilizing a fluorescent probe to scrutinize the damage and variations in microstructures of high polymer materials like hydrogels exhibit great potential.

11.2.2.4.4 In-situ Laser Confocal Microscopy

In a manner akin to the examination of biological tissues, the application of fluorescent dyes enables the staining of distinct tissue structures within the hydrogel. Through the employment of a laser scanning confocal microscope in conjunction with an in-situ loading stage, one is afforded the capacity to observe in real time the microscopic structural changes within the hydrogel during the cyclic loading process. The principal advantage of such an approach is the ability to monitor changes to the hydrogel's original microstructure without causing damage to its intrinsic structure. However, to date, the application of in-situ laser scanning confocal microscopy in fatigue experiment observations has been primarily limited to fiber-reinforced hydrogels (Figure 11.28).

11.2.3 FACTORS AFFECTING THE FATIGUE PROPERTIES OF HYDROGELS

The research on the fatigue properties of hydrogels has just started and there is less relevant literature. This section summarizes the relevant factors affecting the fatigue properties of hydrogels based on recent studies related to the field of hydrogel fatigue.

11.2.3.1 Long-Chain Network

The fatigue threshold of hydrogels is typically explicated by the Lake-Thomas model, which presumes the fatigue threshold solely depends on the primary network or the long-chain network. The model does not account for the entanglement effects or any other effects of polymer interactions, and damage scattered away from the crack tip has no contribution to the threshold. Zhang et al. [42] performed a comparative test on two types of hydrogels with identical long-chain networks and different tougheners (sacrificial bonds), finding that the toughener could effectively enhance the fracture toughness of double-network hydrogels but made little contribution to their fatigue threshold. This might be due to the stress concentration near the crack tip during long-term cyclic loading that causes all short-chain networks to decompose into small clusters, leaving only the elastic long-chain network. As the crack progresses further, it will only activate the non-elastic fracture process of the long chains, as illustrated in Figure 11.23. Zhou et al. [43] found that the long-chain network played a decisive role in the fatigue threshold of a double-network hydrogel [42], and its stiffness was inversely proportional to the length of the short-chain network, and the fatigue threshold was directly proportional to the square root of the

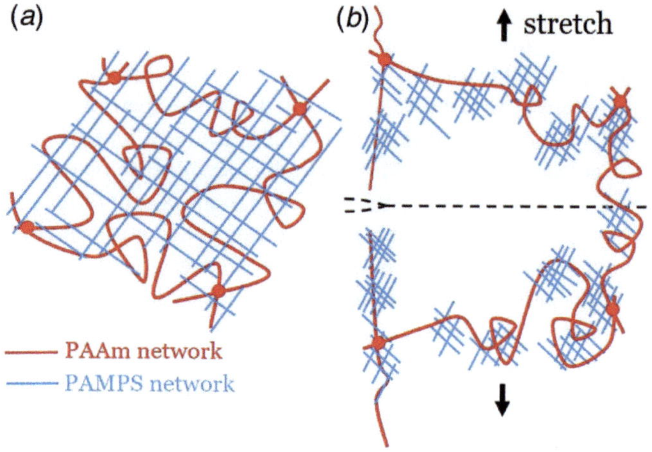

FIGURE 11.23 Schematic diagram of fatigue crack propagation of double network hydrogel [43].

length of the long-chain network. Comparing the fatigue thresholds of different gel systems under actual cyclic loading does not straightforwardly support the hypothesis that the sacrificial bonds has no contributions, and the fatigue threshold of the double-network hydrogel predicted by the Lake-Thomas model is of the same order of magnitude as the actual fatigue threshold, but not the same. The precise impact of sacrificial bonds on the fatigue threshold of hydrogels has not yet been clarified and requires further study.

Based on these studies, there are two methods to improve the fatigue threshold of hydrogels [42]: increasing the length of the long-chain network and introducing a self-healing primary network. Yang et al. [44] studied the effects of PAAm hydrogels cross-linked by disulfide bonds, making the primary network of the hydrogel self-healing. When exposed to UV light, the disulfide bonds break and reform when they interact with radicals that are generated in the hydrogel. The viscoplastic flow produced by this dynamic crosslinking reconfiguration process would reduce the stress at the crack tip and the overall sample, reducing crack extension under dynamic loading conditions and increasing the hydrogel fatigue threshold from 19 to 200 J/m^2.

11.2.3.2 Non-covalent Dynamic Bonds

Double-network hydrogels increase the toughness of materials by introducing sacrificial bonds, but the irreversible breakage of traditional short-chain network sacrificial bonds leads to a significant decline in mechanical performance after bearing cyclic loads. By using covalent networks for elasticity and replacing sacrificial bonds with non-covalent dynamic bonds (e.g., hydrogen bonds [2] and hydrophobic interactions [13]), hydrogel materials can recover their properties after multiple loads and are referred to as self-healing hydrogels. Bai et al. [45] synthesized a self-healing hydrogel composed of a covalently cross-linked polyacrylamide (PAAm) network and un-cross-linked poly(vinyl alcohol) (PVA), in which dynamic reversible hydrogen bonds form between the amide groups on PAAm chains and hydroxyl groups on PVA chains. Fatigue thresholds and crack propagation rates were measured for self-healing hydrogels and hydrogels with the same PAAm network but without PVA, revealing identical fatigue thresholds for both hydrogels. Experimental results showed that self-healing hydrogels were still subject to fatigue failure, and the fatigue threshold depended on the covalent network, with negligible dependence on non-covalent interactions. The reason is that during the stable propagation stage of fatigue cracks, cracks expand by damaging the covalent network. Weaker noncovalent interactions around the crack tip had already failed due to incomplete recovery before reaching this state. Therefore, the contribution of noncovalent interactions to the threshold of fatigue failure can still be ignored. However, it is worth noting that non-covalent dynamic bonds effectively hinder the expansion of fatigue cracks. Li et al. [46] further studied the role of dynamic bonds and found that the fatigue threshold (G_{th}) was weakly correlated with the loading rate in the viscoelastic state but was unrelated to the rate in the elastic state. In the viscoelastic state, dynamic bonds bear the load, protecting the covalent bonds from breaking and increasing G_{th}. By designing dynamic bonds with relatively long relaxation times, hydrogels can bear cyclic loads in the viscoelastic state, thus improving the G_{th}. Zhao et al. [12] studied the effect of hydrogen bonds on hydrogel fatigue damage by comparing hydrogels with different hydrogen bond strengths, finding that both weak (hydrogen bond) hydrogels and strong (hydrogen bond) hydrogels were strengthened by stress-induced hydrogen bond rearrangement in small strain cyclic loading without precut cracks. However, in large strain cyclic loading without precut cracks, both weak and strong hydrogels suffered fatigue damage due to polymer chain sliding. At the same time, strong hydrogen bonds effectively hindered the expansion of fatigue cracks, whereas relatively weaker non-covalent bonds did not significantly improve the fatigue resistance of physical hydrogels, see Figure 11.24.

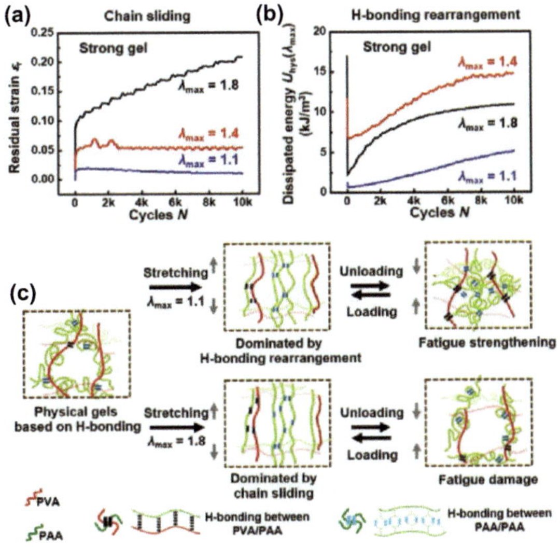

FIGURE 11.24 Fatigue damage mechanism of physical hydrogels based on hydrogen bonding. (Reprinted with permission from Ref. [12], Elsevier.)

11.2.3.3 Entanglements

The entanglements of polymer chains impact the rigidity of polymers and concurrently influence their fatigue performance. Kim et al. [47] achieved a dense entanglement of polymer chains in the acrylamide gel by utilizing a small amount of water, a cross-linker, and an initiator during the synthesis process. They discovered that a hydrogel with high-density entanglement exhibited remarkable elasticity. They applied cyclic strain loads of varying magnitudes at a fixed strain rate, showcasing minimal hysteresis. The dissipated energy (i.e., the area between the loading and unloading curves) versus the applied work (i.e., the area under the loading curve) ratio was below 1%. This near-perfect elasticity is attributed to several factors: Primarily, minimal crosslinking points and dense entanglement help maintain the integrity of the polymer network upon stretching. Secondly, the polymer chains prepared by this method are significantly long and would not rupture before the sample does. Lastly, the low viscosity of water allows for fully swollen hydrogels to have lower inter-chain friction. At the same time, the high-density entanglement significantly improved the fatigue threshold of the hydrogel, which was about four times that of natural rubber and about 20 times that of ordinary hydrogels [11]. Zheng et al. [48] proposed a novel toughening mechanism, near-crack dissipation, to explain the low lag toughness enhancement and fatigue threshold improvement in entangled polymer networks: Once a crack extends within the entangled polymer network, highly entangled polymer chains on the crack plane would be pulled out. Due to the abundant intermolecular interactions between adjacent chains, a significant amount of energy could be dissipated. Moreover, once the entangled chains around the crack top are highly stretched, broken chains can extend to multiple adjacent layers surrounding the crack plane, consuming more energy than single-layer chain rupture, see Figure 11.25.

11.2.3.4 Topological Crosslink

A study conducted by Sun [49] and his team incorporated a meticulous examination of fatigue threshold values, corresponding to an array of PAAm hydrogels synthesized at fluctuating crosslinker concentrations, see Figure 11.26. The data were accordingly plotted on a fatigue threshold-crosslinker concentration graph. Incredibly, a constant phase was identified on this plotted curve where the fatigue threshold maintained its continuity, untouched by the escalating crosslinker concentration. This observation stands as an irrefutable confirmation that the topological associations, coming into existence autonomously during the synthesis process, hold an efficiency equivalent to the chemical crosslinks in fatigue assessments. Such an inference intimates a parallel role shared by both the topological and chemical crosslinks concerning the fatigue threshold of PAAm hydrogels. This elucidation further illuminates the physical implication of the 'characteristic length' in the famed Lake-Thomas model.

11.2.3.5 Crystallization

Compared to synthetic hydrogel counterparts, biological tissues such as cartilage, tendons, muscles, and heart valves exhibit formidable anti-fatigue characteristics. Studies indicate that these anti-fatigue traits may stem from the inherently highly

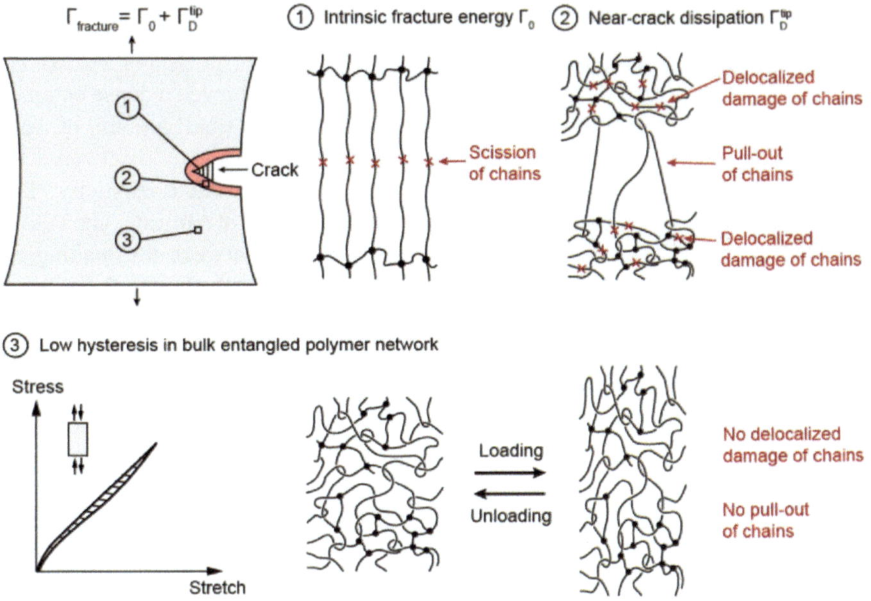

FIGURE 11.25 Schematic illustration of the near-crack dissipation mechanism for entangled polymer network. (Reprinted with permission from Ref. [48], Elsevier.)

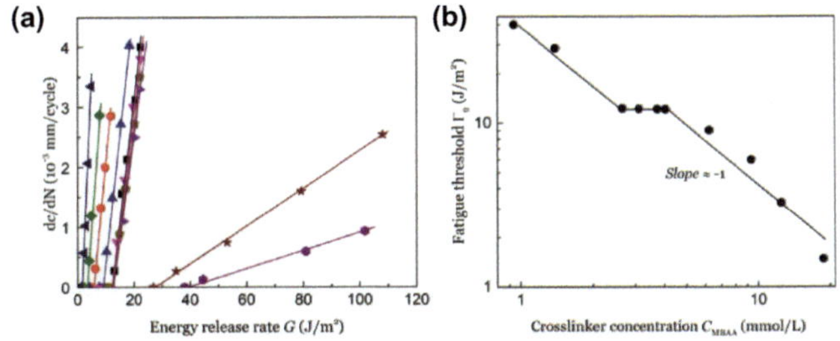

FIGURE 11.26 (a) Crack growth rates for PAAm hydrogels prepared under different crosslinker concentrations and identical monomer concentration, plotted as functions of the applied energy release rates G. (b) Dependence of the fatigue threshold on crosslinker concentration CMBAA. (Reprinted with permission from Ref. [49], Elsevier.)

ordered and partially crystalline structure of collagen fibers within the tissues [50]. Inspired by these fatigue-resistant biological tissues, Lin et al. introduced crystalline inclusions into the hydrogel [51]. By adjusting the annealing time (after freeze-thawing and air-drying) of the PVA hydrogel, hydrogels with different degrees of crystallinity, different sizes of crystalline domain, and different average distances between adjacent domains were prepared, and subjected to fatigue testing. The results revealed that the design of crystalline inclusions in the hydrogels can significantly enhance the fatigue threshold of the hydrogel, with the fatigue threshold increasing as the crystallinity rises. Owing to the presence of dense crystalline domains, fatigue cracks necessitate the breaking of these crystalline domains for propagation. The energy required per unit area to fracture the polymer crystalline domains is much higher than the energy needed to sever a single layer of the same polymer chain [52]. Consequently, PVA hydrogels with a higher degree of crystallinity possess a higher fatigue threshold. Namely, a PVA hydrogel with a crystallinity of 18.9 wt% can reach a fatigue threshold of beyond 1000 J/m^2 in the swollen state. Lin et al. [17] further managed to form neatly arranged nanofibrous structures by inducing nanoparticle crystalline domains via repeated pre-stretching mechanical training, ultimately resulting in a hydrogel with a fatigue threshold as high as 1250 J/m^2. This heightened fatigue resistance is because the arranged nano-precursor fibers are perpendicular to the crack propagation path and puncture the crack. Simultaneously, the nanofibrils at the crack tip are pulled out from the hydrogel but remain connected to the crack tip, with the fracturing of nanofibrils requiring more energy than the breaking of randomly distributed polymer chains. Noteworthily, the cracks pinned by the arranged nanofibrils will neither branch nor deflect under cyclic loading, see Figures 11.27 and 11.28.

11.2.3.6 Fiber Composite

A multitude of natural substances exhibit superior characteristics such as high strength, high toughness, and high fatigue resistance due to their fibrous structures, simultaneously preserving an impressive porosity and water content. Numerous researchers have replicated these distinctive natural structures successfully, integrating them into the production of artificial soft materials [18]. However, conventional rigid fiber reinforcement materials significantly diminish the stretchability and water content of hydrogel materials and are incompatible with pliant matrices. Therefore, the prominence of soft fiber-reinforced hydrogels is on the rise, with the embedded soft fiber ranging from millimeter-scale macrofibres to nanofibers [53,54]. Xiang et al. [54] have prepared a hydrogel possessing high fatigue resistance, high toughness, and minimal hysteresis by combining a rigid elastomer fiber (PDMS) with a soft matrix (PAAm). The fatigue limit of their hydrogel reaches up to 1290 J/m^2, which is an order of magnitude higher than that of its constituent materials (Figure 11.29). The explanation behind the improvement of the fatigue limit of hydrogels through fiber composite is rooted in the extreme disparity in the moduli of the soft matrix and the rigid fiber. When under load, the soft matrix disperses the stress evenly to rigid fibers, which bear most of the load, while the energy required for the fracture of a rigid fiber exceeds that of the matrix, thereby increasing the threshold. Let's consider an extreme case where the matrix is too supple to bear the load and conveys all loads to the fibers, all loads would then be borne by the intact fibers, and the soft matrix would blunt the fibers and no stress concentration would occur. Hence, the fatigue limit of the compound material equals the product of the fiber length and the per unit volume fiber material's endurance energy density. For minuscule characteristic-size fibers, the fatigue limit tends to increase with the decrease in fiber diameter. Zhao et al. [53], following lobster abdomen's Bouligand structure (with a twist angle of 36° between fibrils of two neighboring layers), has proposed a robust and fatigue-resistant nanofiber hydrogel using electrospinning

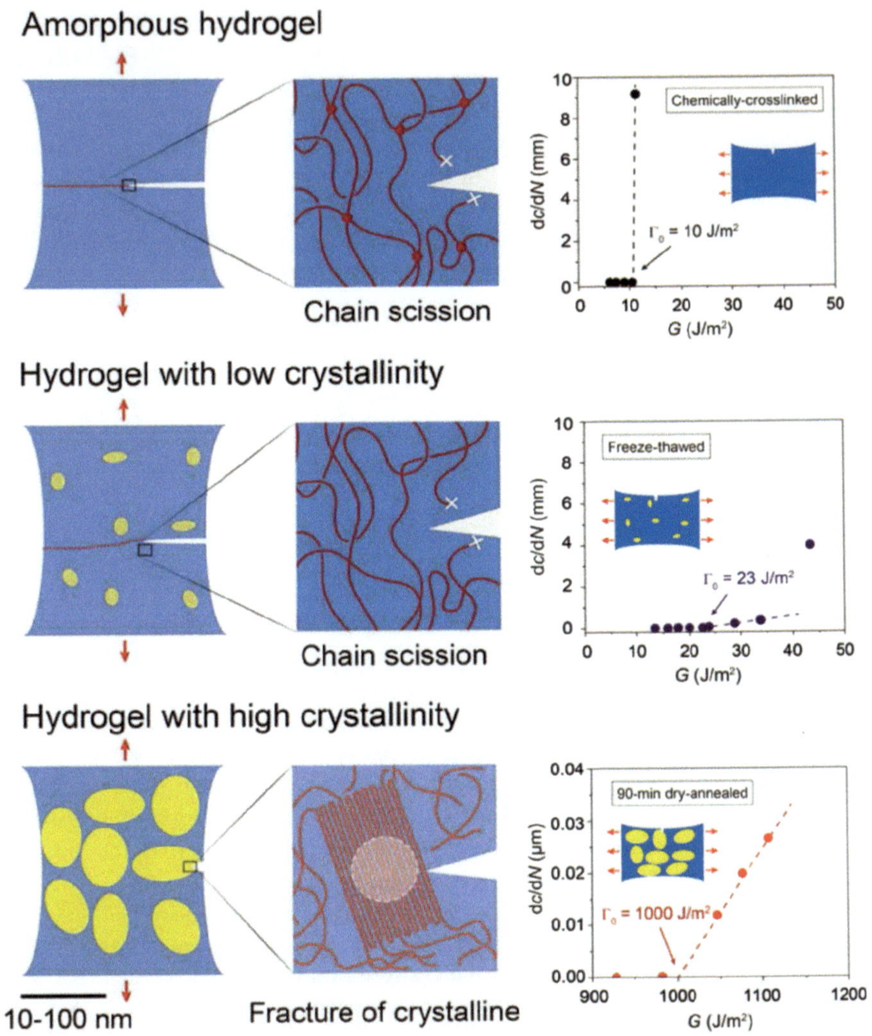

FIGURE 11.27 Illustration of fatigue crack propagation in an amorphous hydrogel and in hydrogels with low and high crystallinities under cyclic loads. The yellow areas represent crystalline domains, and the blue areas denote amorphous domains. (Reprinted with permission from Ref. [51].)

technology, with a fatigue limit reaching up to 770 J/m^2. Preserving the strong interfaces between nanofibers while introducing an ample amount of crystalline domain results in a hydrogel having high strength and stretchability. The reason for the fatigue limit of the nanofiber network is primarily attributed to the fracture energy of a single layer of nanofiber at the crack propagation path. After the introduction of large amounts of the crystalline domain, the nanofibers possess a relatively high fracture energy, which can blunt the tip of the crack and resist crack propagation; the strong interface also enhances the crack resistance capacity of the nanofiber network.

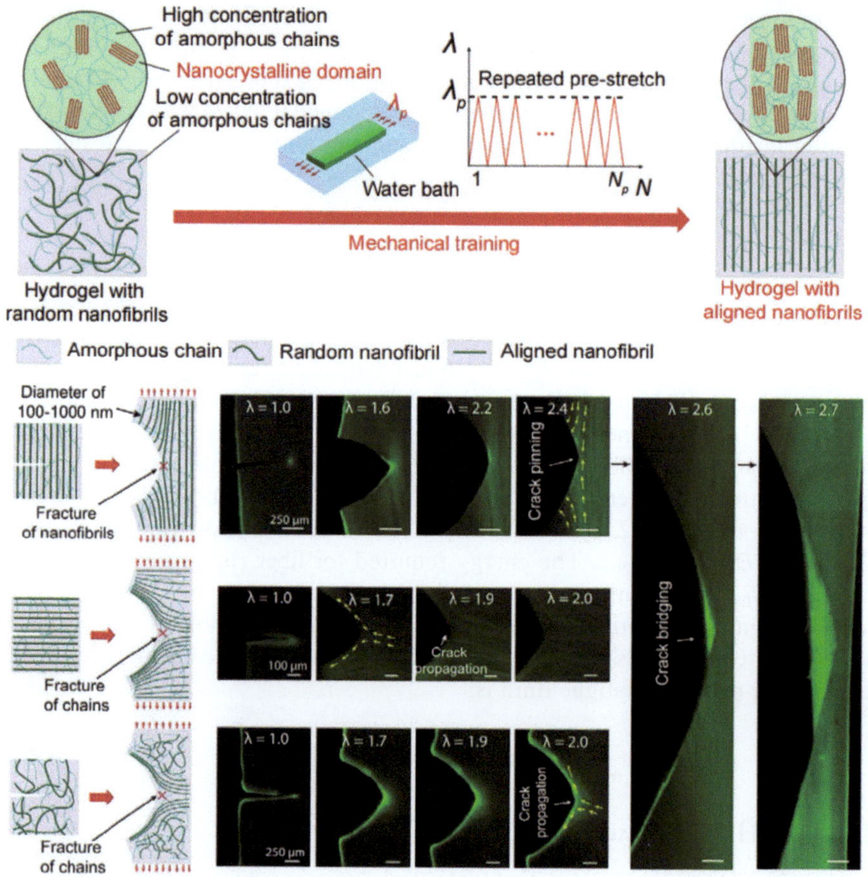

FIGURE 11.28 Schematic illustration of the microstructure of a PVA hydrogel with randomly oriented nanofibrils before mechanical training and a PVA hydrogel with aligned nanofibrils after mechanical training, and mechanisms for high fatigue threshold of pre-stretched PVA hydrogel with aligned nanofibrils [17].

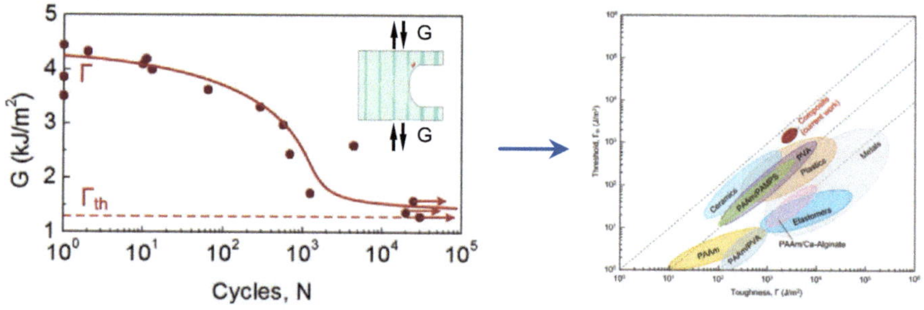

FIGURE 11.29 Stretchable and fatigue-resistant fiber composite hydrogel. (Reprinted with permission from Ref. [54], Elsevier.)

After prolonged cycle loads, nanofibers manifest strain-hardening relations, thereby delaying crack propagation. Additionally, the imitation of biomimetic Bouligand structure further improves its fatigue limit. Following the generation of nanofiber hydrogels through electrospinning technology, the model below was developed and validated for the Lake-Thomas model for nanofiber networks or typical fiber network hydrogels:

$$G_{\text{th}} = \alpha n_{\text{fiber}} U_{\text{fiber}} L_{\text{fiber}}$$

Here, $\alpha > 1$ represents a dimensionless factor, n_{fiber} stands for the quantity of the network of nanofibers (the nanofibers that bear the load) per unit volume in a swollen state, U_{fiber} refers to the energy required for fiber rupture after extended cyclic loads, and L_{fiber} represents the average span of the nanofibers between adjourning cross-linkers in the uncompressed nanofiber network. Assuming there is a negligible number of inactive nanofibers in the nanofiber network, the relationship between the expanded active nanofiber density n_{fiber} and the nanofiber volume fraction Φ is as follows: $\Phi = \pi n_{\text{fiber}} D^2_{\text{fiber}} L_{\text{fiber}}/4$, hence n_{fiber} can be ascertained from the measured values of Φ, D_{fiber} and L_{fiber}. The energy required for fiber rupture upon prolonged cycle loads U_{fiber} can be inferred by the following equation: $U_{\text{fiber}} = \pi W_L D^2_{\text{fiber}} L_{\text{fiber}}/4$, where W_L signifies the utmost strain energy density that a nanofiber can withstand under cyclic tension loads without fatigue failure. Therefore, the ultimate formula of the nanofiber network's fatigue limit is:

$$G_{\text{th}} = \alpha \Phi W_L L_{\text{fiber}}$$

11.2.3.7 Hierarchical Structures

The exceptional fatigue resistance of biological tissues, such as muscles, is linked to their hierarchical arrangement of structures at multiple scales, from the micro to the nano-level. However, it remains elusive how this intricate hierarchical organization contributes to their outstanding anti-fatigue performance. Li et al. [55] employed PA hydrogel as a model material, which contains ionically bonded structures at the 1 nm scale, crosslinked polymer networks at the 10 nm scale, and isotropic bicontinuous hard/soft phase networks at the 100 nm scale (Figure 11.30). They investigated how this multi-scale architecture could inhibit fatigue crack propagation. It was discovered that the fatigue threshold G_{th} was determined by the crosslinked polymer network structure at the 10 nm scale [40], which was constructed by chemical cross-linking and topological entanglement, while the bicontinuous phase network at the 100 nm scale significantly obstructed crack development until the transition G_{tran} exceeded G_{th}. Further insights from in-situ small-angle X-ray scattering analyses revealed that at lower applied strain, the hard network could maintain its integrity, and cyclic loading led to the formation of anisotropicity in the hard network, creating a reinforcement effect akin to composite materials. The fully anisotropic hard phase network evenly bore the majority of the applied stress, efficiently suppressing crack propagation. As the applied strain continued to increase and the energy release rate exceeded G_{tran}, the hard phase network began to fracture, with the soft phase

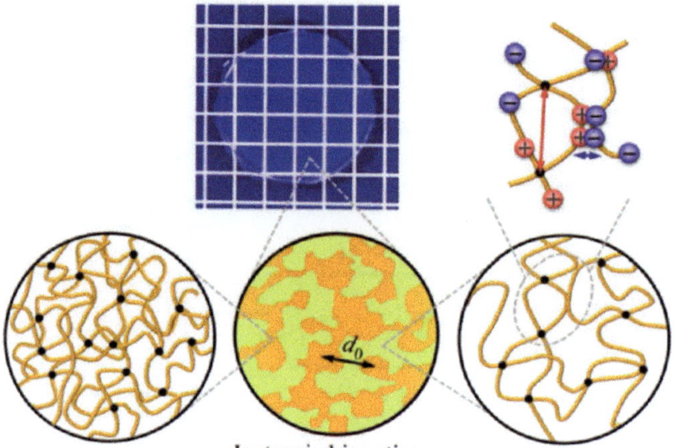

FIGURE 11.30 Illustration for hierarchical structures in PA gels. PA gels are composed of hierarchical structures containing ionic bonds (sacrificial bonds) at the 1-nm scale (indicated by blue arrow), cross-linked polymer network with mesh size at the 10-nm scale (indicated by red arrow), and bicontinuous hard/soft phase networks with mesh size at d_0~100-nm scale (indicated by black arrow) [55].

shouldering the majority of the load, leading to rapid crack propagation. Hua et al. [15] fabricated multi-length-scale hierarchical hydrogel architecture using a freezing-assisted salting-out treatment, achieving a fatigue threshold of up to 10,500 J/m^2. Fibers imbued with a large number of hydrogen bonds and crystalline domains could blunt crack tips and pin cracks; the continuous network formed by the fibers facilitated stress transmission between the fibers and impeded inter-fiber slippage. Consequently, energy dissipation at the crack tip was not limited to the vicinity of the crack tip but rather was extended throughout the network, enhancing energy absorption at multiple scales and rendering the hydrogel highly resistant to fatigue, see Figures 11.30 and 11.31.

11.2.3.8 Crack Tip Softening

Unlike the mainstream strategies of "fiber/matrix Composites" and "crystallization at crack tips" that enhance toughness and fatigue threshold of soft materials, Liu [37] and his team have proposed the concept of crack tip softening (CTS), a method that can improve both toughness and fatigue threshold of single polymer network soft materials. Their research, inducing crack tip softening through UV treatment, compared the fatigue performance of treated and untreated specimens. Results indicate that the fatigue threshold of the specimens with crack tip softening increased from 9.3 to 33.8 J/m^2. The reasons that CTS enhances toughness and fatigue threshold are as follows: Firstly, when the sample undergoes severe tensile deformation (monotonic load and cyclic load), the softened region at the crack tip blunts the crack, inducing stress redistribution. Secondly, soft regions have higher toughness, which elevates

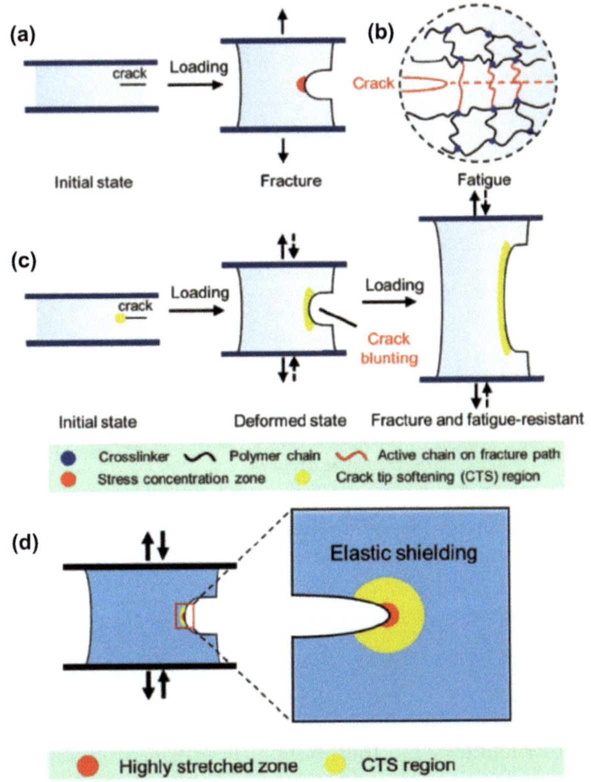

FIGURE 11.31 (a) Schematic illustration of a single-edge notched sample subjected to an external load. (b) The classical Lake-Thomas model assumes that a crack propagates by breaking only a single layer of active chains across the crack plane. (c) CTS makes the crack blunt and resists the crack propagation under both monotonic load and cyclic load. (d) The schematic of elastic shielding makes the sample more fatigue-resistant [37].

crack resistance. Thirdly, due to stress release at the crack tip and the soft region being more elastic compared to the base material, the same stretch does not cause sufficient stress to damage the soft area, implying the soft area provides an elastic barrier for the base material, thus raising the fatigue threshold. Contrarily, other strategies harden materials through a hard phase, whereas the CTS approach adopts a soft phase at the crack tip, and before any other application, modifications or designs are required for the material, whether it be during manufacturing or post-manufacturing training. On another note, the CTS concept also allows materials to toughen during application.

11.2.4 REPRESENTATIVE ANTI-FATIGUE HYDROGELS

The present strategies for enhancing the fatigue performance of hydrogels involve DNs [10,29,33,42,43], organic-inorganic nanocomposites [18,53,54,56], homogeneous hydrophobic crosslinked topological structures [13], slide-ring cross-linkers [57], nanocrystalline domains [51], creating hierarchical anisotropic structures in hydrogels through freeze casting [5,15,58,59] or mechanical training [17], and so forth. Each strategy possesses its unique characteristics, hence, hereafter a brief depiction of certain hydrogels with higher fatigue thresholds is presented.

11.2.4.1 Double-Network Hydrogel

The DN hydrogels, initially proposed by Gong [10], comprise two extensively entangled molecular chain networks. These networks can be either chemically cross-linked, physically cross-linked, or one of each, in which one network (the long chain flexible network) exhibits superior stretchability than the other (the short chain stiff network). The DN hydrogels with high water content (approx. 90% by weight) showcase excellent stiffness (0.1–1 MPa), strength (1–10 MPa), and toughness (100–1000 J/m^2), and its fatigue threshold can reach up to 418 J/m^2 [29]. The high toughness and resistance to crack propagation of DN hydrogels are realized by the synergistic action of two mechanisms that dissipate energy. As the DN hydrogel constitutes two molecular networks, it can concurrently realize robustness and functionality through the structural network and functional network. Nevertheless, as the rupture of the short-chain network is irreversible, there is exhibited hysteresis in the early cycles and a notable decrease in mechanical properties in the later phases, as shown in Figure 11.23.

11.2.4.2 Nanofiber Composite Hydrogel

Zhao and his team [53] rendered a sturdy, fatigue-resistant nanofiber hydrogel mimicking the Bouligand architecture of a lobster's abdomen using electrospinning technology with a fatigue threshold reaching an impressive 770 J/m^2 (Figure 11.32a). Furthermore, Xing et al. [56] synthesized a 3D-printed hydrogel composite, strengthened with aramid nanofiber (ANF) (Figure 11.32b). This material displayed enhanced strength, rupture energy, and fatigue threshold by approximately tenfold in comparison to pure hydrogel at a mere concentration of 0.3 wt% ANF. The augmentation in the fatigue threshold can be attributed to the formation of hybrid polymer networks. Inspired by biological network structures, Xue and his colleagues [60] introduced a hierarchical structure composed of pico-fibers, synthesized through the self-assembly of metal ion-coated peptide β-strands interconnected via flexible hidden lengths (Figure 11.32c). Upon cross-linking these fibers, the ensuing hydrogel demonstrated exceptional mechanical strength (fracture stress ~4.1 MPa), ultrahigh toughness (fracture energy ~25.3 kJ/m^2), and laudable fatigue resistance (fatigue threshold ~424 J/m^2). The phenomenal fatigue performance of the fibers is due to their high mechanical stability, swift reassembly, and hidden lengths. The release of these hidden lengths aids in inhibiting crack propagation, consequently enhancing the fatigue threshold [61].

11.2.4.3 Homogeneous Hydrophobic Cross-Linked Hydrogel

Incorporating hydrophobic association into hydrogels serves as a dynamic physical cross-linking mechanism for energy dissipation within the hydrogel, concurrently enhancing its deformability, beneficial for high extensibility and

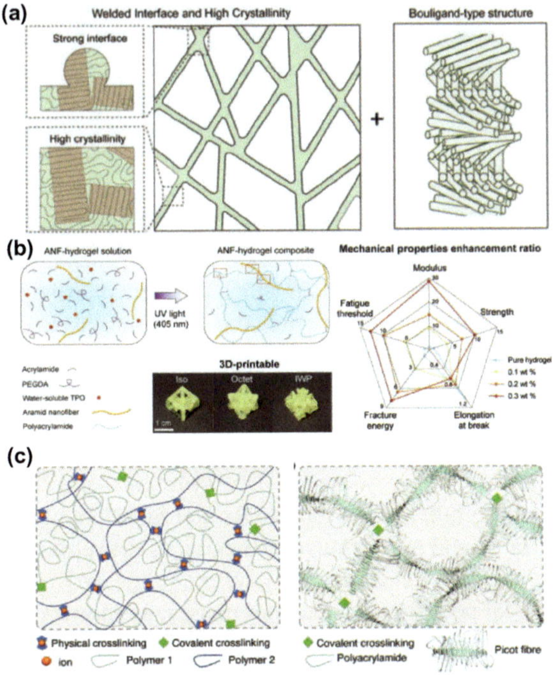

FIGURE 11.32 Nanofiber composite hydrogel. (Reprinted with permission from Refs. [53,56,60], Elsevier.)

self-healing capabilities [13]. A straightforward yet robust homogeneous hydrophobic cross-linking topological structure of hydrogel, exhibiting reversible extreme deformability (10,200% stretchable strain), toughness, and exceptional comprehensive performance in fatigue resistance with a fatigue threshold reaching 2500 J/m^2, has been proposed by Wu et al. [13]. Ordinarily, crack propagation is due to damage in the polymer network within the hydrogel during deformation, leading to fatigue failure. The dynamic association of the introduced hydrophobic nodes dissipates the energy of reversible minor deformation, and the uniformly cross-linked polymer chains unfurl and align to bear external forces, augmenting the structural integrity of the hydrogel. This prevents them from fracturing under cyclical large deformations and inhibits crack propagation. Moreover, the structural integrity of the polymer network during deformation ensures high insensitivity of the polymer to cracks, progressively ordered polymer chains self-augment to prevent further crack propagation during extreme elongation. Hydrogel networks originating from uniform hydrophobic cross-linking provide for hydrophobic dissociation and polymer chain unfurling for energy consumption. Its progressive functioning ensures structural integrity and self-recovery of the hydrogel, which is why a homogeneous hydrophobically cross-linked hydrogel possesses superior mechanical properties, see Figure 11.33.

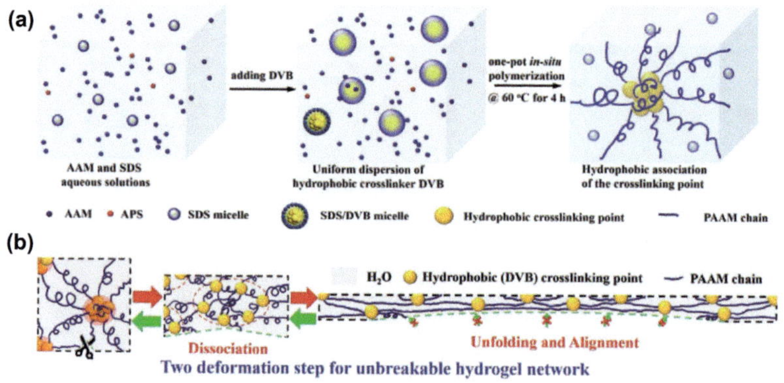

FIGURE 11.33 Homogeneous hydrophobic cross-linked hydrogel [13].

11.2.4.4 Deformable Microsphere-Cross-Linked Isotropic Hydrogels

The robust hydrogels were crafted by harnessing strategies such as physical cross-linking, entanglement networks, and micellar cross-linking, adeptly evading issues such as the accentuation of strain instigated by adopting 'hard' crystalline regions and the decrement of mechanical properties horizontally or vertically, caused by strategies that form anisotropic micro/nanostructures through mechanical training. Feng et al. [62], propounded a generic strategy to form a deformable microsphere-cross-linked isotropic hydrogel only depending on the interpenetrating entanglements of polymer chains in the deformable polymeric microspheres. The mechanical proprieties of the hydrogel hinge on the entanglement of the polymer chains within the deformable polymer microspheres, bestowing a fatigue threshold going as high as 4200 J/m^2. Given the flexibility and adaptability of the microspheres, the interoperable networks within them can instantly convert the hydrogel from isotropy to anisotropy under any load direction, thus, effectively placating stress concentration at the crack tip, dissipating energy, impeding the encroachment of fatigue-induced fractures, and eliminating notch sensitivity. Concurrently, as the stiffness of the microspheres intensifies, the hydrogel exhibits a marked increase in fracture sensitivity and a corresponding ebb in fatigue threshold, see Figure 11.34.

11.2.4.5 Multi-Scale Anisotropic Hierarchical Hydrogel

He et al. [15] proposed a strategy for the preparation of multi-scale anisotropic hierarchical hydrogel structures utilizing a freezing-assisted salting-out treatment, yielding polyvinyl alcohol hydrogels of high anisotropy. These hydrogels exhibit multi-scale hierarchical structures similar to biomaterials, including micrometer-scale honeycomb-like pore walls that house an interconnected network of nanofibers. The water content of these hydrogels ranges from 70% to 95%, and the fatigue threshold of the hydrogel has been amplified to an unprecedented 10,500 J/m^2, matching the performance of natural tendons.

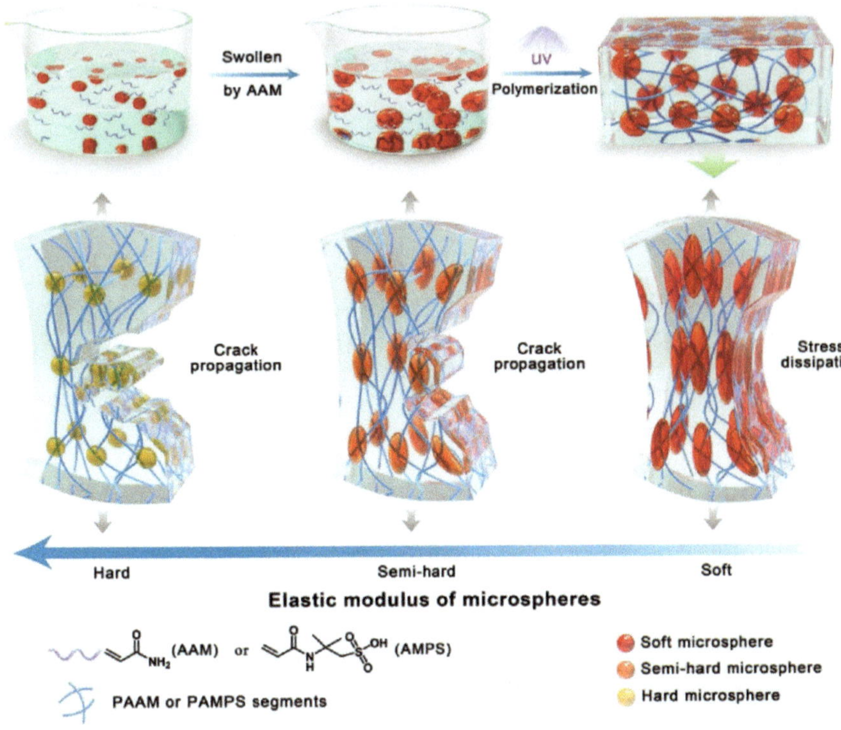

FIGURE 11.34 Deformable polymer microsphere hydrogels [62].

Additionally, Lu et al. [58] introduced a process combining ice-template freeze casting and annealing, thereby preparing anisotropic hydrogels with a fatigue threshold of up to 2740 J/m^2. Meanwhile, Zhai et al. [59] put forth an alternative strategy of Freeze-Assisted Solution Substitution (FASS) to create robust conductive organic hydrogels. These hydrogels possess the distinct traits of anisotropic honeycomb microstructures, with amorphous PVA sub-microscopic hydrophobic aggregates on the honeycomb walls and ion-enhanced PVA chains of nanometer scale, see Figures 11.35 and 11.36.

It becomes evident that multi-scale anisotropic hierarchical hydrogel structures harbor abundant potential in the development of fatigue-resistant hydrogels. Simultaneously, there is a lack of understanding of the role of these multilayered structures in fatigue crack initiation, propagation failure, and their underlying mechanism. Further studies are warranted in this regard. As biomaterials, the in-service performance and damage mechanism of these hydrogels under physiological fluid environments also warrant an exploration. In addition, the anti-fatigue hydrogels with hierarchical anisotropic structures are based on PVA hydrogels, and whether they can be applied to other hydrogels needs to be investigated [63].

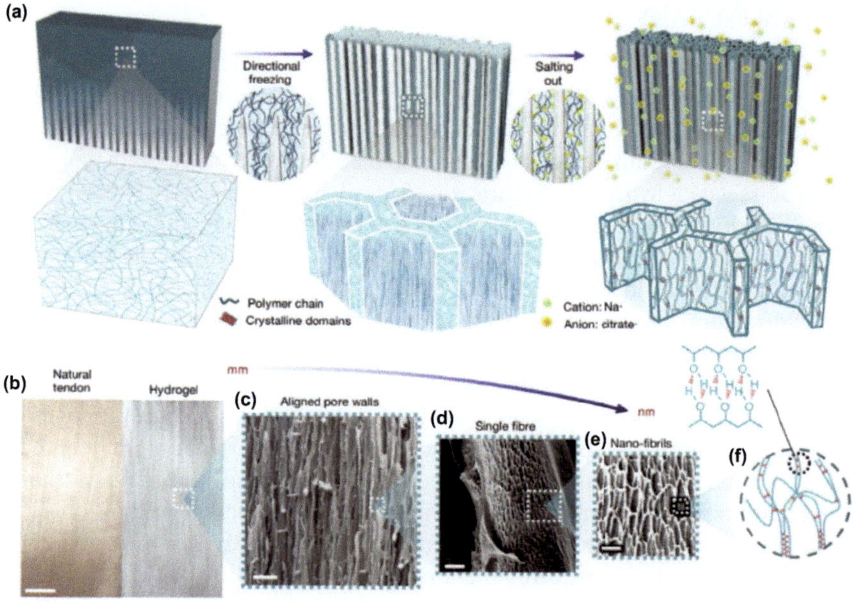

FIGURE 11.35 Preparation of PVA hydrogels with multi-scale anisotropic hierarchical structure by freezing-assisted salting-out treatment [15].

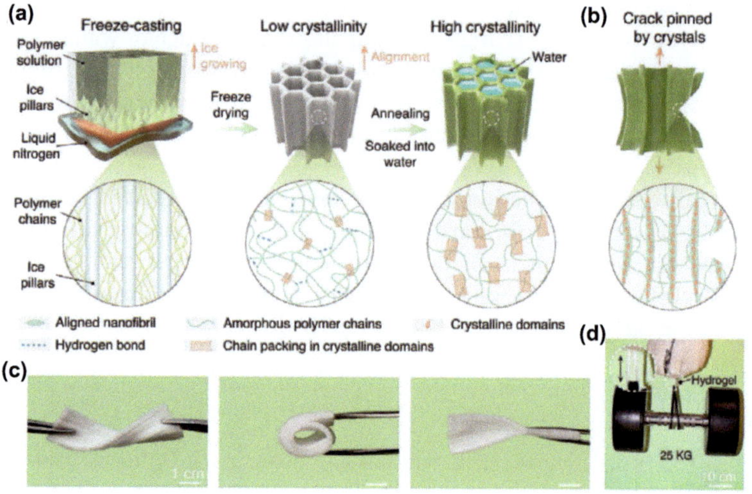

FIGURE 11.36 Anisotropic hydrogels prepared by ice template freeze casting process with annealing [58].

11.2.5 SUMMARY

The present study offers a consolidated review of recent accomplishments in the field of hydrogel fatigue. It delves into topics ranging from the concept of hydrogel fatigue, fatigue experimental approaches, observation and characterization of fatigue damages, and factors impacting hydrogel fatigue performance, to notably representative fatigue-resistant hydrogels. Significant strides have been made to better hydrogel fatigue performance, vastly augmenting its capabilities. Enhancing strategies typically include elastic dissipation mechanisms, multi-scale anisotropic hierarchical structures, crystallization effects, and composite design. Nonetheless, when juxtaposed with natural tissues, synthetic hydrogels invariably reveal performance disparities, and further explorations are mandated for their practical applications.

In terms of fatigue testing, uniaxial loading fatigue testing has garnered widespread application in modern research, yet it is fraught with challenges when applied to hydrogel fatigue testing. Issues stem from a lack of standardization in fatigue testing, significant residual strains leading to buckling, dehydration in humidity chamber testing conditions, the uncertainty of test results, and the lack of microscopic characterization for real-time observations of the internal microevolution within hydrogels. All these quandaries require ameliorations coupled with extensive reviews. In the practical application of hydrogels, they encounter complex stress conditions and environments. There is a need to test the fatigue performance of the hydrogels under these conditions and environments.

Theoretically, the assumptions in the revised Lake-Thomas models necessitate further clarification; the applicability of these models across different hydrogel systems is yet to be scrutinized. The absence of accurate hydrogel lifespan prediction models impedes the evaluation of the damage and longevity of hydrogel use. The extant research on hydrogel fatigue scarcely investigates the influence of water content and solution effect; furthermore, the application of crystallization and multi-scale anisotropic hierarchical structures in fatigue-resistant hydrogels are predominantly based on PVA hydrogels, hence necessitating explorations for their application within other hydrogel types.

To summarize, the construction of a comprehensive system that tests evaluates, and predicts hydrogel fatigue performance is of paramount importance. Simultaneously, it is vital to amalgamate other functional attributes of hydrogel, such as its biocompatibility, swelling resistance, and conductivity, to further enhance fatigue performance and expedite the practical application of hydrogels.

REFERENCES

1. Zhang Y S, Khademhosseini A. Advances in engineering hydrogels. *Science*, 2017, 356(6337).
2. Zhang Y-Z, El-Demellawi J K, Jiang Q, et al. MXene hydrogels: fundamentals and applications. *Chemical Society Reviews*, 2020, 49(20): 7229–51.
3. Yuk H, Wu J, Zhao X. Hydrogel interfaces for merging humans and machines. *Nature Reviews Materials*, 2022, 7(12): 935–52.
4. Zhao Y, Song S, Ren X, et al. Supramolecular adhesive hydrogels for tissue engineering applications. *Chemical Reviews*, 2022, 122(6): 5604–40.

5. Liang X, Chen G, Lin S, et al. Bioinspired 2D isotropically fatigue-resistant hydrogels. *Advanced Materials*, 2022, 34(8): e2107106.
6. Kang M, Lee D-M, Hyun I, et al. Advances in bioresorbable triboelectric nanogenerators. *Chemical Reviews*, 2023, 123(19): 11559–618.
7. Zhao Y, Yang N, Chu X, et al. Wide-humidity range applicable, anti-freezing, and healable zwitterionic hydrogels for ion-leakage-free iontronic sensors. *Advanced Materials*, 2023, 35(22): 2211617.
8. Guo Y, Bae J, Fang Z, et al. Hydrogels and hydrogel-derived materials for energy and water sustainability. *Chemical Reviews*, 2020, 120(15): 7642–707.
9. Xiao K, Mehmet Onur A, Tao Z, et al. Functional tough hydrogels: design, processing, and biomedical applications. *Accounts of Materials Research*, 2022, 4(2): 101–14.
10. Gong J P, Katsuyama Y, Kurokawa T, et al. Double-network hydrogels with extremely high mechanical strength. *Advanced Materials*, 2003, 15(14): 1155–8.
11. Bai R, Yang J, Suo Z. Fatigue of hydrogels. *European Journal of Mechanics - A/Solids*, 2019, 74: 337–70.
12. Zhao X, Wu J, Zhou Y, et al. Fatigue behaviors of physical hydrogels based on hydrogen bonds. *Extreme Mechanics Letters*, 2021, 46.
13. Tan S, Wang C, Yang B, et al. Unbreakable hydrogels with self-recoverable 10200% stretchability. *Advanced Materials*, 2022, 34(40): e2206904.
14. Sun J-Y, Zhao X, Illeperuma W R, et al. Highly stretchable and tough hydrogels. *Nature*, 2012, 489(7414): 133–6.
15. Hua M, Wu S, Ma Y, et al. Strong tough hydrogels via the synergy of freeze-casting and salting out. *Nature*, 2021, 590(7847): 594–9.
16. Li T, Wang Y, Li S, et al. Mechanically robust, elastic, and healable ionogels for highly sensitive ultra-durable ionic skins. *Advanced Materials*, 2020, 32(32): e2002706.
17. Lin S, Liu J, Liu X, et al. Muscle-like fatigue-resistant hydrogels by mechanical training. *Proceedings of the National Academy of Sciences of the United States of America*, 2019, 116(21): 10244–9.
18. Hua M, He X. Soft-fiber-reinforced tough and fatigue resistant hydrogels. *Matter*, 2021, 4(6): 1755–7.
19. Tanaka Y, Fukao K, Miyamoto Y. Fracture energy of gels. *The European Physical Journal E*, 2000, 3(4): 395–401.
20. Tang J, Li J, Vlassak J J, et al. Fatigue fracture of hydrogels. *Extreme Mechanics Letters*, 2017, 10: 24–31.
21. Han Z, Lu Y, Qu S. Design of fatigue-resistant hydrogels. *Advanced Functional Materials*, 2024, 34(21): 2313498.
22. Vivek M, Arockiarajan A, Ganesh T. Influence of water content on the mechanical behavior of gelatin based hydrogels: synthesis, characterization, and modeling. *International Journal of Solids and Structures*, 2021, 233: 111219.
23. Tang J, Li J, Vlassak J J, et al. Fatigue fracture of hydrogels. *Extreme Mechanics Letters*, 2017, 10: 24–31.
24. Illeperuma W R K, Sun J-Y, Suo Z, et al. Force and stroke of a hydrogel actuator. *Soft Matter*, 2013, 9(35): 8504–11.
25. Hu Y, Zhao X, Vlassak J J, et al. Using indentation to characterize the poroelasticity of gels. *Applied Physics Letters*, 2010, 96(12).
26. Karobi S N, Sun T L, Kurokawa T, et al. Creep behavior and delayed fracture of tough polyampholyte hydrogels by tensile test. *Macromolecules*, 2016, 49(15): 5630–6.
27. Thomas A G. Rupture of rubber. V. Cut growth in natural rubber vulcanizates. *Journal of Polymer Science*, 1958, 31(123): 467–80.
28. Paris P C J T E. A rational analytic theory of fatigue. 1961, 13: 9–14.
29. Zhang W, Liu X, Wang J, et al. Fatigue of double-network hydrogels. *Engineering Fracture Mechanics*, 2018, 187: 74–93.

30. Long R, Hui C Y. Fracture toughness of hydrogels: measurement and interpretation. *Soft Matter*, 2016, 12(39): 8069–86.
31. Chen C, Wang Z, Suo Z. Flaw sensitivity of highly stretchable materials. *Extreme Mechanics Letters*, 2017, 10: 50–7.
32. Pyttel B, Schwerdt D, Berger C. Very high cycle fatigue – Is there a fatigue limit?. *International Journal of Fatigue*, 2011, 33(1): 49–58.
33. Zhou Y, Hu J, Zhao P, et al. Flaw-sensitivity of a tough hydrogel under monotonic and cyclic loads. *Journal of the Mechanics and Physics of Solids*, 2021, 153: 104483.
34. Yang F, Zhao J, Koshut W J, et al. A synthetic hydrogel composite with the mechanical behavior and durability of cartilage. *Advanced Functional Materials*, 2020, 30(36): 2003451.
35. Bai R, Yang Q, Tang J, et al. Fatigue fracture of tough hydrogels. *Extreme Mechanics Letters*, 2017, 15: 91–6.
36. Lake G J, Thomas A. The strength of highly elastic materials. *Proceedings of the Royal Society of London Series A, Mathematical Physical Sciences*, 1967, 300(1460): 108–19.
37. Liu B, Yin T, Zhu J, et al. Tough and fatigue-resistant polymer networks by crack tip softening. *Proceedings of the National Academy of Sciences of the United States of America*, 2023, 120(6): e2217781120.
38. Pan Y, Zhou Y, Suo Z, et al. Inelastic zone around crack tip in polyacrylamide hydrogel identified using digital image correlation. *Engineering Fracture Mechanics*, 2023, 289.
39. Glover J D, Yang X, Long R, et al. Creasing in microscale, soft static friction. *Nature Communications*, 2023, 14(1): 2362.
40. Li X, Cui K, Kurokawa T, et al. Effect of mesoscale phase contrast on fatigue-delaying behavior of self-healing hydrogels. *Science Advances*, 2021, 7(16): eabe8210.
41. Zheng Y, Jiang J, Jin M, et al. In situ and real-time visualization of mechanochemical damage in double-network hydrogels by prefluorescent probe via oxygen-relayed radical trapping. *Journal of the American Chemical Society*, 2023, 145(13): 7376–89.
42. Zhang W, Hu J, Tang J, et al. Fracture toughness and fatigue threshold of tough hydrogels. *ACS Macro Letters*, 2018, 8(1): 17–23.
43. Zhou Y, Zhang W, Hu J, et al. The stiffness-threshold conflict in polymer networks and a resolution. *Journal of Applied Mechanics*, 2020, 87(3): 031002.
44. Yang H, Chen X, Sun B, et al. Fracture tolerance induced by dynamic bonds in hydrogels. *Journal of the Mechanics and Physics of Solids*, 2022, 169: 105083.
45. Bai R, Yang J, Morelle X P, et al. Fatigue fracture of self-recovery hydrogels. *ACS Macro Letters*, 2018, 7(3): 312–7.
46. Li X, Gong J P. Role of dynamic bonds on fatigue threshold of tough hydrogels. *Proceedings of the National Academy of Sciences of the United States of America*, 2022, 119(20): e2200678119.
47. Kim J, Zhang G, Shi M, et al. Fracture, fatigue, and friction of polymers in which entanglements greatly outnumber cross-links. *Science*, 2021, 374(6564): 212–6.
48. Zheng D, Lin S, Ni J, et al. Fracture and fatigue of entangled and unentangled polymer networks. *Extreme Mechanics Letters*, 2022, 51: 101608.
49. Sun X, Rao P, He X, et al. Chemically identical gels I – under-crosslinked networks. *Journal of the Mechanics and Physics of Solids*, 2023, 175: 105278.
50. Berisio R, Vitagliano L, Mazzarella L, et al. Crystal structure of the collagen triple helix model [(Pro-Pro-Gly)10]3. *Protein Science*, 2009, 11(2): 262–70.
51. Lin S, Liu X, Liu J, et al. Anti-fatigue-fracture hydrogels. *Science Advances*, 2019, 5(1): eaau8528.
52. Kinloch A J. *Fracture Behaviour of Polymers*. Springer Science & Business Media, 2013.
53. Ni J, Lin S, Qin Z, et al. Strong fatigue-resistant nanofibrous hydrogels inspired by lobster underbelly. *Matter*, 2021, 4(6): 1919–34.

54. Xiang C, Wang Z, Yang C, et al. Stretchable and fatigue-resistant materials. *Materials Today*, 2020, 34: 7–16.
55. Li X, Cui K, Sun T L, et al. Mesoscale bicontinuous networks in self-healing hydrogels delay fatigue fracture. *Proceedings of the National Academy of Sciences of the United States of America*, 2020, 117(14): 7606–12.
56. Xing H, He X, Wang Y, et al. Strong, tough, fatigue-resistant and 3D-printable hydrogel composites reinforced by aramid nanofibers. *Materials Today*, 2023, 68: 84–95.
57. Nian G, Kim J, Bao X, et al. Making highly elastic and tough hydrogels from doughs. *Advanced Materials*, 2022: 2206577.
58. Liang X, Chen G, Lin S, et al. Anisotropically fatigue-resistant hydrogels. *Advanced Materials*, 2021, 33(30): e2102011.
59. Dong X, Guo X, Liu Q, et al. Strong and tough conductive organo-hydrogels via freeze-casting assisted solution substitution. *Advanced Functional Materials*, 2022, 32(31): 2203610.
60. Xue B, Bashir Z, Guo Y, et al. Strong, tough, rapid-recovery, and fatigue-resistant hydrogels made of picot peptide fibres. *Nature Communications*, 2023, 14(1): 2583.
61. Liu J, Yang C, Yin T, et al. Polyacrylamide hydrogels. II. Elastic dissipater. *Journal of the Mechanics and Physics of Solids*, 2019, 133: 103737.
62. Li W, Zheng S, Zou X, et al. Tough hydrogels with isotropic and unprecedented crack propagation resistance. *Advanced Functional Materials*, 2022, 32(43): 2207348.
63. Lu T, Zhang W, Zhou Y. 6-Fatigue of hydrogels. In Li H, Silberschmidt V (eds). *The Mechanics of Hydrogels*. Woodhead Publishing. 2022: 119–38.

Index